AF213824

Promising Techniques for Wastewater Treatment and Water Quality Assessment

Edited by Iqbal Ahmed Moujdin and J. Kevin Summers

Published in London, United Kingdom

IntechOpen

Supporting open minds since 2005

Promising Techniques for Wastewater Treatment and Water Quality Assessment
http://dx.doi.org/10.5772/intechopen.87732
Edited by Iqbal Ahmed Moujdin and J. Kevin Summers

Contributors
Vandana Sakhre, Félix A. López, Francisco Jose Alguacil, Endang Tri Wahyuni, Sameer Al-Asheh, Ahmad Aidan, Markandeya, Vinay Kumar, Pokhraj Sahu, Pramod K. Singh, Nishi K. Shukla, Devendra P. Mishra, Namita Maharjan, Choolaka Hewawasam, Masashi Hatamoto, Takashi Yamaguchi, Nobuo Araki, Hideki Harada, Renato Benintendi, George Yuzhu Fu, Musfiques Salahin, Anass Omor, Karima Elkarrach, Fatima-Zahra ElMadani, Redouane Ouafi, Zakia Rais, Mustafa Taleb, Edward Kwaku Armah, Maggie Manimagalay Chetty, Babatunde Femi Bakare, Jeremiah Adebisi Adedeji, Donald Tyoker Kukwa, Boldwin Mutsvene, Khaya Pearlman Shabangu, M. Amine Didi, Rachid Zegait, Saber Kouadri, Samir Kateb, Mohamed Azlaoui, Yilmaz Yurekli, Luciano Henrique Pinto, Suellen Zucco Bez, Julia Carolina Soares, Thais Francinne, Sabrina Martins Rosa, Aline Mirian Paszuck, Luciana Ferreira Karsten, Hélène Caillet, Laetitia Adelard, Alain Bastide, Nebil Belaid, Yıldırım İsmail İsmail Tosun, Man Djun Lee, Pui San Lee, Khairunnisa Binti Abdul Lateef Khan, Romana Drasovean, Gabriel Murariu, Ruwaya al Kindi, Thies Thiemann, Rana Zeeshan Habib, Emmanuel Edet Etim, Oko Emmanuel Godwin, Ruth Olubukola Ajoke Adelagun

First published in London, United Kingdom, 2021 by IntechOpen
IntechOpen is the global imprint of INTECHOPEN LIMITED, registered in England and Wales, registration number: 11086078, 5 Princes Gate Court, London, SW7 2QJ, United Kingdom
Printed in Croatia

British Library Cataloguing-in-Publication Data
A catalogue record for this book is available from the British Library

Additional hard and PDF copies can be obtained from orders@intechopen.com

Promising Techniques for Wastewater Treatment and Water Quality Assessment
Edited by Iqbal Ahmed Moujdin and J. Kevin Summers
p. cm.
Print ISBN 978-1-83881-900-2
Online ISBN 978-1-83881-901-9
eBook (PDF) ISBN 978-1-83881-902-6

For EU product safety concerns:
IN TECH d.o.o., Prolaz Marije Krucifikse Kozulić 3, 51000 Rijeka, Croatia, info@intechopen.com or visit our website at intechopen.com.

Meet the editors

Iqbal Ahmed, Ph.D., is currently working at the Center of Excellence in Desalination Technology, Department of Mechanical Engineering, King Abdulaziz University, Jeddah, Saudi Arabia. He graduated from the Centre of Hydrogen Energy (ChE), University of Technology Malaysia and has more than sixteen years of research and teaching experience in the field of chemical engineering. Dr. Ahmed specializes in microwave synthesized materials and membrane technology (fabrication to application). He has authored more than sixty journal articles and fifty research papers in a variety of journals and international conferences. He has also contributed more than six book chapters.

Kevin Summers is a Senior Research Ecologist at the Environmental Protection Agency's (EPA) Gulf Ecosystem Measurement and Modeling Division. At present, he is working with colleagues in the Sustainable and Healthy Communities Program to develop an index of community resilience to natural hazards, an index of human well-being that can be related to changes in the ecosystem, social and economic services, and a community sustainability tool for communities with populations less than 40,000. He leads research efforts for indicator and indices development. Dr. Summers is a systems ecologist and began his career at the EPA in 1989 and has worked in various programs and capacities. This includes leading the National Coastal Assessment in collaboration with the Office of Water culminating in the award-winning National Coastal Condition Report series (four volumes between 2001 and 2012), which integrates water quality, sediment quality, habitat, and biological data to assess the ecosystem condition of the estuaries of the United States. He was acting National Program Director for Ecology for the EPA between 2004 and 2006. He has authored approximately 150 peer-reviewed journal articles, book chapters, and reports and has received many awards for technical accomplishments from the EPA and outside the agency. Dr. Summers holds a BA in Zoology and Psychology, an MA in Ecology, and Ph.D. in Systems Ecology/Biology.

Contents

Preface

Rapid population growth and industrialization in emerging countries are contributing to substantial stress on freshwater resources and degrading water quality. Increased pollution of water resources and the reduction of clean water are becoming serious problems. The co-editors would like to articulate their sincere appreciation to the generous number of authors from all over the world for contributing their appropriately distinguished quality work and revising it appropriately at short notice.

This book addresses technologies that are employed in water treatment and decontamination. Researchers in this field directly support the subject as an issue related to liquid/liquid and liquids/solids separations. The book discusses technologies related to pollution control tools that are based upon chemical, biological, and physical techniques of treating and purifying wastewaters. It presents novel and innovative methods for water and wastewater treatment and water reuse. The methods include physical/chemical, physical/biological, and advanced adsorption, absorbance, and oxidation processes tailored to improve the performance of the treatment methods. The book also includes chapters illustrating treatment approaches for the recent pollution of complex toxic materials, organic constituents, nutrients, and emerging micropollutants in numerous water resources. The wastewater treatment methods are defined by the current advanced technologies for water and wastewater treatment and reuse such as membrane separation processes, ozone oxidation, coagulation using flocculation-coagulants, and biological nutrient removal. In this book, the application of the microwave-assisted H2O2 digestions process to treat hazardous sludge has also been addressed. The technique of advanced ozone oxidative process is also investigated for the treatment of biohazardous wastewater.

Membrane technology comprises recent advancements in the removal of challenging substances. Other topics addressed in the book include a downflow hanging sponge system, a self-sustaining option for wastewater treatment, as well as water quality parameters and monitoring and index development approaches.

The co-editors would like to express their sincere appreciation to the authors for contributing their high-quality work.

Iqbal Ahmed Moujdin, Ph.D.
Department of Mechanical Engineering,
Center of Excellence in Desalination Technology,
King Abdulaziz University,
Jeddah, Saudi Arabia

J. Kevin Summers, Ph.D.
Center for Measurement and Modeling,
Office of Research and Development,
U.S. Environmental Protection Agency,
Washington, DC, United States

Section 1

Adsorption

Chapter 1

Evaluation and Quantification of Anionic Surfactant in the Gomti River at Lucknow City, India

Vinay Kumar, Pokhraj Sahu, Pramod K. Singh, Nishi K. Shukla, Devendra P. Mishra and Markandeya

Abstract

In this paper, an attempt has been made to check the level of surfactants particularly in drinking water, which can lead to toxicity in human body system. In this study, a total of 10 locations were selected to enumerate the concentration of surfactants and other physicochemical parameters with metals in the flowing water of river during pre-monsoon 2019. Analyzed result showed that the concentration of surfactants was significantly high and other parameters were also high. It was also found that river at the vicinity of town areas or midstream in the Lucknow city contained high amount of an anionic surfactants due to the nonpoint sources generated by human activities, low concentration was found in upstream, and average concentration was found in downstream, showing natural degradation of surfactants. The values of other parameters were higher than the prescribed limit, which is the serious problem for human being.

Keywords: the river water, detergent, surfactant, sodium dodecyl sulfate, heavy metals

1. Introduction

Water resources have been the most exploited natural system, since man-made and other activity play a negative role on the earth. The pollutants coming as a waste to the water bodies are likely to create nuisance by way of physical, chemical and biological appearance and this contaminated water is also harmful for human utilities. In the last few decades, the significant deteriorations in water and sediment quality throughout the world are as a result of the extensive population growth also increased industrialization with urbanization which increased the demand of river water [1]. Overall, river water is a major source of fresh water. The establishments of industries in river escort are highly responsible to pollute fresh water; it is a global issue that has no respect for National or International boundaries. The Gomti River is considered to originate from a lake "Fulhar Jheel" in the Pilibhit town in Uttar Pradesh, one of the major tributary of the Ganga. The river flows through different district before meeting the Ganga river in Kaithi, Ghazipur bordering Varanasi (at an elevation of 61 m). The distance of Gomti River is 15 km in the Lucknow city and receives improper treated wastewater from point sources such as defense, milk dairy, vegetable, oil, carbon etc. Fifty non-point sources are also discharging wastewater into the river Gomti about 250 MLD from Ghaila bridge upstream to Shaheed Path downstream while three drains Nagaria, Sarkata and Pata were discharged there wastewater without any proper treatment from Daulatganj sewage treatment plant. Three main sources contributed maximum surface water pollution are wastewater discharge from household activity, agriculture practice and from various industries that entire source contribute to detergent in their water. Detergents play a crucial role toward increasing pollution load originated from various sources that come from agriculture runoff in the form of herbicides, residential areas in the form of household detergent and insecticides and from certain industries.

"Detergent" applies to all the products containing soap, surfactant or any substance intended for cleaning and washing process. The uses of detergent have in different forms such as powder, liquid, bar, molded piece and in the form of cake. Several studies reported that the problem of detergents in the river water is growing rapidly in various countries presented in **Table 1**.

River name	Detergent concentration ($g\ m^{-3}$)	Reference
Gomti River (India)	0.1005–0.2758	Present study
Melez Stream (Turkey)	0.01–6.93	[2]
Gediz River (Turkey)	0.023–4.48	[3]
Yuvarlak Stream (Turkey)	0.12	[4]
Bakırçay River (Turkey)	0.01–0.29	[5]
Küçük Menderes River (Turkey)	0–0.93	[6]
Halifax Harbour River (Canada)	0.001–0.2	[7]
England Rivers	0.007–0.173	[8]
England Rivers	0.012–0.08q	[9]
Tama River (Japan)	0.035–0.219	[10]
Hyogo River (Japan)	0.004–2.5	[11]
Tsurumi River (Japan)	0.01–0.29	[12]
Sumida River (Japan)	0.005–0.01	[13]
Saar River (Europe)	0.01–0.09	[14]
Teshiro River (Japan)	0.01–0.27	[15]
Teganuma River (Japan)	0.019–1.4	[16]
Yodo River (Japan)	0.043–0.089	[17]
Miami River (America)	<0.05	[18]
Oohori River (Japan)	0.5–1.6	[19]
Mississippi River (America)	0.01–0.3	[20]
Litheos River (Mediterranean)	0.1	[21]
Yorkshire River (England)	0.05–0.25	[22]
Laguna Bay (Filipinler)	0.002–0.102	[23]
Itter River (Germany)	0.007–0.011	[24]
Gediz River	0.084–5.592	[25]

Table 1.
Comparison of detergent concentrations in water of different rivers.

However, surfactant; any organic substance has surface-active properties due to hydrophilic and hydrophobic groups with capable of reducing surface tension of water which forming water-air interface and make a emulsions at water solid interface. The surfactants are also responsible not only for causing foam in rivers and creating procedural problem during wastewater treatment but also responsible for reduction of water quality [26]. Surfactants are classified in several category as anionic, cationic, non-anionic and zwitter ionic or amphoteric surfactant by their ionic activity (special nature) in water [27], if the head of an ionic surfactant carries a positive charge called cationic surfactant, head carries negative charge called anionic surfactant, center attach with both positive and negative charge called amphoteric or zwitter surfactant and the hydrophobic part bonded with oxygen-containing hydrophilic groups (without charge) in non-ionic surfactant. The non-anionic and anionic surfactant is used in laundry detergent due to their high cleaning efficiency as compare to cationic surfactant. The anionic surfactants are carrying negative charge at their head, the negative compound is group of alkyl sulfates and another compound is also using in phosphate and carboxylates. The natural soaps have poor ability in hard water and production of soaps was uneconomic because of lacking of food and oil, after First World War competition of soap industry increase with food and oil industry. German Scientists experimented with synthetic detergent and developed second-grade detergent by using alkyl-naphthalene sulfate. In 1930s the petroleum industries growing rapidly and refineries waste converted into long alkenes chain then reached with benzene and sulfuric acid and developed alkylbenzene sulfonate after neutralization with sodium hydroxide. The alkylbenzene sulfonate has become one of the best laundry detergents and their sale increased a thousand fold as compare to soap. In 1950s, river beginning to foam due to over exploiting of the detergent in Europe and America while the concentration of detergent in potable water of 32 ppb American city during 1959 found 15 to 34 ppb [23]. The Legal action was followed by German Detergent Act of 1962. The United Kingdom passed legal requirement mandatory of laboratory test for bio-degradation of anionic surfactant in 1973. The alkylbenzene sulfonate have replaced with linear alkylbenzene sulfonate (LAS) due to high biodegradability. LAS is a group of anionic surfactant that is found in drinking water, as well as in domestic and industrial wastewaters in bulk form than any other groups because of their ease and low cost [28–30]. They have also adverse effects on aquatic species at 0.005 LAS/L and may cause histological degradation in fish species [31, 32]. They can further induce severe damage to vital organs, even hematological, hormonal and enzyme disturbances, growth and development of the plankton constituents [33, 34], as well as toxic to aquatic life at ≤0.025 LAS/L. The alkyl sulfates as sodium dodecyl sulfate (SDS) have found economic and safe ingredient using in food processing (USEPA) and it also called sodium lauryl sulfate. Sodium dodecyl sulfate is less negative effect and inexpensive as compare to LAS so it is using huge after 1980s.

Water quality assessment and seasonal variations in physicochemical parameters of the Gomti River were studied by several researchers [35–40] but still, the Gomti River water study about surfactants level was not studied. The present investigation reveals the level of detergents in the Gomti River which generates base line data and help to government and policy maker.

2. Materials and methods

2.1 Study area

The Gomti River is one of the major tributaries of the Ganga, situated 26.8°N, 80.9°E in Lucknow city after 250 km flowing river from the origin point. The river

Sample code	Sampling location	Latitude	Longitude (north)	Distance (km)
W/S-1	Ghaila bridge	25°30.58′ 7	8°44.74′	1.5
W/S-2	Gau Ghat	25°29.43′	78°46.54	1.8
W/S-3	Kudiya Ghat	25°30.99′	78°46.71′	1.8
W/S-4	Daliganj Bridge	25°32.25′	78°47.20′	3.0
W/S-5	Hanuman Setu	25°32.25′	78°44.27	3.4
W/S-6	Khatu Shyam	25°32.09′	78°47.14	3.7
W/S-7	Nisatganj Bridge	25°30.56′	78°43.09′	4.1
W/S-8	Gomti Barraj	25°27.56′	78°46.44′	5.1
W/S-9	Dilkusha Garden	25°28.78′	78°42.89′	6.3
W/S-10	Shaheed Path	25°27.90′	78°39.09′	7.9

Table 2.
Description and geographical position of sampling sites in the Gomti River.

serves as one of the major source of drinking water for the Lucknow city and the population of about more than 4.5 million helped by the river in a different way. The study area covers 10 different locations presented in **Table 2**. In this study, selected total 10 reasonable locations in the Lucknow city being an over loaded of pollutants source in stream and identified the levels of toxic heavy metals and other characteristics.

2.2 Criteria for the river water sampling

Sampling stations were selected considering the population load, demographical distribution of river and sources of wastewater. One of the main goals of this study is to collect the samples that are representative of the site conditions, so that an accurate assessment can be made with a minimum number of samples. Predetermined sampling protocols have been chosen for the present investigation. Two pre-washed polythene bottles having 1 and 2 L capacities with stopper were used for the samples collection as per CPCB guideline and store for further analyses.

2.3 Method used for the river water quality assessment

In this study essential physicochemical parameters were recorded during pre-monsoon 2019 and analyzed in the laboratory according to standard procedures [41]. Various physicochemical parameters like temperature, pH, turbidity, total dissolved solids (TDS), total suspended solids (TSS), total hardness, biochemical oxygen demand (BOD), dissolved oxygen (DO), carbon dioxide (CO_2), electrical conductivity (EC), chloride, sulfate, nitrate, phosphate, total alkalinity, chemical oxygen demand (COD), fluoride (F^-), cadmium (Cd), iron (Fe), lead (Pb), chromium (Cr), copper (Cu), manganese (Mn), cobalt (Co), nickel (Ni) and zinc (Zn) were analyzed.

2.4 Determination of anionic surfactant through methylene blue active substances method (MBAS)

The sample 20 mL was transferred into a 40 mL vial (vial A) equipped with a screw cap and teflon liner. The alkaline buffer (2 mL), neutral methylene blue

Metals	Wave length (nm)	Detection limit (ppm)	Sensitivity (ppm)
Cd	228.8	0.004	-0.009
Pb	217.0	0.1	0.06
Cr	357.9	0.03	0.05
Cu	324.7	0.01	0.025
Ni	232.0	0.09	0.04
Zn	213.9	0.005	0.008
Fe	248.3	0.05	0.05
Mn	279.5	0.15	0.02

Table 3.
Operating condition of atomic absorption spectrophotometer (AAS).

solution (1 mL) followed by chloroform (5 mL), were added to vial A in successive order. The vial was tightly closed using a holed screw-cap and teflon liner before being vigorously shaken for 2 min using a vortex mixer. After being shaken, the screw-cap was loosen to release the pressure inside and awaited the phase to separate. Once the two phases were separated, a Pasteur pipette was used to transfer the bottom (chloroform) layer into the new vial (vial B) that contained ultra-pure water (22 mL) and acid methylene blue solution (1 mL). Vial B was then shaken using a vortex mixer for 2 min. The cap was then loosened for few seconds and re-tightened. After the chloroform had completely separated from the water (after 2 min), the chloroform layer was collected using a Pasteur pipette and placed in a 10 mm quartz cell. The absorbance of the chloroform phase was measured by using a UV spectrometer at a wavelength of 650 nm. The concentration of anionic surfactants as MBAS is calculated from the calibration curve established with appropriate reference material such as sodium dodecyl sulfate (SDS) (0.05, 0.1, 0.5, 1.0, 1.5 and 2.0 μM).

2.5 Quality assurance and quality control

Appropriate quality assurance and procedures were followed to ensure the reliability and reproducibility of the results. Glassware were dipped in chromic acid for overnight and washed 2–3 times with de-ionized water before use. Analytical grade standard chemicals and reagent blanks were used during analyses to check chemical impurities and other environmental contaminations. Always, freshly prepared solutions were standardized against primary standard for actual strength. Triplicate samples were analyzed to check precision of the analytical method and instrument. For elemental analysis, certified reference material (CRM) multi element standard solution IV (CertiPUR® 1.11355.0100 Lot. No. HC081563, Merck) were used and the operating condition are given in **Table 3**.

3. Result and discussion

The river is characterized by sluggish flow throughout the year, except during the monsoon season, when heavy rainfall causes a manifold increase in the runoff causing more turbid of water. The summaries of result for physicochemical and heavy metals are presented in **Figures 1–3**.

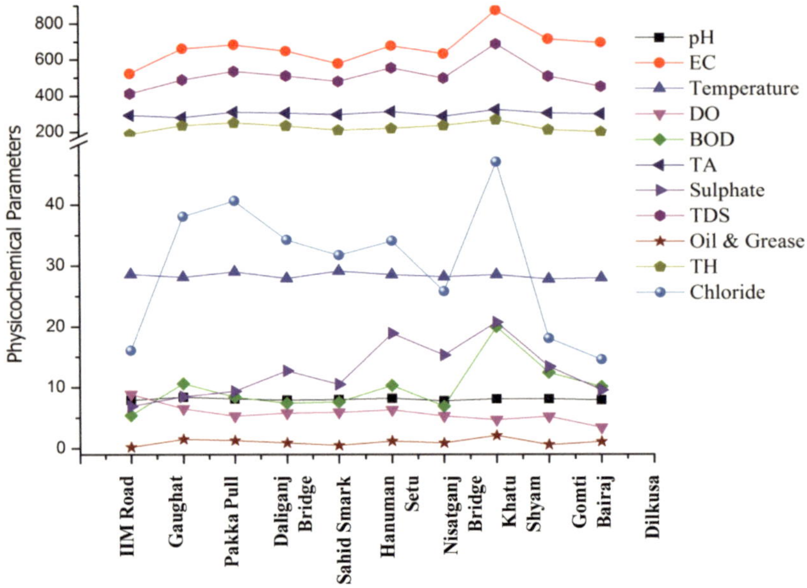

Figure 1.
Graphical representation of physicochemical parameters in the Gomti River water [in mg/L except pH and EC (µS/cm)].

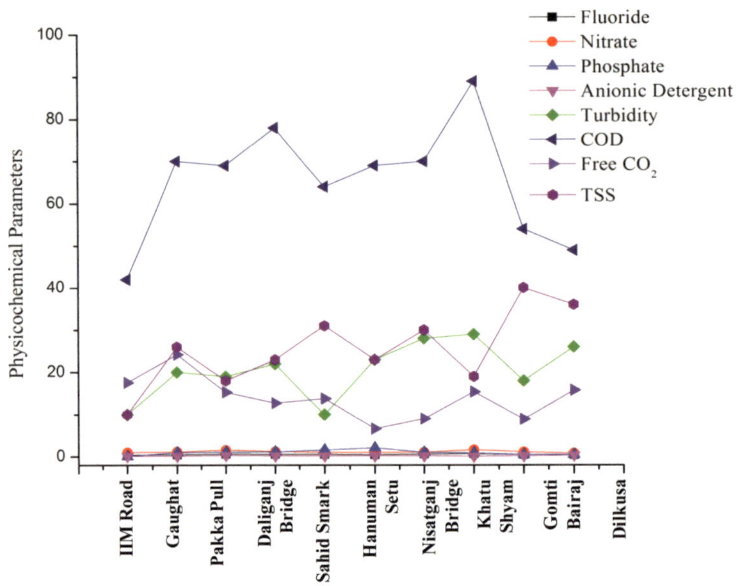

Figure 2.
Graphical representation of physicochemical parameters in the Gomti River water (in mg/L).

3.1 Physical parameter in the Gomti River water

Electrical conductivity varied from 523 to 874 µS/cm and it depended upon the presence of cations and anions, mobility, valence and temperature of water which was a good measure of total amount of salt present in water. Spot temperature were measured in the ranged from 27.8 to 29.1°C (**Figure 1**). Temperature is one of the critical and influencing physical parameter of water quality because it influences the

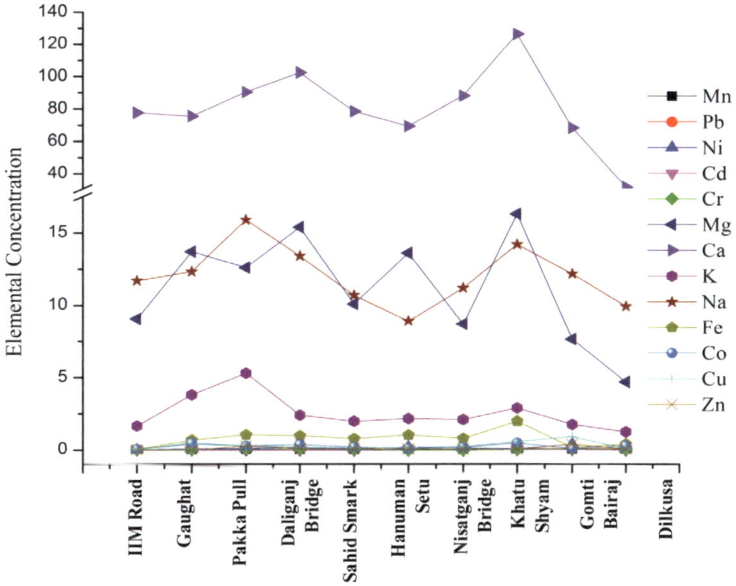

Figure 3.
Graphical representation of trace elements in the Gomti River water (in mg/L).

aquatic life by altering the dissolve oxygen (DO) and other concentration in the water making oxygen less available for respiration and metabolic activity of aquatic organisms [42, 43]. Turbidity was measured by the Turbidity meter and average value is recorded between minimum 10.0 NTU at IIM Road and maximum 29.0 NTU at Gomti Barraj. Obtained turbidity was very high in all the studied locations and found above the prescribed limits by BIS (10500-2012) [44]. Comparatively high turbidity in Gomti Barraj may be due to the mixing of large amount of sewage water, disposal ash and organic pollutants from the surrounding locality. The increase in turbidity by organic pollutants resulting eutrophication of water bodies which consequently diminish the light transmission into water and thus gradually condense overall productivity.

The average value of TSS in river was recorded 25.6 mg/L, which was well within the standard limit (150 mg/L) prescribed by WHO in terms of inland surface water. The amount of total dissolve solids (TDS) in water indicates salinity of water and may also be used as an indicator for rapid plankton growth and sewage contamination. In this study average TDS value is measured value were ranged from 412 to 687 mg/L at IIM Road (**Figure 1**).

The lowest and highest values of Hardness were observed 188.0 mg/L at IIM road and 268.0 mg/L at Gomti Barraj location. Hardness is caused due to presence of cations like Ca^{+2}, Mg^{+2}, Fe^{+3} etc. This is the property of water to precipitate soap by formation of complex with calcium, magnesium present on water. The calcium and magnesium are exists in surface and ground water mainly as carbonates and bicarbonates. River water contributed calcium as due to higher proportion of calcium in the surrounding rocks and soils which is essential for plant precipitation of lime, bone building etc. The main source of magnesium is sewage inflows and minerals generate from soil erosion and are important for enzyme activation, growth of chlorophyll and phytoplankton [45]. In the present study calcium content were found to 126 mg/L and magnesium 11.18 to 16.30 mg/L. The lowest values of calcium and magnesium were observed 31.62 and 4.68 mg/L at Dilkusha, while highest 126.3 and 16.3 mg/L at Khatu Shyam location respectively.

3.2 Chemical parameters in the Gomti River water

The pH value of all the studied water samples is measured by digital pH meter and average pH is recorded between 7.8 and 8.4 which was found very approximate to the high limit (6.5 to 8.5) prescribed by the BIS (10500-2012) [44]. As per measurement, pH values were found lowest (7.8) in Khatu Shyam location and highest (8.4) found in Gau Ghat location (**Figure 2**). pH that maintains the acidic or alkaline property is a vital characteristic of any aquatic ecosystem since all the biochemical activities and retention of physicochemical attributed to the water are greatly depend on pH of the surrounding water [46]. It also prevails in biological and physicochemical attributes of surrounding water.

DO was measured between 3.4 and 8.9 mg/L whereas the highest concentration at IIM Road was found and lowest concentration at Dilkusha. The aquatic life distressed when DO levels drop from 4 to 2 mg/L [47] and as DO level falls undesirable changes in odor, taste and color reduce the usefulness of water [43]. Biochemical oxygen demand (BOD) is also an important parameter of water quality which measures the quantity of oxygen consumption by microorganisms during decomposition of organic matter present in the water. BOD of the water river was measured by using method APHA 5210. For drinking purpose, BOD of water should be zero. High BOD than the prescribed value was found in the studied river where the minimum average value was found 5.4 mg/L at IIM Road and highest was found in 19.9 mg/L at Gomti Barraj (**Figure 2**). The alkalinity express the buffering capacity of the water which appreciably maintain the pH by absorbing access H^+ ions and protects the water body from pH fluctuation. Whereas carbonates and bicarbonates associated with sodium and potassium contribute only alkalinity not hardness because of incapability of sodium and potassium to form complex with electron donor ligands. In this study, lowest and highest alkalinity was found to be 280.0 mg/L at Gau Ghat and Gomti Barraj respectively whereas the average value of alkalinity was found to be 300.8 mg/L. The alkalinity values were found higher than the prescribed limit by IS-10500, 2012 standard. The average value of potassium was recorded 2.53 mg/L were ranged from 1.24 mg/L at Dilkusha to 5.30 mg/L at Pakka Pul. In this study, the Sodium content is measured between 8.90 and 15.90 mg/L whereas the highest concentration at Pakka Pul is found and lowest concentration at Hanuman Setu. In this study, the oil and grease was observed between 0.6 and 2.1 mg/L, whereas the highest concentration at Khatu Shyam, while lowest concentration at Gomti Barraj. The chemical oxygen demand (COD) was analyzed by using open reflex digestion method and COD varied from 42.0 to 89.0 mg/L with an average value was 13 mg/L. The maximum value of COD was found to be 89.0 mg/L at Khatu Sham station. Mixing and accumulation of sewage discharges from nonpoint sources resulted high COD at this location. The Hanuman Setu and Gau Ghat are contributing lowest and highest value of CO_2 concentrations were 6.6 to 24.2 mg/L respectively and the average values was found to be 13.9 mg/L. The activity of the microorganisms increase due to organic load of water results the increase the level of CO_2 in river water.

Chloride is not harmful to humans but high concentration of chloride increase the corrosive property of water. The chloride content was within permissible limit as prescribed by BIS and average values were recorded as lowest and highest (14.5 to 47.1 mg/L) at Shaheed Smarak and Gomti Barraj, whereas the average values of chloride was 30.0 mg/L. The value of sulfate were recorded between 6.9 and 20.7 mg/L. Comparatively lowest and highest amount of sulfate was recorded at IIM Road (6.9 mg/L) and at Gomti Barraj (20.7 mg/L), which is might be due to discharge of municipality sewage and dumping of domestic waste into the lakes. The average value of nitrate and phosphate were found 1.20 and 1.0 mg/L respectively

(**Figure 2**). High level of nitrates is found in rural areas because of extensive application of nitrogenous fertilizers in agriculture. In urban areas sewage water rich in nitrates contaminate surface water thus increases the nitrate amount [48].

3.3 Anionic surfactant level in the Gomti River water

Detergents make our cloth brighter or whiter by emitting blue light after absorbing UV light. In this study, the anionic surfactant is measured between 0.1005 and 0.2758 mg/L whereas the highest concentration was analyzed at Pakka Pul and lowest concentration at Khatu Shyam station. The average value (0.1832 mg/L) was found within the prescribed standard of Indian standard, while the value were exceed in Pakka Pul, Daliganj Bridge and Dilkusha sampling station, at this location the activity of washer man is higher to washing the cloths that can creating the emerging level of anionic detergent (**Figure 2**). Different type of detergent are using for cleaning of cloth such as Surf Excel, Active Wheel, Ariel, Tide Nirma, Saheli, Ghari, Rin Supreme and Fena in the powder form and many others in liquid form. This washing substance made up with many different chemical components such as surfactants, enzyme, bleaches, optical brightness, perfume, color and builders, and anti-redeposition agent and filter. They are dissolved in water and react with dirt and remove from cloth and other materials. The discharging of that wastewater into the river water without proper treatment created serious issue such as foaming, decreasing light penetration into different depth of the river making aerobic zone, eutrophication and several eco-toxic effects on aquatic animals. The concentration of anionic surfactant from 1 to 20 mg/L showed LC50 in fish [49, 50]. Koç and Güven [51] studied the single and progressive dose of 12.5 to 35.0 mg/L in *Oncorhynchus mykiss*. Many researchers have shown that activated carbon is an effective adsorbent for treating water with high concentrations of organic compounds [44]. Various biotechnological techniques have recently described an important group of bio surfactants using lipases or enzymes that can reduce environmental load of detergent products as the chemicals used in conventional methods; they are biodegradable and non-toxic, and also leave no harmful residues [52].

3.4 Heavy metals in the Gomti River water

The Gomti River receives industrial as well as domestic waste from various drains in the city. The analyzed concentration of Mn, Pb, Ni, Cd, Cr, Fe, Co, Cu and Zn were represented in **Figure 3** and the average value were found to be 0.045, 0.057, 0.134, 0.007, 0.036, 0.792, 0.250, 0.303, and 0.110 mg/L respectively.

The results were found that Cd and Pb in higher amount which is very toxic for environment as well as aquatic ecosystem. The level of Mn, Fe, Co, Cu and Zn were higher in Khatu Shyam. Higher concentration of metal in water and sediment could be due to the industrial/agricultural/domestic runoff coming into the river. The maximum concentration of Pb, Ni, Cd and Cr were observed at Gomti Barraj. Intake of excessive concentration of heavy metals can pose carcinogenic effect on human and water quality assessment is necessary where the water using as a drinking purpose because of human actives creating huge quantities of wastes that can cause environmental pollution. The concretions of As, Cu, Fe and Cd in the Gomti River water were 0.029 to 0.079, 0.0145 to 0.061, 0.077 to 1.685 and 0.0144 to 0.0244 mg/L respectively. Kumar et al. [53] also found the similar result in the Gomti River. Industrialization and urbanization is growing rapidly with population and untreated waste disposal also increased due to mismanagement and overload wastes discharges from household into aquatic system [54]. That untreated wastewater and sewage loaded with heavy metals and organic compound into the water stream.

4. Conclusions

River is an important for every human being. History showed that major civilization was developed on the bank of river. But due to modern civilization, increased in population and changing lifestyle of human, the quality and quantity of river is gradually reduced. The present study concluded that the level of total alkalinity, total dissolved solid, total hardness, turbidity, Pb, Ni, Cd, Ca, Fe and Cu was found beyond the acceptable limits of drinking water quality standard. The level of anionic detergent in drinking water is 0.2 mg/L recommended by Indian government but the value of anionic surfactant were exceed in Pakka Pul, Daliganj Bridge and Dilkusha sampling point. The emerging level of water pollution caused due to continuously discharging of improper and untreated sewage and wastewater. This study strongly recommended to awareness is need to people of this city and good operation practice to operator of STP.

Acknowledgements

The first author is highly grateful to Dr. G.C. Kisku, Scientist CSIR-IITR, Lucknow.

Conflict of interest

There is no conflict of interest.

Author details

Vinay Kumar[1], Pokhraj Sahu[1], Pramod K. Singh[1], Nishi K. Shukla[2], Devendra P. Mishra[3] and Markandeya[4*]

1 Department of Environmental Science, Babu Banarsi Das University, Lucknow, India

2 Environment Department, Wimpey Laboratories LLC, Abu Dhabi, UAE

3 Department of Applied Science and Humanities, Rajkiya Engineering College, Ambedkarnagar, UP, India (affiliated to Dr. A.P.J. Abdul Kalam Technical University, Lucknow, India)

4 Department of Civil Engineering, Indian Institute of Technology (BHU) Varanasi, India

*Address all correspondence to: mktiwariiet@gmail.com

References

[1] Liyanage PC, Yamada K. Impact of population growth on the water quality of natural water bodies. Sustainability. 2017;**9**:1405. DOI: 10.3390/su9081405

[2] Izgoren SF. Detergent pollution in Melez Stream and correlation with nutrients [master thesis]. Turkey: Ege University, The Institute of Natural Sciences; 1992

[3] Tuğrul G. Investigation of anionic detergent pollution in Gediz River system [master thesis]. Đzmir, Turkey: Ege University, Biology Department; 1992

[4] Balık S, Ustaoğlu MR, Egemen O, Cirik S, Eltem R, Sarı HM, et al. The Formulation of a Action Plan Sustainable Usage for Yuvarlak Stream. Ankara, Turkey: TC Ministry of the Environment, Private Environmental Protection Establishment; 2002. p. 182

[5] Başaran AK. Pollution parameters in Bakırçay Delta and with relation Candarlı Bay [PhD dissertation]. Đzmir, Turkey: Ege University The Institute of Natural Sciences; 2004

[6] Egemen Ö, Ustaoğlu MR, Önen M, Hakarerler H, Sarı HM, Tanrıkul T, et al. Water quality of Küçük Menderes River and investigation of interaction with ecosystem. Ege University Scientific Investigation Project Report, 2005. p. 65

[7] Gagnon MJ. Monitoring anionic surfactants at a sea outfall, Halifax Harbour, Canada. Water Research. 1983;**17**:1653-1659

[8] Waters J, Garrigan JT. An improved microdesulphonation/gas liquid chromatography procedure for the determination of linear alkylbenzene sulphonates in UK rivers. Water Research. 1983;**17**:1549-1562

[9] Gilbert PA, Pettigrew R. Surfactants and the environment. International Journal of Cosmetic Science. 1984;**6**:149-158

[10] Yoshikawa S, Sano H, Harada T. Determination of LAS in river water by high-performance liquid chromatography. Suishitsu Odaku Kenkyu. 1984;**7**:191-194

[11] Kobuke Y. Concentration of linear alkylbenzene sulfonates (LAS) and its composition in the river waters of Hyogo Prefecture. Japanese Journal of Limnology. 1985;**46**:279-286

[12] Yoshikawa S, Sano H, Harada T. Distribution of LAS in river water and sediment at Tsurumi River. Japanese Journal of Water Pollution and Research. 1985;**8**:755-758

[13] Kikuchi M, Tokai A, Yoshida T. Determination of trace levels of linear alkylbenzene sulfonates in the marine environment by high performance liquid chromatography. Water Research. 1986;**20**:643-650

[14] Matthijs E, De Henau H. Determination of LAS. Tenside, Surfactants, Detergents. 1987;**24**:193-199

[15] Kojima S. Pollution by fluorescent whitening agents (FWA) and linear alkylbenzene sulfonates (LAS) in environmental water. Announce Rep. Environ. Poll. Res. Inst. Nagoya. 1989;**19**:61-68

[16] Nonaka K, Hattori Y, Nakamoto M. Determination of LAS in environmental and domestic wastewaters. Suishitsu Odaku Kenkyo. 1989;**12**:194-200

[17] Nonaka K, Hayashi Y, Nakamoto M. The distribution and fate of LAS in some river waters of Osaka. Announce Rep. Environment Pollution Control Cent. Osaka Prefect. Gov. 1990;**12**:68-79

[18] Hand VC, Rapaport RA, Pittinger CA. First validation of a model for the adsorption of linear alkylbenzene sulfonate (LAS) to sediment and comparison to chronic effects data. Chemosphere. 1990;**21**:741-750

[19] Amano K, Fukushima T, Nakasugi O. Fate of linear alkylbenzene sulfonates in a lake estuary. Water Science and Technology. 1991;**23**:497-506

[20] McAvoy DC, Eckhoff WS, Rapaport RA. Fate of linear alkylbenzene sulfonate in the environment. Environmetal Toxicology and Chemistry. 1993;**12**:977-987

[21] Dassenakis M, Scoullos M, Foufa E, Krasakopoulou E, Pavlidou A, Kloukiniotou M. Effects of multiple source pollution on a small Mediterranean river. Applied Geochemistry. 1998;**13**:197-211

[22] Fox K, Holt M, Daniel M, Buckland H, Guymer I. Removal of linear alkylbenzene sulfonate from a small Yorkshire stream: Contribution to GREAT-ER project 7. The Science of the Total Environment. 2000;**251**(252):265-275

[23] Eichhorn P, Flavier ME, Paje ML, Knepper TP. Occurrence and fate of linear and branched alkylbenzene sulfonates and their metabolites in surface waters in the Philippines. The Science of the Total Environment. 2001;**269**:75-85

[24] Wind T, Werner U, Jacob M, Hauk A. Environmental concentrations of boron, LAS, EDTA, NTA and Triclosan simulated with GREAT-ER in the river Itter. Chemosphere. 2004;**54**:1135-1144

[25] Orkide M, Meral Ö, Özdemir E, Ersin M. Detergent and phosphate pollution in Gediz River, Turkey. African Journal of Biotechnology. 2009;**8**(15):3568-3575

[26] Vinod D, Neha S, Shalini S, Archna S, Aparna P. Effect of detergent use on water quality in Rewa city of India. Journal of Applied Chemistry. 2012;**1**(4):28-30

[27] Dehghani M, Mahvi A, Najafpoor A, Azam K. Investigation the potential of using acoustic frequency on the degradation of linear alkylbenzene sulfonates from aqueous solution. Journal of Zheijang University Science A. 2007;**8**:1462-1468

[28] Makwana SA, Patel CG, Patel TJ. Physico-chemical analysis of drinking water of Gandhinagar District. Archives of Applied Science Research. 2012;**4**(1):461-464

[29] Salvato JA, Nemero NL, Agardy FJ. Environmental Engineering. 5th ed. New York: John Wiley and Sons, Inc.; 2003

[30] Shafqat A, Shailendra Y, Tamheed F. Spectrophotometric determination of anionic detergents in the river Sai at Jaunpur. Paripex - Indian Journal of Research. 2012;**1**(12):72-73

[31] Fujii S, Polprasert C, Tanaka S, Lien NPH, Qiu Y. New POPs in the water environment: Distribution, bioaccumulation and treatment of perfluorinated compounds - A review paper. Journal of Water Supply: Research and Technology-AQUA. 2007;**56**:313-326

[32] Misra VG, Chawla G, Kumar V, Lal H, Viswanathan PN. Effect of linear alkyl benzene sulfonate in skin of fish fingerlings (Cirrhinamrigala): Observations with scanning electron microscope. Ecotoxicology and Environmental Safety. 1987;**13**:164-168

[33] Mukherjee B, Nivedita M, Mukherjee D. Plankton diversity and dynamics in a polluted eutrophic lake, Ranchi. Journal of Environmental Biology. 2010;**31**(5):827-839

[34] Ogundiran MA, Fawole OO, Adewoye SO, Ayandıran TA. Pathologic lesions in the gills of *Clarias gariepinus* exposed to sublethal concentrations of soap and detergent effluents. Journal of Cell and Animal Biology. 2009;**3**(5):78-82

[35] Kumar V, Singh PK, Kumar P, Sahoo P, Shukla NK, Markandeya KGC. Characterization of the river bed sediment profile and evaluation of urbanization pollutants at Lucknow city area. International Journal of Advanced Research. 2017a;**5**(8):370-380

[36] Kumar V, Singh PK, Sahoo P, Kumar P, Shukla NK, Markandeya K, et al. Status assessment of physicochemical parameters in Gomti River water quality at Lucknow city area, Uttar Pradesh, India. International Journal of Applied Research and Technology. 2017b;**2**(4):225-236

[37] Richards A. Effects of detergent use on water quality in Kathmandu, Nepal [master thesis]. Nepal: Massachusetts Institute of Technology; 2003

[38] Singh KP, Mohan D, Singh VK, Malik A. Studies on distribution and fractionation of heavy metals in Gomti river sediments - A tributary of the Ganges. Indian Journal of Hydrology. 2005;**312**:14-27

[39] Singh MK, Ruhela M, Kumar V, Bhatnagar VK. To study of water quality from river of Lucknow and effect on human health. International Archive of Applied Sciences & Technology. 2017;**8**(2):51-55

[40] Tiwari M, Kisku GC. Impact assessment of Gomti river water quality after immersion of idols during Durga Utsav. Biochemistry and Analytical Biochemistry. 2016;**5**:1-5. DOI: 10.4172/2161-1009.1000287

[41] APHA, AWWA, WEF. Standard Methods for Examination of Water and Wastewater. 22nd ed. Washington: American Public Health Association; 2012. p. 1360. ISBN 978-087553-013-0

[42] Jalal FN, Sanalkumar MG. Hydrology and water quality assessment of Achencovil river in relation to pilgrimage season. International Journal of Scientific and Research Publications. 2012;**2**(12):1-5

[43] Tank SK, Chippa RC. Analysis of water quality of Halena blocks in Bharatpur. International Journal of Scientific and Research Publications. 2013;**3**:1-6

[44] Bureau of Indian Standards. Indian Standard: Drinking Water – Specification (Second Revision), IS 10500. New Delhi, India; 2012

[45] Verma P, Chandawat D, Gupta U, Solanki H. Water quality analysis of an organically polluted lake by investigating different physical and chemical parameters. International Journal of Research in Chemistry and Environment. 2012;**2**(1):105-110

[46] Jalal FN, Sanalkumar MG. Water quality assessment of Pampa river in relation to pilgrimage season. International Journal of Research in Chemistry and Environment. 2013;**3**(1):341-347

[47] Francis-Floyd R. Dissolved Oxygen for Fish Production. Institute of Food and Agriculture Sciences, University of Florida; 2003. p. 3

[48] Taneja P, Labhasetwar P, Nagarnaik P. Nitrate in drinking water and vegetables: Intake and risk assessment in rural and urban areas of Nagpur and Bhandara districts of India. Environmental Science and Pollution Research. 2019;**26**:2026-2037. DOI: 10.1007/s11356-017-9195-y

[49] Spehar RL, Holcombe GW, Carlson RW, Drummod RA, Yount JD,

Pickering QH. Effects of pollution on fresh water fish. Journal - Water Pollution Control Federation. 1979;**51**:1616-1684

[50] Weith GD, Konasewitch DE. Structure-activity correlations in studies of toxicity and bioconcentration with aquatic organisms. In: Proceedings of a Symposium Held in Burlington, Ontario. Vol. 11. 1975. p. 347

[51] Koç H, Güven KC. Toxicity and identification of LAS in tissue of rainbow trout (*Oncorhynchus mykiss*). Turkish Journal of Pharmaceutical Sciences. 2002;**8**:197-208

[52] Fariha H, Shah AA, Javed S, Hameed A. Enzymes used in detergents: Lipases. African Journal of Biotechnology. 2010;**9**(31):4836-4844. DOI: 10.5897/AJBx09.026

[53] Kumar D, Verma A, Dhusia N, More N. Water quality assessment of River Gomti in Lucknow. Universal Journal of Environmental Research and Technology. 2013;**3**(3):337-344

[54] Ajmal MR, Uddin R, Khan UK. Heavy metals in water, sediments plants and fish of Kali Nadi U.P. (India). Environment International. 1988;**14**:515-523

Chapter 2

Adsorption Processes in the Removal of Organic Dyes from Wastewaters: Very Recent Developments

Francisco Jose Alguacil and Felix A. Lopez

Abstract

The problem of the treatment of contaminated wastewaters is of the upmost worldwide interest. This contamination occurs via the presence of inorganic or organic contaminants of different nature in relation with the industry they come from. In the case of organic dyes, their environmental impact, and thus, their toxicity come from the air (releasing of dust and particulate matter), solid (scrap of textile fabrics, sludges), though the great pollution, caused from dyes, comes from the discharge of untreated effluents into waters, contributing to increase the level of BOD and COD in these liquid streams; this discharge is normally accompanied by water coloration, which low the water quality, and caused a secondary issue in the wastewater treatment. Among separation technologies, adsorption processing is one of the most popular, due to its versatility, easiness of work, and possibility of scaling-up in the eve of the treatment of large wastewater volumes. Within a miriade of potential adsorbents for the removal of organic dyes, this work presented the most recent advances in the topic.

Keywords: organic dyes, adsorbents, wastewaters, water quality

1. Introduction

Nowadays, one of the most important environmental issues is related to water scarcity, water contamination and water quality. In many cases, water contamination is due to pollution by means of the presence of inorganic and/or organic compounds; particularly, the presence of organic dyes in waters, or in general, in effluents must be avoided since they are toxic (mutagenic, i.e. Azure B or Disperse Red 1, and carcinogenic, i.e. Basic Red 9 and Crystal Violet) against life and humans. Under the word "dye," is included substances which.

produced a color on materials of diverse nature, being the textile industry (54%) which produced the highest quantity of dyes-bearing effluents, followed by the dyeing industry (21%), paper and pulp industry (10%), tannery and paint industry (8%) and the own dye manufacturing industry (7%).

These compounds can be classified on structural bases, i.e. basic, acid, azo, disperse, anthraquinone or metal complex. The international standard of dye effluent discharge into the environment considered the next tolerable limits [1]:

i) biological oxygen demand: under 30 mg/L, ii) chemical oxygen demand: below 50 mg/L, iii). color: under 1 ppm; .iv) pH value in the 6–9 range, v) suspended solids: less than 20 mg/L, vi) temperature: below 42° C, and vii) toxic pollutants: completely avoid. However and despite all prohibitions, still today there is an important number of illegal dyes (i.e. Solvent Yellow 4, a member of the azolipophilic compounds) used in the textile industry. There are a number of useful technologies to eliminate these organic dyes from waters-effluents [2, 3], and among them, adsorption processes gained a paramount interest due to their manipulation easiness, scaling up, and the possibility of using a countless number of potential adsorbents.

The present work, and due to space constrains, reviewed the most recent results (January–April 2020) about the removal of organic dyes by adsorption processes.

2. Adsorbents and organic dyes: the adsorption results

Being a plant practically present around the world, cactus, and due to their chemical composition and biological and nutritional properties, find various applications, being one of them, their use as adsorbents for toxic metals and organic dyes [4]. Thus, different cactus parts: fruit seeds, peel, clodades, among others, had been investigated in the topic of the removal of organic dyes from waters. Often, these bioadsorbents were subject, before use, to some type of treatment, such as heat treatment, chemical treatment, sun-dehydration. Some results about the adsorption capacity of these adsorbents are given in **Table 1**.

These cactus-based adsorbents presented maximum adsorption capacities in close relation with those derived from other materials of different origins (**Table 2**).

Others bioadsorbents used in the removal of organic dyes (maximum capacity) are: leaves of *Lawsonia sp.* (malachite green (no data)) [5], *Platanus orientalis* (rhodamine B (557 mg/g), methyl orange (327 mg/g)) [6], *α-chitin* (methylene blue

Adsorbent	Dyes (maximum capacity, mg/g)
Cladodes of *Tacinga palmadora*	Crystal violet (229)
Fruit peels of *O. ficus indica*	Crystal violet (312). Metyhylene blue (416)
Natural cladodes of *O. ficus indica*	Methylene blue (3.5)
Palm cactus	Crystal violet (173-220)
O. ficus indica	Basic Blue 9 (35-278)

Table 1.
Use of cactus in dyes adsorption.

Adsorbent	Dyes (maximum capacity, mg/g)
Leaves of *Tacinga Palmadora*	Crystal violet (229)
Tarjuna sawdust waste	Crystal violet (46)
Rice husk	Crystal violet (293)
Cactus cladodes (activated carbon)	Methylene Blue (750)
Walnut shell	Methylene Blue (315)
Wood apple rind	Methylene Blue (40)

Table 2.
Crystal violet and methylene blue adsorption onto different adsorbents.

(95 mg/g)) [7], pods of *Clitoria fairchildiana* (rhodamine 66 (571 mg/g)) [8], and diatomite waste (methylene blue (25 mg/g), acid orange (35 mg/g)) [9].

Nanomaterial-based adsorbents are another type of materials that, due to their properties and adsorption capacities, have applications in the removal or organic dyes from waters. Including in these nanomaterials, carbon nanotubes (CNTs), grapheme sheets (GS), and metal oxides (MO) are found [10].

Carbon nanotubes presented a sp^2 allotropic carbon of graphite structure in cylindrical or tube shaped sheets. Based on the number of these sheets presented in the adsorbent, CNTs can be found as single-walled carbon nanotubes (SWCNTs) or multi-walled carbon nanotubes (MWCNTs). Typically, SWCNTs presented a diameter in the 0.4–10 nm range, whereas MWCNTs have a diameter in the 10–100 nm range and spacing between sheets in the 0.34–0.38 nm range.

The adsorptive effectiveness of these carbon nanotubes can be improved by functionalizing them or modifying some of their characteristics: specific area, charge density, porosity, and hydrophilicity. These modifications can be done by acid/oxidant treatment, combination with metals/MO, and grafting special functional groups, such as polymer and surfactants.

These carbon materials presented four characteristics adsorption sites in their surfaces. Thus, the adsorption process occurred at i) the external, and/or ii) internal surface of the nanotubes, iii) the interstitial pathways between individual nanotubes sheet, and iv) the external groove sites. In the case of MWCNTs, the space between the sheets can also be used to adsorb organic dyes.

Organic dyes uptake onto these nanotubes responded very often to the Langmuir and Freundlich isotherm models, and the adsorption kinetics is best fitted to the pseudo-second-order kinetic model.

Graphene is formed by a single layer of sp^2 allotropic carbon atoms arranged in a two-dimensional hexagonal honeycomb lattice structure.

Similarly to carbon nanotubes, single-layer (SLG) or multiple-layer graphene (MLG) materials can be yielded in a 2D structure from a graphite-based material. Other derived materials, such as graphene oxide (GO) and reduced graphene oxide (RGO), with enhanced adsorptive characteristics, can be produced by chemical oxidation of graphite and reduction of grapheme oxide, respectively. These two last materials, presented better adsorption characteristics than the above grapheme materials.

A graphene/wastepaper composite [11], had been used in the removal of methylene blue and Congo red from waters, with maximum capacities of 58 and 90 mg/g, respectively. Nanoribbons of graphene were used in the adsorption of methylene blue and orange II dyes [12], in this case the maximum capacities, presented for the adsorbent, were of 280 and 265 mg/g, respectively. Others graphene-based materials had been recently used in the adsorption of crystal violet (69 mg/g) [13], rhodamine B (963 mg/g) [14], etc.

It was found that for selected organic dyes, graphene-based adsorbent had an average 2–5 times higher dye adsorption capacity than carbon nanotubes and metal oxides.

MOs adsorbents applied for the treatment of organic dyes-bearing waters included, iron oxide (Fe_3O_4), zinc oxide (ZnO), titanium dioxide (TiO_2), magnesium oxide (MgO), alumina oxide (Al_2O_3)), and zirconium oxide (ZrO_2). Among them, iron oxide nanoparticles presented good properties i.e. high specific surface area, to adsorb organic dyes, and they are magnetic. This characteristic facilitated the dispersion of the nanoparticles in the aqueous solution, and their removal from it, when an external magnetic field is applied [15].

Other investigations described the use of nanohybrids of $Cu_xO/Fe_2O_3/MoC$ as materials used in the adsorption of reactive red 195A and reactive yellow

84 (maximum capacities 435 and 278 mg/g, respectively) [16], also the use of Cr-doped ZnO in the adsorption of methyl orange (19 mg/g) and methylene blue (41 mg/g), the adsorption of methyl orange (833 mg/g) by a magnetic composite [17], and Fe_3O_4/PPy composites (eosin Y, methyl orange and brilliant green: 212, 149 and 264 mg/g, respectively) [18]. It was described in the literature [19], the usefulness of Ag_2O as adsorbent of Congo red (181 mg/g), acid orange 7 (125 mg/g) and amido black 10B (83 mg/g), however this investigation, as many others, did not give any information about the desorption step.

The list of nanoparticles or nanomaterials used to remove organic dyes from waters seemed not to end [20, 21], considering that a series of materials such as biomass, clay minerals, different wastes, etc., when modified with magnetic nanoparticles enhanced their respective adsorption capacity towards organic dyes, because they increased their surface area and porosity and with the addition of the magnetic nanoparticles, they adopt a new property, as is the magnetic character, which improve their separation from the treated water. Moreover, by the addition of adequate functional groups to these nanoadsorbents, basically on their surface, they further improve their respective capacities on the treatment of waters contaminated with organic dyes. Not being exhaustive, **Table 3** summarized some of the results encountered in this field.

Polyaniline, polypyrrole and other conducting polymers, had been also used in the removal of organic dyes from waters. In fact, these polymers reacted with organic dyes due to the similarity of the conjugated molecular structures, of both types of compounds, which enhanced the reactivity between them. The use of these organic dyes as templates in the conducting polymer synthesis may affect both the conductivity and morphology control of the end product, specially in the case of polypyrrole.

It was described [22], how conducting polymers and organic dyes reacted:

i. π-π interaction between the aromatic rings,

ii. electrostatic ionic interactions,

iii. hydrogen bonding, and.

iv. hydrophobic interactions.

Adsorbent	Organic dye (removal efficiency, mg/g)
Magnetic baker's yeast biomass	methyl violet (61)
Modified multi-walled carbon nanotubes	alizarin yellow R (45)
Magnetic peach gum bead	methylene blue (232)
Magnetic polyacrylamide micropheres	methylene blue (1990)
Malachite@clay nanocomposite	Congo red (238)
ZnO nanorods loaded activated carbon	brilliant green (58)
Sorel's cement nanoparticles	methyl orange (21)
Alkaline treated timber sawdust	methylene blue (694)
Ultrathin $MoSe_2$ nanosheets	rhodamine B (133)
Cellulose nanocrystal-reinforced keratin	reactive black 5 (1201)

Table 3.
Adsorbents and adsorbed organic dyes.

Some dyes used in the preparation of polyaniline are: methyl orange and green GS, whereas in the case of polypyrrole, the list included both mentioned above and Congo red, Thymol blue, cresol red and rhodamine B among others.

In the conducting role, polyaniline and polypyrrole are polycations, thus, it is expected that they normally reacted with anionic dyes, however, experimentally it was found that both cationic and anionic dyes reacted with these conducting polymers; it seemed that electrostatic ionic interactions are not the most important factor to explain this reactivity.

As it is mentioned above, these conducting polymers had been used, alone or in the composite forms, in the removal of numerous organic dyes from waters, and basically their success is due to that they have a relative low production cost.

In the case of polyaniline, the list of investigations related to the removal of organic dyes from waters included: Congo red, eosin Y, rose bengal, indigo carmine, etc. Polyaniline composites are classified along the non-conducting component, and included composites containing (in various forms): aluminum, bismuth, carbon, iron, silicium, natural polymers (cellulose), synthetic polymers, etc.

Pristine polypyrrole had been investigated in the removal of methyl orange, methylene blue, etc. Polypyrrole composites again contained in various forms: aluminum, carbon, titanium, zinc, etc.

Table 4 presented a comprehensive (but not exhaustive) list of organic dyes adsorbed by polyaniline and polypyrrole composites.

The use of polyaniline-related materials, such as aniline oligomers and copolymers, polyaniline chemically modified, etc., had a further interest in the removal of organic dyes from waters. The use of polypyrrole-related materials in this environmental role has a significant minor development.

Due to their environmentally friendly and availability, polymer derivatives based on polysaccharides are also of interest in the removal of organic dyes from contaminated waters. Thus, a variety of modified polysaccharides, i.e. chitosan, starch, dextran, cellulose, have been investigated as adsorbents in this role; however, and despite some of pullulan characteristics such as: high solubility and flexibility of the backbone when compared with other polysaccharides, they are not amply used in waters purification [27].

	Composite	**Organic dye (maximum capacity, mg/g)**	**Ref.**
Polyaniline	TiO_2	Methyl orange (62)	[23]
	TiO_2	Methylene blue (83) Congo red (20) Crystal violet (50) Rhodamine 6G (57)	[24]
	SiO_2	Methylene blue (55) Congo red (33) Crystal violet (25) Rhodamine 6G (2)	[24]
	resin	Methylene blue (40)[a] Methylene blue (28)[b]	[25]
	PSMA	Methyl orange (148)	[26]
Polypyrrole	Fe_3O_4	Eosin Y (712) Methyl orange (149) Brilliant green (264)	[18]

[a]*Mixed solutions of Cr(VI) and methylene blue.*
[b]*Solutions containing only methylene blue.*

Table 4.
Organic dyes adsorbed onto polyaniline and polypyrrole composites.

Pullulan, having a chemical formula $(C_6H_{10}O_5)_n$, is a linear, non-ionicpolysaccharide consisting of maltotrioseunits:α-(1 → 6)-linked(1 → 4)-α-d-triglucosides. The known pullulan derivatives are: i) soluble ionic pullulan derivatives: this type of compounds can be synthesized by chemical modification of the polysaccharide. It can include i.i) various content and length of grafted chains, and i.ii) various content of tertiary amine groups, ii) pullulan microspheres: they can be formed by suspension cross-linking of the previously grafted pullulan with cationic moieties (P-g-pAPTAC); iii) nonionic thermosensitive pullulan copolymer: it was prepared by graft-polymerization of p(N-isopropylacrylamide) onto the pullulan. Here, cerium(IV) was used as initiator. The resulted thermosensitive material has the (P-g-pNIPAAm) acronysm; iv) nonionic pullulan-graft-polyacrylamide hydrogel: this pullulan was synthesized by free radical polymerization in presence of a cross-linking agent and calcium carbonate.

Pullulan derivatives showed high removal efficiency of organic dyes contaminants, i.e. P-g-APTAC microspheres were used in the adsorption of azocarmine B (maximum capacity: 114 mg/g), acid orange 7B (65 mg/g) and methyl orange (55 mg/g); pullulan-graft-polycrylamide hydrogel was used in the adsorption of methylene blue and reactive blue with maximum capacities (70° C) of 399 and 356 mg/g, respectively.

The properties of metal–organic frameworks (MOFs), made of them interesting materials for the removal of organic dyes from waters. Some of these properties are: thermal stability, high surface area and porosity, nanosized cavities, etc. The metal centers of these materials provided additional coordination locations aimed to fixing organic dyes, whereas one step ahead in the practical use of these MOFs is provided them with magnetic properties *via* the incorporation of magnetic materials within the framework. Some examples of the use of these materials are given in **Table 5**.

Similarly to the number of materials used to adsorb organic dyes, the number of these chemical compounds investigated to be adsorbed, seemed to be countless. Besides all the mentioned along this work, below, it is summarized further investigations and results (maximum capacities, mg/g):

i. methylene blue: modified Cs-ZnS (502) [33], Cr-doped ZnO nanorods (41) [34], attapulgite derivative (115) [35], mesoporous Zr-based polymer (60) [36], diatomite waste (32) [37], iron-carbon nanosheets (185) [38], graphene oxide derivative (1370) [39], zeolite/CeO_2 nanocomposite (2.5) [40], pyridine derivative (175) [41], cellulose nanocomposite (2067) [42], cellulose/carbon aerogel (1179) [43], MgO modified biochar (475) [44], MoS_2/WO_3 (228) [45],

ii. methyl orange: Cr-doped ZnO nanorods (16) [34], TOCN/CGG hydrogel (134) [46],

Adsorbent	Organic dye (maximum adsorption, mg/g)	Reference
Cd-MOF	Congo red (192)	[28]
Zn-MOF	methylene blue (116)	[29]
HSO_3-MOF	methylene blue(833)	[30]
LDH-MOF	orange II (1173)	[31]
aramid nanofibrils-MOF	methyl violet (114)	[32]

Table 5.
MOFs and adsorption of organic dyes.

iii. Congo red: iron-carbon nanosheets (532) [38], $Ba_5Si_8O_{21}$ microspheres (1239) [47], cellulose/carbon aerogel (585) [43], MIL-100(Fe) (1791) [48],

iv. bromophenol blue: *Hermetia illucens* larvae (571) [49],

v. acid chrome blue K: MIL-100(Fe) (926) [50],

vi. crystal violet: attapulgite derivative (69) [37], vii) thioflavin T: TOCN/CGG hydrogel (430) [46],

vii. methyl blue: nitrogen-doped carbon derivative (1054) [50],

viii. eriochrome black T: carbon derivative (166) [51],

ix. rhodamine B: WS_2/WO_3 (237) [52],

x. reactive yellow: diatomite waste (33) [37],

xi. indigo carmine: N and S organic framework (547) [53],

xii. direct red 31: nanoporous composites (526) [54],

xiii. acid blue 92: nanoporous composites (556) [54].

3. Conclusions

During the first quarter of 2020 year, an important number of dye adsorption procedures have been proposed in a series of published papers, all of them, claiming successful dye removal. However, in the opinion of the authors of the present review some criticism to the papers must be accounted for:

i. comparison of the maximum dye capacity presented different numbers, see **Table 6**. Thus, the successfulness of some of these adsorbents can be questioned if compared with the figure presented by other materials,

ii. near 52% of the published papers included some data about the desorption step, that is, the other 48% of the published papers are uncompleted in terms of the overall adsorption–desorption process. Authors must aware that as important is the adsorption step than the desorption one, and in all the cases in which the latter is included, the authors do not give any comment about what to do with the now dye-bearing desorption solution. In the manuscripts, organic dyes simply go to one solution (the feed one) to another solution (the desorption one) (except in the case of degradation occurs),

iii. all the reviewed manuscripts lacked the investigation of a sometimes key variables on adsorption processes and in batch mode, as the stirring speed and how the phases mixed can be. This is because, with the correct stirring speed, the.

iv. thickness of the aqueous film layer reached a minimum and the adsorption reached a maximum. Together with the above, the form in which the phases are mixed also ensured the best contact, and thus the best solute transfer, between the phases involved in the process.

Organic dye	Adsorbent	Maximum capacity (mg/g)	Ref.
Methylene blue	functionalized-organic polymer	2740	[55]
	cellulose hydrogel	756	[56]
	modified grapheme oxide	257	[57]
	functionalized lignosulfate	63	[58]
Methyl orange	NiAlTi	1250	[59]
	MIL-101-NH_2	462	[60]
	polyaniline composite	148	[26]
Crystal violet	fly ash	433	[61]
	polyaniline-metal oxides	50	[24]
Congo red	MIL-100(Fe)	1791	[31]
	cellulose/carbon aerogels	585	[43]
	polyaniline-metal oxides	33	[24]
Rhodamine B	Ni-graphene composite	963	[14]
	MoS_2 nanoflowers	365	[62]
	polyaniline-metal oxides	20	[24]

Table 6.
Difference in organic dyes adsorption capacity of various adsorbents.

It is needed to mention here, that in Refs. [63, 64] a heavy scientific fault was detected: the authors of the manuscripts investigated, besides the adsorption of organic dyes, the adsorption of Cr^{6+}, written as such, when this element in the VI oxidation state, never exists as a cation in aqueous solutions. This fault also is responsibility of the corresponding reviewers and of the Editors.

In practical terms, the real bottlenecks in the usefulness of these adsorbents are:

i. lack of information about the desorption step,

ii. evident loss of adsorption capacity under continuous adsorption–desorption cycles,

iii. adsorbent cost and possibility of production at large scale to scale-up the adsorption–desorption process,

iv. environmental friendship of the own adsorbent,

v. lack of information on the use of the purification of real waters bearing organic dyes.

Acknowledgements

To the CSIC Spanish Agency for support. Authors also acknowledge support of the publication fee by the CSIC Open Access Publication Support Initiative through its Unit of Information Resources for Research (URICI).

Author details

Francisco Jose Alguacil* and Felix A. Lopez
National Center for Metallurgical Research (CENIM), Spanish National Council for Scientific Research (CSIC), Madrid, Spain

*Address all correspondence to: fjalgua@cenim.csic.es

References

[1] Katheresan V, Kansedo J, Lau SY. Efficiency of various recent wastewater dye removal methods: A review. Journal of Environmental Chemical Engineering. 2018;6:4676-4697. DOI: 10.1016/j.jece.2018.06.060

[2] Ismail M, Akthar K, Khan, MI, Kamal T, Khan A, Asiri AM, Seo J, Khan SB. Pollution, toxicity and carcinogenicity of organic dyes and their catalytic bio-remediation. Current Pharmaceutical Design. 2019; 25:3653-3671. DOI: 10.2174/13816128256661910 21142026

[3] Lellis B, Favaro-Polonio CZ, Pamphile JA, Polonio JC. Effect of textile dyes on health and the environment and bioremediation potential of living organisms. Biotechnology Research & Innovation. 2019. 3:275-290. DOI: 10.1016/j.biori.2019.09.001

[4] Amari A, Alalwan B, Eldirderi MM, Mnif W, Ben Rebah F, Cactus material-based adsorbents for the removal of heavy metals and dyes: a review. Materials Research Express. 2020;7:012002. DOI: 10.1088/2053-1591/ab5f32

[5] Sharmila S, Dinesh M, Kowsalya E, Kamalambigeswari R, Rebecca LJ. Biosorption of dye using Lawsonia sp. as adsorbent. Research Journal of Pharmacy and Technology. 2020;13:1651-1654. DOI: 10.5958/0974-360X.2020.00299.1

[6] Jiang X, Sun P, Xu L, Xue Y, Zhang H, Zhu W. Platanus orientalis leaves based hierarchical porous carbon microspheres as high efficiency adsorbents for organic dyes removal. Chinese Journal of Chemical Engineering. 2020;28:254-265. DOI: 10.1016/j.cjche.2019.03.030

[7] Abou-Zeid RE, Salama A, Al-Ahmed ZA, Awwad NS, Youssef MA. Carboxylated cellulose nanofibers as a novel efficient adsorbent for water purification. Cellulose Chemistry and Technology. 2020;54:237-245. DOI: 10.35812/CELLULOSECHEMTECHNOL.2020.54.25

[8] da Silva AMB, Serrão NO, de Gusmão Celestino G, Takeno ML, Antunes NTB, Iglauer S, Manzato L, de Freitas FA, Maia, PJS. Removal of rhodamine 6G from synthetic effluents using *Clitoria fairchildiana* pods as low-cost biosorbent. Environmental Science and Pollution Research. 2020;27:2868-2880. DOI: 10-1007/s11356-019-07114-6

[9] Liu N, Wu Y, Sha H. Magnesium oxide modified diatomite waste as an efficient adsorbent for organic dye removal: adsorption performance and mechanism studies. Separation Science and Technology. 2020;55:234-246. DOI: 10.1080/01496395.2019.1577456

[10] Awad AM, Jalab R, Benamor A, Nasser MS, Ba-Abbad MM, El-Naas M, Mohammad AW. Adsorption of organic pollutants by nanomaterial-based adsorbents: An overview. Journal of Molecular Liquids. 2020;301:112335. DOI: 10.1016/j.molliq.2019.112335

[11] Feng C, Ren P, Li Z, Tan W, Zhang H, Jin Y, Ren F. Graphene/waste-newspaper cellulose composite aerogels with selective adsorption of organic dyes: Preparation, characterization, and adsorption mechanism. New Journal of Chemistry. 2020;44:2256-2267. DOI: 10.1039/c9nj05346h

[12] Gao Y, Liang X, Han S, Wu L, Zhang G, Qin C, Bao S, Wang Q, Qi L, Xiao L. High-efficiency adsorption for both cationic and anionic dyes using graphene nanoribbons formed by atomic-hydrogen induced single-walled carbon nanotube carpets. Carbon Letters. 2020;30:123-132. DOI: 10.1007/s42823-019-00089-x

[13] Czepa W, Pakulski D, Witomska S, Patroniak V, Ciesielski A, Samorì P. Graphene oxide-mesoporous SiO2 hybrid composite for fast and efficient removal of organic cationic contaminants. Carbon. 2020;158:193-201. DOI: 10.1016/j.carbon.2019.11.091

[14] Tao X, Wang S, Li Z. Ultrasound-assisted bottom-up synthesis of Ni-graphene hybrid composites and their excellent rhodamine B removal properties. Journal of Environmental Management. 2020;255:109834. DOI: 10.1016/j.jenvman.2019.109834

[15] Luyen NT, Linh HX, Huy TQ. Preparation of Rice Husk Biochar-Based Magnetic Nanocomposite for Effective Removal of Crystal Violet. Journal of Electronic Materials. 2020;49:1142-1149. DOI: 10.1007/s11664-019-07798-z

[16] Bastami TR, Khaknahad S, Malekshahi M. Sonochemical versus reverse-precipitation synthesis of CuxO/Fe2O3/MoC nano-hybrid: removal of reactive dyes and evaluation of smartphone for colorimetric detection of organic dyes in water media. Environmental Science and Pollution Research. 2020; 27:9364-9381. DOI: 10.1007/s11356-019-07368-0

[17] Natarajan S, Anitha V, Gajula GP, Thiagarajan V. Synthesis and Characterization of Magnetic Superadsorbent Fe3O4-PEG-Mg-Al-LDH Nanocomposites for Ultrahigh Removal of Organic Dyes. ACS Omega. 2020;5:3181-3193. DOI: 10.1021/acsomega.9b03153

[18] Zhang M, Yu Z, Yu H. Adsorption of Eosin Y, methyl orange and brilliant green from aqueous solution using ferroferric oxide/polypyrrole magnetic composite. Polymer Bulletin. 2020;77:1049-1066. DOI: 10.1007/s00289-019-02792-1

[19] Wang Y, Bi N, Zhang H, Tian W, Zhang T, Wu P, Jiang W. Visible-light-driven photocatalysis-assisted adsorption of azo dyes using Ag2O. Colloids and Surfaces A: Physicochemical and Engineering Aspects. 2020;585:124105. DOI: 10.1016/j.colsurfa.2019.124105

[20] Tara N, Siddiqui SI, Rathi G, Chaudhry SA, Inamuddin, Asiri AM. Nano-engineered adsorbent for the removal of dyes from water: A review. Current Analytical Chemistry. 2020;16:14-40. DOI: 10.2174/1573411015666190117124344

[21] Lu F, Astruc D. Nanocatalysts and other nanomaterials for water remediation from organic pollutants. Coordination Chemistry Reviews. 2020;408:213180. DOI: 10.1016/j.ccr.2020.213180

[22] Stejskal J. Interaction of conducting polymers, polyaniline and polypyrrole, with organic dyes: polymer morphology control, dye adsorption and photocatalytic decomposition. Chemical Papers. 2020;74:1-54. DOI: 10.1007/s11696-019-00982-9

[23] Cionti C, Della Pinal C, Meroni D, Falletta E, Ardizzone S. Photocatalytic and oxidative synthetic pathways for highly efficient PANI- TiO_2 nanocomposites as organic and inorganic pollutant sorbents. Nanomaterials. 2020;10:441. DOI:10.3390/nano10030441

[24] Maruthapandi M, Eswaran L, Luong JHT, Gedanken A. Sonochemical preparation of polyaniline@TiO2 and polyaniline@SiO2 for the removal of anionic and cationic dyes. Ultrasonics Sonochemistry. 2020;62:104864. DOI: 10.1016/j.ultsonch.2019.104864

[25] Ding J, Pan Y, Li L, Liu H, Zhang Q, Gao G, Pan B. Synergetic adsorption and electrochemical classified recycling of Cr(VI) and dyes in synthetic dyeing wastewater. Chemical Engineering Journal. 2020;384:123232. DOI: 10.1016/j.cej.2019.123232

[26] Liu Y, Song L, Du L, Gao P, Liang N, Wu S, Minami T, Zang L, Yu C, Xu X. Preparation of polyaniline/ emulsion microsphere composite for efficient adsorption of organic dyes. Polymers. 2020;12:167. DOI: 10.3390/ polym12010167

[27] Ghimici L, Constantin M. A review of the use of pullulan derivatives in wastewater purification. Reactive and Functional Polymers. 2020;149:104510. DOI: 10.1016/j. reactfunctpolym.2020.104510

[28] Ghomshehzadeh SG, Nobakht V, Pourreza N, Mercandelli P, Carlucci L. A new pillared Cd-organic framework as adsorbent of organic dyes and as precursor of CdO nanoparticles. Polyhedron. 2020;176:114265. DOI: 10.1016/j.poly.2019.114265

[29] Hong YS, Sun SL, Sun Q, Gao EQ, Ye M. Tuning adsorption capacity through ligand pre-modification in functionalized Zn-MOF analogues. Materials Chemistry and Physics. 2020;243:122601. DOI: 10.1016/j. matchemphys.2019.122601

[30] Saghian M, Dehghanpour S, Sharbatdaran M. Unique and efficient adsorbents for highly selective and reverse adsorption and separation of dyes: Via the introduction of SO3H functional groups into a metal-organic framework. RSC Advances. 2020;10:9369-9377. DOI: 10.1039/ c9ra10840h

[31] Soltani R, Marjani A, Shirazian S. A hierarchical LDH/MOF nanocomposite: Single, simultaneous and consecutive adsorption of a reactive dye and Cr(VI). Dalton Transactions. 2020;49:5323-5335. DOI: 10.1039/d0dt00680g

[32] Zong L, Yang Y, Yang H, Wu X. Shapeable aerogels of metal-organic-frameworks supported by aramid nanofibrils for efficient adsorption and interception. ACS Applied Materials and Interfaces. 2020;12:7295-7301. DOI: 10.1021/acsami.9b22466

[33] Ali, H. Ternary system from mesoporous CdS–ZnS modified with polyaniline for removal of cationic and anionic dyes. Research on Chemical Intermediates. 2020;46:571-592. DOI: 10.1007/s11164-019-03968-0

[34] Chen J, Xiong Y, Duan M, Li X, Li J, Fang S, Qin S, Zhang R. Insight into the Synergistic Effect of Adsorption-Photocatalysis for the Removal of Organic Dye Pollutants by Cr-Doped ZnO. Langmuir. 2020;36:520-533. DOI: 10.1021/acs.langmuir.9b02879

[35] Cui M, Mu P, Shen Y, Zhu G, Luo L, Li J. Three-dimensional attapulgite with sandwich-like architecture used for multifunctional water remediation. Separation and Purification Technology. 2020;235:116210. DOI: 10.1016/j. seppur.2019.116210

[36] Liang C, Ren J, El Hankari S, Huo J. Aqueous Synthesis of a Mesoporous Zr-Based Coordination Polymer for Removal of Organic Dyes. ACS Omega. 2020;5:603-609. DOI: 10.1021/ acsomega.9b03192

[37] Ma T, Wu Y, Liu N, Wu Y. Hydrolyzed polyacrylamide modified diatomite waste as a novel adsorbent for organic dye removal: Adsorption performance and mechanism studies. Polyhedron. 2020;175:114227. DOI: 10.1016/j.poly.2019.114227

[38] Manippady SR, Singh A, Basavaraja BM, Samal AK, Srivastava S, Saxena M. Iron-Carbon Hybrid Magnetic Nanosheets for Adsorption-Removal of Organic Dyes and 4-Nitrophenol from Aqueous Solution. ACS Applied Nano Materials. 2020;3:1571-1582. DOI: 10.1021/ acsanm.9b02348

[39] Neskoromnaya EA, Burakov AE, Melezhik AV, Babkin AV, Burakova IV,

Kurnosov DA, Tkachev AG. Synthesis and evaluation of adsorption properties of reduced grapheme oxide hydro- and aerogels modified by iron oxide nanoparticles. Inorganic Materials: Applied Research. 2020;11:467-475. DOI: 10.1134/S2075113320020264

[40] Nyankson E, Adjasoo J, Efavi JK, Yaya A, Manu G, Kingsford A, Abrokwah RY. Synthesis and kinetic adsorption characteristics of Zeolite/ CeO2 nanocomposite. Scientific African. 2020;7:e00257. DOI: 10.1016/j.sciaf.2019.e00257

[41] Rajendran N, Samuel J, Amin MO, Al-Hetlani E, Makhseed S. Carbazole-tagged pyridinic microporous network polymer for CO2 storage and organic dye removal from aqueous solution. Environmental Research. 2020;182:109001. DOI: 10.1016/j.envres.2019.109001

[42] Wang G, Zhang J, Lin S, Xiao H, Yang Q, Chen S, Yan B, Gu Y. Environmentally friendly nanocomposites based on cellulose nanocrystals and polydopamine for rapid removal of organic dyes in aqueous solution. Cellulose. 2020;27:2085-2097DOI: 10.1007/s10570-019-02944-6

[43] Yu Z, Hu C, Dichiara AB, Jiang W, Gu J. Cellulose nanofibril/carbon nanomaterial hybrid aerogels for adsorption removal of cationic and anionic organic dyes. Nanomaterials. 2020;10:169. DOI: 10.3390/nano10010169

[44] Zheng Y, Wan Y, Chen J, Chen H, Gao B. MgO modified biochar produced through ball milling: A dual-functional adsorbent for removal of different contaminants. Chemosphere. 2020;243:125344: DOI: 10.1016/j.chemosphere.2019.125344

[45] Zheng Y, Wang J, Wang Y, Zhou H, Pu Z, Yang Q, Huang W. The combination of MoS2/WO3 and its adsorption properties of methylene blue at low temperatures. Molecules. 2020;25:2. DOI: 10.3390/molecules25010002

[46] Dai L, Cheng T, Xi X, Nie S, Ke H, Liu Y, Tong S, Chen Z. A versatile TOCN/CGG self-assembling hydrogel for integrated wastewater treatment. Cellulose. 2020;27:915-925. DOI: 10.1007/s10570-019-02834-x

[47] Xu L, Sun P, Jiang X, Chen J, Wang J, Zhang H, Zhu W. Hierarchical quasi waxberry-like Ba5Si8O21 microspheres: Facile green rotating hydrothermal synthesis, formation mechanism and high adsorption performance for Congo red. Chemical Engineering Journal. 2020;384:123387. DOI: 10.1016/j.cej.2019.123387

[48] Guo XZ, Han SS, Yang JM, Wang XM, Chen SS, Quan S. Effect of Synergistic Interplay between Surface Charge, Crystalline Defects, and Pore Volume of MIL-100(Fe) on Adsorption of Aqueous Organic Dyes. Industrial and Engineering Chemistry Research. 2020;59:2113-2122. DOI: 10.1021/acs.iecr.9b05715

[49] de Souza PR, do Carmo Ribeiro TM, Lôbo AP, Tokumoto MS, de Jesus RM, Lôbo I.P. Removal of bromophenol blue anionic dye from water using a modified exuviae of *Hermetia illucens* larvae as biosorbent. Environmental Monitoring and Assessment. 2020;192:197. DOI: 10.1007/s10661-020-8110-z

[50] He Y, Fu X, Wu H, Zhu T, Li S, Na B, Peng C. Highly efficient removal of methyl blue from aqueous solution using a novel nitrogen-doped porous magnetic carbon, Desalination and Water Treatment. 2020;173:409-419. DOI: 10.5004/dwt.2020.24658

[51] Khan A, Wei D, Khuda F, Ma R, Ismail M, Ai Y. Comparative adsorption capabilities of rubbish

tissue paper–derived carbon-doped MgO and CaCO3 for EBT and U(VI), studied by batch, spectroscopy and DFT calculations. Environmental Science and Pollution Research. 2020;27:13114-13130. DOI: 10.1007/s11356-020-07796-3

[52] Li G, Wang Y, Bi J, Huang X, Mao Y, Luo L, Hao H. Partial oxidation strategy to synthesize WS2/WO3 heterostructure with enhanced adsorption performance for organic dyes: Synthesis, modelling, and mechanism. Nanomaterials. 2020;10:278. DOI: 10.3390/nano10020278

[53] Pan X, Qin X, Zhang Q, Ge Y, Ke H, Cheng, G. N- and S-rich covalent organic framework for highly efficient removal of indigo carmine and reversible iodine capture. Microporous and Mesoporous Materials. 2020;296:109990. DOI: 10.1016/j.micromeso.2019.109990

[54] Hasanzadeh M, Simchi A, Shahriyari Far H. Nanoporous composites of activated carbon-metal organic frameworks for organic dye adsorption: Synthesis, adsorption mechanism and kinetics studies. Journal of Industrial and Engineering Chemistry. 2020;81:405-414. DOI: 10.1016/j.jiec.2019.09.031

[55] Yan KY, Chen JY, Li XY, Wang Q, Kuang GC. Carboxylic Acid Enriched Porous Organic Polymer as a Platform for Highly Efficient Removal of Methylene Blue from Aqueous Solution. Macromolecular Chemistry and Physics. 2020;221:1900553. DOI: 10.1002/macp.201900553

[56] Li Y, Hou X, Pan Y, Wang L, Xiao H. Redox-responsive carboxymethyl cellulose hydrogel for adsorption and controlled release of dye. European Polymer Journal. 2020;123:109447. DOI: 10.1016/j.eurpolymj.2019.109447

[57] Wang Z, Gao M, Li X, Ning J, Zhou Z, Li G. Efficient adsorption of methylene blue from aqueous solution by graphene oxide modified persimmon tannins. Materials Science and Engineering C. 2020;108:110196. DOI: 10.1016/j.msec.2019.110196

[58] Abboud M, Sahlabji T, Haija MA, El-Zahhar AA, Bondock S, Ismail I, Keshk SMAS. Synthesis and characterization of lignosulfonate/amino-functionalized SBA-15 nanocomposites for the adsorption of methylene blue from wastewater. New Journal of Chemistry. 2020; 44:2291-2302. DOI: 10.1039/d0nj00076k

[59] Rathee G, Singh N, Chandr, R. Simultaneous Elimination of Dyes and Antibiotic with a Hydrothermally Generated NiAlTi Layered Double Hydroxide Adsorbent. ACS Omega. 2020;5:2368-2377. DOI: 10.1021/acsomega.9b03785

[60] Zhang W, Zhang RZ, Yin Y, Yang JM. Superior selective adsorption of anionic organic dyes by MIL-101 analogs: Regulation of adsorption driving forces by free amino groups in pore channels. Journal of Molecular Liquids. 2020;302:112616. DOI: 10.1016/j.molliq.2020.112616

[61] Sočo E, Papciak D, Michel MM. Novel application of mineral by-products obtained from the combustion of bituminous coal-fly ash in chemical engineering. Minerals. 2020;10:66. DOI: 10.3390/min10010066

[62] Xiao X, Wang Y, Cui B, Zhang X, Zhang D, Xu X. Preparation of MoS2 nanoflowers with rich active sites as an efficient adsorbent for aqueous organic dyes. New Journal of Chemistry. 2020;44:4558-4567. DOI: 10.1039/d0nj00129e

[63] Chen X, Cui J, Xu X, Sun B, Zhang L, Dong W, Chen C, Sun D. Bacterial cellulose/attapulgite magnetic composites as an efficient adsorbent for heavy metal ions and dye

treatment. Carbohydrate Polymers. 2020;229:115512. DOI: 10.1016/j.carbpol.2019.115512

[64] Li D, Tian X, Wang Z, Guan Z, Li X, Qiao H, Ke, H, Luo L, Wei, Q. Multifunctional adsorbent based on metal-organic framework modified bacterial cellulose/chitosan composite aerogel for high efficient removal of heavy metal ion and organic pollutant. Chemical Engineering Journal. 2020;383:123127. DOI: 10.1016/j.cej.2019.123127

Chapter 3

Isolation and Identification of Carbazole Degrading Bacteria from Lake Water

Khairunnisa Binti Abdul Lateef Khan

Abstract

Heterocyclic hydrocarbon compounds have been identified as one of the major components of water pollution that occurs as a result of urbanization. It has been known that the presence of these compounds is hazardous and remain in the environment for a long period of time. This study was conducted to isolate and identify heterocyclic hydrocarbon degrading bacteria from lake water by genomic DNA extraction and sequencing as well as measure the degradation rate of the bacteria using Gas Chromatography Flame-Ionization Detector (GC-FID). The water sample was collected from west campus lake of Universiti Malaysia Sarawak where six strains of bacteria that has degrading ability was isolated using sub-culturing technique on MSM double layer agar plates. The genomic DNA of bacteria designated as strain IM1, IM2, IM3, IM4, IM5 and IM6 were extracted and amplified using Polymerase Chain Reaction (PCR). The isolates were then sequenced and were identified *as Bradyrhizobium sp., Ochrobactrum sp., Pseudomonas aeruginosa sp. and Burkholderia sp.* All six isolates possessed the ability to utilize carbazole as sole carbon and energy source as the degradation rate of carbazole was measured using GC-FID analysis. After 12 days of incubation, IM2 showed 96.37% degradation while the other five isolates were able to degrade 100% of the carbazole. Thus, bacteria isolated from this study may provide great benefit for bioremediation.

Keywords: heterocyclic hydrocarbon, carbazole, degrading bacteria, bioremediation

1. Introduction

Heterocyclic hydrocarbon has been known to manifest various cytotoxic effect such as severe toxicity, cytotoxicity, mutagenicity, carcinogenicity and photo-induced toxicity [1–3]. Gas, petroleum and coal are the three main natural resource of hydrocarbons. One of the examples of hydrocarbon causing toxic affect to the environment is the accidental crude oil spill in seawater [4]. It was reported that a black bag containing 15 to 20 liters of oil was dumped in the seawater of Redang island in Terengganu [5]. The marine life along the shore was affected, especially turtles landing on the island to lay eggs. Moreover, the toxic chemicals found in the crude oil could also affect human health especially kidney system, liver and sensory system. Some of the hydrocarbons that have their degradation product detected in industrial waste water, contaminated sites are carbazole, biphenyl, dibenzofuran, dibenzothiophene and fluorine.

Carbazole is one of the heterocyclic hydrocarbons that is highly mutagenic and toxic especially to aquatic organism as documented in a recent study [1, 2, 6]. However, this compound has also been widely used in the raw industry for various purpose such as dye productions, medicines and plastics. Carbazole compound is a derivation of shale oil, crude oil and creosote and it has been proven as non-human carcinogen although however its derivatives such as N-methylcarbazole, which are detected in automobile emission and cigarette smokes are highly genotoxic and are now classified as IARC Group 2B carcinogens [1, 7].

Biodegradation has been known to be a technology that provides various benefits for organic pollutants. As it has been understood for so long, microorganism was able to degrade pollutant, bioremediation sets a goal to transform organic pollutant to various type of nontoxic or environmental friendly metabolites or mineralize it into water and carbon dioxide [8]. Many researches have been done using carbazole as sole carbon source where various bacteria with degradable properties has been isolated from different type of environment by enrichment culture method. Most carbazole degrading bacteria are gram negative and aerobic such as *Microbacterium esteraromaticum* strain SL6 [9].

2. Objectives

1. To isolate and identify carbazole degrading bacteria strains from lake water environment that are able to contribute in bioremediation.

2. To assess degradation ability and degradation rate of isolated degrading bacteria strains using gas chromatography.

3. Literature review

3.1 Aromatic hydrocarbon

Aromatic compounds are organic molecules that contains one or more aromatic rings. The three major categories of aromatic compounds are polycyclic aromatic hydrocarbons (PAHs), heterocyclic, and substituted aromatics. Polycyclic aromatic hydrocarbon (PAH) has carcinogenic and mutagenic properties that are connected to some chemical properties such as dipole moment, electrophilic potency, intramolecular and subcellular binding, electronegativity or L- and K- region reactivity, hydrophobicity and others.

Whereas heterocyclic hydrocarbon which is also known as heterocycles are compound that contains at least one atom other than carbon and some or all the atoms are joined in rings. Heterocyclic hydrocarbon has high boiling and melting point, low water solubility and low vapor pressure. These characteristics proves that heterocyclic hydrocarbon can stay in the environment for a long period of time [10]. **Figure 1** shows the molecular structure of aromatic heterocyclic organic compound known as carbazole.

Heterocyclic hydrocarbon is by products of incomplete combustion of organic materials such as coal, petroleum, tar and gas [11]. According to Bamforth and Singleton [12], heterocyclic hydrocarbon can be found easily in groundwater, soil and sediments and also the atmosphere. Some areas such San Diego Bay, California,

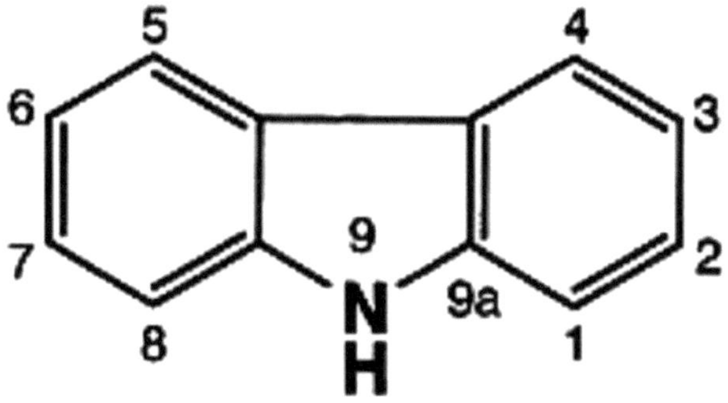

Figure 1.
Molecular structure of carbazole.

Central Pacific Ocean, intertidal sediments, gas works site soils, sewage sludge-contaminated soils, aquifers and groundwater and atmospheric air pollution have shown presence of high concentration of heterocyclic hydrocarbon. Heterocyclic hydrocarbon also originates from two major sources which are natural and anthropogenic sources. Heterocyclic hydrocarbon does not only occur during partial combustion of organic material but also occur with fossil fuels such as petroleum and coal. Carbazole and dibenzothiophene which are heterocyclic compounds often exist together with PAH and some other aromatic compounds as they are components of crude oil, shale oil and creosote [13].

3.2 Health and environmental concerns

Heterocyclic hydrocarbon is a major concern in urban and industrial areas as some of them have been classified as carcinogenic, mutagenic and teratogenic [14, 15]. The properties of heterocyclic hydrocarbon are that they are relatively insoluble in water and they are also highly lipophilic. Another found property is that this compound ban be degraded by some bacteria in soil [14, 15]. Whereas in the atmosphere, heterocyclic hydrocarbon can react with different kind of pollutants such as ozone, nitrogen oxides and sulfur dioxide.

Studies have shown that heterocyclic hydrocarbon are strongly bio concentrated or metabolized [16]. Human are exposed to most PAH through their eating diet that consist of marine lives. For heterocyclic hydrocarbon such as carbazole, tobacco smoking and breathing from polluted air are the routes of exposure [17, 18]. Carbazole has also been classified as “benign tumorigen” by Nojiri and Omori [19]. Moreover, heterocyclic hydrocarbons are potential carcinogens that can produce tumors. Benzo(a)pyrene, a common heterocyclic hydrocarbon, is shown to cause lung and skin cancer in laboratory animals. When ingested, heterocyclic hydrocarbons are absorbed very fast into the gastrointestinal tract. This is because of their high lipid solubility [12, 20]. In general, the more number of benzene rings present, the more harmful the heterocyclic hydrocarbon would be.

3.3 Bioremediation

Bioremediation is defined as a process where microorganisms or their enzymes are used to degrade contaminants to its original condition. Microbial degradation is natural mechanism to demolish hydrocarbon pollutants (and crude oil) from the environment [21, 22]. Research has shown that, in order to induce the ability of certain microbes to degrade or transform toxic or pollutants, bioremediation and biotransformation have been the most successful method chosen [23, 24]. The main objective or aim in bioremediation is to destroy or remove contaminants by using microbes or degrading bacteria and stimulate them with nutrients and other chemicals to aid the degradation process. The presence of suitable microbes and the ideal environmental conditions such as the right amount of nutrient, oxygen and suitable pH and temperature for microbes to grow are some of the factors that also lead to bioremediation [25].

In this process of bioremediation, different microbes will act upon in parallel or sequence in order to degrade the compound. Two approaches are widely used which is in-situ where contaminants are treated at the site and ex-situ approach where contaminants are removed to be treated elsewhere. The ability of variety of microbes to degrade different kinds of pollutants proves that bioremediation is a crucial and important technology that can be used in different conditions [26]. Previous study by Alvarez et al. [27] stated that bioremediation has been proven to be an environmental-friendly technique as the degrading agents are microbes that can be easily decomposed. Therefore, this method appears to fulfill the characteristics of the demanding, growing industry as it is sustainable, easy to implement and cost effective. **Table 1** shows some carbazole degrading bacteria that has been reported in previous literature that are able to contribute to bioremediation.

3.4 Degradation of heterocyclic hydrocarbon

According to Surajit Das from National Institute of Technology, Rourkela Odisha, India, the toxicity, mutagenic properties as well as high carcinogen of PAH in nature has cause a great environmental concern to scientist. Therefore, researchers have collected many marine bacteria that has potential in bioremediation. A research done by Latha and Laithakumari [28] has also shown on how the efficiency of degradation can be increased when a catabolic plasmid from *Pseudomonas putida* that has genotype of hydrocarbon degradation is inserted in a marine bacterium. Some examples of marine bacteria that has been taken and used for bioremediation are *Neptunomonas naphthovorans, Lutibacterium anuloederans*, and *Cycloclasticus spirillensus* [23, 29].

According to research by Nojiri and Omori [19], the structural analog of dioxin and some carbazole-degrading enzymes that plays the same role as dioxin degrading enzymes are the factors that made carbazole known as model compound for bioremediation and led to more study on bacterial degradation of carbazole. There are three main degradation pathways of carbazole that has been identified which are the angular deoxygenation, lateral deoxygenation and hydroxylation pathway of carbazole that can be catalyzed with different enzymes. **Figure 2** shows lateral deoxygenation carbazole at C3 and C4 [9, 30–32].

The common reaction by carbazole degrading bacteria would be hydroxylation as identified by Lobastova et al. [30], who studied on hydroxylation of Aspergillus *flavus* VKM F-1024 by carbazole, where hydroxycarbazole were produced as major product. Nojiri et al. [33] explained on the angular deoxygenation where oxidation

Bacteria Strain	Medium	Product
Achromobacter sp. SL1	Carbon	Anthranilic acid, catechol
Pseudomonas sp. SL4	Carbon	Anthranilic acid, catechol
Microbacterium esteraromaticum SL6	Carbon	Anthranilic acid, catechol
Sphingomonas sp. GTIN11	Nitrogen	Anthranilic acid

Table 1.
Carbazole degrading bacteria, adapted from Salam et al. [9].

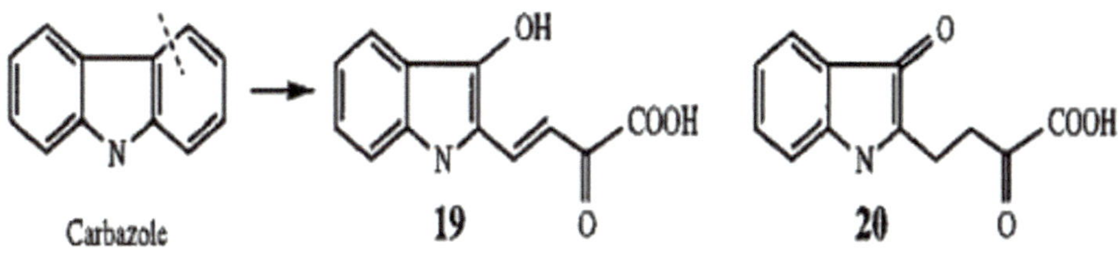

Figure 2.
Lateral deoxygenation of carbazole at C3 and C4.

occur at the ring-fused position of carbazole which is induced by a multicomponent enzyme known as carbazole 1,9a- dioxygenase (CARDO) in an angular position. Based on literature, carbazole degradation pathway has been well reported and many types of carbazole degrading bacteria has been identified successfully.

4. Materials and methods

4.1 Sample collection

Five hundred ml of water sample was collected from lake water from west campus at University Malaysia Sarawak (UNIMAS). The location was at latitude 1.468616 and longitude 110.430536. The samples obtained were stored in falcon tube and transported to the laboratory at room temperature (**Figure 3**).

4.2 Preparation of minimal salt media (MSM) broth and agar

A standard formulation was used to prepare the Minimal Salt Media (MSM). All the chemical compounds were placed in 1 L conical flask after weighing using precision digital balance. Than the pH was adjusted to 7.2 0.2 at 25 °C. The chemical compounds used to prepare the media are as in the **Table 2**: Minimal Salt Media mixture;

The MSM double layer agar was prepared by adding 3 g of Bacto agar (Difco) into 200 ml prepared MSM broth in two conical flasks. Next, the MSM agar solution was autoclaved at 121 °C/15 atm for 1 hour and 30 minutes. Then, one of the conical flasks containing MSM agar solution was supplemented with 2 ml of 10X carbazole

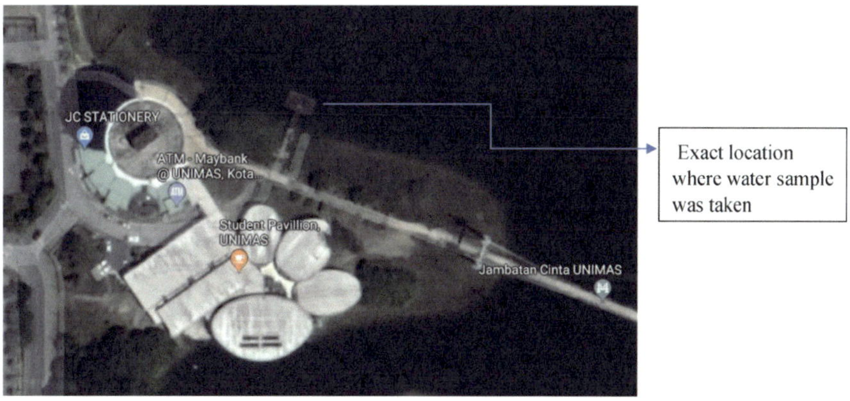

Figure 3.
Location of sampling from a lake located at UNIMAS west campus, Sarawak.

Chemical Compound	Weight
KH2PO4,	1.36 g
(NH4)2SO4,	0.5 g
MgSO4.7H2O	0.2 g
CaCI.2H2O	0.01 g
FeSO4.7H2O	0.005 g
MnSO4. 7H2O	0.0025 g
NaMoO4.2H2O	0.0025 g
Na2HPO4	0.00213 g

Table 2.
Minimal salt media mixture [34].

stock and stirred to mix it well. Finally, the agar solution was poured into agar plates forming double layer and stored in fridge at 4 °C for further use.

4.3 Enrichment culture and isolation of degrading bacteria

4.3.1 First enrichment culture

Enrichment cultures were undertaken where 2 ml of sample were added into test tube with 8 ml of Minimal Salt Media(MSM) and 0.05%(w/v) substrate(carbazole). The test tubes were then incubated at 30 °C and shaken at 200 rpm until color change was observed indicating heterocyclic hydrocarbon degradation.

4.3.2 Second enrichment culture

For the second enrichment culture, 2 m1 of bacteria samples was transferred from the first enrichment culture into another 8 ml of MS media and 0.1% (w/v) substrate (Carbazole) as the sole carbon source. Next, the samples were incubated at 30 °C and shaken at 200 rpm for another two to four weeks until color change observed.

4.3.3 Isolation of pure colony

After second enrichment, the bacteria were inoculated on MS double layer agar plate. Heterocyclic hydrocarbon which is carbazole was added as the sole carbon source before incubating them for several days. Then the bacteria were sub cultured until pure colony is observed. Colonies that shows clear zone were picked by inoculating needles and purified by streaking several times on a fresh new plate. The growing bacteria colonies were observed as the pure cultures are obtained to be used in the next procedure.

4.4 Glycerol stock preparation and test

The pure colony of carbazole degrading bacteria isolated were inoculated into the MSM broth and incubated for 24 hours. Next, the bacteria culture was transferred into sterile 2 ml.

Eppendorf tube and sterile glycerol were added to make the final concentration 20% and 40%. Finally, the glycerol stocks were stored in −80 °C. After 3 weeks stored, the bacteria from glycerol stock was streaked on MSM agar where growth were observed.

4.5 Gram staining

A single colony from culture plate of isolated degrading bacteria was picked using sterile inoculation loop and mixed with a few drops of distilled water to form a smear on microscopic slide. Then, the smear was air dried and heat fixed by passing through the flame on bunsen burner for a few seconds. Next the primary stain, crystal violet was applied on the smear for 1 minute and then rinsed with water. Then the smear was covered with iodine for one minute before rinsing with water and ethyl alcohol was used to decolorize the slide for.

30 seconds. After decolorizing, safranin was added for 30 seconds before rinsing with water. Then the slides were observed under light microscope to identify and distinguish between gram positive and gram-negative bacteria.

4.6 Molecular characterization

4.6.1 Genomic DNA extraction of isolated bacteria

For genomic DNA extraction, the bacteria isolated, that has pure colony on a plate was washed with 2 ml of TE buffer. Then, 1 ml of the mixture was transferred into new Eppendorf tube and centrifuged at 10000 rpm for 2 minutes in order to obtain pellet. Next, the pellet will be resuspended with 500 μl TE buffer, vortex until pellet dissolved and divided into two tubes with 750 μl in each tube. Then, 65 μl of 10% sodium dodecyl sulphate (SDS) and 10 μl proteinase K were added before incubating for 1 hour in incubator and flipped every 10 minutes. Next, 825 μl of phenol/chloroform/isoamyl alcohol (25:24:1) was added and placed on shaker for 45 minutes. After that, the tubes were centrifuged for 15 minutes at 10000 rpm until double layer is formed. Then the upper layer of the mixture was extracted into new sterile 1.5 ml tubes. Next, 200 μl of isopropanol will be added before spinning for 15 minutes at 10000 rpm and the tubes were flipped a few times. Once the pellet of DNA was obtained, the supernatant was removed, and the pellet was air dried before adding 50 μl TE buffer to dissolve the pellet and stored at −20 °C for further analysis. Then, the extracted genomic DNA was also analyzed by using agarose gel electrophoresis to confirm presence of DNA.

4.6.2 Polymerase chain reaction

Polymerase Chain Reaction (PCR) was carried out by using universal primers which is forward primer 27F and reverse primer 1492R to amplify 16S rRNA gene. **Table 3** shows the sequence for PCR.

All PCR reagents were pipetted into 0.5 μl Eppendorf tube and mixed gently to make sure all mixtures were collected at the bottom of the tube. **Table 4** shows the reaction mixture used for polymerase chain reaction. Next, the tubes containing

Primer	Sequences
27F	5' AGAGTTTGATCCTGGCTCAG 3'
1492R	5' TACGGCTACCTTGTTACGACTT 3'

Table 3.
Polymerase chain reaction universal primer [35].

PCR Reagents	Volume (µl)
Sterile distilled water (dH2O)	9.5
Master Mix	12.5
Forward primer(27F)	1.5
Reverse primer(1492R)	1.5
DNA template	1.5
Silicon oil	10

Table 4.
Reaction mixture for polymerase chain reaction.

Phase	Temperature and Duration
Initial denaturation	95 °C for 5 minutes
Denaturation	95 °C for 30 seconds
Annealing	50 °C for 30 seconds
Extension	72 °C for 1 min 30 seconds

Table 5.
PCR profiles.

PCR mixture were placed in DNA thermal cycler machine and performed according to parameters shown in **Table 5**.

4.6.3 Agarose gel electrophoresis (AGE), DNA purification and sequencing

Agarose gel 1%was prepared for gel electrophoresis where 0.75 g of agarose powder was added to 50 ml of distilled water in a conical flask which is then heated for about 40 seconds in microwave until agarose powder is fully dissolved. Next, 1 ml of 50X TAE buffer was added to the mixture and the mixture was poured into agarose gel electrophoresis tray with comb to let it solidify. After about 30 minutes when the gel is solidified, the PCR products and was loaded into respective wells and the gel was run at 100 V,400 mA for 30 minutes. After 30 minutes, the gel was stained with ethidium bromide for 10 minutes and then rinsed with distilled water for 5 minutes. Then, the gel is viewed under the UV light to observe the DNA band.

Once bands were observed, the PCR products were purified using DNA purification kit. Firstly, 20 µl of PCR mixture was mixed with 35 µl of binding buffer and the mixture was then transferred into a high pure filter tube that had been inserted into a collection tube. Next, the tube was centrifuged for 15 seconds at 5000 rpm and then the flowthrough in the collection tube was discarded. Then, 800 µl of washing buffer was added into the high pure filter tube which was then centrifuged at 5000 rpm for 15 seconds. The flowthrough in the collection tube was then discarded. Washing using buffer was done twice to enhance the result. Next the collection tube was replaced and the high pure filter tube was inserted into another sterile collection tube. Then, 30 µl of elution buffer was inserted into high pure filter tube and centrifuged for 15 seconds at 5000 rpm. The flowthrough, which was the purified PCR product in the collection buffer was collected and transferred into new 1 mL tube. Then, the purified DNA was sent to First BASE Laboratories Sdn. Bhd. Company for sequencing and the sequence data obtained were analyzed using MEGA software and BLAST from NCBI [36].

4.7 Gas chromatography-flame ionization detector (GC-FID) analysis

The gas chromatography-flame ionization detector (GC-FID) was conducted to measure the degradation rate of the bacteria strains. The bacteria strains were grown back using enrichment culture method in different tubes and each tube was harvested on day 0, day 3, day 6, day 9 and day 12. Next, 1 ml of ethyl acetate was added to 1 ml of each sample and the mixture is vortex for around 10 seconds until a double layer could be observed from the mixture. Then the upper layer was pipetted into a vial tube [35]. The samples were then analyzed using gas chromatography-flame ionization detector machine and the data was collected. The degradation percentage were calculated using the formula given below. The percentage of residual substrate are calculated first using the Eq. 1. Once the percentage is obtained, the degradation percentage are calculated using the formula in Eq. 2.

$$\text{Percentage of residual substrate} : (\text{average peak area average control}) \times 100 = \% \quad (1)$$

$$\text{Degradation percentage} : 100 - \text{Percentage of residual substrate} = X\% \quad (2)$$

5. Result

5.1 Enrichment cultures and streaking on MSM agar plates

After preparing media, enrichment cultures were done where 2 ml of lake water was added into a test tube containing 8 ml of MSM prepared before and 0.1% of carbazole was also added as sole carbon source. The cultures are then sub cultured two times and color change was observed and compared with the control. Another test tube was also prepared but without any water samples as a control. **Table 6** shows change in color observed on first, second and third enrichment medium for the lake water sample collected from Unimas lake. In tube C, it could be observed major color change from cloudy to greenish yellow after three weeks. On the second and third enrichment the substrate residue become smooth and the color change to pale yellow compared to the control which is clear after two weeks.

Control media **Tube A**	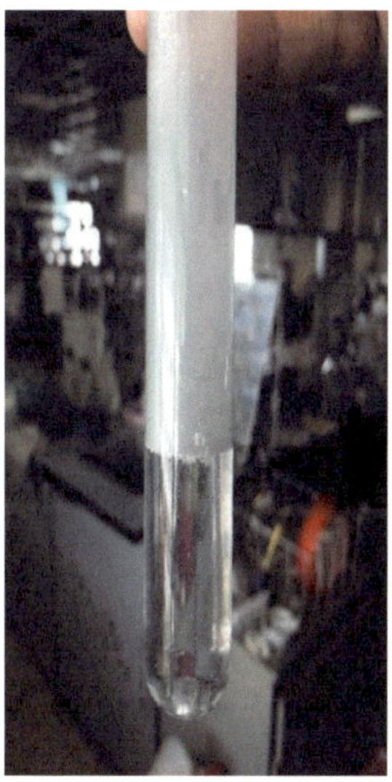

First enrichment **Tube B (before) Tube C (after 3 weeks)**	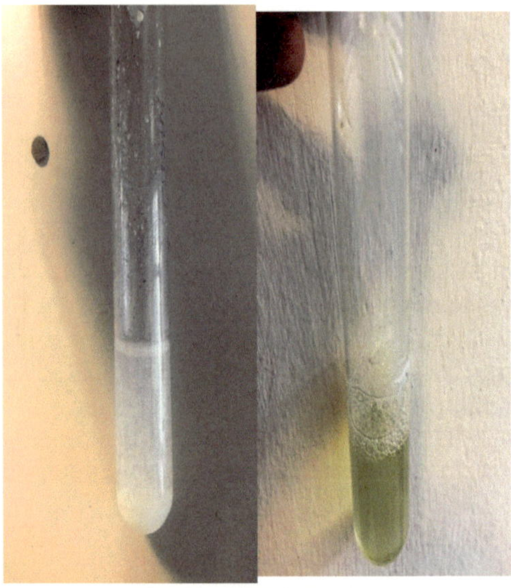
Second enrichment Tube C (after 2 weeks)	
Third enrichments **Tube D and Tube E** **(after 2 weeks)**	

Table 6.
Observation on first, second and third enrichment.

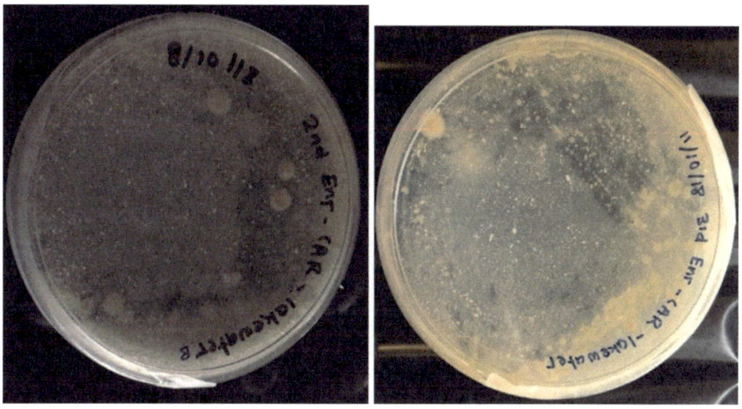

Figure 4.
Streaking results of second enrichment and third enrichment of lake water samples.

The **Figure 4** below shows bacteria growth observed on MSM double layer agar after streaking with second and third enrichment cultures. Both plates are able to form clear or hollow zone proving degradation of substrate carbazole.

5.2 Isolation of carbazole degrading bacteria and gram staining

The **Figure 5** below shows the growth of bacteria isolated from second and third enrichment plates which was cultured on MSM broth supplemented with 0.1% carbazole as sole carbon source. It could be observed that IM2 and IM3 turned to dark yellow while IM6 is cloudier but still turned yellow in color. IM1, IM3 and IM5 became light yellow in color compared to control from **Table 6**.

After isolating six different bacteria by streaking on MSM double layer agar plates supplemented with 0.1% carbazole, it could be observed that not all of them produce clear or hollow zones. Despite that, all the isolates showed growth on agar plate proving that carbazole was utilized efficiently by the bacteria. Next, gram staining was done for all isolates to distinguish between Gram positive and Gram-negative groups of bacteria before proceeding with DNA extraction and sequencing (**Table 7**).

Figure 5.
Six bacteria isolated, from right, IM1, IM2, IM3, IM4, IM5 and IM6.

Pure colony of isolated bacteria	Gram staining results
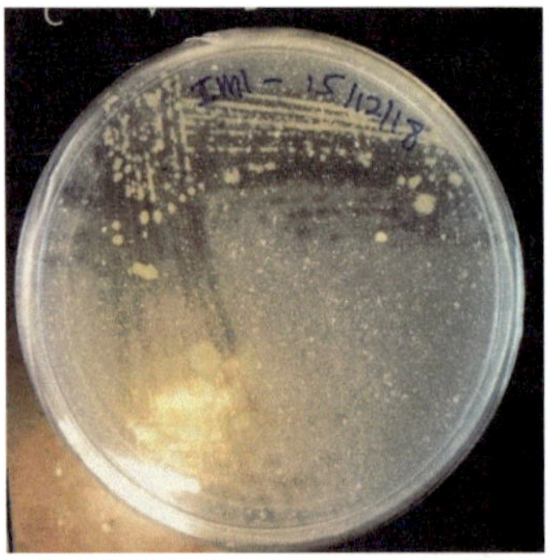 IM1	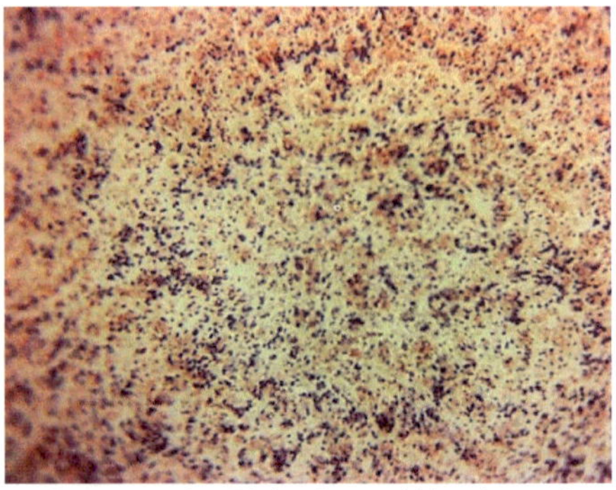 Gram negative: Pink color could be observed
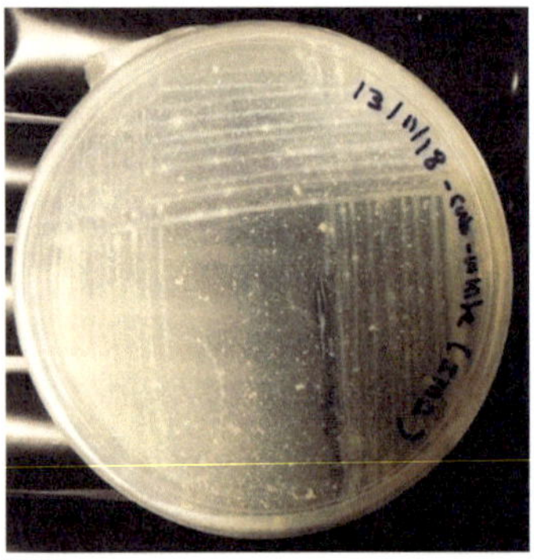 IM2	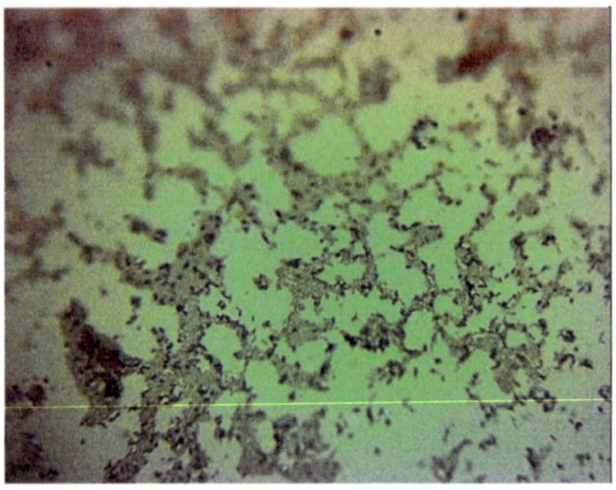 Gram negative: Pink color could be observed
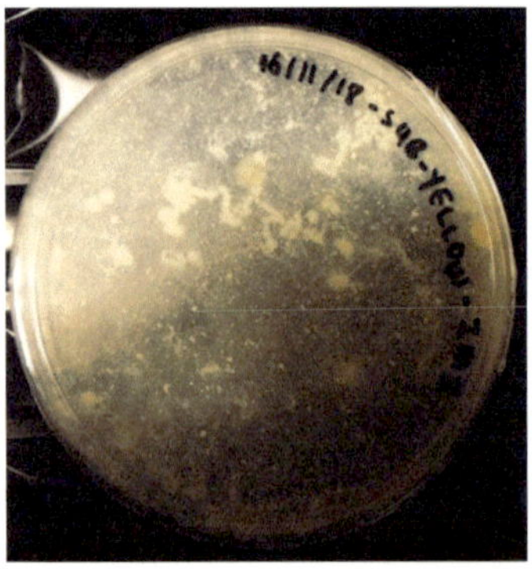 IM3	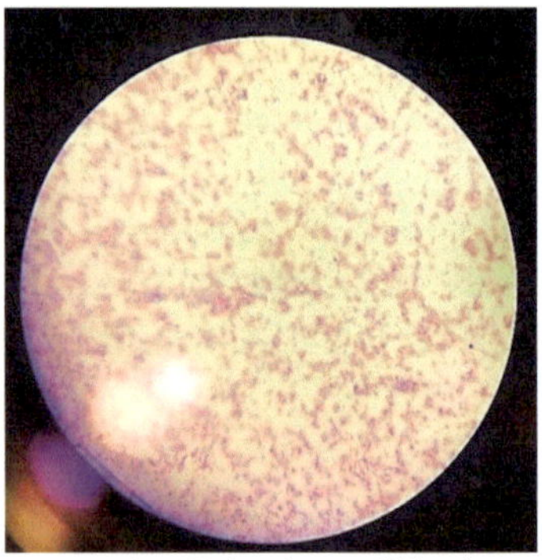 Gram negative: Pink color could be observed
 IM4	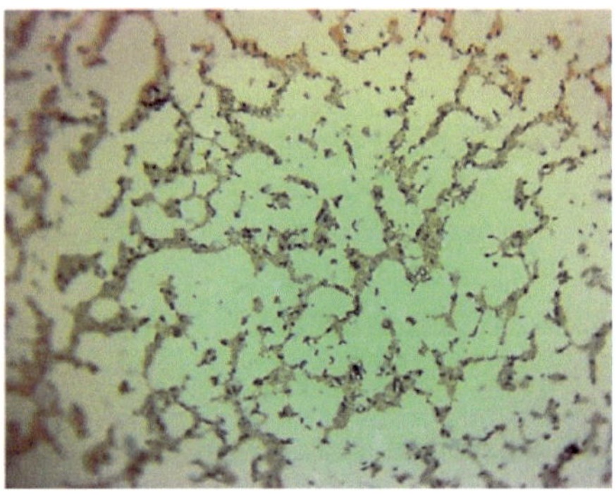 Gram negative: Pink color could be observed

Pure colony of isolated bacteria	Gram staining results
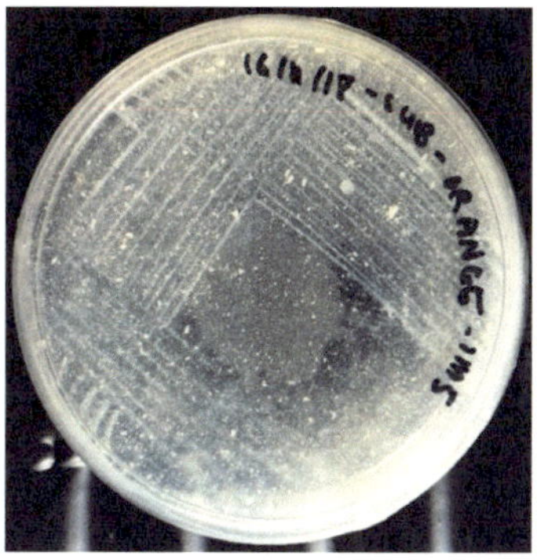 IM5	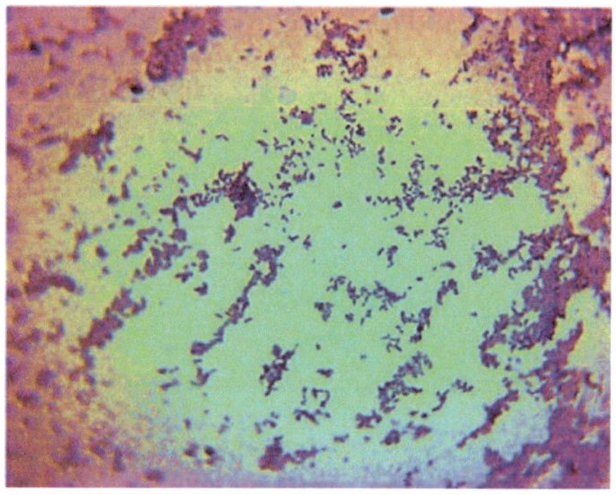 Gram negative: Pink color could be observed
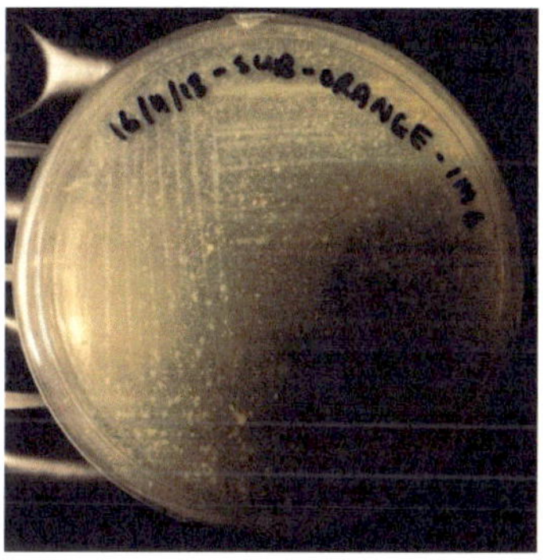 IM6	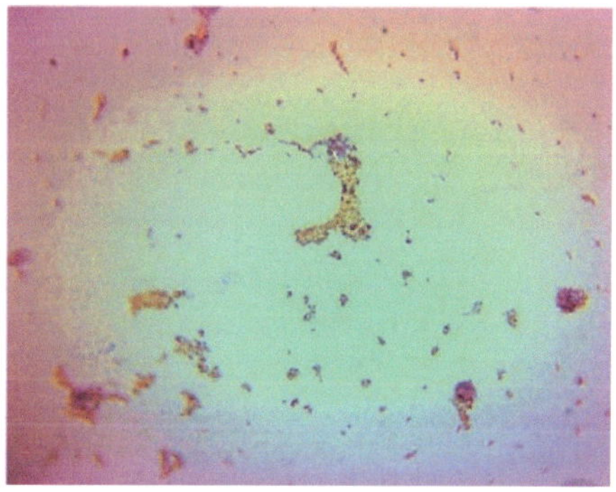 Gram negative: Pink color could be observed

Table 7.
Table of isolated bacteria and gram staining results.

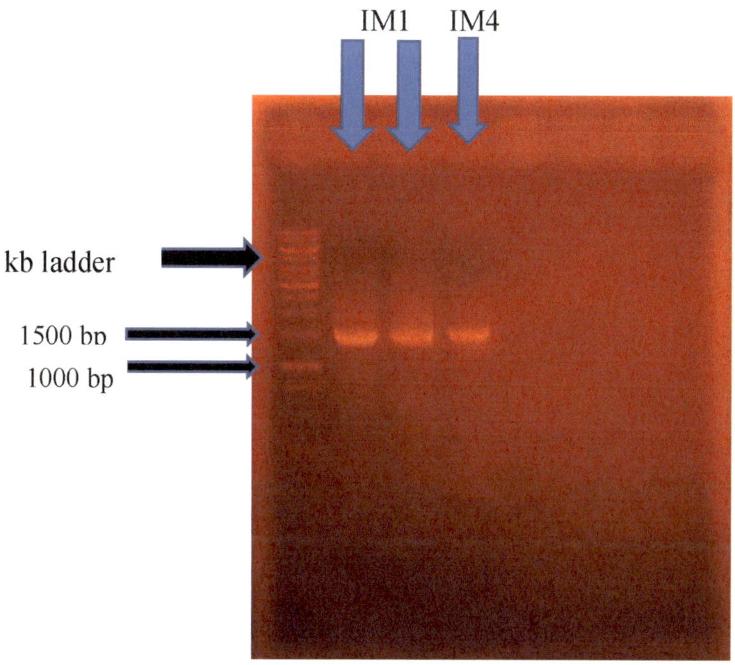

Figure 6.
Clear band on IM1 and IM4 genomic DNA fragment at 1500 bp.

5.3 Molecular identification

5.3.1 Agarose gel electrophoresis (AGE)

After the carbazole degrading bacteria were isolated from lake water sample, DNA Extraction was done continued with PCR and Agarose Gel Electrophoresis

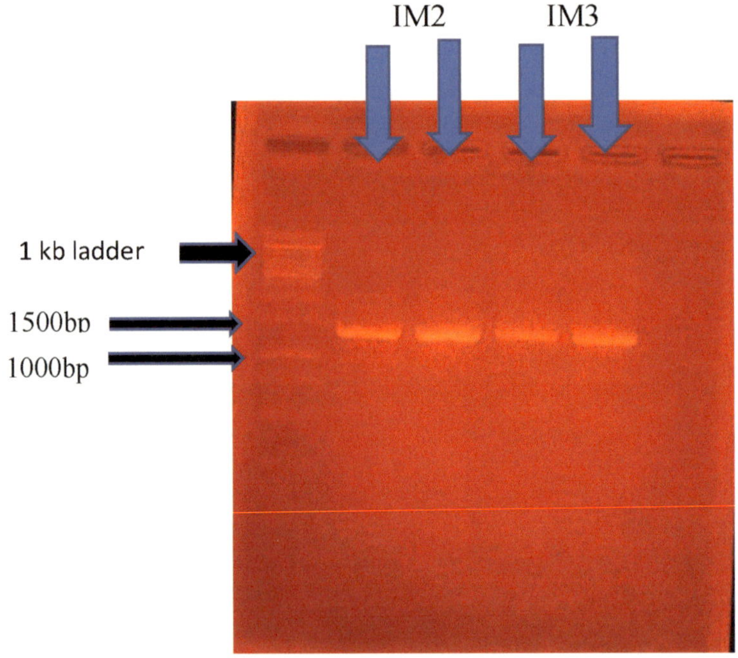

Figure 7.
Clear band on IM2 and IM3 genomic DNA fragment at 1500 bp.

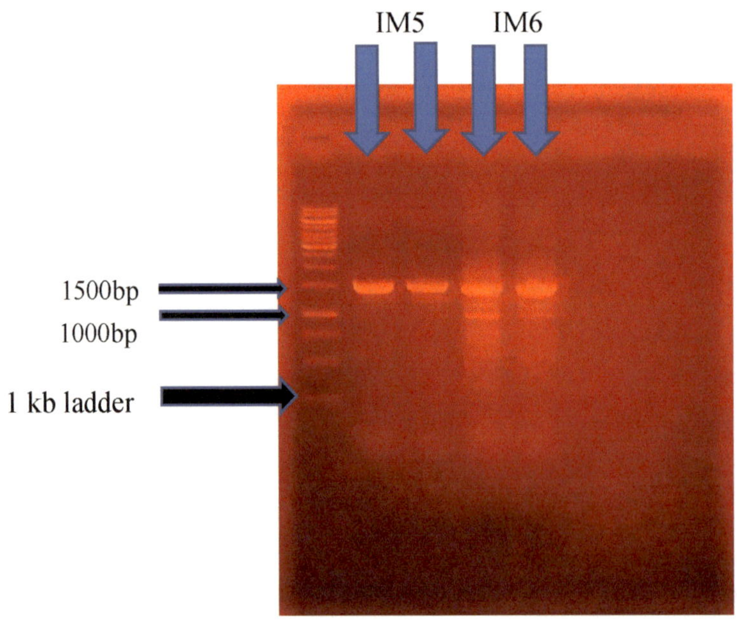

Figure 8.
Clear band on IM5 and IM6 genomic DNA fragment at 1500 bp.

(AGE) to identify the bacteria. All lanes produced intact, clear bands. The band for each isolated bacterium from IM1 to IM6 are shown at 1500 bp as in **Figures 6–8**.

5.3.2 Sequencing analysis

The nucleotide sequence retrieved from the company was processed using MEGA 6 software. The nucleotide sequence with weak signals was cut off leaving only the nucleotide sequence with strong signal which was then analyzed using BLAST tool. From the analysis, one of the six isolated bacteria showed weak,

Isolated strain	Identification	Percentage of Similarity
IM1	*Bradyrhizobium sp.*	90%
IM2	*Ochrobactrum sp.*	99%
IM3	*Pseudomonas aeruginosa sp.*	96%.
IM4	*Burkholderia sp.*	100%
IM5	*Burkholderia sp.*	100%
IM6	*Ochrobactrum anthropic sp.*	100%

Table 8.
Sequencing analysis results.

Days	Control	3	6	9	12
Average peak area	4432.86	481.64	215.34	101.7	0
Percentage of residual substrate (%)	100	10.865	4.858	2.294	0
Degradation percentage (%)	0	89.13	95.14	97.71	100

Table 9.
Summary of GC-FID analysis of IM1 samples.

Days	Control	3	6	9	12
Average peak area	6688.91	6311.12	1078.71	849.48	309.38
Percentage of residual substrate (%)	100	94.35	16.13	12.7	4.63
Degradation percentage (%)	0	5.65	83.87	87.3	95.37

Table 10.
Summary of GC-FID analysis of IM2 samples.

Days	Control	3	6	9	12
Average peak area	4590.37	4432.86	2777.48	2647.27	0
Percentage of residual substrate (%)	100	96.57	60.51	57.67	0
Degradation percentage (%)	0	3.43	39.49	42.33	100

Table 11.
Summary of GC-FID analysis of IM3 samples.

Days	Control	3	6	9	12
Average peak area	4432.86	623.3	0	0	0
Percentage of residual substrate (%)	100	14.06	0	0	0
Degradation percentage (%)	0	85.94	100	100	100

Table 12.
Summary of GC-FID analysis of IM4 samples.

Days	Control	3	6	9	12
Average peak area	4432.86	1704.71	1635.69	308.38	0
Percentage of residual substrate (%)	100	38.46	36.9	6.96	0
Degradation percentage (%)	0	61.54	63.1	93.04	100

Table 13.
Summary of GC-FID analysis of IM5 samples.

Days	Control	3	6	9	12
Average peak area	4432.86	389.16	0	0	0
Percentage of residual substrate (%)	100	8.78	0	0	0
Degradation percentage (%)	0	91.22	100	100	100

Table 14.
Summary of GC-FID analysis of IM6 samples.

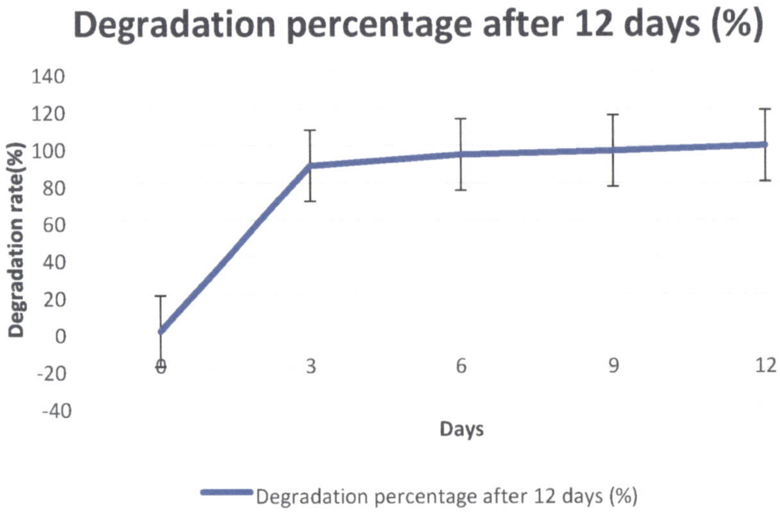

Figure 9.
The degradation percentage of IM1 strain on CAR after 12 days incubation.

overlapping signals giving result with lower percentage of identification. **Table 8** shows the sequencing analysis results of six isolated bacteria strains.

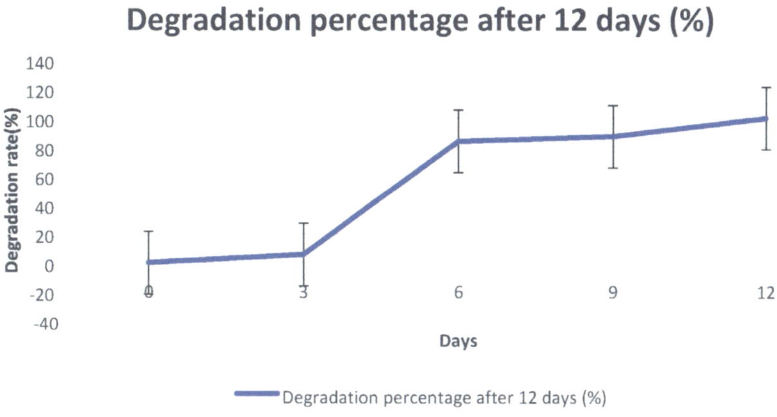

Figure 10.
The degradation percentage of IM2 strain on CAR after 12 days incubation.

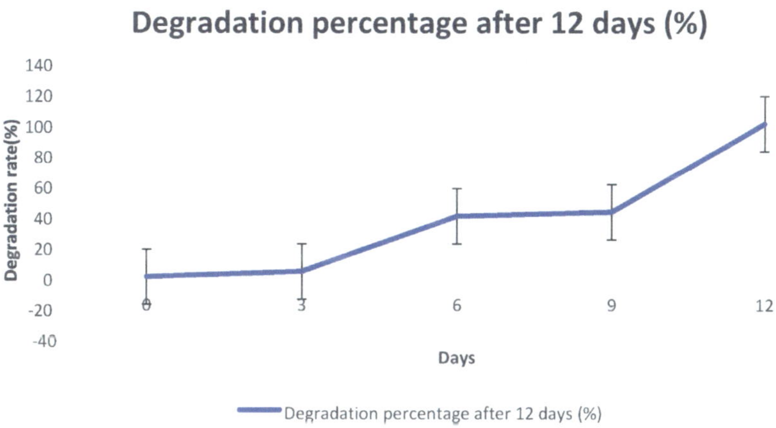

Figure 11.
The degradation percentage of IM3 strain on CAR after 12 days incubation.

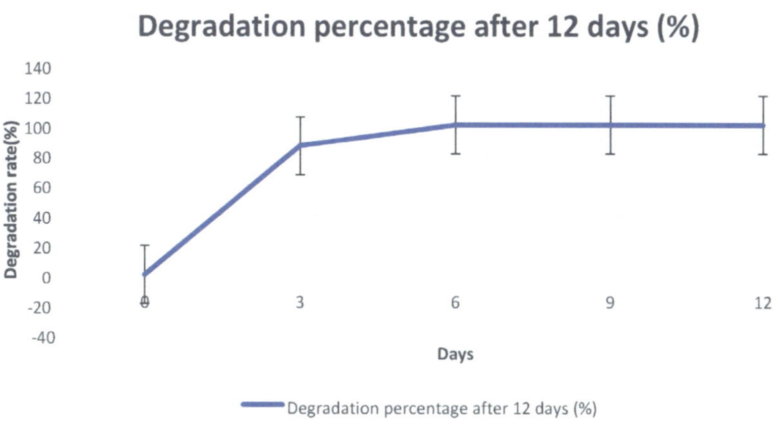

Figure 12.
The degradation percentage of IM4 strain on CAR after 12 days incubation.

5.4 Carbazole degradation analysis

5.4.1 Gas chromatography-flame ionization detector (GC-FID) analysis

The isolated bacteria strains degradation rate on CAR was analyzed from day 0, day 3, day 6, day 9 and day 12 using GC-FID analysis. From the chromatogram obtained, the degradation rate of six isolated strains could be summarized as shown in **Tables 9–14**. **Figures 9–14** shows the percentage of degradation after 12 days for six isolated strains.

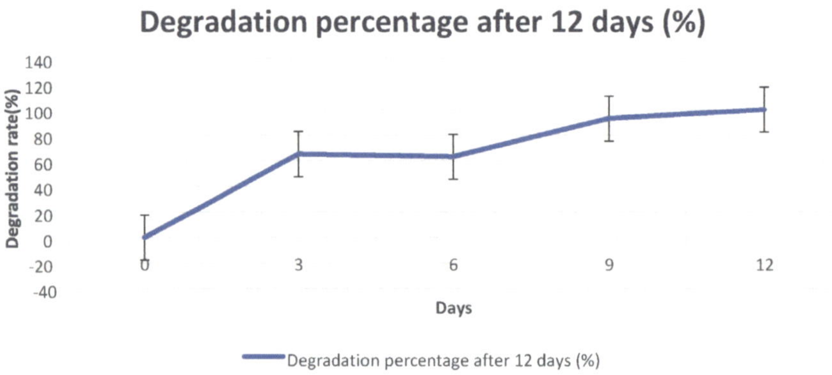

Figure 13.
The degradation percentage of IM5 strain on CAR after 12 days incubation.

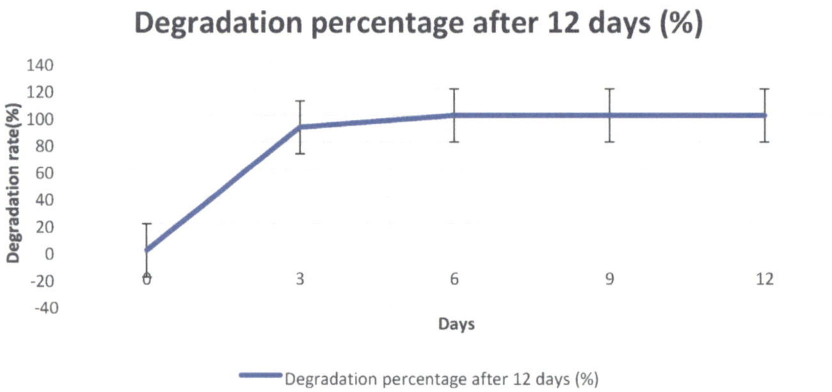

Figure 14.
The degradation percentage of IM6 strain on CAR after 12 days incubation.

6. Discussion

6.1 Enrichment and isolation of carbazole degrading bacteria

Many different types of bacteria exist in our natural environment according to Zulkharnain and Taka [37]. Although most research focus on isolating heterocyclic hydrocarbon degrading bacteria from sewage or site with high probability of contamination, these degrading bacteria could be found anywhere in the environment and not just in certain area. Therefore, lake environment is chosen to discover different type of novel carbazole degrading bacteria. The project started by taking lake water from Unimas west campus lake, Sarawak to isolate heterocyclic hydrocarbon mainly carbazole degrading bacteria. Sampling was done at the side of the lake closer to land as the number of bacteria are higher at coastal, shallow area [38]. Once, sample was taken, enrichment was done using minimal salt media (MSM) with carbazole as sole carbon source where the degradation of carbazole could be observed in **Table 6**. After around three to four weeks of the first enrichment, color change could be observed where the enrichment turned to clear yellow from cloudy white before. Here, based on study by Zulkharnain and Taka [37], it proves that the bacteria present in the enrichment had utilize the carbazole as sole carbon source and degrade it causing color change. It also explains that the accumulation and production of different metabolites during degradation caused the color change [37]. Another research by Stope et al. [39] and Maeda et al. [40] explained that the change in color of the enrichment was due to the meta cleavage product formation which are the intermediates. When second and third enrichment was done, color change could be observed in a faster rate within two weeks as degradation rate increases. According to the concept where 3-ring structure is more stable and recalcitrant than 2d-structure, bacteria takes about 10 days to degrade fully [37].

From the enrichment, MSM agar plates supplemented with carbazole were prepared and used in this experiment to provide a selective condition and avoid growth of any other bacteria [41]. Double layer agar was used in this experiment to observe the clear zones easily. Streaking was done to observe the growth and degradation on MSM agar plates. After two weeks, clear or hollow zones could be observed on the plates in **Figure 4**. According to Takahashi et al., [42], clear zones that appear on the bacterial colony is because of the mineralization of insoluble CAR by bacteria. As clear zone could be observed, this proves the ability or capability of bacteria to degrade the substrate and utilize it as sole energy source.

Once clear zone was observed, sub-culturing was done to isolate pure colony of bacteria that shows degradation where six bacteria were isolated and designated as IM1, IM2, IM3, IM4, IM5 and IM6 strains. Based on the result shown, all the bacteria were able to grow on MSM agar supplemented with carbazole plate. Although, not all the six isolated showed clear zones or halo formation, the isolates were able to grow on the plates proving the capability of using the substrate. To observe the production of intermediate product and confirm the degradation, the isolated bacteria were inoculated on MSM liquid media as in liquid media it is easier to degrade the substrate because the bacteria becomes free moving cell and it mimics the original or natural condition of the bacteria in lake water. The MSM culture with bacteria was placed on orbital shaker to make sure the bacteria get enough nutrient as the shaker moves in circular motion.

6.2 Gram staining

The isolated bacteria were then characterized morphologically by gram staining. Based on results in **Table 7**, all isolated bacteria (IM1, IM2, IM3, IM4, IM5, IM6) were identified as Gram negative, rod shaped bacteria after viewing under light microscope with 100X magnification with immersion oil. To differentiate between Gram-positive and Gram-negative bacteria, it has been known that gram staining is the best and easiest method so far. Bacteria that can retain the primary strain which is crystal violet (purple color) are Gram-positive bacteria while Gram-negative bacteria classifications are the ones that could not retain the primary strain, where it will stain pink with safranin instead [43]. Based on a few studies, some Gram-negative bacterium such as *Pseudomonas sp.* LD2 [44], *Achromobacter sp.* Strain CAR1389 [45] and *Sphingomonas sp.* GTIN1 1 [46] has been reportd to utilize and degrade carbazole. This shows that most or many of the heterocyclic hydrocarbon degrading bacteria are Gram-negative.

6.3 Molecular characterization

The molecular characterization was done firstly by DNA extraction procedure based on method in 3.6.1. Detergent which is SDS and proteinase K was used to break the cells and digest contaminated protein in the cell. The digestion of protein process was done under optimal condition of 37°C and incubated where the tubes were flipped every 10 minutes to increase digestion. Once the mixture became viscous, it shows that the digestion process has ended. To take up the DNA in aqueous phase, phenol: chloroform: isoamyl (25:24:1) was added and placed on rotary shaker to maximize the bacterial DNA take up. Centrifugation was done to form the double layer of organic and aqueous phase and to separate the upper layer containing DNA into new tubes. Next, isopropanol was added before centrifuging leaving the precipitate of DNA at the side and bottom of the tube. For further amplification using PCR machine, the DNA precipitate was redissolved in TE buffer.

According to method in 3.6.2. amplification was done using thermal cycler before 16S rDNA sequencing to identify the bacteria. In the PCR amplification, three main process were involved which are denaturation, annealing and extension phase which is all repeated for 30 cycles, therefore producing a million copies in around 2 hours. As shown in AGE analysis in the results, all six isolates (IM1, IM2, IM3, IM4, IM5, IM6) was successfully amplified using 27F as forward and 1492R (reverse) primers showing clear, intact band which is located at 1500 bp after being compared with 1Kb DNA ladder in lane 1. According to Angeline [47], the total of two primers and the sequence of the target DNA is the size of the PCR products.

One of the most common technique to separate DNA was used in this experiment which is gel electrophoresis. In this experiment, according to method in 3.6.3., 1% of gel was used with 1 ml of 50X TAE and 50 ml of distilled water. As the resolution of DNA band depends on the concentration of gel, most concentration of the gels are made in range 0.7% to 2% [48]. The four major factors that influence the migration rate of DNA, according to Cheng and Zhang [48] are, the concentration of agarose where lower concentration will produce better results for large DNA fragments. Then, the size of DNA is also a major factor where smaller fragments tend to migrate in a faster rate compared to larger fragments. Furthermore, the DNA conformation where DNA in supercoiled form is faster than in linear form. Lastly, during electrophoresis, the voltage supplied plays important role as well as the lower the voltage the slower the migration rate. After the bands of PCR were viewed under UV light, the remaining products were purified using the right protocol before sending for sequencing to identify the isolated carbazole degrading bacteria.

6.4 The identification of isolated bacteria

The sequence data for strain IM1, IM2, IM3, IM4, IM5 and IM6 was obtained from the First BASE Laboratories Sdn Bhd which are shown in **Table 8**. All the sequence data were analyzed using MEGA software before the species were determined molecularly using BLAST programming from NCBI (http://blast.ncbi.nlm.nih.gov/Blast.cgi).

From the analysis, it is shown that the signals for IM1 and IM3 signs were not strong enough resulting in lower identification percentage. For IM1strain, the BLAST result showed that the bacteria belong to the genus of *Bradyrhizobium sp.* with 90% identity compared to the database. *Bradyrhizobium sp.* is a gram negative, nitrogen fixing bacteria that is rod in shape. It is most commonly found in soil and it is slow growing compared to *Rhizobium sp.* A study by Qu and Spain [49], showed the first clear evidence for the initial steps in biodegradation of nitroanilines by *Bradyrhizobium JS329.* Although not many studies could be found on degradation of carbazole by *Bradyrhizobium sp.*, a study on isolation and characterization of species that degrades polycyclic and heterocyclic aromatic compounds under extremely low oxygen conditions, found that Shinorhizobium sp. C9 (AF227756), Mesorhizobium amorphae (AJ271899), Mesorhizobium sp. SH2851 (AY141983) and Rhizobium ciceri (U07934) has the ability to degrade heterocyclic hydrocarbons [35].

For IM2 strain, it was discovered that the bacteria belong to the genus *Ochrobactrum sp.* with 99% identity to the database while IM6 was also discovered to be another type of *Ochrobactrum* species with a specified name which is *Ochrobactrum anthropi sp.* with 100% identity. This species is known to be gram negative, rod shaped, non-pigmented, aerobic bacteria. This explains how the two isolates could be from the same species as it is hard to differentiate when the bacteria are not pigmented. Novel strain of *Ochrobactrum anthropi* HM-1 has been isolated from oil-contaminated soil where the degradation potential has been reported that contributes to bioremediation of used engine oil polluted sites [50].

With similarity of 97% to the database, IM3 strain was found to be *Pseudomonas aeruginosa sp. this* species is a pathogenic bacterium that could cause disease to plant, animal and even human. This bacterium is also a gram negative, rod shaped bacteria that could be found widely in environment, especially in soil, water and plants. In an experiment where potent biodegradation of crude oil was assessed, *Pseudomonas aeruginosa sp. have* been reported to degrade at the percentage of 58% of crude oil with direct or indirect assistance of glycerol or rhamnolipid. As *Pseudomonas aeruginosa sp. has* the ability to utilize vegetable oil or glycerol as sole carbon source [51]. Crude oil is mainly involved in biodegradation as the major component of crude oil are volatile hydrocarbon that needs to undergo biodegradation.

IM4 and IM5 strain has been confirmed to belong to the genus of *Burkholderia sp.*

with 100% similarity to the database after the 16S rDNA sequencing was done. *Burkholderia sp.* are obligate aerobic, gram-negative bacteria that is well known for their antibiotic resistance. Some family of *Burkholderia sp.* such as *Burkholderia mallei sp.* are known to be pathogenic to mostly horses and other related animals. Based on a study by Inoue [52], 27 carbazole utilizing bacteria was isolated from environment where three of them were *Burkholderia sp. strain* NE-7, *Burkholderia sp.a and Burkholderia sp. strain* NW-1. When hybridization was done, CAR gene was found in these isolates proving the ability to utilize carbazole as sole carbon source.

6.5 Gas chromatography-flame ionization detector (GC-FID) analysis

As the rate of degradation cannot be determined through direct observation, GC-FID analysis was used in order to obtain the result of the degradation rate of carbazole by the six (IM1, IM2, IM3, IM4, IM5, IM6) isolated bacteria. The GC-FID machine was used by analyzing the residual substrate after 12 days incubation period of the bacteria supplemented with carbazole where pattern of degradation was observed and compared. In previous study by Okoh et al. [53], the degradation of crude oil by Pseudomonas aeruginosa was determined by measuring the reduction rate of crude oil. Hedlund et al. [54] had also done analysis using GC-FID where the degradation rate of substrate was calculated by looking at the disappearance of substrate.

In this experiment, the sample were analyzed every two days until day 12. Based on the chromatogram observed from the analysis result in 4.4, the peak of substrate, CAR was detected at the retention time from 14.0 to 15.0 minutes and the pattern of CAR utilization are revealed to vary among all isolates. This could be due to the sundry nature or environment of hydrocarbon present at the location from which the isolates were taken [55]. For IM1 strain, the percentage of residual carbazole was seen to be decreasing from 10.87% on day 3 to 0% on day12 while the degradation percentage showed gradual increase from 89.13% on day 3, 95.14% on day 6, 97.71% on day 9 and 100% on day 12. This shows that *Bradyrhizobium sp.* possess the ability to degrade carbazole gradually with highest percentage on day 12 resulting on zero amount of carbazole left.

For IM2 strain, the percentage of residual of CAR was high on day 3 with 94.35% but reduced greatly to 16.13% on day 6 and subsequently to 12.7% on day 9 and 4.63% on day 12. For the degradation rate, on the third day, the degradation rate was low at 5.65% but increase sharply on day 6 up to 83.87 and 87.3% on day 9 followed by 95.37 on day 12. Therefore, the degradation rate was slower on the first 3 days and increased a lot from the sixth day. However, it could be observed that this *Ochrobactrum sp.* does not totally degrade CAR after 12 days as there was still some residual substrate left. Thus, indication could be made by observing the plotted degradation graph that the strain might need more time to completely degrade the substrate compared to other bacteria.

The percentage of residual substrate for IM3 isolate, decreased from 96.57% on day 3 followed by 60.51% on day 6 and 57.67% on day 9. Whereas the degradation percentage on day 3 was low at 3.43% and increased to 39.49% on day 6, 42.33% on day 9 and finally to 100% on day 12. The *Pseudomonas aeruginosa sp. which* had been widely analyzed as a degrader showed that the strain took 12 days to completely utilize and degrade carbazole as sole carbon source based on the analysis result.

For IM4 strain, it could be observed from the result that the degradation rate was one of the highest compared to other isolates. The percentage of residual carbazole reduced to 14.06% on day 3 and was fully utilize leaving 0% from day 6 onwards. The degradation rate went down to 85.94% on the third day and increase fully to 100% on day 6. For IM5 strain which was identified as the same species with IM4, the rate of degradation however was different from IM4 strain with residual substrate percentage of .38.46% on day 3 and gradually falling to 36.9% on day 6, 6.96% on day 9 and was totally utilized by day 12. The degradation rate on day 3 was 61.545 which was higher compared to IM4 strain and gradually decreased to 93.04% on day 9 and 100% on day 12. Thus, both IM4 and IM5 strain which was identified as *Burkholderia sp.* could have been different type of stains showing different level of degrading ability.

The IM6 strain which was identified as *Ochrobactrum anthropi sp.* has the highest degradation rate based on the GC-FID analysis result where it was shown that the degradation rate on day 3 was up to 91.22% and the substrate, CAR was totally degraded by day 6 onwards. The residual substrate left on day 3 was 8.78% from 100% on the first day and by day 6 there was no substrate left. Based on study by Abulgasem Alenabi [56], the degradability of bacteria on CAR or other substrate was believed to be influenced by the adaptability of isolates to utilize the substrate as sole carbon source. The optimum growth factors such as temperature, pH and presence of nutrients also plays an important role on the capability of bacteria to degrade substrate [57].

7. Conclusion and recommendations

As a conclusion, all the objectives were successfully achieved. The appearance of clear zone on MSM agar plates and color change in MSM broth of all six isolates proved to be carbazole degrading bacteria as they were able to utilize the substrate carbazole as sole carbon source. All six species of bacteria isolated from Unimas lake water was successfully identified using 16S rDNA sequencing which was found to be *Bradhyrhizobium sp., Ochrobactrum sp., Burkholderia sp.* and *Pseudomonas aeruginosa sp. all* these bacteria were found to be gram negative bacteria through gram staining analysis. As for the second objective of measuring the degradation rate, all the six isolates were able to degrade carbazole after 12 days of incubation when tested using GC-FID analysis where *Burkholderia sp.* and *Ochrobactrum sp.* showed high degradation rate. Thus, this proves that all the six species are a good candidate for bioremediation.

For the recommendations, the bacterial growth parameter such as pH, agitation and temperature can be manipulated to increase the degradation rate on the substrate used. The degrading ability of isolated bacteria strains could also be tested with other heterocyclic compound such as dibenzofuran and dibenzothiophene. Next, when preparing and handling samples for GC-FID analysis, precautions should be applied in order to gain more accurate results. Moreover, in order to fully understand on the degradation pathway of isolates, experiment on detection of metabolites produced can be done in the future.

A. APPENDICES

A.1 DNA Sequence for six isolates

IM1:

GGCAGTGGTGCTGCTTACCATGCAAGTCGAGCGCCCCCTGGGGTAGCGG
CTTACTGGTGAGTAACGCGTGGGATCTCCCTTTTGCTACAGAATAACATACG
GAAACTTGTGCTAATACAGTATGTGCCCTTTGGGGGAAAGACTTATGGGCA
AAGGATGAGCCCGCGTTGGATTACCTAGTTGGTGGGGTAAATGCCTACCAA
GGCGACTATCCATATCTGGGCTGAGAGGATGATCAACCCCGTTGTGACTGA
AACACAGCCCAAACTCCTACGGGAGGCAGCAACGGGGAATATTGGACAATG
GGTCCAAGCCTGAGCCATTCGTGCCCTGAGAGAGATGAAGGTCTTACGGTT
GAAAAGCTCTTTCACCGGCGAAGATAATGACGGTAACCGGAGAAGAAACC
CCGGCTAACTGCGTGCCAACCTCCGCG GAAACACGAAGGGGGCTAACGTTG
CTCGGATTTACTGGGTGTAAACCGCACGGACGACGATCGATTAGTCAGGGA
TGAAATCCCACGGCTCACCCCTGGAGCTGCCTTTGAGACTGTATATCTGGAG
TATGTATCAGGTGAGTG GAGTTCCGATTGCAAAGTTGACATTCTCATATAT
CTGTCGGAACACTA TAGGTGAAAGCTGATAGGACATGACCTAGTGAGACAG
AGGTGCA CAAGCGTGGGGAGCTATCTGGAGCACGTACGTTGGAAGTTG
ACTGCATT GACCATCATGGCTGCGCGCACGTCCGTATACTGCCTATGTG
TCTCT GATGCCGGACGAACTCAACTCTCTGGTGAGGACGTTCTCGCGAGAC
TAGCCTCTTTTCTACTCGAACCGTTCACTCACGCGCGATAGCAG GATG
TTTCTTCATTGCTAAGTGACCCTCTAGACGCTGCTCCACACGTGACC TATTG
ACTGTCGCATGAAGCAGATGCTATCTCATCTACCGCTCTTAGCAGTT CACG
TAGCCCTTAAGCTCGATCAAAAATCTGAGATCTTTTGTCTAGTTTGA
CATTATTAGCTTATCCAATGCTTGGCACGTGAATCGTCTGTCTAGTTCGC
CATCAAAAGCGACTTTCGGCTATCATACATCTTAGGAGAGATGCTGGCG G
ACTGAACGCCCTTTATGCATCGATCAGCCAGTTCCACT GAAATGCCTTGTG
CGCAGCCTTGACTGCTAGTATTCCAGAACTTTCACGC GAGACCTCAGATTCG
ATTAGTGAACTGTGTAATCAGGCATCGATGACGCCG.

IM2:

CNNNNNNNGNGGNNGCTNANNNATGCAAGTCGAGCGCCCCGCAAGGG
GAGCGGCAGACGGGTGAGTAACGCGTGGGAACGTACCTTTTGCTACGGAA
TAACTCAGGGAAACTTGTGCTAATACCGTATGTGCCCTTCGGGGGAAA
GATTTATCGGCAAAGGATCGGCCCGCGTTGGATTAGCTAGTTGGTGAGG

TAAAGGCTCACCAAGGCGACGATCCATAGCTGGTCTGAGAGGATGATCAGC
CACACTGGGACTGAGACACGGCCCAGACTCCTACGGGAGGCAGCAGTGGG
GAATATTGGACAATGGGCGCAAGCCTGATCCAGCCATGCCGCGTGAGTGAT
GAAGGCCCTAGGGTTGTAAAGCTCTTTCACCGGTGAAGATAATGACGG
TAACCGGAGAAGAAGCCCCGGCTAACTTCGTGCCAGCAGCCGCGGTAATAC
GAAGGGGGCTAGCGTTGTTCGGATTTACTGGGCGTAAAGCGCACG
TAGGCGGACTTTTAAGTCAGGGGTGAAATCCCGGGGCTCAACCCCG
GAACTGCCTTTGATACTGGAAGTCTTGAGTATGGTAGAGGTGAGTG
GAATTCCGAGTGTAGAGGTGAAATTCGTAGATATTCGGAGGAA
CACCAGTGGCGAAGGCGGCTCACTGGACCATTACTGACGCTGAGGTGC
GAAAGCGTGGGGAGCAAACAGGATTAGATACCCTGGTAGTCCACGCCG
TAAACGATGAATGTTAGCCGTTGGGGAGTTTACTCTTCGGTGGCGCAGC
TAACGCATTATACATTCCGCCTGGGGAGTACGGTCGCAAGATTAAAA
CTCATAGGAATTGACGGGGGCCCGCACAAGCGGTGGAGCATGTGGTT
TAATTCGAAGCAACGCGCAGAACCTTACCAGCCCTTGACATACCGGTCGCG
GANNCANAGATGTGTCTTTCAGTTCGGCTGGACCGGATACAGGTGCTG
CATGGCTGTCGTCAGCTCTNGTCGGGAGATGTTGGGTTAAGTCCG
CAACGAGGGCAACCCTNGCCCTTAGTTGCCAGCATTTAATTGGGNNNTC
TAAGGGACTGCCGGGGATAACCCACAGAAAGGGGGGGATGACGG
CAAGNCCCNAGGGCCCTTACGGGCTGGGTNACACNGGGTTANAATGGGGG-
GAANGGGGGGCACNAACCCCNGAGGGGGAGCTATTTCCCAN NACNCTNC
CAATTCGGAATGCNCTTGGAACCCGGGGCCTAAAATG GAAACCNTT.

IM3:

CCTGCATTGGGGGCAGCTACCATGCAAGTCGAGCGGATGCAACGG
GAGTTTGCTCCGGGGTTCAGCGGCGGACGGGTGAGTAATGCCTAG
GAACCTGCCCGGTAGCGGGGGATAACTTCCGGAAACGGGCGCTAATACCG
CATACGCCCTGAGGGAGAAAGTGGGGGATCTTCGGACCTCACGCTTTGGAT
GAGCCTATGTCGGATTAGCTAGTTGGTGGGGAAAGGCCTACCAAGGCGAC
GATCCATAACTGGTCTGAGAGGATGATCAGCCACACTGGAACTGAGA
CACGGCCCAGACTCCTACGGGAGGCAGCAGTGGGGAATATTGGA
CAATGGGCGAAACCCTGATCCAGCCATGCCGCGTGTGTGAAGAAGGT
CTTCGGATTGTAAAGCACTTTTGGTTGGGAGGAAGGGCAGCGAGTTAAT
ACCTTGCTGATTTGACGGTACCTGCAGAATAAGCACCGGCTAACTTCG
TGCCAGCAGCCGCGGTAATACGAAGGGTGCAAGCGTTAATCGGAA
TTACTGGGCGTAAAGCGCGCGCAGGTGGTTCAGAAAGATGGATGTGA
AATCCCCGAGGCTCAACCTGGGAACTGCATTTTAACTACTGAGCTAGAG
TGCGGTAGAGGGGAGGTGGGAATTCCGCTGTGTAGCAGTGAAATGCGT
AGATATGCGGAGGAACACCGATGGCGAAGCAGACTCCTGGGATAACAC
TGACGCTCATGCACGAAAGCGTGGGGAGCAAACAGGATTAGATACCC
TGGTAGTCCACGCCCTAAACGATGTCAACTAGCTGTTGGGGACTTCTGA
GCTTTGAAGCGCAGCTAACACGTGAAATTGACCGCCTGGGGAGTACAG
TCGCGAGATTATAACTCTCAAGGAGTTGACACGGGACCCCACAGACGC
TGGGATGATGTGTATTATATCGATGAGACGCGCAAAACTTACCTACCG
CTTTACATGTCTGAATGCTTACAGAATTTGATTGGTTCTACGAGACTC
GAACACAGTGCTGCATGCTGTCGTCAGCTCGGTCTGGATGTGGGTAG
TTCCGTACGAGCGCACTGGCATAGTGCTACCACTGGACTCGTACTGT
ACTGGCAGTGACAGTCGAGCTGTGGAGTGAGTAAGGCTCTAGTACCTA
GGCTAAGGCTAGCTCATACGTCAGTCAGTACCTGTCGCATGAAGCG
GATCTCTATCCGATGACG ATCG

IM4:

NNNNNNNNNGCATGCTTACNNTGCAGTCGAACGGCAGCACGGGT
GCTTGCACCTGGTGGCGAGTGGCGAACGGGTGAGTAATACATCGGAAC
ATGTCCTGTAGTGGGGGATAGCCCGGCGAAAGCCGGATTAATACCGC
ATACGATCTACGGATGAAAGCGGGGGACCTTCGGGCCTCGCGCTATAGGG

TTGGCCGATGGCTGATTAGCTAGTTGGTGGGGTAAAGGCCTACCAAG
GCGACGATCAGTAGCTGGTCTGAGAGGACGACCAGCCACACTGGGACT
GAGACACGGCCCAGACTCCTACGGGAGGCAGCAGTGGGGAATTTTGGA
CAATGGGCGAAAGCCTGATCCAGCAATGCCGCGTGTGTGAAGAAGGC
CTTCGGGTTGTAAAGCACTTTTGTCCGGAAAGAAATCCTTGGCTCTAATA
CAGTCGGGGGATGA CGGTACCGGAAGAATAAGCACCGGCTAACTACG
TGCCAGCAGCCGCGGTAATACGTAGGGTGCAAGCGTTAATCGGAATTAC
TGGGCGTAAAGCGTGCGCAGGCGGTTTGCTAAGACCGATGTGAAATC
CCCGGGCTCAACCTGGGAACTGCATTGGTGACTGGCAGGCTAGAGTA
TGGCAGAGGGGGGTAGAATTCCACGTGTAGCAGTGAAATGCGTAGAGAT
GTGGAGGAATACCGATGGCGAAGGCAGCCCCCTGGGCCAATACTGACGC
TCATGCACGAAAGCGTGGGGAGCAAACAGGATTAGATACCCTGGTAGT
CCACGCCCTAAACGATGTCAACTAGTTGTTGGGGATTCTTTCCTTAGTAAC
GTAGCTAACGCGTGAAGTTGACCGCCTGGGGAGTACGGTCGCAAGATTA
AANNCNAAGGACT.

IM5:

NNNNNNNNGNNNGGCATGCTTACNNTGCAAGTCGAACGGCAGCA
CGGGTGCTTGCACCTGGTGGCGAGTGGCGAACGGGTGAGTAATACATC
GGAACATGTCCTGTAGTGGGGGATAGCCCGGCGAAAGCCGGATTAA
TACCGCATACGATCTACGGATGAAAGCGGGGGACCTTCGGGCCTCGC
GCTATAGGGTTGGCCGATGGCTGATTAGCTAGTTGGTGGGGTAAAGG
CCTACCAAGGCGACGATCAGTAGCTGGTCTGAGAGGACGACCAGCCACAC
TGGGACTGAGACACGGCCCAGACTCCTACGGGAGGCAGCAGTGGGGAATT
TTGGACAATGGGCGAAAGCCTGATCCAGCAATGCCGCGTGTGTGAA
GAAGGCCTTCGGGTTGTAAAGCACTTTTGTCCGGAAAGAAATCCTTGGC
TCTAATACAGTCGGGGGATGACGGTACCGGAAGAATAAGCACCGGCTAA
CTACGTGCCAGCAGCCGCGGTAATACGTAGGGTGCAAGCGTTAATCGG
AATTACTGGGCGTAAAGCGTGCGCAGGCGGTTTGCTAAGACCGATGTG
AAATCCCCGGGCTCAACCTGGGAACTGCATTGGTGACTGGCAGGCTAGA
GTATGGCAGAGGGGGGTAGAATTCCACGTGTAGCAGTGAAATGCGTAGA
GATGTGGAGGAATACCGATGGCGAAGGCAGCCCCCTGGGCCAATACTG
ACGCTCATGCACGAAAGCGTGGGGAGCAAACAGGATTAGATACCCTG
GTAGTCCACGCCCTAAACGATGTCAACTAGTTGTTGGGGATTCATTTCCTT
AGTAACGTAGCTAACGCGTGAAGTTGACCGCCTGGGGAGTACGGTCGCA
AGATTAAAACTCAAAGGAATTGACGGGGACCCGCACAAGCGGTGGATG
ATGTGGATTAATTCGATGCAACGCGAAAAACCTTACCTACCCTTGACA
TGGTCGGAATCCTGCTGAGAGGTGGGAGTGCTCGAAAGAGAACCGGCGC
ACAGGTGCTGCATGGCTGTCGTCAGCTCGTGTCGTGAGATGTTGGGTTAA
GTCCCGCAACGAGCGCAACCCTTGTCCTTAGTTGCTACGCAAGACACTCTA
AGGAGACTGCCGGTGACAAACCGGAGGAAGGTGGGGATGACGTCAATCC
TCNTGGCCCTTATGGGTAGGGCTTCCACGTCNNACATGGTCCGAACC
AAGGTTTCCANCCCCCGAGGGGAACTATCCCAAAAACNANTCTAAT
CCGAATGCCCTGCAACCCAGGGCTAAACTGGATCCTTAANCCCGANNNNT
GCCCGGGGAAACTTCCCGGTTTTTACCCCCCNNCCCCCGGGAGGGGTTT
TCCAAAGGGGGTNNAC.

IM6:

NNNNNNNNNNNGNNGCTTACCATGCATGTCGAGCGCCCCGCAAGGGGA
GCGGCAGACGGGTGAGTAACGCGTGGGAACGTACCTTTTGCTACGGAA
TAACTCAGGGAAACTTGTGCTAATACCGTATGTGCCCTTCGGGGGAAAGA
TTTATCGGCAAAGGATCGGCCCGCGTTGGATTAGCTAGTTGGTGAGG
TAAAGGCTCACCAAGGCGACGATCCATAGCTGGTCTGAGAGGATGA
TCAGCCACACTGGGACTGAGACACGGCCCAGACTCCTACGGGAGGCAG
CAGTGGGGAATATTGGACAATGGGCGCAAGCCTGATCCAGCCATGCC
GCGTGAGTGATGAAGGCCCTAGGGTTGTAAAGCTCTTTCACCGGTG

AAGATAATGACGGTAACCGGAGAAGAAGCCCCGGCTAACTTCGTGCCAG
CAGCCGCGGTAATACGAAGGGGGCTAGCGTTGTTCGGATTTACTGGGC
GTAAAGCGCACGTAGGCGGACTTTTAAGTCAGGGGTGAAATCCCGGGG
CTCAACCCCGGAACTGCCTTTGATACTGGAAGTCTTGAGTATGGTAGA
GGTGAGTGGAATTCCGAGTGTAGAGGTGAAATTCGTAGATATTCGG
AGGAACACCAGTGGCGAAGGCGGCTCACTGGACCATTACTGACGCTGAG
GTGCGAAAGCGTGGGGAGCAAACAGGATTAGATACCCTGGTAGTCCACG
CCGTAAACGATGAATGTTAGCCGTTGGGGAGTTTACTCTTCGGTGGCG
CAGCTAACGCATTAAACATTCCGCCTGGGGAGTACGGTCGCAAGATTAA
AACTCATAGGAATTGACGGGGGCCCGCACAAGCGGTGGAGCATGTGGTT
TAATTCTAAGCAACGCGCAGAACCTTACCAGCCCTTGACATACCGGTCG
CGGACACAGAGATGTGTCTTTCAGTTCGGCTGGACCGGATACAGGTG
CTGCATGGCTGTCGTCAGCTCGTGTCGTGAGATGTTGGGTTAAGTCCC
GCAACGAGGGCAACCCTCNCCCTTAGTTGCCAGCATTTAGTTGGGGA
CTCTAAGGGGACTGCCGGTGATAANCCAAAAGAAAGGGGGGATGANN
NCAAGTCTCCTGGGCCTTACGGCTGGGGTACCACCGGGCTACAATGGG
GGGGACNGGGGGCACCGAGACCCCNAGGGGGNCTNNTTNCCAAANCN
NNNCAGTCGGATTGCNNNNCNACNCCGGGGCTGAAGTGGAATCCGNG
GAATCCGGANNCCATGCCCGGGGAANNCTTCCGGGCTTNTAACCCCCC
CGTCNCCCTGGGGATTGGTTTTCCCAAAGGNTTGTTTACCCCAGGGG
GGGGACCCCCTGGGGNNCACCTGGGGGATTTCAAAAGGTCNNGA

A.2 Overlay of GC-FID results

IM1:

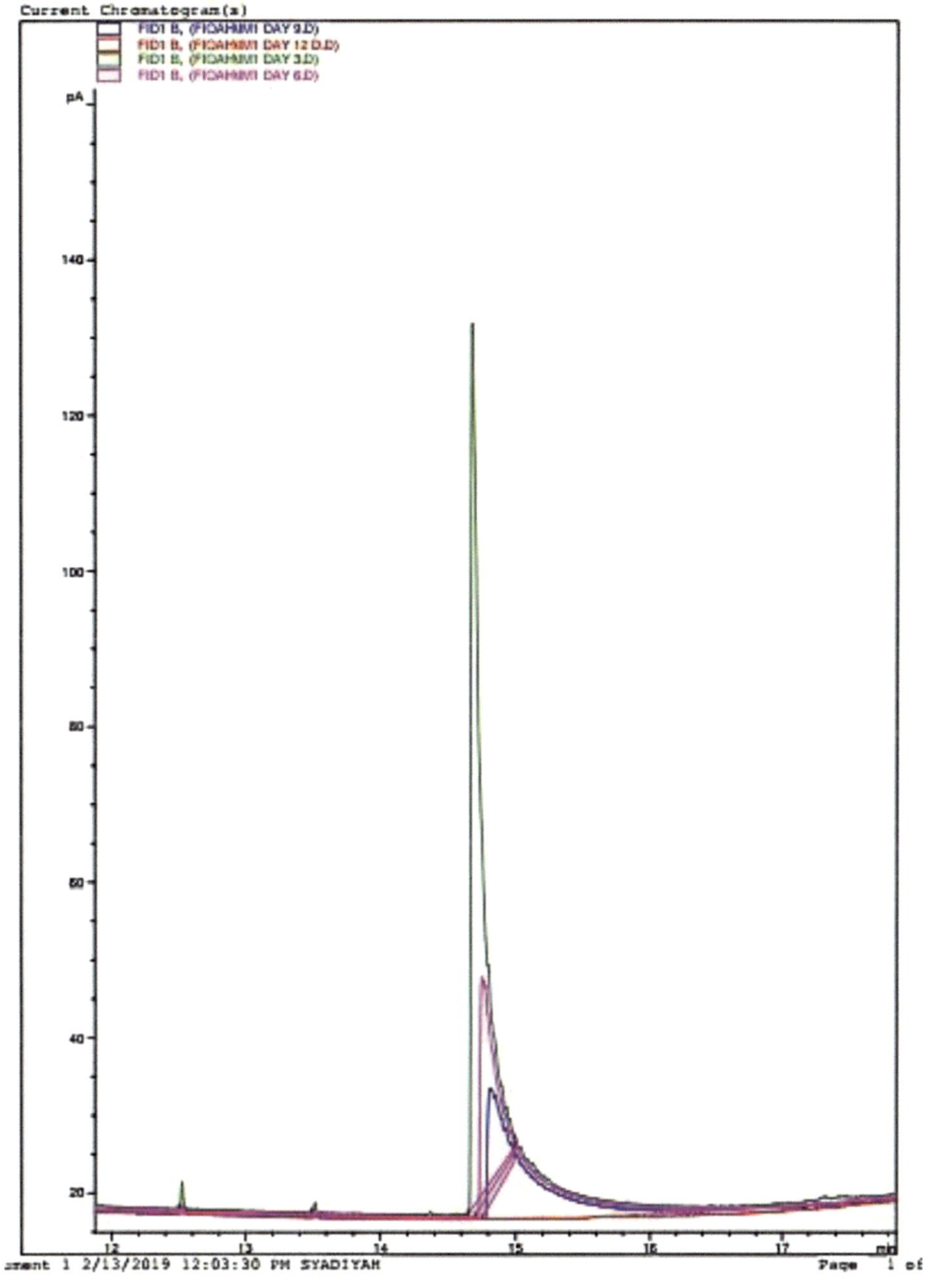

IM2:

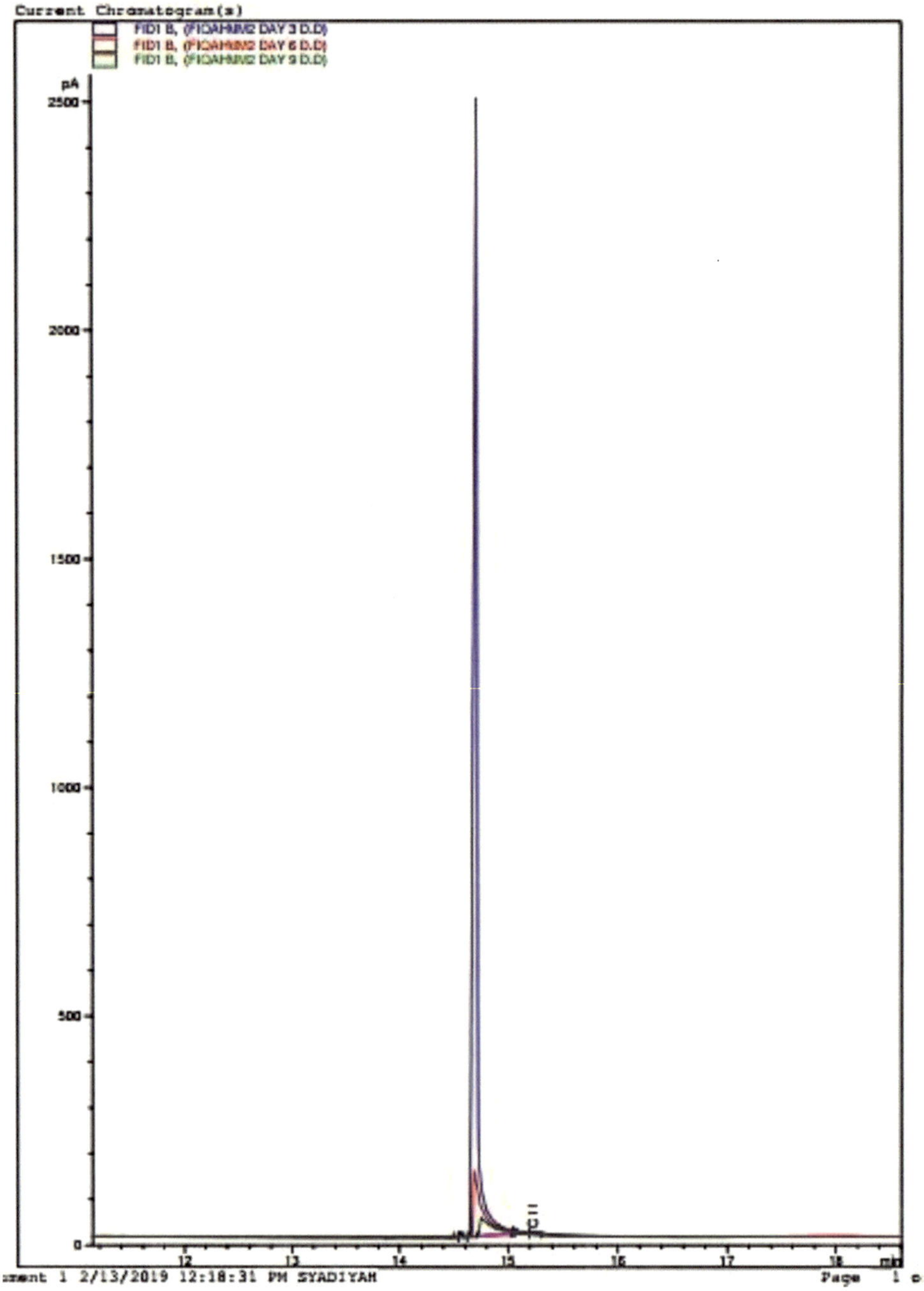
Current Chromatogram(s)
pA
2500
2000
1500
1000
500
0
12
13
14
15
16
17
18
min
2/13/2019 12:18:31 PM SYADIYAH
Page

IM3:

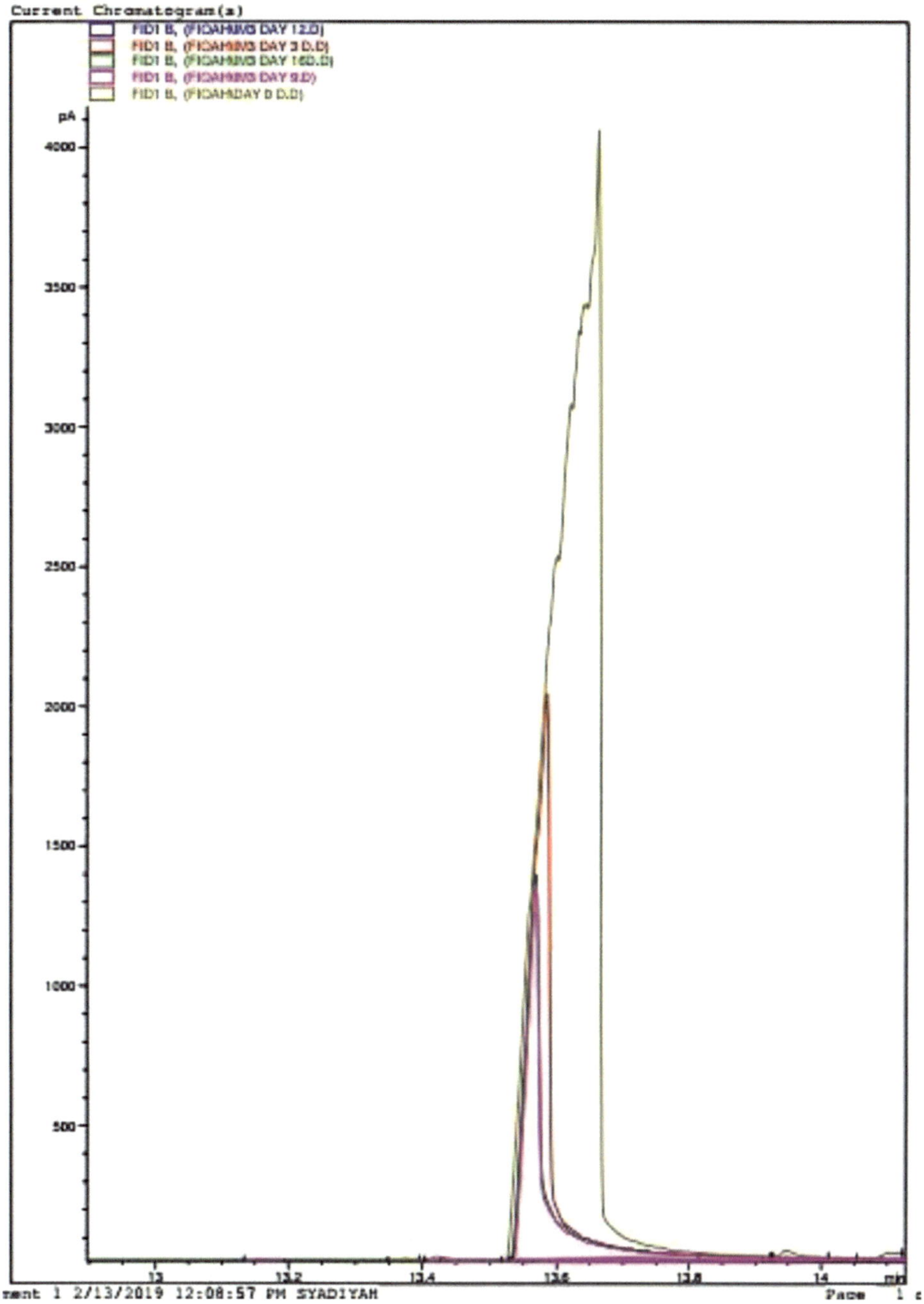
Current Chromatogram(s)
pA
4000
3500
3000
2500
2000
1500
1000
500
13
13.2
13.4
13.6
13.8
14
min
ment 1 2/13/2019 12:08:57 PM SYADIYAH
Page 1

IM4:

IM4:

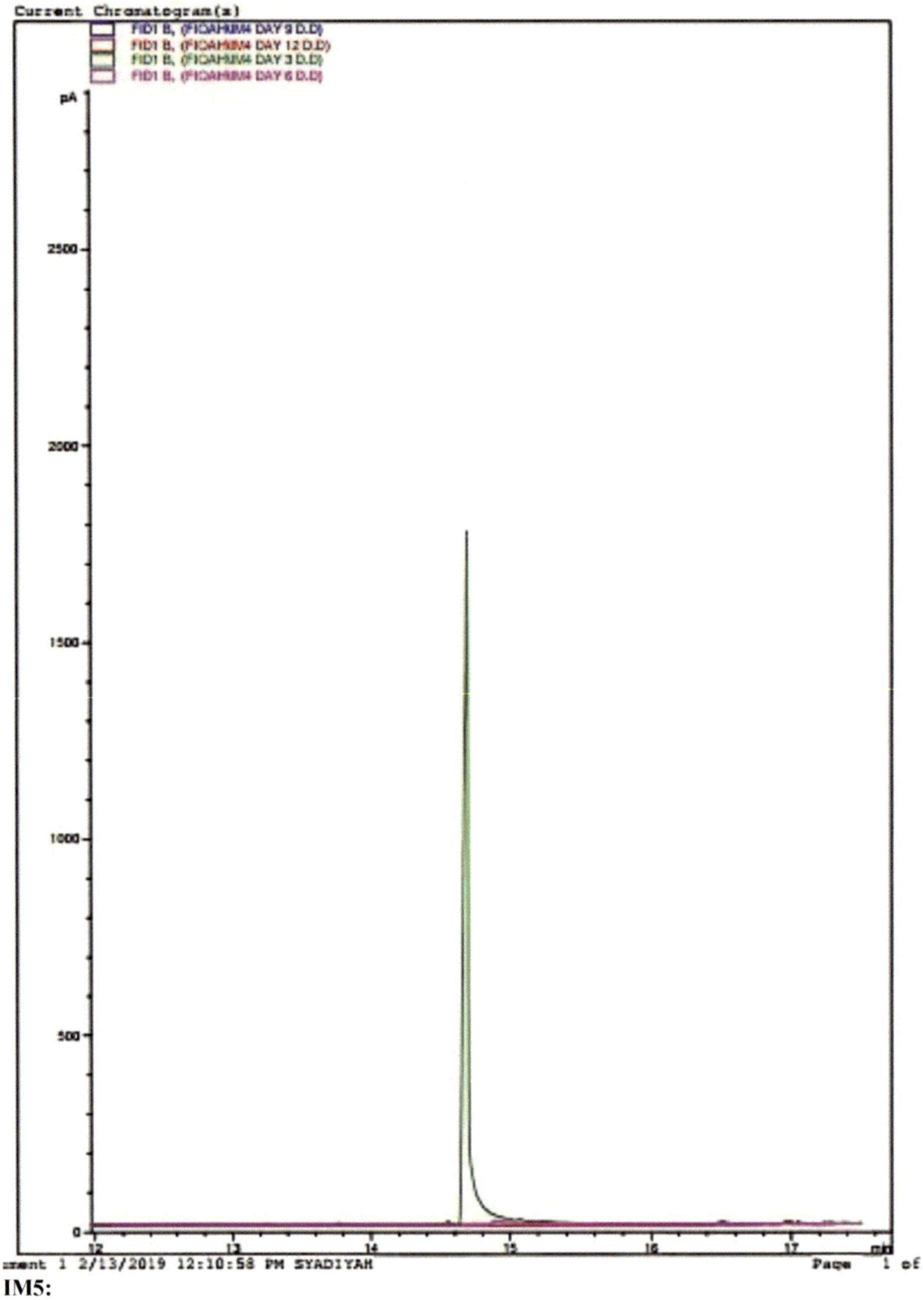
Current Chromatogram(s)
pA
2500
2000
1500
1000
500
0
12
13
14
15
16
17
min
2/13/2019 12:10:58 PM SYADIYAH
Page 1 of

IM5:

IM5:

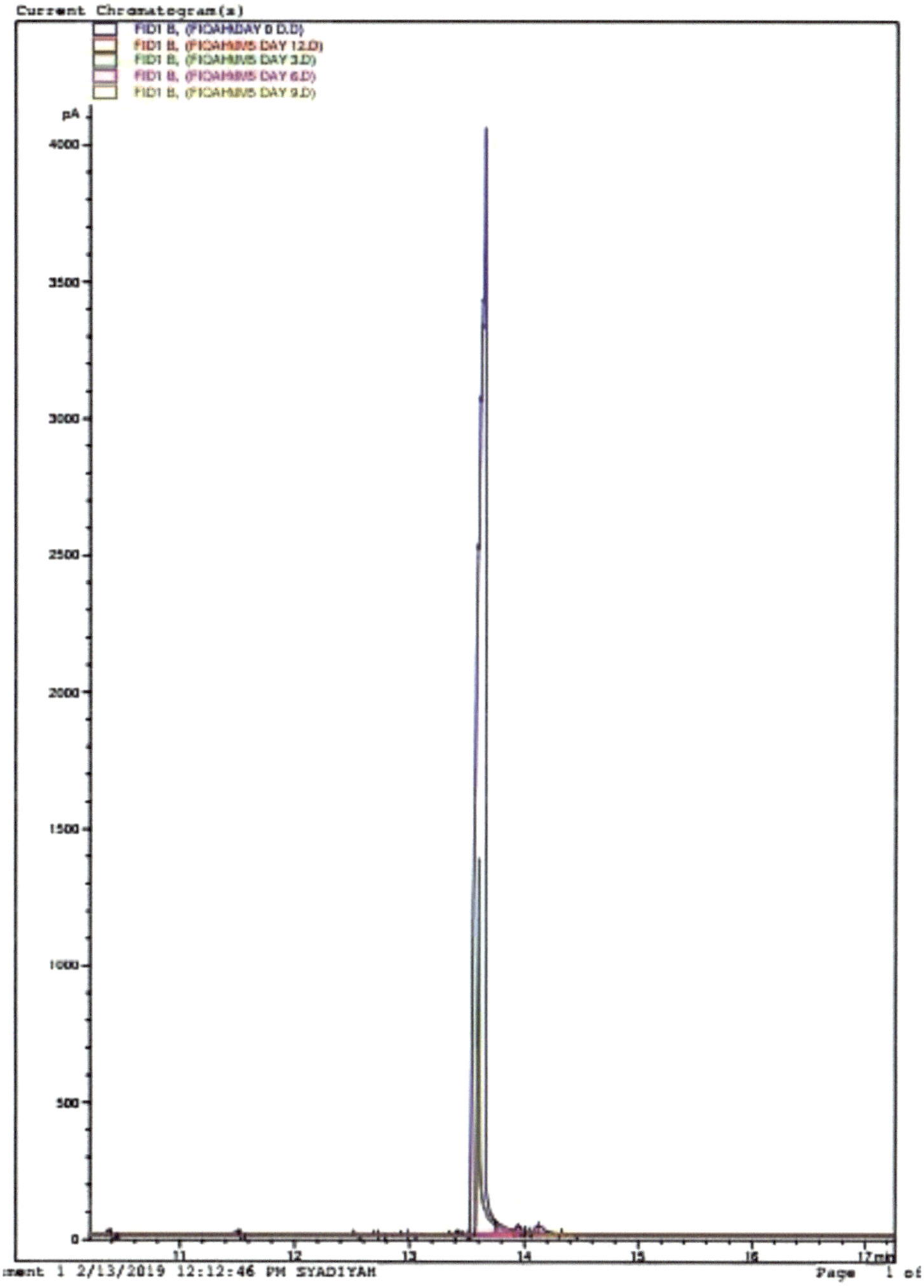
Current Chromatogram(s)
pA
4000
3500
3000
2500
2000
1500
1000
500
0
11
12
13
14
15
16
17 min
2/13/2019 12:12:46 PM SYADIYAH
Page 1 of

IM6:

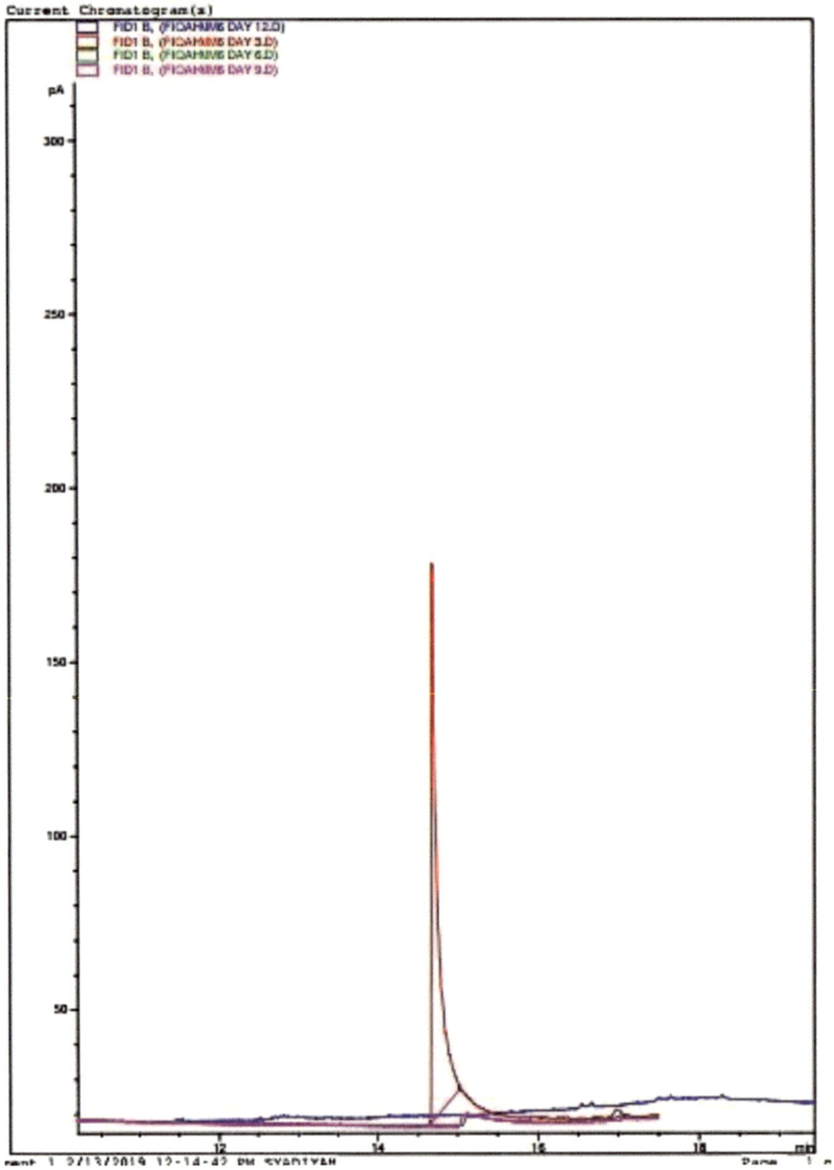

Author details

Khairunnisa Binti Abdul Lateef Khan
Faculty of Resource Science and Technology, Universiti Malaysia Sarawak, Malaysia

*Address all correspondence to: khairunnisakhan97@gmail.com

References

[1] Salam, L. B., Ilori, M. O., & Amund, O. O. (2017). *Properties, environmental fate and biodegradation of carbazole*. 3 Biotech, 7(2). https://doi.org/10.1007/s13205-017-0743-4.

[2] Isentraeger A, Brinkman C, Hollert H, Sagner A, Tiehm A, Neuwoehner J. (2008). Heterocyclic compounds: toxic effects using algae, daphnids, and the Salmonella/microsome test taking methodical quantitative aspects into account. Environ Toxicol Chem 27: 1590–1596.

[3] Ferraro DJ, Gakhar L, Ramaswamy S (2005) Rieske business: structure-function of Rieske non-heme oxygenases. Biochem Biophys Res Commun 338:175–190.

[4] Abha, S., & Singh, C.S. (2012). *Hydrocarbon pollution: Effects on living organisms, remediation of contaminated environments, and effects on heavy metals contamination on bioremediation*. In L. Romero-Zeron (Ed.), Introduction to enhanced oil recovery (EOR) process and bioremediation of oil contaminated sites (pp. 159–206). Croatia: Intech.

[5] Chu GJ, Murad A. A case of ethanol-induced systemic allergic dermatitis. Contact Dermatitis. 2017 Mar;76(3):182-184. doi: 10.1111/cod.12668. PMID: 28220569.

[6] Peddinghaus S, Brinkmann M, Bluhm K, Sagner A et al. (2012). Quantitative assessment of the embryotoxic potential of NSO-heterocyclic compounds using zebrafish (Danio rerio). Reproduc Toxicol 33: 224–232.

[7] Smith CJ, Hansch C. (2000). The relative toxicity of compounds in mainstream cigarette smoke condensate. Food Chem Toxicol 38: 637–646.

[8] Alexander, M. (1999). *Biodegradation and bioremediation*. Academic Press. San Diego, USA; pp. 325–327.

[9] Salam L.B., Ilori M.O., Amund O.O. et al (2014). *Carbazole angular dioxygenation and mineralization by bacteria isolated from hydrocarbon-contaminated tropical African soil*. Environ Sci Pollut Res 21:9311–9324.

[10] Masih, J., Singhvi R., Kumar K., Jain V.K., Taneja A. (2012). *Seasonal variation and sources of polycyclic aromatic hydrocarbons (PAHs) in indoor and outdoor air in a semi arid tract of northern India*. Aerosol Air Qual Res, 12, pp. 515–525.

[11] Fetzer, J.C. 2000. The Chemistry and Analysis of the large polycyclic aromatic hydrocarbon. Polycyclic aromatic compound (New York: Wiley) 27, 143, Food Safety Authority of Ireland; 2006, Investigation into levels of polycyclic AromaticHydrocarbons (PAH) in food on the Irish market. Grifoll.

[12] Bamforth, M., Singleton. (2005). *Bioremediation of polycyclic aromatic hydrocarbons: Current knowledge and future directions*. Journal of Chemical Technology and Biotechnology. Vol 80; pp.723–736.

[13] Max Nestler, F.H. Characterization of wood-preserving coal-tar creosote by gas-liquid chromatography. Anal. Chem. 1974, 46, 46-53.

[14] Igwe, J. C. (2015). Environmental Effects of Polycyclic Aromatic Hydrocarbons. 5(7), 117–132.

[15] Sims RC, Overcash MR. (1983). Fate of polynuclear aromatic compounds (PNAs) in soil-plant systems. Residue Reviews 88:1–66

[16] Bleeker, E., Wiegman, A.S., Droge, S.T.J., Kraak, M.H.S., van Gestel, C.A.M.

(2003). *Towards an improvement of the risk assessment of polycyclic (hetero) aromatic hydrocarbons*. Report 2003–01 of the department of Aquatic Ecology and Ecotoxicology, University of Amsterdam, Amsterdam, the Netherlands, pp. 58.

[17] Abdel-Shafy, H. I., & Mansour, M. S. M. (2016). *A review on polycyclic aromatic hydrocarbons: Source, environmental impact, effect on human health and remediation*. Egyptian Journal of Petroleum, 25(1), 107–123.

[18] IARC (International Agency for Research on Cancer). (1983). *Polynuclear Aromatic Compounds, Part 1, Chemical, Environmental and Experimental Data, IARC Monographs on the Evaluation of the Carcinogenic Risk of Chemicals to Humans*. Vol. 32, Lyon, France, pp. 33–91.

[19] Nojiri, H., Omori, T. (2007). *Carbazole metabolism by Pseudomonads*. In: Ramos J-L,Filloux A (eds) Pseudomonas, vol 5. Springer, New York, pp. 107–145.

[20] Chung, W. H. (2013). *Isolation of Heterocyclic Hydrocarbon from Seawater*. (Unpublished degree thesis). Universiti Malaysia Sarawak, 14–17.

[21] Sufinas, S.F.N. (2012). *Isolation and Characterization of carbazole degrading bacteria from soil. mangrove environment*. (Unpublished degree thesis). Universiti Malaysia Sarawak.

[22] Cristol SJ, Zaki Ali M. Photochemical transformations. 36. Intramolecular electron transfer from photoexcited benzene rings to carbon-mercury bonds. Stereochemical fate of zwitterionic biradicals leading to Wagner-Meerwein rearranged products. *Tetrahedron Letters*. 24: 5839-5842.

[23] Grison, C., & Biton, J. (2018). *Heterocyclic Compound from Phytoextraction to Green Chemistry and Vice Versa via*, 1–7.

[24] Karigar, C. S., & Rao, S. S. (2011). *Role of microbial enzymes in the bioremediation of pollutants: a review*. Enzyme Research, 2011, 805187. https://doi.org/10.4061/2011/805187

[25] Meike, A., & Stroes-Gascoyne, S. (2000). *Review of microbial responses to abiotic environmental factors in the context of the proposed Yucca mountain repository*. Springfield: Lawrence Livermore National Laboratory.

[26] Sylvia, D. M., Fuhrmann, J.F., Hartel, P.G., and D.A Zuberer (2005). "*Principles and Applications of Soil Microbiology*." New Jersey, Pearson Education Inc.

[27] Alvarez, V.M., Santos, S.C., Casella, R.C., Vital, R.L., Sebastian, G.V. & Seldin, L. (2008). *potential of a tropical soil contaminated with a mixture of crude oil and production water*. Journal of Microbiology and Biotechnology, 18 (12), 1966–1974.

[28] Latha, K., Lalithakumari, D. Transfer and expression of a hydrocarbon-degrading plasmid pHCL from Pseudomonas putida to marine bacteria. *World Journal of Microbiology and Biotechnology* 17, 523–528 (2001). https://doi.org/10.1023/A:1011917408368

[29] Chung, W. K. and King, G. M. (2001). 'Isolation, Characterization, and Polyaromatic Hydrocarbon Degradation Potential of Aerobic Bacteria from Marine Macrofaunal Burrow Sediments and Description of Lutibacterium anuloederans gen. nov., sp. nov., and Cycloclasticus spirillensus sp. nov.', *Applied and Environmental Microbiology*. American Society for Microbiology Journals, 67(12), pp. 5585–5592. doi: 10.1128/AEM.67.12.5585-5592.2001.

[30] Lobastova, T.G., Sukhodolokaya, G. V., Nikolayeva, V.M., Baskunov, B.P.,

Turchin, K.F.,Donova, M.V. (2004). *Hydroxylation of carbazoles by Aspergillus flavus VKM F1024*. FEMS Microbiol Lett 235:51–56.

[31] Nojiri, H. (2012). *Structural and molecular genetic analyses of the bacterial carbazole degradation system.* Biosci Biotechnol Biochem 76(1):1–18.

[32] Grifoll M, Selifonov SA, Chapman PJ. (1995). Transformation of substituted fluorenes and fluorine analogs by Pseudomonas sp. strain F274. Appl Environ Microbiol 57:3462–3469

[33] Nojiri H, Nam JW, Kosaka M, Morii K-I, Takemura T et al. (1999). Diverse oxygenations catalyzed by carbazole 1,9a-dioxygenase from Pseudomonas sp. strain CA10. J Bacteriol 181(10):3105–3113.

[34] Gutierrez, T., Biddle, J.F., Teske, A., Aitken, M.D. (2015). *Cultivation-dependent and cultivation-independent characterization of hydrocarbon-degrading bacteria in Guaymas Basin sediments*. Frontiers in Microbiology, 6.

[35] Hirano, S., Kitauchi, F., Haruki, M., Imanaka, T., Morikawa, M. and Kanaya, S. (2004). *Isolation and characterization of Xanthobacter polyaromaticivorans sp. nov. 127W that degrades polycyclic and heterocyclic aromatic compounds under extremely low oxygen conditions*. Bioscience, biotechnology, and biochemistry, 68(3) 557–564.

[36] Grosser RJ, Warshawsky D, Vestal JR. (1991). Indigenous and enhanced mineralization of pyrene, benzo(a) pyrene and car- bazole in soils. Appl Environ Microbiol 57:3462–3469

[37] Zulkharnain, A., & Taka, J.S. (2011). *Preliminary studied on heterocyclic hydrocarbon degrading bacteria isolated from Southwest Coast of Borneo*. International Congress of the Malaysian Society for Microbiology, 2, 397–340.

[38] Reuszer HW. (1933). Marine bacteria and their role in the cycle of life in the sea: III. The distribution of bacteria in the ocean waters and muds about Cape Cod. Biological Bulletin, 65, 480.

[39] Stope, M.B., Becher, D., Hammer, E., & Schaeur, F. (2002). *Cometabolic ring fission of dibenzofuran by Gram negative and Gram-positive biphenyl-utilizing bacteria.* Applied and Environmental Microbiology, 59, 62–67.

[40] Maeda R, Nagashima H, Widada J, Iwata K, Omori T. (2009a). Novel marine carbazole-degrading bacteria. FEMS Microbiol Lett 292:203–209

[41] Zheng, K. *et al.* (2014). 'A unique carbazole–coumarin fused two-photon platform: development of a robust two-photon fluorescent probe for imaging carbon monoxide in living tissues', *Chem. Sci.* The Royal Society of Chemistry, 5(9), pp. 3439–3448. doi: 10.1039/C4SC00283K.

[42] Takahashi, T. *et al.* (2009). 'Climatological mean and decadal change in surface ocean pCO2, and net sea–air CO2 flux over the global oceans', *Deep Sea Research Part II: Topical Studies in Oceanography*, 56(8), pp. 554–577. doi: https://doi.org/10.1016/j.dsr2.2008.12.009.

[43] Beveridge, T.J. (2001). *Use of the gram staining in microbiology*. Biothechnic Histochemistry, 76, 111–118.

[44] Gieg, L.M., Otter, A., & Fedorak, P. M. (1996). *Carbazole degradation by Pseudomonas sp.LD2: Metabolic characteristics and identification of some metabolites*.Environmental Science Technology, 30, 575–585.

[45] Farajzadeh, Z., & Karbalaei-Heidari, H.R. (2012). *Isolation and characterization of a new Achrombacter sp. strain CAR1389 as a carbazole*

degrading-bacterium. World Journal Microbiology Biotechnology, 28, 3075–3080.

[46] Kilbanne, J.J., Draram, A., Abbasian, J., & Kayser, K.J. (2002). *Isolation and characterization of Sphingomonas sp. GTIN11 capable of carbazole metabolism in petroleum*. Biochemical and Biophysical Research Communication, 297, 242–248.

[47] Angeline, W.S.L. (2006). *Broad Range of PCR amplification of bacteria DNA in various clinical samples for the rapid diagnosis of infections.* Published thesis for degree of masters. Kuala Lumpur, Malaysia: University of Malaya.

[48] Cheng, L., & Zhang, D.Y. (2008). *District Laboratory Practice in Tropical Countries.* (2nd). Cambridge: Cambridge University Press.

[49] Qu, Y. and Spain, J. C. (2010) 'Biodegradation of 5-nitroanthranilic acid by Bradyrhizobium sp. strain JS329.', *Applied and environmental microbiology*. American Society for Microbiology (ASM), 76(5), pp. 1417–22. doi: 10.1128/AEM.02816-09.

[50] Ibrahim, H. M. M. (2016). 'Biodegradation of used engine oil by novel strains of Ochrobactrum anthropi HM-1 and Citrobacter freundii HM-2 isolated from oil-contaminated soil.', *3 Biotech*. Springer, 6(2), p. 226. doi: 10.1007/s13205-016-0540-5.

[51] Zhang, G. *et al.* (2005). 'Biodegradation of crude oil by Pseudomonas aeruginosa in the presence of rhamnolipids.', *Journal of Zhejiang University. Science. B*. Zhejiang University Press, 6(8), pp. 725–30. doi: 10.1631/jzus.2005.B0725.

[52] Inoue, K. *et al.* (2005) 'Diversity of carbazole-degrading bacteria having the car gene cluster: Isolation of a novel gram-positive carbazole-degrading bacterium', FEMS Microbiology Letters. Oxford University Press, 245(1), pp. 145–153. doi: 10.1016/j.femsle.2005.03.009.

[53] Okoh, O.B., Ilori, M.O., Akinyemi, J. O., and Adebusoye, S.A. (2006). *Hydrocarbon degrading potential of bacteria isolated from a Nigerian bitumen (Tarsand) deposit.* Nature and Science, 4 (3): 51–57.

[54] Hedlund, B.P., Geiselbrecht, A.D., Bair, T.J. and Staley, J.T. (1999). *Polycyclic aromatic hydrocarbon degradation by a new marine bacteriu, Neptunomonas naphthovorans gene. Nov., sp. nov.* Applied and Environmental Microbiology, 65; 251–259.

[55] Leahy, J.G., Colwell, R.R. (1990). *Microbial degradation of hydrocarbons in the environment*. Microbiol Rev 54:305–315.

[56] Alennabi, K. A. A. (2012) 'Study on biodegradation of Miri and Masila crude oil and used car oil by microorganisms isolated from Malaysian soil and the effect of aeration and NPK addition on biodegradation process', (June).

[57] Olowomofe, T. & Babalola, Toyin & Oluyide, Oluwabusayo & Adedayo, O. (2019). Microbial Assessment of In-door Air and Equipment Used in Banks within Ekiti State University, Ado-Ekiti, Ekiti State, Nigeria. Annual Research & Review in Biology. 1-13. 10.9734/arrb/2019/v33i530134.

Section 2

Membrane

Chapter 4

Emerging Trends in Wastewater Treatment Technologies: The Current Perspective

Edward Kwaku Armah, Maggie Chetty, Jeremiah Adebisi Adedeji, Donald Tyoker Kukwa, Boldwin Mutsvene, Khaya Pearlman Shabangu and Babatunde Femi Bakare

Abstract

The quality of freshwater and its supply, particularly for domestic and industrial purposes are waning due to urbanization and inefficient conventional wastewater treatment (WWT) processes. For decades, conventional WWT processes have succeeded to some extent in treating effluents to meet standard discharge requirements. However, improvements in WWT are necessary to render treated wastewater for re-use in the industrial, agricultural, and domestic sectors. Three emerging technologies including membrane technology, microbial fuel cells and microalgae, as well as WWT strategies are discussed in this chapter. These applications are a promising alternative for manifold WWT processes and distribution systems in mitigating contaminants to meet acceptable limitations. The basic principles, types and applications, merits, and demerits of the aforementioned technologies are addressed in relation to their current limitations and future research needs. The development in WWT blueprints will augment the application of these emerging technologies for sustainable management and water conservation, with re-use strategies.

Keywords: contaminants, membrane technology, microalgae, microbial fuel cell, wastewater treatment

1. Introduction

The modern-day world has seen a boom in industrial activities. Due to extensive manufacturing activities taking place, large volumes of waste are produced, including wastewaters which are of major interest for re-use due to the scarcity of potable water in most countries. The wastewater produced poses serious environmental problems in its disposal. Because of new products that are emerging and being manufactured, so are new and recalcitrant wastes produced in production lines. Convectional wastewater technologies may be limited to process these contaminants, further exacerbating the problems the world is already facing with respect to potable water. Hence, there is a dire need to develop new methods to mitigate wastewater's effect on the already degrading environment. On the other hand, clean, fresh potable water has become scarce especially in most African countries due to contamination by intensive industrial activities. To date over one hundred technologies for the treatment of organic and inorganic wastewater streams have been documented; several of these technologies have been emerging and these range from chemical and physical to biological methods. This book chapter focuses on the emerging trends of wastewater treatment technologies, with respect to membrane and biological methods.

Exhibiting high levels of novelty in purification technologies, membranes have been widely used and serve a crucial role in various fields, such as fatty and oily industrial water treatment [1–3].

Microalgae-based technologies are autotrophic in nature and microalgae is a highly potential atmospheric carbon fixation technology. After upstream treatment processes, microalgae technology is usually employed as secondary or tertiary treatment process for effluents that are laden with inorganic components such as nitrogen and phosphorus which cause eutrophication and more long term challenges that are caused by organic material and heavy metals in disposed of wastewater. Microalgal processes then chip in to offer at attractive dimension for the treatment of wastewater coupled with the generation of possibly biomass of high value which can further used for various purposes. Microalgae has minimal risk of production of secondary pollution because of its ability to use inorganic nitrogen and phosphorus for their growth; and their ability to remove heavy metals and toxic organics [4–6].

Another powerful, emerging treatment methodology is the Microbial fuel cells (MFCs) technology which capitalizes on the bioelectrical catalytic activity of microorganisms to generate electric power by oxidizing the organic matter and sometimes inorganic material in wastewater. MFC technology offers a dual goal as it allows for energy recovery and wastewater treatment in a single configuration [7, 8].

2. Wastewater contaminants

The term wastewater is said to be water containing contaminants mainly due to human use. It emanates from diverse sources such as domestic, commercial, agricultural, or infiltration and storm run-off, with most wastewater being 99.9% water and the rest solids [9]. The characteristics of wastewater are usually determined by the chemical components and flow conditions, as this is used in the design of each wastewater treatment plant [10]. The flow conditions of wastewater are based on the seasons and it is mainly the wet season which will result in an inflow of storm run-offs. The organic and inorganic constituents of wastewater are used as an indicator of the chemical quality of wastewater. The following parameters are usually considered when measuring the chemical characteristics of wastewater; biochemical oxygen demand (BOD), chemical oxygen demand (COD), total solids (TS), volatile solids (VS), total nitrogen (TN), total phosphorus (TP), pH and alkalinity [11]; among others.

2.1 Chemical oxygen demand (COD)

This is usually a representative of the contaminants in wastewater as the higher the COD content in wastewater, the higher the degree of contamination. The COD content in industrial wastewater is usually higher when compared to that of domestic/municipal wastewater as presented in **Table 1**. It gives an indication of the degree of biodegradation in wastewater when compared with BOD as the ratio of BOD to COD higher than 0.5 makes the wastewater biologically treatable [16]. It is

Parameters	Brewery	Abattoir	Cane Sugar	Oil refinery	Coke Oven	Tannery	Textile
BOD_5, mg/L	1609.34-3980.61	476-3850	350-2750	100-500	510-1360	1000-2000	50-500
COD, mg/L	1096.41-8926.08	935-6600	1000-4340	150-800	930-3120	2000-4000	250-8000
TSS	530.67-3728.02	750-4400	760-800	130-600	19-3330	2000-3000	100-700
pH	4.6-7.3	6.85-8.19	5-6.5	2-6	6.8-8.2	11-12	5.6-9

Table 1.
Characteristics of raw industrial wastewater [12–15].

measured as the quantity of oxygen required to stabilize the carbonaceous organic matter chemically. It is used to quantify the organic matter, nitrite, sulphide and ferrous salts present in wastewater [17].

COD in wastewater could either be readily biodegradable matter, active autotrophic and heterotrophic biomass, soluble inert organic matter, inert inorganic matter [18]. Generally, the COD content in wastewater is either soluble or particulate (suspended). Classification of domestic wastewater based on COD include low (300-500 mg/L), medium (500-750 mg/L) and high (700 – 1200 mg/L) strength wastewater [19]. According to Henze and Comeau [19], the degradable COD content of a typical medium strength is 90% for soluble COD, 66% for particulate COD and 76% for total COD while the remaining percent are the inert component. The use of membrane technology only is very effective for low-strength wastewater [20] but the efficiency can be increased when combined with other technologies for treatment of high strength wastewater such as seen in the study by Matheus et al. [21] where microfiltration and nanofiltration was preceded by coagulation and flocculation to achieve a 96% COD removal (from 4610 mg/L to 184 mg/L) for dairy wastewater. Wastewater with high COD content usually causes fouling for the membrane [21], therefore, the use of biological treatment techniques such as microalgae and microbial fuel cell are more appropriate for high strength wastewater [22, 23].

2.2 Biochemical oxygen demand (BOD)

This is the quantity of oxygen required by microorganisms for the decomposition of organic matter under aerobic conditions. As stated for COD, BOD is also an indication of the degree of contamination, it affects the amount of dissolved oxygen required by aquatic organisms, and if lower than 6 mg/L could lead to their death. The typical BOD value of domestic wastewater with minor industrial wastewater in it ranges from 100 – 200 mg/L, 200 – 300 mg/L and 300 – 560 mg/L for low, medium and high strength wastewater [19]. The relationship between BOD and dissolved oxygen is inversely proportional, as a low dissolved oxygen indicates a high BOD content in wastewater [24]. However, as the organic biodegradable content of water increases, the BOD increases also [25]. Since increase in biodegradable organic pollutants is an increase in the BOD, therefore, most biological treatment processes such as microalgae or microbial fuel cell technique can remove the BOD content in wastewater. Zhang et al. [26], indicated a 98.6% BOD removal using MFC while Marassi et al. [27] reported a 96-97% efficiency using a tubular MFC. The use of microalgae has also been reported to have effectively reduce the BOD content of wastewater by generation of O_2 during photosynthesis [28] and 87% removal efficiency [29].

2.3 Total solid (TS)

This is the organic and inorganic matter; suspended and dissolved solids; settleable and volatile solid content of wastewater. Though physical separation techniques easily remove most suspended solids, some still find their way into the environment. The dissolved and volatile solid (VS) contents are a representative of the degradable content in wastewater; therefore, some treatment techniques do account for the number of volatile solids removed. The VS content of wastewater, likewise, indicate its strength as higher VS indicate high strength wastewater and vice versa. The more the VS content of wastewater, the greater the impact on the treatment plant as it is an indication of the organic solid content. Total dissolved solids (TDS) are composed of inorganic salts and small quantities of organic matter dissolved in water. TDS in wastewater increases due to chemicals either from washing, cleaning, and production processes [30].

2.4 Total nitrogen and phosphorus

These are plant nutrients that are present in wastewater as either nitrates or ammonia, and fertilizer manufacturing companies usually generate them, agricultural sectors and industries that utilize corrosion inhibitors. Total nitrogen is the combination of both the inorganic and organic nitrogen, and ammonia in wastewater, it exists as either nitrate, nitrite, ammonium, and organic dissolved compounds such amino acids, urea, and organic nitrogen composites. In aquatic ecosystems, phosphorus is also present as phosphates such as orthophosphates, condensed phosphates and phosphates organically bound [25].

Nitrogen and phosphorus in wastewater cause eutrophication in water bodies which can lead to the death of aquatic habitats, if discharged without treatment [31]. High removal rate of nitrogen and phosphorus have been achieved using microalgae treatment process with industrial application of this technique been reported to achieve between 87 and 93% removal [32].

2.5 Metals

Metals are generally found in wastewater, mainly from the manufacturing, mining, and textile industries. Metals such as arsenic, iron, chromium, lead, copper, tin, sodium, potassium, mercury, aluminum, and nickel are common pollutants in industrial wastewaters [33]. Industries such as iron and steel, mining, micro-electronics, and textiles often generate wastewater with heavy metals therein. Metals in wastewater lead to an increase in the treatment costs, and they are known to cause varying environmental problems such as distortion in plant growth, algal bloom, death of aquatic biota, debris formation and sedimentation [34]. Human related health effects include carcinogenicity, chronic asthma, skin related problems, depression, internal organ damage, coughing and nervous system-related diseases [35].

The presence of metal in wastewater in low concentration (1-3 mg/L) is toxic because metals are non-biodegradable and some metals do accumulate overtime [33, 36]. Although some metals which are essential to human, animal and plants may still be tolerated in minimal quantities such as copper, zinc, chromium but above the limit required can be toxic. An example is the reproduction of water flea Daphnia affected by exposure to 0.01 mg hexavalent chromium/L, therefore, the lethal chromium level for several aquatic and terrestrial invertebrates has been reported to be 0.05 mg/L. Some elements, however, such as arsenic, lead, cadmium, mercury is known to be toxic to living beings at any concentration and are not required to be taken into the body even at ultra-trace level [33].

2.6 Viruses and bacteria

The occurrence of human pathogenic viruses in wastewater is a usual occurrence, and newly discovered cases that were not associated with wastewater previously, are now considered as wastewater pollutants. Viral and bacterial infections from waterborne outbreaks are usually connected with environments associated with the discharge of wastewater [37, 38]. Enteric viruses are known to cause gastroenteritis infections, hepatitis, and respiratory tract infections. Enteric viruses such as noroviruses, rotaviruses, enteroviruses, sapoviruses, astroviruses, bocaviruses, hepatitis A virus, hepatitis E virus, Aichi virus, Human polyomaviruses (PyVs), papillomaviruses, a plant virus called pepper mild mottle virus (PMMoV), and enteric bacteria such as bacteriophages, fecal coliforms and *Escherichia coli* are found in wastewater, and the full details of their occurrence and concentration in untreated and treated wastewater by continents can be seen in a review by Farkas et al., [37]. The emergence of severe acute respiratory syndrome coronavirus 2 (SARS-CoV-2), the virus related to the COVID-19 pandemic has been discovered in wastewater with entry through human feaces into sewer systems as other viruses [39, 40]. Research is on-going on the effect of SARS-CoV-2 on aquatic habitats and its resulting long-term effect.

2.7 Pharmaceutical compounds

These compounds are part of the emerging pollutants in wastewater since their long-term effect on human and aquatic habitats are unknown. Compounds such as analgesics, antibiotics, anticonvulsants, anti-cancer agents, beta-blockers, contrast agents, hormones, lipid-regulators and antidepressants are pharmaceutical compounds that have recently been found in wastewater [41]. This is because human drugs are excreted either in original or metabolized form after administration. Though most pharmaceutical compounds are biologically degradable, but some product is seen in the effluent of wastewater treatment plant [41]. In effluents from a sewage treatment plants about 2 μg/L of tetracycline, ibuprofen, contrast products, caffeine, and codeine were found [42]. Likewise, Clara et al. [43] reported the presence of antibiotics (such as metronidazole, norfloxacin, and dextromethorphan (DMP)) at concentrations below 0.05 μg/L in another effluent. Studies indicate that the removal rate of antibiotic is around 50% and Bisphenol A 71%, that of analgesics, anti-inflammatory drugs, and beta-blockers is within 30–40% because of their resistant to treatment [41].

3. Emerging trends in wastewater treatment technologies

One of the primary reasons that has driven the inception of new or improved wastewater treatment technologies is the legislation and hefty fines that are attracted when the disposal of wastewater does not meet the set discharge limits. This impact on the financial wellbeing of factories and industries has fueled the emergence of new or improved treatment technologies.

Anaerobic and aerobic technologies have been popular lately in the treatment of organic wastewater because of their friendliness to the environment and cost-effectiveness. Anaerobic technologies are, however, a cut above other technologies because of the low energy consumption.

The nature of the wastewater primarily dictates the choice of technology to be adopted, and thus it is crucial to characterize streams to determine key wastewater characteristics, such as COD, TS, VS, and salt content, among others. The main thrust of this chapter is premised on three emerging technologies, that is, membrane, microalgal, and microbial fuel cell (MFC) technologies. These technologies can be employed independently or in series as a treatment mechanism.

3.1 Membrane technology

Membrane technology (MT) encompasses the related engineering and scientific approaches for the transport of components, species, or substances through or by membranes [44]. This technique is generally adopted to explain the mechanical processes for the separation of gas or liquid streams. Membranes are classified as a thin layer barrier for size differential separation, which are usually integrated with chemical and biological treatments, or as a standalone system in secondary treatment of wastewater [44, 45]. For a typical membrane mechanism, there is usually a driving force such as a semi-permeable barrier which controls the rate of movement of components by fractional permeation, and rejection through pores of different sizes as depicted in **Figure 1**. The permeation and selective rejection are a function of the membrane pore size and chemical affinity, allows for a product stream devoid of target components. Some advantages and drawbacks are presented in **Figure 2**.

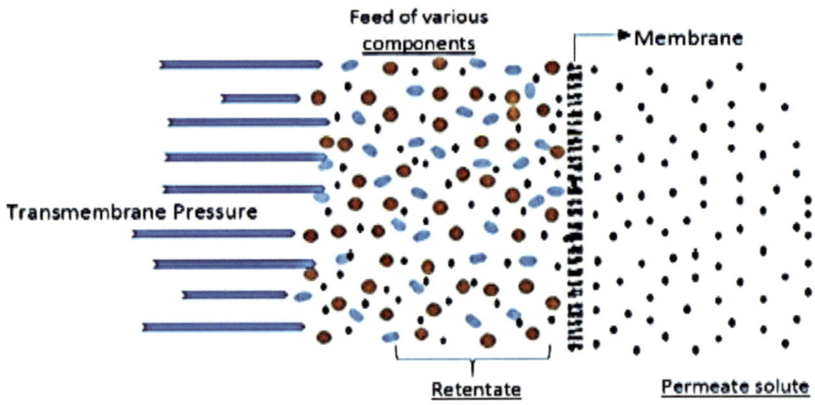

Figure 1.
Membrane selective permeation for various solutes adapted from Tetteh et al., 2019 [45].

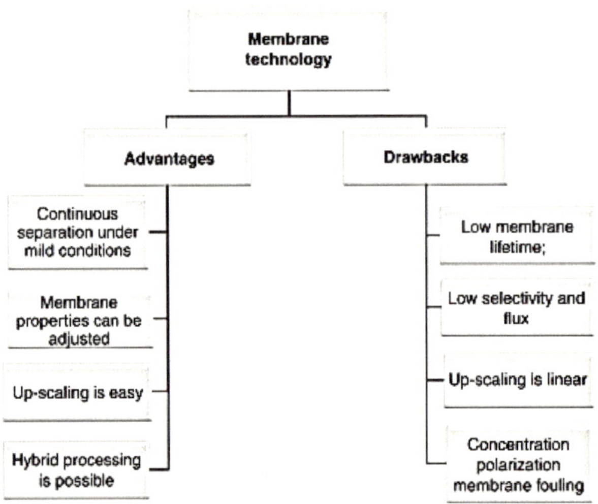

Figure 2.
Some advantages and drawbacks of membrane technology. Adapted from Burggraaf 1996 [46].

3.1.1 Classification of membranes

3.1.1.1 Microfiltration

Microfiltration (MF) employs a sieving mechanism to retain macromolecules or particles larger than 0.1 μm, or more specifically, in the range of 0.1–10 μm [45]. Unlike ultrafiltration (UF), reverse osmosis (RO) and nanofiltration (NF), the transmembrane pressure (TMP) for both sides of the membrane is low as a result of the retention of smaller particles. Thus MF requires a relatively small TMP, that is, lower than 2 bars but it may vary from 0.1 to 2 bar [47]. Larger pore sizes of MF membranes limit the removal of suspended solids, bacteria, viruses, protozoan cysts and on a lesser extent, organic colloids within the region [48].

3.1.1.2 Ultrafiltration

The performance of ultrafiltration (UF) processes are currently receiving increasing recognition as a pretreatment for desalination and membrane bioreactor applications. UF like MF uses physical sieving as a separation mechanism. The pore size, molecular weight cut-off (MWCO) and pressure for a UF membrane filtration ranges from 0.05 μm to 1 nm, 1–500 kDa and an operating pressure of 1–7 bar [47]. In effect, UF with a definite MWCO are impermeable to compounds with molecular weights exceeding the MWCO and have demonstrated a 3–6 log removal of chlorine resistant protozoan cysts, colloids, viruses, and coliform bacteria. The use of MF and UF as pretreatment to reverse osmosis (RO) has progressively arose at an industrial scale. Both could serve as pretreatment strategies for NF and RO processes for the reduction of membrane fouling, which is applied as a post treatment to chemical precipitation of organic chemical removal, pH adjustment, and phosphorus, hardness, and metal removal. Fouling is extremely distinguished in UF applications, due to the high molecular weight of fractions retained in relation with the small osmotic pressure differentials, and liquid phase diffusivity. However, this does not negatively influence the demand for UF's, as any design, configuration and application will be fouled [49, 50]. The configuration for application could be influenced by the mechanical stability, hydrodynamic requirement, and cost implications.

3.1.1.3 Ion exchange membranes

Membranes are classified as anion exchange membrane (AEM) if the polymer matrix is embedded with fixed positive charge groups, and vice versa, for cation exchange membranes (CEM) [51], which involves the permeation of anions/cations, and rejection of cations/anions in the effluent. Electrodialysis (ED), reverse electrodialysis (RED), diffusion dialysis (DD) and the Donnan membrane process (DMP) are examples of such, which usually involves the exchange of ions between solutions across the membrane as shown in **Figure 3**. The application of these processes is usually based on the type of effluent which is usually reported as an energy resourceful mechanism of separation by potential gradient.

3.1.1.4 Reverse and forward osmosis

Reverse osmosis (RO) is often referred to as a tight membrane and has been widely used in brackish and WWT. Its effectiveness in desalination was found to be more effective than conventional thermal multistage flashing [49]. High external pressures of 15 to 150 bars is a result of the hypertonic feed and is usually greater than the osmotic pressure which is applied to retain dissolved solute, and prevent

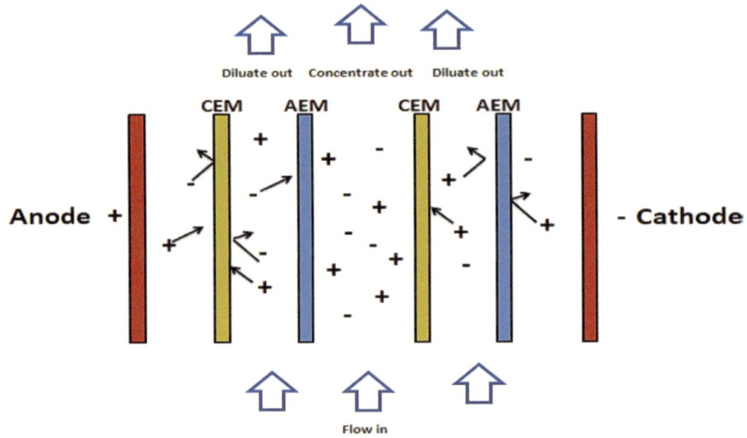

Figure 3.
Schematic diagram of ED adapted from Obotey 2020 [47].

and allow for solvent permeation at a MWCO around 100 Da through diffusion mechanism [47]. Some advantages of the RO system that have been reported in previous studies include low energy consumption, simple configuration and operation, low membrane fouling tendencies and high rejection of a wide range of contaminants. With a concentration gradient as the driving force, the separation and concentration in forward osmosis (FO) occurs as the concentrated solution (e.g. salts such as NaCl) draws water from a less concentrated feed solution. The use of FO operates at ambient conditions, hence irreversible fouling is low. However, to attain the desired process flow and optimum configuration, ROs are arranged in stages and passes. The sequence of the stages has the concentrate stream of the first stage as the feed inlet to the second stage. In addition, the permeate streams from both stages are directed into one discharge channel.

3.1.1.5 Electro-dialysis (ED) and electro-dialysis reversal (EDR)

These processes combine the principles of electricity generation and ion-permeable membranes in the separation of dissolved ions from water [45]. A difference in electric potential leads to a transfer of ions from a dilute solution to a concentrated solution through an ion-permeable membrane. During electro dialysis, two types of ion exchange membrane are used as shown in **Figure 3**. One is permeable to anions and rejects cations, while the other is permeable to cations and rejects anions. There are also two streams which are the concentrate and the diluate (feed). When an electric current is passed through the system, ions from the diluate migrate into the concentrate through oppositely charged membranes (cations migrate to the cathode whiles anions migrate to the anode). The cations are then retained by the positively charged anion-exchange membrane (AEM). Likewise, the anions are retained by the cation-exchange membrane (CEM). The outcome of this is a feed stream depleted of ions, while the concentrate stream becomes rich in ions [44].

3.1.2 Applications of membrane technology (MT)

A wider scope of industrial and environmental applications of MT are based on its advantages such as (1) clean technology, (2) energy saving (in most cases) and (3) ability to replace conventional processes; such as filtration, distillation, ion exchange, and chemical treatment systems [52]. A schematic representation of the applications of MT is depicted in **Figure 4**. Other advantages are (4) its ability

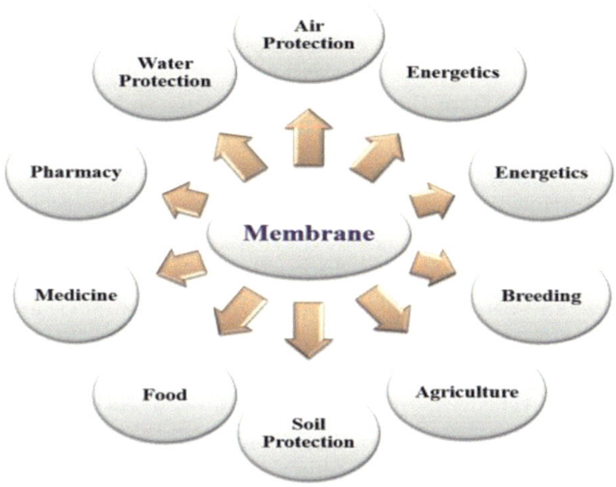

Figure 4.
Application of membrane processes adapted from Obotey 2020 [47].

to produce high-quality products and (5) its flexibility in system design. Because of its multidisciplinary application, this technique is applied in several industries, including water treatment for domestic and industrial water supply, chemical, pharmaceutical, biotechnological, beverages, food, metallurgy, and various separation processes.

3.2 Microalgal wastewater treatment (MWWT)

Water-security is a perspective which defines the reliable availability of an acceptable quality and quantity of water for health, livelihoods and production; coupled with an acceptable level of water-related risks [53, 54]. However, population dynamics and the proliferation of industrial set-ups have induced an imbalance in the water-resource equation. Domestic use of water and the demand for water in the production sector of the economy, coupled with commercial services and the agricultural sector, have surpassed the supply capacity of potable water sources [54]. The unethical discharge of wastewater from some of these sources results in serious social, health, and environmental problems. In addition, freshwater-scarce nations have the growing need to encourage strategies for water reuse, because of inadequate precipitation and lack of capacity to harvest rainwater,, which in turn aims to reduce effluent wastewater disposal. Functional wastewater treatment plants (WWTP) for municipalities across the globe have proven to be highly demanding to run in terms of chemical input and energy. Although the basic stages of treatment are primary, secondary, and tertiary, the effluent from these plants contribute to secondary pollution as they are unable to meet the green-drop guidelines [54]. Phytoremediation is a green strategy that sequesters residual pollutants from wastewater and renders it potent for re-injection into the water supply system. The use of microalgae-based WWT systems has received serious scrutiny in the research community; and in synergy with industry, various wastewater technologies and strategies have been developed to address specific needs in the sector [55].

3.2.1 Microalgal intervention

Standard culture media have been optimized for specific microalgae strains and are subsequently modified to cultivate many other strains. These are then used as templates to define wastewater characteristics and to select the microalgal strain or microalgae consortium that would best be able to treat a given wastewater source. The microalgae intervention protocol (MAIP) is mainly designed to rid the effluent wastewater from WWTP of the residual pollutants and concurrently produce high value products, thereby meeting the green-drop requirements [2, 3]. MAIP is therefore integrated into regular WWTP and upgrades it to advanced WWTPs (AWWTPs). This in turn confers the ability to sequester nitrates and orthophosphates, which, if unsuccessful will result in eutrophication to be induced and propagated in the receiving waters [3]. The need to regulate nitrogen and phosphorus discharge to the environment is born out of the following: (i) as free ammonia, ammonia-nitrogen is harmful to fish and other aquatic biota, (ii) ammonia consumes dissolved oxygen (DO) and therefore presents the potential of DO depletion, (iii) both phosphorus and nitrogen are plant nutrients and therefore contribute to eutrophication, (iv) is the NO_3- ion, nitrate-nitrogen reacts and combines with hemoglobin, which contributes to infant mortality. In addition, nitrate-nitrogen can be reduced to mutagenic nitrosamines in the gastrointestinal tract thereby posing more hazards to infants [56]. Various research teams [57–60] reported the presence of emerging pollutants (EP) in WW and the possible undesirable effects many of them can have on the environment and living organisms. These EP include, among others, pesticides, pharmaceuticals, and personal care products; and some technologies have been proposed for their removal; such as physico-chemical and biological treatment strategies. EP removal using pure microalgae strains has been proven to be effective. However, microalgae-based EP removal technologies have not received appreciable attention in the global research community.

The advocacy for employing microalgae to sequester wastewater nutrients, as a treatment option has attracted global acceptance. However, there are skepticisms in employing wastewaters for microalgal cultivation to produce biomass and bio-products. This is primarily due to the reality that wastewaters are of a wide variety of sources and therefore have a wide range of properties whose stability is in question. Pre-treatment is therefore a necessary stage for microalgal WWT, which imposes on the economy of the process. This brings to bear the necessity to adopt the integrated microalgal WWT protocol [61–64].

3.2.2 Microalgal WWT strategies

Aside from the ability of microalgae to sequester NH3-N, NO3- -N and PO43-, microalgae also removes heavy metals as well as organic carbon from wastewater, while preventing secondary pollution. However, previous research has indicated that microalgae can rarely grow in undiluted wastewater due to high concentrations of ammonium and other compounds frequently present in wastewater. Different microalgae species present different growth indices in each wastewater treatment application. It is therefore paramount to select a suitable microalgal strain to treat a given wastewater source. Ungureanu and co-workers [63, 65] reported that the microalga *C. mexicana* recorded the highest removal of N, P and C from piggery wastewater compared with five other species (*C. vulgaris*, *M. reisseri*, *Nitzschia cf. pusilla*, *O. multisporus* and *S. obliquus*). On cultivation of the microalga C. zofingiensis with piggery wastewater using different dilution ratios, 79.84% of COD, 82.70% of total N and 98.17% of total P were removed [63]. In another study with *V. vulgaris*, 60–70% of COD and 40–90% of $NH4^{+}$ -N were removed from diluted piggery wastewater [65]. The highest removal percentage was obtained with 20-fold diluted wastewater. Whilst tertiary treatment of municipal wastewater effluent and remediation of animal waste streams are an additional technological and economic pressure on municipalities and farms that threatens their economic sustainability, but at the same time it also presents an opportunity [63]. However, there are several challenges with current microalgae growth systems. For example, algae grown in an open pond or raceway system are suspended in the water in the presence of soluble and suspended waste and can be extremely difficult to harvest because oilagenous microalgae are approximately 5–10 micrometers in diameter. Many of the highly productive microalgae cannot be easily filtered and harvested through centrifugation which is an expensive unit operation. Algae can be harvested by sedimentation; however, this is a slow process and requires considerable floor space. Metal salts can be used as flocculants to facilitate sedimentation; however, this results in water contamination. Algal pond systems are also susceptible to washout, where algae leaves the system and enters surface waters [63, 65]. Integrated microalgal WWT systems are examples of green technology, which incorporates both the conventional WWTP and the microalgal WWT protocol which is primarily considered to address imperative issues such as global warming and climate change. The microalgal biomass generated during wastewater treatment, represents a carbon sink, and thus mitigates the negative effect of CO_2 by photosynthetic sequestration of this greenhouse gas [66].

3.2.2.1 Open ponds

Open ponds are grouped into natural systems, artificial ponds, and containers. Natural systems include the lakes and lagoons; artificial ponds which are either unmixed open ponds, circular open ponds mixed with a center pivot mixer, or raceway ponds; and containers. The commonly used forms include raceway ponds, circular ponds, and tanks, of which raceway ponds have received the most attention [64].

Waste stabilization ponds are used for wastewater treatment by tens of thousands of small communities around the world. These ponds are low cost, simple to operate and provide effective wastewater treatment in terms of organic carbon and pathogen removal. However, phosphorus removal in waste stabilization ponds is often low, generally between 15 and 50% [62, 64]. Because of this, there is increasing pressure from regulators to upgrade pond systems to prevent eutrophication of receiving water bodies. The problem is that current upgrade options often involve the use of chemical dosing which contributes to secondary pollution that makes recovery and reuse of the phosphorus very difficult, and in some cases almost impossible. What is needed is a sustainable low-cost solution to remove phosphorus from the wastewater and ideally allow the phosphorus to be recovered and reused. A potentially emerging environmental process technology has been identified whereby microalgae in waste stabilization pond systems may be triggered to excessively accumulate phosphorus within their cells. While microalgae in lakes can store polyphosphate there is the potential of using this natural phenomenon to optimize for phosphorus removal in algal wastewater treatment ponds [62, 63].

Figure 5(A) Is the raceway pond that uses a motorized paddle wheel (PW) to initiate and sustain movement and mixing of the microalgal cell (MCs,) thereby preventing them from settling to the reactor bed. It enhances exposure of the MC to light and nutrients and promotes interphase mass transfer. However, while the mixing energy requirement of a PW is relatively low, efficiency of gas transfer is also low. In some instances, aerators are used to supplement CO_2 to improve microalgae growth, and hence promote nutrient sequestration from the broth. The pond operates at the prevailing temperature and light intensity depends on the incoming solar insolation [68]. **Figure 5(B)** is a rectangular open unmixed pond (ROP). The MCs here do not have the privilege of equal exposure to light. The MCs that are near the bottom are shielded from light by those above, thereby creating blind zones to photosynthetic activities resulting in reduction in cell density (CD)

(A) Raceway pond (mixed)

(B) Rectangular open pond (unmixed)

(C) Open circular containers (unmixed)

(D) Circular open pond system (mixed)

Figure 5.
Microalgal open pond systems [66–68].

and productivity. **Figure 5(C)** shows open circular containers (OCC) which are unmixed. **Figure 5(D)** shows circular open pond systems (COPS) equipped with mixers [15, 16].

3.2.2.2 Closed bioreactor (CBR) systems

Closed photobioreactor systems are characterized by (i) efficient photosynthetic activities associated with adequate control of the operational variables, (ii) lower risk of contamination and (iii) minimization of water loss by evaporation, which is a serious concern in open systems. However, closed systems are more expensive, since they must be constructed with transparent materials, and are more complicated to operate and challenging to scale up. Closed photobioreactors vary in configuration, and the main types are bubble columns, airlift reactors, tubular (loop) and stirred tank reactors. Photobioreactors employing microalgae to treat wastewater and produce biomolecules have (i) elevated efficiency in the use of light energy, (ii) an adequate mixing system, (iii) ease of control of the reaction conditions, (iv) reduced hydrodynamic stress on the cells [69–71].

Figure 6 gives a pictorial view of photobioreactor scenarios for bubble column, airlift, and annular configurations. A bubble column reactor is basically a cylindrical vessel with a gas distributor at the bottom. The gas is sparged in the form of bubbles into either a liquid phase or a liquid–solid suspension without mechanical agitation. During operation, mixing and CO_2 mass transfer are carried out through the action of the spargers with an external light supply. The configuration of a gas sparger is important since it determines the properties of bubbles; such as bubble size, which in turn affects gas hold-up and other hydrodynamic parameters associated with bubble columns. Photosynthetic efficiency depends on the gas flow rate, which further depends on the photoperiod as the liquid is circulated regularly from central dark zones to external photic zones. This exposes more MCs to the nutrients in the medium, which in the context of this chapter, is wastewater. Photosynthetic efficiency can be increased by increasing the gas flow rate (≥ 0.05 m/s), which in turn leads to shorter photoperiods [69, 70]. This type of reactor has advantages of higher mass transfer rates; and low operational and maintenance costs due to

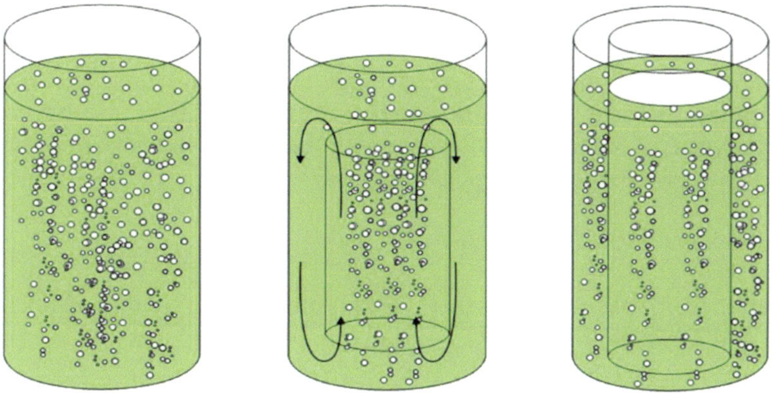

Figure 6.
Bubble column reactors Płaczek et al., 2017 [47].

fewer moving parts. However, back-mixing and coalescence have been identified as major challenges for these reactors. There is an upper limit for increasing the flow rate, beyond which the heterogeneous flow formed will eventually cause the back-mixing of gas components. Scalability and economics of microalgae cultivation using photobioreactors remain the challenges that have to be overcome for large-scale microalgae production.

Hom-Diaz and co-workers [57], in an outdoor pilot 1200 L microalgal photobioreactor (PBR) used toilet wastewater (WW) and evaluated the PBR's ability to remove pharmaceutically active compounds (PhACs). Nutrients (ammonia-nitrogen, nitrate-nitrogen and total phosphorous) were removed and chemical oxygen demand (COD) was efficiently reduced to the extent of 80%, whilst as much as 48% of the pharmaceutical residues were removed, thereby satisfying the green-drop requirement.

Airlift photobioreactors comprise of two interconnecting zones, called the riser and the down-comer, in an annular setup. Generally, there are two types of airlift photobioreactor: (i) the internal-loop and (ii) the external-loop [19]. For the internal-loop airlift reactor, the two regions are separated by either a draft tube or a split cylinder, whilst for an external-loop airlift reactor, the riser and down-comer are separated physically by two different tubes. Mixing is done by bubbling the gas through a sparger in the riser tube, with no mechanical agitation. A riser is synonymous with bubble column, where sparged gas moves upward randomly and haphazardly, and decreases the density of the riser making the liquid move upward. Gas hold up in the down-comer significantly influences the fluid dynamics of the airlift reactor thus forcing the liquid downwards The external-loop which is a draft tube confers certain advantages to the airlift bioreactor, namely, preventing bubble coalescence by directing them in one direction; distributing shear stresses more evenly throughout the reactor. This exposes more MCs to the nutrients, minerals, volatile organic compounds and a host of other pollutants for sequestration and for cell growth; enhancing the cyclical movement of fluid, thus increasing mass and heat transfer rates [71–73].

Fully closed tubular photobioreactors are potentially attractive for large-scale axenic culture of microalgae and is one of the more suitable types for outdoor mass culture. Tubular photobioreactors consist of an array of straight, coiled, or looped transparent tubes that are usually made of transparent plastic or glass. Algae are circulated through the tubes by a pump, or airlift technology [21].

Many factors contribute to the inability of microalgae to remove nutrients and produce biomass. Some minerals, such as calcium, iron, silica, magnesium, manganese, potassium, copper, sulfur, cobalt, and zinc, also influence microalgae development in wastewater, along with pH, temperature, light, mixing, and dissolved oxygen, which influence development rates and chemical composition of microalgae in wastewater treatment systems [74, 75].

3.2.3 Benefits of microalgal WWT

Molinuevo-Salces and co-workers [76] pointed out the benefits of microalgal-based WWT systems to include:

1. treating diverse kinds of wastewater including domestic, commercial, agricultural, and industrial wastewater

2. reducing pollutants and pathogens

3. recovering nutrients as biomass

4. mitigating CO_2 gas emissions

5. recovery of metabolites and

6. energy savings

Starch-based textile de-sizing wastewater (TDW) was treated with the microalgae, *Scenedesmus sp*. to remove organic carbon with lab-scale reactors, which achieved 92.4% color removal, reduction in chemical oxygen demand (COD) by 89.5%, carbohydrates by 97.4% and organic acids by 94.7% [22]. Phasey and co-workers [23] averred that cultivation of microalgae using municipal and agricultural wastewater in high rate algal ponds (HRAP) partitions nutrients into microalgal biomass, which can be recovered and reused.

3.2.4 Microalgal WWT challenges

In spite of all the advantages, some challenges have to be surmounted before the microalgal WWT protocol can be applied. The challenges include (1) land requirement, (2) effect of wastewater characteristics, (3) environmental and operational condition influence and (4) biomass harvesting and valorization [14]. However, limitations such as algae biomass separation from water, process efficiency in cold climates and limited ability of the algae biomass to reduce micropollutant content in wastewater discourages full-scale use [77].

3.3 Microbial fuel cells for wastewater treatment

In order to build a sustainable platform for the future society needs to substantially reduce its reliance on fossil fuels. This reduction can then minimize the global scale of pollution. As has been discussed in this chapter, these two global challenges could be concurrently addressed through the application of wastewater treatment technologies which reduce pollution and provide the starting blocks for biofuels. In recent years, a paradigm shift has occurred where wastewater, which can also be referred to as waste matter, is being used by industries generate electricity. In particular, studies have illustrated that a number of biological processing methods can be used to produce bioenergy or bio-chemicals while treating industrial wastewater. Specifically, brewery wastewater treatment has been highlighted for the application of microbial fuel cells (MFCs) [78]. One such instance of this method is using MFCs to simultaneously treat wastewater and produce bioenergy which is most referred to as bioelectricity. Production of these bio-products happens from simply converting the organic and chemical energy contained in wastewater to electrical energy. To further explore these possibilities, this section first describes an MFC, second it discusses applications of MFCs in wastewater treatment, and thirdly it reviews the different techniques and operations that use MFCs to treat wastewater while concurrently producing electricity. In addition, it also describes other applications and bioenergy products of this technique, its advantages and disadvantages, further promising applications of the MFC technology in wastewater treatment. An MFC is a device that converts organic matter to electricity using microorganisms as the biocatalyst. Typical MFCs have three major components: electrodes, separator, and electrogens. All MFCs contain two electrodes, which, depending on the design, can either be separated into one or two chambers. These chambers operate as completely mixed reactors. As illustrated in **Figure 7** below, each electrode is placed on each side of the membrane, which can either be a proton exchange membrane (PEM) or a cation exchange membrane (CEM). The anode faces the chamber that contains the liquid phase, and the cathode faces the chamber that only contains air [79].

Aforementioned literature proposed the use of carbon, graphite, and metal-based materials as electrodes. For example, materials made from carbon cloth, carbon paper, carbon felt [80], graphite granules, carbon mesh [81], platinum, platinum black and activated carbon with single or tubular or multi-electrode configurations are suitable as electrodes [82]. These electrodes should have properties which render them biocompatible and stable In addition high electrical conductivity, and large surface area is recommended [83, 84]. The cathode can be exposed to air or other additional electron acceptors like permanganate, chromium hexacyanoferrate and azo dye, etc. [85]. The separator is either a cation exchange membrane [86] or a salt bridge [87] which is used to keep the chamber. The potential difference generated between the two chambers drives the electrons to move through the circuit while microbial degradation of wastewater acts as the substrate to generate bioelectricity [88]. MFCs were first considered to be used to treat wastewater as early as 1991 [89]. Municipal wastewater contains a multitude of organic compounds that can fuel MFCs. The amount of power generated by MFCs in the wastewater treatment process can potentially halve the electricity demand in a conventional treatment process which consumes a significant amount of electric power

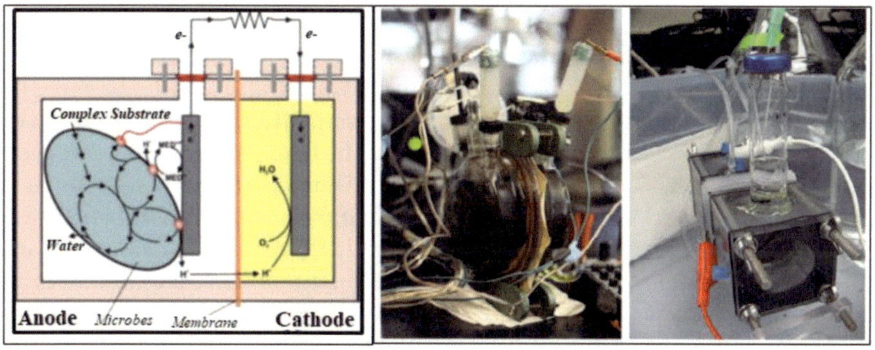

Figure 7.
Schematic diagram and pictures of a typical double-chamber microbial fuel cell (MFC), sourced from Logan et., 2006 [78].

for aerating the activated sludge. MFCs yield 50–90% less solids to be disposed of than conventional activated sludge treatment methods. Anaerobic digesters, are sometimes integrated with aerobic sequencing batch reactors to overcome the challenges of sludge disposal [90]. Furthermore, organic molecules such as acetate, propionate and butyrate can be thoroughly broken down to CO2 and H2O. A hybrid MFC incorporating both electrophiles and anodophiles are especially suitable for wastewater treatment because more organics can be biodegraded by a variety of organics. MFCs using certain microbes display a special ability to remove sulphide as necessary in wastewater treatment [91]. MFCs can enhance the growth of bio electrochemically active microbes during wastewater treatment, thus enabling operational stabilities. Continuous flow, single-compartment MFCs and membrane-less MFCs are favored for wastewater treatment amidst concerns in scale-up of other technologies [92–94]. Sanitary waste, food processing wastewater, swine wastewater and corn stover are all favorable biomass sources for MFCs because they are rich in organic matters [95–97]. Up to 80% of the Chemical Oxygen Demand (COD) can be reduced in some cases [96, 98] and a columbic efficiency as high as 80% was obtained by Kim et al. [99].

MFC technologies are a promising yet novel strategy in wastewater treatment, as the treatment process itself becomes a method to capture energy in the form of electricity or hydrogen gas, rather than being a net consumer of electrical energy. In the early 1990s Kim and colleagues illustrated that bacteria could be used in a biofuel cell as an indicator of the lactate concentration in water [80], which in turn supports electricity generation [81]. Although the power generation was low, it was not apparent whether the technology would have much impact on reducing wastewater strength. In 2004 this changed, and the link between electricity generation with MFCs and wastewater treatment was clearly forged when it was proven that domestic wastewater could be treated to practical levels while simultaneously producing electricity [82]. The amount of electricity produced in this study, while low (26 mW/m2), was considerably higher than previously obtained with other wastewater types. Research conducted prior to 2004 had shown that organic and inorganic matter in marine sediments could be used in a novel type MFC design [83], making it apparent that a wide variety of substrates, materials and system architectures could be used to generate electricity from organic content with bacterial biomass. Still, power levels in all these applications were relatively low. The final development that raised the current interest in MFCs was peaked when power densities of two orders of magnitude greater was produced in an MFC with the addition of glucose [84]. This application had no need for exogenous chemical mediators or catalyst thus ensuring this operation was purely biological.

Following these demonstrations, the competition was on to advance a rather practical approach to MFC applications. The first objective being the development of a scalable approach and design of the MFC for various wastewater treatment types [78]. While the energy that could be harnessed from the wastewater may not be enough to power a typical city, it has been reported that a substantial amount of energy can be used to power the WWTPs. As can be observed in the few studies discussed above on MFC technology, the per capital basis of the energy is not particularly substantial and impressive. Also, it can be noted that the most significant energy savings associated with the use of MFC for wastewater treatment, besides generation of electricity and removal of high strengths pollutants form these recalcitrant substrates, is savings in expenses for aeration and solids handling in typical WWTPs. The main operating costs for wastewater treatment are, aeration, sludge treatment and pumping. It has been argued that aeration alone can account for half of the operational costs at a typical WWTP [85]. Reducing this cost can also ensure that WWTPs become net producers of energy if MFCs are integrated with other treatment technologies.

3.3.1 Applications of microbial fuel cells in wastewater treatment

Applications of MFCs in wastewater treatment include a variety of advantages like long-term sustainability, use of renewable resources, degradation of organic and inorganic waste, bio-hydrogen production, and removal of compounds like nitrates, etc. [86]. The electrochemical active microbial community requires an in-depth understanding of its solution chemistry to engage in full-scale implementation and exploitation of MFC technology for electricity generation. [9]. Under ideal laboratory conditions, these systems have produced power densities of 2 to 20 mW/m^2 [87]. However, the amount of biomass-based energy produced by microbial processes is very low. It has yet to reach to its full potential to work in pilot scale units. It has also been noted that the success of specific MFC applications in wastewater treatment will depend on the concentrations and biodegradability of the organic matter in the effluent, the wastewater temperature, and the absence of toxic chemicals [78]. One of the first applications could be the development of a pilot-scale reactor at industrial locations where a high quality and reliable influent is available. Food processing wastewaters and digester effluents are good candidates. Moreover, decreased sludge production could substantially decrease the payback time. In the long term, dilute substrates, such as domestic sewage, could be treated with MFCs, thus decreasing society's need to invest substantial amounts of energy in their treatment. A varied array of alternative applications could also emerge, ranging from biosensor development and sustained energy generation from the seafloor, to bio-batteries operating with various biodegradable fuels. While full scale, and highly effective MFCs are not yet within our reach, the technology holds considerable promise, and major hurdles will undoubtedly be overcome by engineers and scientist in the near future [88]. The growing pressure on our environment, and the call for renewable energy sources will further stimulate development of this technology, to full scale plant operation. As part of the aforementioned applications of MFC in wastewater treatment, potential for application of this technology it as a typical sensor for pollutant strength analysis for in situ process monitoring and control [89]. The proportional correction between the columbic efficacy of MFCs and the strength of the wastewater can propose MFCs to be potential biological oxygen demand (BOD) sensors [80]. An accurate method to measure the BOD value of a liquid is to calculate its Columbic yield. A number of works, namely [80, 90] showed a strong linear relationship between the Columbic yield and the strength of wastewater in BOD concentration range. MFC-type BOD sensors are advantageous because they have excellent operational stability, and good reproducibility and accuracy. An MFC-type BOD sensor constructed with the microbes can be kept operational for over five years without extra maintenance [80]. These biological sensors promise a longer service life than ordinary versions of BOD sensors reported in literature.

3.3.2 Promising techniques of MFCs in wastewater treatment and electricity valorisation

Waste biomass is a cheap and relatively abundant source of electrons for microbes capable of producing electrical current outside the cell [85]. Rapidly developing microbial electrochemical technologies, such as microbial fuel cells, are part of a diverse platform of future substantial energy and chemical production technologies. In this section, we discuss the key advances that will enable the use of exo-electrogenic micro-organisms to generate biofuels, hydrogen gas, methane, and other valuable inorganic and organic chemicals. Moreover, this section will scrutinize the crucial challenges for implementing these systems and compare them to similar renewable energy technologies. Although commercial development is already underway in several different applications, ranging from wastewater treatment to industrial chemical production, further studies are still required regarding efficiency, scalability, system lifetimes and reliability of MFCs in the field of wastewater treatment and bioenergy production [85].

Power generation using domestic wastewater in the flat plate system was developed and found to be capable of continuously generating electricity from the organic matter in the wastewater while undergoing treatment [82]. Following an acclimation period of approximately 1-month, constant power generation from wastewater was obtained with the Flat Plat Microbial Fuel Cell (FPMFC) over a period of five months. For wastewater containing 2463 mg COD/L, an average power density of 560 mW/m^2 was obtained with a hydraulic retention time (HRT) of 2.0 h (0.22 mL/min flow rate; 164 mg/L log mean COD) and an air flow rate of 2 mL/min with a 470 *ohms'* resistor. Under these operating conditions, the COD removal rate was 1.2 mg/L min (58% COD removal), and the maximum power density was achieved at a flow rate of 0.22 mL/min. This power density was about 10% higher than that obtained under typical operating conditions with a 470 *ohms'* resistor.

Continuous wastewater treatment and electricity generation using a Single Chamber Microbial Fuel Cell (SCMFC) was successfully piloted with feasible results [82, 91]. It was found that the system could generate 26 mW/m^2 at the maximum power density while reducing 80% of the COD. In a specially designed, smaller batch system by Liu et al. [92] showed that up to 28 mW/m^2 of power could be generated with domestic wastewater. It was further demonstrated that by removing the proton exchange membrane (PEM), they could generate a maximum of 146 mW/m^2 of power. In these systems, the anode was separated from the PEM/cathode or plain cathode in a large chamber, but the anode chamber was not mixed except by the flow of liquid into the system. In other MFCs, the anode chamber was often mixed in [93–95] . In hydrogen fuel cells, the electrodes are usually combined into a single strip separated by a PEM. This is necessary to keep the two electrodes near to enhance proton conduction between the two electrodes. However, PEMs such as nafion are permeable to oxygen, resulting in the transfer of small amounts of oxygen from the cathode chamber to the anode chamber.

Domestic wastewater treatment was examined under two different temperature gradients, (23 ± 3°C and 30 ± 1°C) and flow modes (fed-batch and continuous) using a single-chamber air–cathode microbial fuel cells (MFCs) in view of the effect of operating parameters which affect the production of electricity [94]. Temperature was an important parameter which influenced efficiency and power generation. The highest power density of 422 mW/m^2 (12.8 W/m^3) was achieved under continuous flow and mesophilic conditions, at an organic loading rate of 54 g COD/L-d with reduction of COD by 25.8%. Energy recovery was found to depend significantly on the operational conditions (flow mode, temperature, organic loading rate, and Hydraulic Retention Time (HRT)) as well as the reactor architecture. The results demonstrate that the main advantages of using temperature gradients, in series MFC configurations for domestic wastewater treatment are power savings, low solids production, and higher treatment efficiencies.

A study on MFCs used to produce electricity from different compounds sources, including acetate, lactate, and glucose has proven its ability in high efficiencies and versatility in applications for wastewater treatment [96]. Clearly, the possibility to produce electricity in a MFC from domestic wastewater, while at the same time accomplishing biological wastewater treatment (reduction of COD) was emphasized. Tests were conducted using SCMFC containing eight graphite electrodes (anodes) and a single air cathode. The system was operated under continuous flow conditions with primary clarifier effluent obtained from a local wastewater treatment plant. The prototype SCMFC reactor generated electrical power (*maximum of 26* mW/m^2) while reducing the COD by about 80%. The power output was proportional to the hydraulic retention time over a range of 3 to 33 h, and to the influent wastewater strength over a range of 50–220 mg/L for COD. Current generation was controlled primarily by the efficiency of the cathode. Optimal cathode performance was obtained by allowing passive air flow rather than forced air flow (4.5–5.5 L/min). The Columbic efficiency of the system, based on COD reduction and current generation, was <12%, indicating that a substantial fraction of the organic matter was not accessible to the microorganisms thus limiting the current generation. Bioreactors based on power generation in MFCs may represent a completely new approach to wastewater treatment. If power generation in these systems can be increased, MFC technology may provide a new method to offset wastewater treatment plant operating costs, whilst making advanced wastewater treatment more affordable for both developing and industrialized nations.

The development of electric power from MFCs was initially investigated for its potential contribution to applications in space research [97]. It was discussed that one of the determining factors in MFC technology was the use of applied microbial cultures, which are responsible for converting electric energy from the chemical bonds in the substrates. In the last decade, despite the intensive development there is a knowledge gap regarding electricity production from microbes and the screening for electricity production. The fast screening method was based on microbial iron (III) – reduction, and do not require any MFC infrastructures. The method is suitable for the evaluation of numerous microbe species or strains simultaneously; and in this way there is possibility to extend the range of potential MFC biocatalysts and be able to predict the electricity generation from the chosen cultures. The knowledge which was generated from this study concerning the growth – iron (III) – reduction, substrate utilization, adhering and biofilm forming properties, extracellular conductive proteins and redox mediator production measurements is essential for the utilization of *G.toluenoxydans* and *S. xiamenensis* species for the different types of MFC applications (wastewater treatment and/or energy production). This information is vital for further strain-improvement and to create an efficient MFC design for electricity production. *S.xiamenensis* DSMZ 22215 species can catalyze *maltose* or *maltodextrine* efficiently. This ability makes the microbes available to be useful in MFC systems for the treatment of starch-based wastewaters *(e.g. Brewery wastewater, starch wastewater and the pulp and paper industry).*

Simultaneous wastewater treatment for biological electricity generation, through the membrane electrode assembly air-cathode MFC in starch processing wastewater (SPW) as substrate, was proven in this study [82]. Over the entire experimentation time, it was perceived that the optimum voltage output of 490.8 mV and power density of 293.4 mW/m^2 was ascertained with a current density of 893.3 mA/m^2. An internal resistance of 120 ohms was also recorded within the third cycle of experiments. Removal efficiencies for COD and $NH_4^+ - N$ increased with time, with a maximum of 98.0% and 90.6%, respectively. This was higher than most reported works on MFC operations. High values of nitrate removal might have been a result of both biological and physiochemical processes. Columbic Efficiency (CE) was not high (maximum 8.0%) and was mainly caused by other electron acceptors in the SPW, and oxygen diffusion during long operation periods. SEM revealed the presence of biofilm on the anode, in which short rod-shaped bacillus might have been the dominating bacteria responsible for MFC operation. This study demonstrated the feasibility of using MFC technology to generate electricity and simultaneously treat SPW with high removals of COD and $NH_4^+ - N$, thus providing an attractive alternative to reduce the cost of wastewater treatment whilst generating electricity from a renewable resource.

3.3.3 Advantages and disadvantages (Limitations) of MFCs in WWT

MFCs present several advantages and disadvantages (**Table 2**), both operational and functional in comparison to currently implemented wastewater treatment technologies for both high organic pollutant removals in the form of CODs and for the valorization of bioenergy in the form of electricity [98]. The generation of bioenergy from wastewater treatment is mostly considered to be the green or blue energy aspect of MFCs [92]. Electricity is generated in a direct way from biomass and organic matter, hence chemical energy is directly converted to electrical energy. The direct conversion of wastewater substrates to bioenergy has also been reported to be a third of the input during the thermal combustion of biogas [85]. Due to the harvesting of electrical energy, the bacterial growth yield in a MFC is considerably lower than the sludge output of an aerobic process [85, 99]. Generally the off-gas of an anaerobic process has a high content of nitrous gases together with the targeted hydrogen and methane [78]. The off-gases of MFCs has less economic viability, since the energy contained in the substrate was previously directed towards the anodic chamber of the MFC during processing [78]. The gas produced in the anodic chamber of the MFC can be literally discharged, considering no large amounts or other odorous compounds are present, and in addition no aerosols with noxious or undesired bacterial contents are liberated into the environment. Power generation from MFCs have improved considerably and reached the level of primary power target, at least in small scale systems, but the scale up is still a big challenge and a major limitation of the application of MFC technologies. The high cost of cation exchange membranes, the potential for biofouling and associated high internal resistance restrain the power generation and limit the practicality and commercial application of this technique [100].

Domestic wastewater is organic matter with embedded energy content of almost 10 times the energy needed for treatment [101]. While emerging techniques are promising, none of the processes available today can yet fully extract all the energy available in wastewater without further investment in their research and development [100]. A major setback of MFC applications is associated with the process start up time, and sequence which may be between 4 to 103 days depending on the inoculum, electrode materials, reactor design and operating conditions (temperature, external loading rates etc.), but it is largely affected by the type of substrate being fed into the MFC system [96]. Another vital impediment in scaling up of MFCs for wastewater treatment is the shortage of buffer capacity of electrolytes. This might require some external mediators, or chemical substance to maintain and stabilize the hydrogen potential of the anodic and cathodic chambers. This has to enhance the wastewater treatment process but still favor the valorization of bioenergy within the MFC system.

Advantages	Disadvantages (Limitations)
• Generation of energy from Wastewater / Biomass	• Low Power Densities
• Direct Conversion of Substrate Energy to Bio Electricity	• High Design and Fabrication Costs
• Minimal Sludge Production	• Electricity Up-Scaling Problems
• Less Gas Emissions / Treatment	• Activation Losses
• Low Aeration Costs	• Ohmic Losses
	• Bacterial Metabolic Losses
	• Concentration Losses

Table 2.
List of advantages and disadvantages of MFCs, sourced from Quach-Cu et al., 2018 [61].

4. Conclusion

In this chapter, we have reviewed the use of the MT, Microalgae and MFC technology, particularly focusing on their strengths and limitations in treating wastewater while producing bioenergy and other viable products. In the case of membrane distillation, continuous studies are needed to adequately understand the concept of temperature polarization and, accordingly, a suitable membrane should be developed to make the process viable for large scale application. Microalgal WWT achieves a dual purpose of reducing wastewater of their pollutants and producing biomass of value. It also adds the benefit of mitigating global warming as microalgae biofix anthropogenic carbon dioxide. Microalgal WWT by the airlift bioreactor technology application has advantages over other available reactor technologies as it maximizes carbon dioxide and oxygen gas mass transfer with high remediation potentials. Presently, MFC technology is at research stage hence more research and practical attempts are a necessity for its commercial viability and applications practically at large scale. Although some of the basic knowledge has been gained in MFC research, there is still a lot to be learned in the scale-up of MFC for real plant application and commercialization.

Author details

Edward Kwaku Armah[1*], Maggie Chetty[1], Jeremiah Adebisi Adedeji[1], Donald Tyoker Kukwa[1], Boldwin Mutsvene[1], Khaya Pearlman Shabangu[2] and Babatunde Femi Bakare[2]

1 Department of Chemical Engineering, Durban University of Technology, Durban, South Africa

2 Department of Chemical Engineering, Mangosuthu University of Technology, Durban, South Africa

*Address all correspondence to: edwardkarmah@gmail.com

References

[1] Khanzada NK, Farid MU, Kharraz JA, Choi J, Tang CY, Nghiem LD, et al. Removal of organic micropollutants using advanced membrane-based water and wastewater treatment: A review. J. Memb. Sci. 2020.

[2] Yue X, Zhang T, Yang D, Qiu F, Li Z, Zhu Y, et al. Oil removal from oily water by a low-cost and durable flexible membrane made of layered double hydroxide nanosheet on cellulose support. J Clean Prod. 2018;

[3] Zhang Y, Wei S, Hu Y, Sun S. Membrane technology in wastewater treatment enhanced by functional nanomaterials. J Clean Prod. 2018;

[4] Abdel-Raouf N, Al-Homaidan AA, Ibraheem IBM. Microalgae and wastewater treatment. Saudi J. Biol. Sci. 2012.

[5] Gupta S kumar, Bux F. Application of Microalgae in Wastewater Treatment Volume 1: Domestic and Industrial Wastewater Treatment. Appl. Microalgae Wastewater Treat. Springer; 2019.

[6] Sukla LB, Pradhan D, Subbaiah T. Future Prospects of Microalgae in Wastewater Treatment. In: Sukla LB, Subudhi E, Pradhan D, editors. Role Microalgae Wastewater Treat. Springer Singapore; 2019. p. 129-35.

[7] Tang RCO, Jang JH, Lan TH, Wu JC, Yan WM, Sangeetha T, et al. Review on design factors of microbial fuel cells using Buckingham's Pi Theorem. Renew. Sustain. Energy Rev. 2020.

[8] Yaqoob AA, Khatoon A, Setapar SHM, Umar K, Parveen T, Ibrahim MNM, et al. Outlook on the role of microbial fuel cells in remediation of environmental pollutants with electricity generation. Catalysts. 2020.

[9] Yongabi K. Biocoagulants for water and waste water purification: a review. Int Rev Chem Eng. 2010;

[10] Qasim SR. Wastewater treatment plants: Planning, design, and operation, second edition. Wastewater Treat. Plants Planning, Des. Oper. Second Ed. 2017.

[11] Von Sperling M. Wastewater Characteristics, Treatment and Disposal. Water Intell Online. 2015;

[12] Chang EE, Hsing HJ, Chiang PC, Chen MY, Shyng JY. The chemical and biological characteristics of coke-oven wastewater by ozonation. J Hazard Mater. 2008;

[13] Fazal T, Mushtaq A, Rehman F, Ullah Khan A, Rashid N, Farooq W, et al. Bioremediation of textile wastewater and successive biodiesel production using microalgae. Renew. Sustain. Energy Rev. 2018.

[14] Enitan AM, Adeyemo J, Kumari S, Swalaha FM, Bux F. Characterization of Brewery Wastewater Composition. Int J Environ Ecol Eng. 2015;

[15] Munter R. Industrial wastewater characteristics. Water Use Manag. 2000;

[16] Hu Z, Grasso D. Water Analysis - Chemical Oxygen Demand. Encycl Anal Sci Second Ed. 2004.

[17] Li D, Liu S. Water Quality Detection for Lakes. Water Qual Monit Manag. 2019.

[18] Yu P, Cao J, Jegatheesan V, Du X. A real-time BOD estimation method in wastewater treatment process based on an optimized extreme learning machine. Appl Sci. 2019;

[19] Henze M, Comeau Y. Wastewater characterization. Biol Wastewater Treat

Princ Model Des [Internet]. 2008 [cited 2020 Sep 3]. p. 33-52. Available from: https://books.google.com/books?hl=en&lr=&id=41JButufnm8C&oi=fnd&pg=PA33&ots=nSK7g0wE4j&sig=wKmtnw4CGaLlOlm5I-pfjmYkcvc

[20] Buscio V, Crespi M, Gutiérrez-Bouzán C. Application of PVDF ultrafiltration membranes to treat and reuse textile wastewater. Desalin Water Treat [Internet]. Taylor and Francis Inc.; 2016 [cited 2020 Sep 3];57:8090-6. Available from: http://www.tandfonline.com/doi/full/10.1080/19443994.2015.1021854

[21] Mateus GAP, Formentini-Schmitt DM, Nishi L, Fagundes-Klen MR, Gomes RG, Bergamasco R. Coagulation/Flocculation with *Moringa oleifera* and Membrane Filtration for Dairy Wastewater Treatment.

[22] López Velarde Santos M, Rodríguez Valadéz FJ, Mora Solís V, González Nava C, Cornejo Martell AJ, Hensel O. Performance of a microbial fuel cell operated with vinasses using different cod concentrations. Rev Int Contam Ambient. Centro de Ciencias de la Atmosfera, UNAM; 2017;33:521-8.

[23] Liu F, Sun L, Wan J, Shen L, Yu Y, Hu L, et al. Performance of different macrophytes in the decontamination of and electricity generation from swine wastewater via an integrated constructed wetland-microbial fuel cell process. J Environ Sci (China). 2020;

[24] US EPA O of W. 5.2 Dissolved Oxygen and Biochemical Oxygen Demand.

[25] Islam MMM, Shafi S, Bandh SA, Shameem N. Impact of environmental changes and human activities on bacterial diversity of lakes. Freshw Microbiol Perspect Bact Dyn Lake Ecosyst. 2019.

[26] Zhang L, Fu G, Zhang Z. Electricity generation and microbial community in long-running microbial fuel cell for high-salinity mustard tuber wastewater treatment. Bioelectrochemistry. 2019;

[27] Marassi RJ, Queiroz LG, Silva DCVR, Silva FT da, Silva GC, Paiva TCB d. Performance and toxicity assessment of an up-flow tubular microbial fuel cell during long-term operation with high-strength dairy wastewater. J Clean Prod. 2020;

[28] Fard FA, Yengejeh RJ, Ghaeni M. Efficiency of Microalgae Scenedesmus in the Removal of Nitrogen from Municipal Wastewaters. Iran J Toxicol [Internet]. 2019 [cited 2020 Sep 3];1-6. Available from: http://www.ijt.ir

[29] Ogwueleka TC, Samson B. The effect of hydraulic retention time on microalgae-based activated sludge process for Wupa sewage treatment plant, Nigeria. Environ Monit Assess [Internet]. Springer; 2020 [cited 2020 Sep 3];192:1-16. Available from: https://link.springer.com/article/10.1007/s10661-020-8229-y

[30] Canada H. Total Dissolved Solids (TDS).

[31] USEPA. Biological Nutrient Removal Processes and Costs. J Exp Psychol Gen. 2007;

[32] Wollmann F, Dietze S, Ackermann JU, Bley T, Walther T, Steingroewer J, et al. Microalgae wastewater treatment: Biological and technological approaches. Eng. Life Sci. 2019.

[33] Baysal A, Ozbek N, Akm S. Determination of Trace Metals in Waste Water and Their Removal Processes. Waste Water - Treat Technol Recent Anal Dev [Internet]. InTech; 2013 [cited 2020 Sep 3]. Available from: http://dx.doi.org/10.5772/52025

[34] Akpor OB. Heavy Metal Pollutants in Wastewater Effluents: Sources,

Effects and Remediation. Adv Biosci Bioeng. 2014;

[35] Burakov AE, Galunin E V., Burakova I V., Kucherova AE, Agarwal S, Tkachev AG, et al. Adsorption of heavy metals on conventional and nanostructured materials for wastewater treatment purposes: A review. Ecotoxicol. Environ. Saf. 2018.

[36] Al-Saydeh SA, El-Naas MH, Zaidi SJ. Copper removal from industrial wastewater: A comprehensive review. J. Ind. Eng. Chem. 2017.

[37] Farkas K, Walker DI, Adriaenssens EM, McDonald JE, Hillary LS, Malham SK, et al. Viral indicators for tracking domestic wastewater contamination in the aquatic environment. Water Res. 2020.

[38] Osuolale O, Okoh A. Human enteric bacteria and viruses in five wastewater treatment plants in the Eastern Cape, South Africa. J Infect Public Health. 2017;

[39] Ahmed W, Angel N, Edson J, Bibby K, Bivins A, O'Brien JW, et al. First confirmed detection of SARS-CoV-2 in untreated wastewater in Australia: A proof of concept for the wastewater surveillance of COVID-19 in the community. Sci Total Environ. 2020;

[40] Alpaslan Kocamemi B, Kurt H, Hacioglu S, Yarali C, Saatci AM, Pakdemirli B. First Data-Set on SARS-CoV-2 Detection for Istanbul Wastewaters in Turkey. medRxiv. 2020;

[41] Deblonde T, Cossu-Leguille C, Hartemann P. Emerging pollutants in wastewater: A review of the literature. Int J Hyg Environ Health. 2011;

[42] Dargnat C, Teil M-J, Chevreuil M, Blanchard M. Phthalate removal throughout wastewater treatment plant. Sci Total Environ. 2009;

[43] Clara M, Windhofer G, Hartl W, Braun K, Simon M, Gans O, et al. Occurrence of phthalates in surface runoff, untreated and treated wastewater and fate during wastewater treatment. Chemosphere. 2010;

[44] Díez B, Rosal R. A critical review of membrane modification techniques for fouling and biofouling control in pressure-driven membrane processes. Nanotechnol Environ Eng. 2020;5:1-21.

[45] Tetteh EK, Rathilal S, Chetty M, Armah EK, Asante-Sackey D. Treatment of water and wastewater for reuse and energy generation-emerging technologies. IntechOpen; 2019.

[46] Burggraaf AJ, Cot L. Fundamentals of inorganic membrane science and technology. Elsevier; 1996.

[47] Obotey Ezugbe E, Rathilal S. Membrane Technologies in Wastewater Treatment: A Review. Membranes (Basel). 2020;10:89.

[48] Pabby AK, Rizvi SSH, Requena AMS. Handbook of membrane separations: chemical, pharmaceutical, food, and biotechnological applications. CRC press; 2015.

[49] Jiang K, Song B, Shi X, Song T. An overview of membrane computing. J Bioinforma Intell Control. 2012;1:17-26.

[50] Gai H, Zhang X, Chen S, Wang C, Xiao M, Huang T, et al. An improved tar–water separation process of low–rank coal conversion wastewater for increasing the tar yield and reducing the oil content in wastewater. Chem Eng J. 2020;383:123229.

[51] Drioli E, Stankiewicz AI, Macedonio F. Membrane engineering in process intensification—An overview. J Memb Sci. 2011;380:1-8.

[52] Zheng X, Zhang Z, Yu D, Chen X, Cheng R, Min S, et al. Overview of

membrane technology applications for industrial wastewater treatment in China to increase water supply. Resour Conserv Recycl. 2015;105:1-10.

[53] Department of Environmental Affairs D. SOUTH AFRICA ENVIRONMENT OUTLOOK Executive summary | 1 2 nd South Africa Environment Outlook Executive summary A report on the state of the environment. Pretoria; 2016.

[54] Sershen, Rodda N, Stenström TA, Schmidt S, Dent M, Bux F, et al. Water security in South Africa: Perceptions on public expectations and municipal obligations, governance and water re-use. Water SA [Internet]. South African Water Research Commission; 2016 [cited 2020 Aug 14];42:456-65. Available from: http://dx.doi.org/10.4314/wsa.v42i3.11Availableonwebsitehttp://www.wrc.org.

[55] Chalivendra S. Bioremediation of wastewater using microalgae. 2014.

[56] Kendrick M. Algal Bioreactors for nutrient removal and biomass production during the tertiary treatment of domestic sewage[Internet]. Loughborough University; 2011 Jan. Available from: https://repository.lboro.ac.uk/articles/thesis/Algal_bioreactors_for_nutrient_removal_and_biomass_production_during_the_tertiary_treatment_of_domestic_sewage/9456713

[57] Hom-Diaz A, Jaén-Gil A, Bello-Laserna I, Rodríguez-Mozaz S, Vicent T, Barceló D, et al. Performance of a microalgal photobioreactor treating toilet wastewater: Pharmaceutically active compound removal and biomass harvesting. Sci Total Environ. Elsevier B.V.; 2017;592:1-11.

[58] Fagan R, McCormack DE, Dionysiou DD, Pillai SC. A review of solar and visible light active TiO2 photocatalysis for treating bacteria, cyanotoxins and contaminants of emerging concern. Mater. Sci. Semicond. Process. 2016.

[59] Gimeno O, García-Araya JF, Beltrán FJ, Rivas FJ, Espejo A. Removal of emerging contaminants from a primary effluent of municipal wastewater by means of sequential biological degradation-solar photocatalytic oxidation processes. Chem Eng J. 2016;

[60] Xiong JQ, Kurade MB, Abou-Shanab RAI, Ji MK, Choi J, Kim JO, et al. Biodegradation of carbamazepine using freshwater microalgae Chlamydomonas mexicana and *Scenedesmus obliquus* and the determination of its metabolic fate. Bioresour Technol. 2016;

[61] Quach-Cu J, Herrera-Lynch B, Marciniak C, Adams S, Simmerman A, Reinke RA. The effect of primary, secondary, and tertiary wastewater treatment processes on antibiotic resistance gene (ARG) concentrations in solid and dissolved wastewater fractions. Water (Switzerland). 2018;

[62] Brown N, Shilton A. Luxury uptake of phosphorus by microalgae in waste stabilisation ponds: Current understanding and future direction. Rev. Environ. Sci. Biotechnol. 2014.

[63] Monfet E, Unc A. Defining wastewaters used for cultivation of algae. Algal Res. 2017.

[64] Shahot K, Idris A, Omar R, Yusoff HM. Review on biofilm processes for wastewater treatment. Life Sci J. 2014;

[65] Houser JB, Venable ME, Sakamachi Y, Hambourger MS, Herrin J, Tuberty SR. Wastewater Remediation Using Algae Grown on a Substrate for Biomass and Biofuel Production. J Environ Prot (Irvine, Calif). 2014;

[66] Mambo MP. Integrated Algae Pond Systems for the Treatment of Municipal Wastewater. 2014.

[67] Powell N. Biological Phosphorus Removal by Microalgae in Waste Stabilisation Ponds. [Pamerston North]: Massey University; 2009.

[68] Institute of Environmental Biotechnolgy. Evaluation of integrated algae pond systems for municipal wastewater treatment. !he Belmont Volley ww w P1lot..Scole lAPS Case Study • • 0 • [Internet]. Republic of South Africa: Water research Commission; 2016 [cited 2020 Aug 14]. Available from: www.wrc.org.za

[69] Jacob-Lopes E, Teixeira T. Microalgae-Based Systems for Carbon Dioxide Sequestration and Industrial Biorefineries. Biomass. 2010.

[70] Molina Grima E, Belarbi EH, Acién Fernández FG, Robles Medina A, Chisti Y. Recovery of microalgal biomass and metabolites: Process options and economics. Biotechnol Adv. 2003;

[71] Płaczek M, Patyna A, Witczak S. Technical evaluation of photobioreactors for microalgae cultivation. E3S Web Conf. 2017.

[72] Cozma P, Gavrilescu M. Airlift reactors: Applications in wastewater treatment. Environ Eng Manag J. 2012;

[73] Elawwad A, Karam A, Zaher K. Using an algal photo-bioreactor as a polishing step for secondary treated wastewater. Polish J Environ Stud. 2017;

[74] Lin CY, Nguyen MLT, Lay CH. Starch-containing textile wastewater treatment for biogas and microalgae biomass production. J Clean Prod. 2017;

[75] Phasey J, Vandamme D, Fallowfield HJ. Harvesting of algae in municipal wastewater treatment by calcium phosphate precipitation mediated by photosynthesis, sodium hydroxide and lime. Algal Res. 2017;

[76] Molinuevo-Salces B, Riaño B, Hernández D, García-González MC. Microalgae and wastewater treatment: Advantages and disadvantages. Microalgae Biotechnol Dev Biofuel Wastewater Treat. 2019.

[77] Lavrinovičs A, Juhna T. Review on Challenges and Limitations for Algae-Based Wastewater Treatment. Constr Sci. 2018;

[78] Logan BE, Hamelers B, Rozendal R, Schröder U, Keller J, Freguia S, et al. Microbial fuel cells: methodology and technology. Environ Sci Technol. 2006;40:5181-92.

[79] Min B, Cheng S, Logan BE. Electricity generation using membrane and salt bridge microbial fuel cells. Water Res. 2005;39:1675-86.

[80] Kim HJ, Park HS, Hyun MS, Chang IS, Kim M, Kim BH. A mediator-less microbial fuel cell using a metal reducing bacterium, Shewanella putrefaciens. Enzyme Microb Technol. 2002;30:145-52.

[81] Kim JR, Min B, Logan BE. Evaluation of procedures to acclimate a microbial fuel cell for electricity production. Appl Microbiol Biotechnol. 2005;68:23-30.

[82] Liu H, Logan BE. Electricity generation using an air-cathode single chamber microbial fuel cell in the presence and absence of a proton exchange membrane. Environ Sci Technol. 2004;38:4040-6.

[83] Prasad J, Tripathi RK. Scale up sediment microbial fuel cell for powering Led lighting. Int J Renew Energy Dev. 2018;7:53.

[84] Rabaey K, Verstraete W. Microbial fuel cells: novel biotechnology for energy generation. TRENDS Biotechnol. 2005;23:291-8.

[85] Logan BE. Microbial Fuel Cells [Internet]. Hoboken, N.J.: Wiley-Interscience; 2008. Available from: http://search.ebscohost.com/login.aspx?direct=true&db=nlebk&AN=219803&site=eds-live

[86] Lovley DR. Microbial fuel cells: novel microbial physiologies and engineering approaches. Curr Opin Biotechnol. 2006;17:327-32.

[87] Ieropoulos IA, Greenman J, Melhuish C, Hart J. Comparative study of three types of microbial fuel cell. Enzyme Microb Technol. 2005;37:238-45.

[88] Nishika GVYSA, Mongia SAADG, GulatiMarwah IKDR. A Short Review on Microbial Fuel Cell Technology and A Proposed approach for Generation of Electricity using Waste Water Treatment. 2015;

[89] Moon H, Chang IS, Kang KH, Jang JK, Kim BH. Improving the dynamic response of a mediator-less microbial fuel cell as a biochemical oxygen demand (BOD) sensor. Biotechnol Lett. 2004;26:1717-21.

[90] Zhang Y, Min B, Huang L, Angelidaki I. Electricity generation and microbial community response to substrate changes in microbial fuel cell. Bioresour Technol. 2011;102:1166-73.

[91] Lu N, Zhou S, Zhuang L, Zhang J, Ni J. Electricity generation from starch processing wastewater using microbial fuel cell technology. Biochem Eng J. 2009;43:246-51.

[92] Liu H, Ramnarayanan R, Logan BE. Production of electricity during wastewater treatment using a single chamber microbial fuel cell. Environ Sci Technol. 2004;38:2281-5.

[93] Wei J, Liang P, Huang X. Recent progress in electrodes for microbial fuel cells. Bioresour Technol. 2011;102:9335-44.

[94] Wen Q, Kong F, Ma F, Ren Y, Pan Z. Improved performance of air-cathode microbial fuel cell through additional Tween 80. J Power Sources. 2011;196:899-904.

[95] Ren H, Rangaswami S, Lee H-S, Chae J. A micro-scale microbial fule cell (MFC) having ultramicroelectrode (UME) anode. 2013 IEEE 26th Int Conf Micro Electro Mech Syst. IEEE; 2013. p. 869-72.

[96] Wang H, Ren ZJ. A comprehensive review of microbial electrochemical systems as a platform technology. Biotechnol Adv. 2013;31:1796-807.

[97] Szöllősi A, Rezessy-Szabó JM, Hoschke Á, Nguyen QD. Novel method for screening microbes for application in microbial fuel cell. Bioresour Technol. 2015;179:123-7.

[98] Kalathil S, Khan MM, Lee J, Cho MH. Production of bioelectricity, bio-hydrogen, high value chemicals and bioinspired nanomaterials by electrochemically active biofilms. Biotechnol Adv. 2013;31:915-24.

[99] Kim I-S, Chae K-J, Choi M-J, Verstraete W. Microbial fuel cells: recent advances, bacterial communities and application beyond electricity generation. Environ Eng Res. 2008;13:51-65.

[100] Hu Z. Electricity generation by a baffle-chamber membraneless microbial fuel cell. J Power Sources. 2008;179:27-33.

[101] Shi CY. Mass flow and energy efficiency of municipal wastewater treatment plants. IWA Publishing; 2011.

Chapter 5

A Review on AI Control of Reactive Distillation for Various Applications

Vandana Sakhre

Abstract

In this chapter, previous studies on reactive distillation process control including control using conventional as well as soft sensor control, membrane assisted reactive distillation design and simulation, estimation and control are discussed. The review of literature in different dimensions is carried out to explore the opportunities in the field of research work. The chapter is focused on dynamics and control of Reactive distillation, its control using Conventional Techniques, Model Predictive Control MPC), Reactive Distillation using Soft Sensors/Soft Controllers, Membrane assisted reactive distillation, Biodiesel in Reactive Divided Wall Column: Design and Control and Membrane reactive divided wall column. These control techniques are proposed and analyzed by many researchers. These techniques have potential use in process industries to have better soft sensor control of nonlinear processes.

Keywords: control, divided wall column, model predictive control, reactive distillation

1. Introduction

Reactive distillation is a part of process intensification technique which aims at carrying out reaction to form products which are then separated in the same column based on difference in boiling points. Dynamic study of any chemical process is the most crucial part which should be performed at early design stage as the dynamic nature leads to nonlinearity. This nonlinearity introduces sluggish behavior in the system which will lead to distract the process output out of the range. The dynamic behavior or nonlinearity was usually studied by rigorous modeling or using simulation software to predict the nature of system. For this advanced control systems are designed and implemented to get effective control strategies. Neural network based soft controller are suitable for reactive distillation system as these are suitable to overcome with multiplicity issues. The dynamic performance of reactive divided wall distillation column with nonlinear control structures is proven to be cost effective. The use of reactive distillation with membrane separation techniques are emerged as effective separation techniques which give corresponding recoveries as well.

2. Reactive distillation dynamics and control

Non linearity of reactive distillation was first reported by Ciric and Gu et al. [1] using simulation of reactive distillation processes involving simultaneous solution of material and energy balances and stoichiometric relationship which corresponds to the solution of a considerable large set of non-linear modeling equations of reactive distillation column. They have used a mixed integer nonlinear programming (MINLP) approach was used to synthesize an optimum reactive distillation column. The MINLP minimizes the total annual cost subject to a MESH model. The solution of this MINLP yields the optimal number of trays, the tray holdups, the feed tray locations, and their feed distribution. Amte et al. [2] presented a MINLP optimization technique that would assist to identify a suitable configuration for selectivity maximization at conceptual design level. Results obtained through MINLP gives a good agreement with those obtained by performing independent simulation using ASPEN PLUS. In this work, authors have considered feed and catalyst tray location, reflux as the variables for the maximization of selectivity. Thus, MINLP optimization process proves conceptual design for the selectivity engineering with reactive distillation. Doherty et al. [3] have given pioneering contributions to the analysis and design of the reactive distillation and developed thermodynamically based approach for analyzing equilibrium limited, thermally neutral reactive distillation systems. This work employed a novel composition coordinate system to transform the problem into a form completely analogous to nonreactive distillation.

Subawalla and Fair et al. [4] worked on dynamic study of TAME synthesis in reactive distillation column by considering different design parameters such as number of trays, feed flow rate etc., resulting into nonlinear interaction of these input parameters. He has also provided some intuitive guidelines about coupling of these design parameters order. Schenk et al. [5] have presented equilibrium and non-equilibrium models for predicting the steady state and dynamic behavior of RDC based on a rate-based model in which mass transfer rates between liquid and vapor phase are considered explicitly, based on the Maxwell-Stefan equations. A switching policy makes it possible to switch from one model to the other, based on the knowledge gained, by following the Gibbs free energy as a function of time. Cardoso et al. [6] have proposed a new simulation/optimization model for the Mixed Integer Nonlinear Programming (MINLP) formulation of reactive distillation columns as used by Ciric and Gu [1], where the simulation algorithm is based on conventional distillation and optimization is performed by stochastic algorithm. Estimate of the initial composition profile in the column was obtained by relaxation method and material balance equations are solved by Newton–Raphson method, to compute the composition profile. Vora and Daoutidis et al. [7] represented work on dynamics of reactive distillation by proposing a different feed configuration for two feed and two product system. They have also presented control of an ethyl acetate reactive distillation system to achieve higher conversion than the previously proposed configuration which involves single feed reactive distillation column. Taylor and Krishna et al. [8] have presented work on modeling of homogeneous and heterogeneous reactive distillation processes by considering equilibrium stage models, non-equilibrium stage modeling, the conventional NEQ model, NEQ modeling, hydrodynamics, and mass transfer. Dynamic simulation for NEQ model without back mixing was also presented by Kataria et al. [9] in which model equation for reactive distillation without back mixing was developed for open loop system. For the NEQ model with partial mixing, bifurcation behavior with stability analysis and simulation were carried out which confirmed the existence of oscillation in the NEQ model as well. Once the nonlinear nature of reactive distillation is confirmed, various control strategies were reported by the authors to make the system linear. Linearization was carried out to make system to oscillate around the set points of desired parameters which makes the system easy to operate without much affecting the performance over the time. Mehran et al. [10] have employed Takagi-Sugeno fuzzy model to express the local dynamics of each fuzzy implication by a linear system model to plot the resultant error of surface with time and reported that although system's non linearity can be linearize to some extent but still many control problems involve uncertainties in the model. In view of this, Gruner et al. [11] have taken an industrial case of reactive distillation and proposed a nonlinear control scheme for the system operated by Bayer AG.

Interaction of various parameters in a reactive distillation leads to higher order nonlinearity and dynamics like interaction between number of trays, reflux ratio, reboiler duty, etc. This interaction will lead to multiple steady states and multiplicity in the system. The type of multiplicity will limit the values which should be broken by application of suitable controllers. Heath et al. [12] studied the interactions of design variables and applied control schemes for an ethyl glycol reactive distillation system and assumed that the process variables are fixed i.e. no revamping is done. Author has taken process economics as a major issue which ultimately depends on the number of variables interacting in a system and its control was achieved as per the desired objective. For the same assumptions, Schenk et al., [13] and Georgiadis et al. [14] have worked on interaction of design variables in a reactive distillation column and proposed advanced optimization technique for the control of a reactive distillation system to smoothen the operability and increase the system efficiency.

Figure 1 shows general control scheme of reactive distillation column. The common controls used are Flow Control (FC), Level Control and Composition Control (CC). Level control is given in condenser and reboiler while composition control is provided to condenser, reboiler and one of the feed flow rates to get tighter composition analysis. Flow control is provided to feed flow rate at the top of the reactive section.

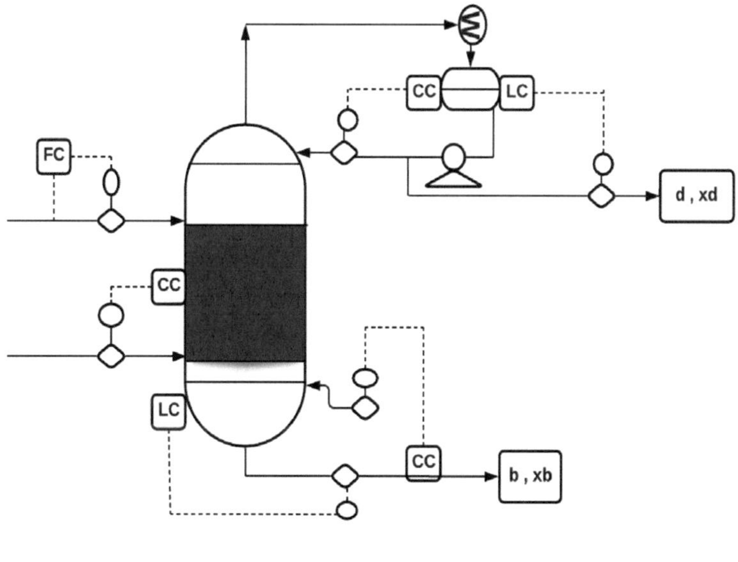

Figure 1.
Reactive distillation control.

2.1 Control using conventional techniques

Simplified models clarify process dynamics but cannot represent the process under wide range of operating condition. Thus, various mathematically driven controllers like proportional, integral, derivative or a combination of theseareusedas hardware sensor-controller system which are mathematically functioned or scripted. Furthermore, the RD process contains a large degree of uncertainties, which cannot be well described using single mathematical expression. Therefore, techniques without using exact process models are more attractive for RD control. Monroy-Loperena et al. [15] studied the control problem of reactive distillation system for ethyl glycol by proposing a robust PI control configuration. They have also revealed the existence of input multiplicity in the system and proposed a first order output feedback control system to regulate the product composition. The control design involves interpretation of error signal whose dynamic is again constructed based on available data. Sneesby et al. [16] moved in process control by highlighting an integrated control scheme by taking ETBE as the case column. He proposed to change control objective to reflect changing economic variables like starting from optimum purity, minimum number of trays, optimum reflux ratio, etc. In view of this author has also presented a rigorous MESH based modeling to represent the main chemical reaction. Al-Arfaj & Luyben et al. [17] applied different conventional PI, PID and other conventional controller scheme for reactive distillation as well as for simple reactor. They compared the control of both these systems to produce methyl acetate. Various control strategy was proposed in this paper, the first one was for a composition and temperature control while second scheme was based on tow temperature controllers. A comparison between these schemes shows that different scheme corresponds to over design or under design system hence proper balancing of degree of freedom of a system is equally important.

Various tuning methods are proposed in the literature to calculate the ultimate period or ultimate gain. Chandra et al. [18] have calculated ultimate gain and ultimate period using Ziegler Nicolas tuning rule for an ARX model structure. The objective was to control the desired product purity in distillate stream. In this work, an ARX model structure relates the plant output with present and past plant input output to formulate a predictive control. Recursive least square estimator was used which provides updated parameters to ARX model. Goyal et al. [19] have presented support vector regression to tune a PID controller. Model gain scheduling was included in one of the control strategies for reactive distillation. Temperature control was given priority because to balance the stoichiometry, temperature of feed trays can be used to adjust the fresh feed streams. For this the gain of controller was define as the change in temperature with respect to the feed flow rate. Nizami et al. [20] have constructed one or two loop composition PID controller, however, the conventional controllers applied for the control of reactive distillation was not capable of actually control the simultaneous interacting parameters because of occurrence of reaction and separation in single column. Lei et al. [21] have described the design and optimization of reactive distillation column for the synthesis of Tert-Amyl Ethyl Ether (TAEE), the temperature–composition cascade strategy was proposed to control the Reactive Distillation (RD) process for the synthesis of TAEE. In those optimized conditions, the proposed control strategy was introduced to manage the RD process by changing the sensitive variables. Dimian et al. [22] have carried out thermodynamic analysis in residue curve map and simulation of reactive distillation column. Process dynamic and control was considered in detail to design the column. The feasibility of fatty acid esterification with individual alcohol was studied by means of residue curve maps. By sensitivity analysis it was found that the reflux of heavy alcohol is the key manipulating variable for controlling the water content in the reactive liquid phase.

The controllability of the system could be studied using tools from control theory. Feedback control of inventory is measured, and a feedback control loop is implemented with the fresh feed as manipulating variable. Konakom et al. [23] have proposed to control distillate rate subject to a given product purity constraints. A conventional batch reactive distillation model described by a system of differential algebraic equations is formulated and solved using an optimal control algorithm. In open loop simulation of production of industrial grade ethyl acetate of 90% purity, dynamic optimization programming was implemented which increases the purity as per the product specification.

2.2 Model predictive control and other control techniques

Due to the complexity of the process dynamics involved in reactive distillation, conventional control technologies, e.g. PI, PID control, cannot provide satisfactory control performance, while the application of modern control technology requires good process models. A reasonable process model as described by Sneesby et al. [24] contain hundreds of equations for the RD process which is to be controlled. Pattern-based predictive control (PPC) is such a method that does not rely on exact process models while providing improved control performance for complex processes over conventional, e.g. PI, control algorithms. Some progress has been made in this direction, like for time delay compensation, Zhao et al. [25] worked for dynamic models that reject disturbances. Since various chemical processes possess time delays and uncertainties, for example, the flowing fluid in a pipe was taken as time delay variable. To represent such system the proposed model works on first order lag dynamics to compensate for uncertainties. Bode plots were also constructed to show that pattern based fuzzy predictive allows a trade-off between robustness and the performance. Seem et al. [26] have proposed a novel predictive scheme by considering a proportional integral controller in which the gain and integral time is calculated automatically and hence they have given such system a name of self- regulating system. However, this scheme was based on the pattern reorganization methodology, but author has asserted that this scheme requires less memory and is more efficient as compared to the conventional techniques. Jang et al. [27] worked on fuzzy predictive control which does not depends on exact process model but considers a pattern predictive control that provides improved performance in both set point tracking and disturbance rejection, shown for the RD process. Local optimum was identified to minimize prediction error and global optimum was then identified through various subsystems. The nonlinear transformation, feature pattern extraction, and PP design was discussed in detail by Tian et al. [28] who have designed a pseudo input–output linear process gain, which needs only a rough and easily obtained knowledge of the steady state characteristics of the process. Author has worked on one-point control strategy i.e. control of bottom purity by considering reflux ratio and reboiler duty as the variables. They also used state estimation approaches for measurement of sample at multiple rates.

Bansal et al. [29] have developed an algorithm for solving Mixed Integer Dynamic Optimization (MIDO) problems. This algorithm is different from other conventional algorithm as they do not depend on the use of a particular primal dynamic optimization method. However, overthe last decades, many versions of Extended Kalman filter (EKF) has coined that deals with the measurements at multiple rates. Patwardhan et al. [30] have worked on output feedback system for the case of plant mismatch. The EKF for nonlinear systems has been explored by Beccera et al. [31]. The concept involves the use of a time varying linearization of differential algebraic equation in which estimation was performed using a simplified EKF that was integrated with the differential algebraic equation model which accommodate measurements obtained only from the differential states. This technique has serious limitations. This limitation has been overcome by Mandela et al. [32]. They presented formulations for EKF and Unscented Kalman Filter, in which measurements of differential as well as algebraic states were recorded. To achieve more rigorous control of reactive distillation nonlinear system over a wide operating range, various successive linearization based nonlinear predictive control scheme was developed by Huang et al. [33]. They proposed error feedback scheme that introduces integral action in the controller for controlling a multi rate sampled data system. Author has not implemented the conventional linear feedback but adopted a novel variable feedback concept that effectively reduces the noise making the system quite robust for the designer. Akesson et al. [34] have developed a scheme in which control objective was to keep the output close to a specified reference trajectory in such a way that large control signal variations are avoided and possible hard constraints on the state and inputs was satisfied. Main control objective of this work was to minimize the cost. Procedure was adopted by the authors in which controller was trained directly to minimize the cost for a data set, without having to compute the optimal MPC control signals by off-line optimizations.

MPC are classified to various type such as dynamic matrix control (DMC), quadratic dynamic matrix control (QDMC), robust multivariable predictive control technology (RMPCT), generalized predictive control (GPC), and other advanced control techniques and was reviewed in brief by Sharma et al. [35]. They also presented work on comparison of conventional strategies with MPC and neural network predictive control by considering a TAME reactive distillation column for different load changes and proved that NNPC and MPC provide much accurate result as compared to conventional PIDs. **Figure 2** shows the general MPC structure.

2.3 Control of reactive distillation using soft sensors/soft controllers

Design of a soft sensor for a reactive distillation column includes three steps: first, selection of secondary measurement of the process, second, moving data collection and processing this data, and the last, modeling of process based on selected

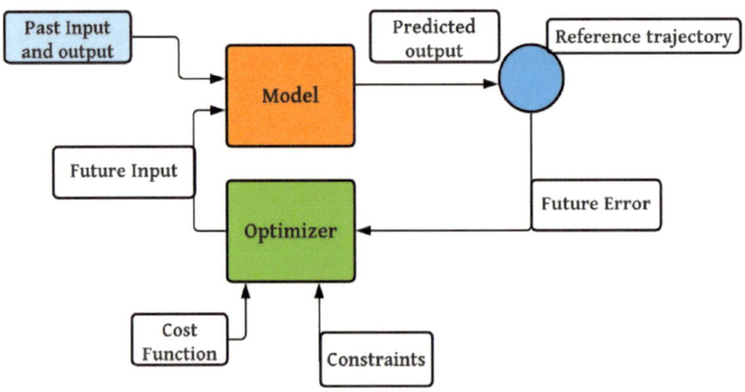

Figure 2.
General MPC structure.

secondary measurements and processing of data. Generally, the differential algebraic equations (DAE), describes the dynamics of a reactive distillation column. Soft computing techniques are methods in which real practical situation could be dealt in the same way as human deals them i.e. based on intelligence, common sense, reasoning, analogies, etc. Fuzzy logic is the oldest control schemes used not only in process industries but in vast area of other engineering applications also. In view of this, Babusk et al. [36] have first coined up the detailed concept of fuzzy logic in both static and dynamic system and proved fuzzy system as an interactive method, facilitating the active participation of the user in a computer-assisted modeling session. The fuzzy model proposed by Takagi and Sugeno described by fuzzy 'IF-THEN' rules represents local input–output relations of a nonlinear system. Rico et al. [37] applied the fuzzy control technique to control the process output from distillation column inthe desired range for different input disturbances. As an initiation to fuzzy logic, industries as well researchers move toward the field of soft sensing and control using soft computing techniques. These techniques were initially based on local optimization such as given by Pekkanen et al. [38] for a stage by stage specification of reactive distillation. They initiated the control procedure from each column ends i.e., from top as well the bottom while making the design specifications at each stage.

Soft controllers are also known by an alternative name known as intelligent control technique or inferential control as these controllers can estimate and control the process based on past experiences. Use of soft computing approaches take its first start up with the launch of natural evolution based algorithms like genetic algorithm, ant bee search algorithm etc. and merged toward more rigorous approach by combining one algorithm with other such as artificial neural networks or ANN's and fuzzy logic as a black box technique to model systems and gained substantial interest in different areas of engineering. These are also known as hybrid techniques which consist of a frame- work of dynamic mass and energy balances, supplemented with fuzzy models. The hybrid models have shown that the use of fuzzy logic in hybrid modeling introduces flexibility, which enables the description of complex behavior with a pre defined, interpretable overall model structure. Araromi et al. [39] designed a continuous RD using hybrid Fuzzy Hammerstein (FH) model consisting nonlinear fuzzy model and linear state space model was then developed. The developed model was compared with linear autoregressive input exogenous (ARX) and nonlinear autoregressive input exogenous (NARX). Sumana et al. [40] investigated the use of gramian covariance matrices for sensor configuration in continuous multi component reactive distillation by applying extended Kalman Filter to obtain the instantaneous composition information from temperature sensors data of reactive distillation column. The sensors configurations were further evaluated by IAE criteria incorporating the measurements suggested by the state estimator. **Figure 3** shows general soft sensor control.

Soft computing or inferential computing is the most advanced stage of control schemes. At present, neural network is one of the most demanded intelligent controllers which works on the imitation of working of neurons in human brain. Wang et al. [41] have taken a case-based modeling program with an industrial example of distillation column. The basic features of this case base modeling described in brief were discreteness, nonlinearity, contradiction and complexity. Theyreported that neural network is promising in process control and fuzzy distributed neural network can be used to design a soft sensor for a high purity distillation column. Multilayer neural network was utilized when creating a system inverse neural model. In view of this, Zilkova et al. [42] have developed three-layer feed forward neural network with one hidden layer elected to approximate nonlinear function. The first subsystem served for desired current component reconstruction and second system

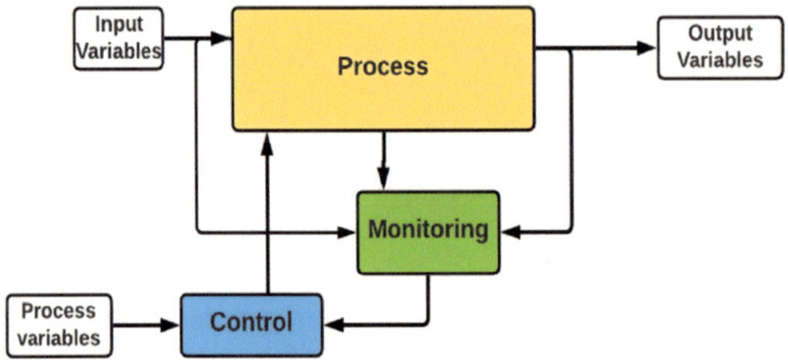

Figure 3.
General structure of soft sensor control.

serves for corresponding voltage components reconstruction for PWM converter. This was carried out using Simulink model. Simulation result verified the effectiveness of proposed controller. Akesson et al. [34] have taken control objective to keep the output close to a specified reference trajectory in such a way that large control signal variations were avoided and possible hard constraints on the state and inputs were satisfied. Main control objective of this work was to minimize the cost. For a data set, without having to compute the optimal MPC control signals by off-line optimizations. Raghavan et al. [43] have developed a Recurrent Neural Network (RNN) based inferential state estimation scheme for an ideal reactive distillation column. The performance of the estimator for both open loop and close loop was compared with that of kalman filter in terms of qualitative and quantitative indices and concluded that RNN has better level of inferential control over the conventional suggested methods. Prakash et al. [44] proposed an artificial neural network based nonlinear control algorithm for simulated batch reactive distillation column by considering a homogenously catalyzed esterification reaction which was controlled using ANN based state predictor. The open loop dynamic was presented in detail for this system and the proposed law was tested against gain scheduling controller to compare the performance. Bahar et al. [45] presented an Elman neural network to control the product composition from distillation column using temperature measurements inferentially. The main limitation of the neural network controller was that substantial offline computations may be needed in order to train it properly, and for some choices of cost functions it may not even be feasible to achieve satisfactory accuracies.

Other soft sensing techniques like genetic algorithm, particle swarm optimization, ant colony optimization etc. are although very old but are still being employed for inferential control which is based on natural evolution theory. Nithya et al. [46] have used real world experimentation in which pneumatic control valve is used to control the flow of water in and out of the tank. Using the black box modeling, the transfer function has been derived which was used to design a PI controller to find the values of gain and transfer function. Fuzzy logic controller has been designed for spherical tank considering nonlinear system. For tuning of the PI and Fuzzy Logic controller, genetic algorithm was used and the responses for servo and load disturbance were observed. Idris et al. [47] have considered the case of methyl acetate production in a continuous reactive distillation column in which tracking index term has been coined which was define as squared sum of differences between the predicted outputs and set points change over the prediction horizon. The control algorithms were applied in gPROMS against various tuning parameters and concluded that optimizing controller can be easily applied to a simulated complex model of reactive distillation to enable real time dynamic optimization.

Sujatha et al. [48] have considered a MIMO system defining integration of manipulated variable and control variable. Various novel techniques such as relative gain array, Niederlinski index, singular value decomposition, Morari Resiliency Index, dynamic relative gain array, hankel interaction index array, participation matrix and H_2-norm were studied for the interactions and subsequently for input output pairing.

Differential evolution is one of the latest coined soft controlling techniques which aims at approaching toward global minimum. Subudhi et al. [49] have developed a differential approach based on a jumping rate fittest approach in which individual were selected from the union of current population and opposite population. The aim was to approach toward global minimum and then LM was used to move forward achieving fast convergence. LM is a gradient based algorithm to increase convergence speed. Opposition based learning improves the chance of starting with better initial population by checking the opposite solutions. Lu et al. [50] have developed a differential algorithm based stochastic search technique which is a powerful and global optimizer. The author has proposed a Modified Differential Evolution Fuzzy Neural Network (MDEFNN) which consists of a FNN identifier, a MDE estimator, a computation controller, and a hitting controller. Lawrynczuk et al. [51] have reported model predictive control strategy for a high purity, high pressure ethylene ethane distillation column. In this study, multi-layer perceptron neural network was applied with one hidden layer and a linear output. In the MPC-NPL algorithm, the nonlinear neural model was linearizing using eight repetitive steps such as estimation, approximation, control increment, and iteration. Kandanapitiya [52] has recently reported Modeling of Reactive Distillation for Acetic Acid Esterification. The mathematical model considered material balance equations, equilibrium relationships, summation equations and energy balance equations. The model was simulated for acetic acid and ethanol esterification reaction. Wierschem [53] has worked on design of Continuous Enzymatic Reactive Distillation with Immobilized Enzyme Beads. Based on kinetic and thermodynamic data, a detailed rate-based model of the ERD is developed. Simulation results and experimental data of the ERD setup are in good agreement. Zhang [54] has designed centralized and decentralized stochastic adaptive fuzzy output feedback control by using dynamic surface control method. Fuzzy systems are used to approximate unknown nonlinear continuous function. For the online process, the young's inequality norm of fuzzy basis vector is adjusted. Peng [55] has shown that two riccati equations were employed with relaxing the agent dynamics for uncertain nonlinear systems. The author has developed a cooperative output feedback adaptive control (COFAC). The NN identification of the individual uncertain dynamics is decoupled from the network topology, which is useful for practical implementations since the uncertain nonlinear dynamics can be suppressed by local NN. Cui [56] deals with the problem of adaptive decentralized NN control by combining Lyapunov-Razamikhin functional approach, minimum learning parameter algorithm and back stepping design technique. Here only adaptive parameter needs to be estimated for each subsystem which shows that all signal in close loop system are uniformly ultimately bounded in probability.

2.4 Membrane assisted reactive distillation

Membrane not only plays the role of a separator, but also used in the reaction itself. Membrane assisted reactive distillation has emerged as a novel technique of hybrid process intensification to achieve higher efficiency and yield in the production of bulk chemicals. In several cases, non-ideal aqueous-organic mixtures are formed which tend to form azeotropes. They can be overcome using membrane separations like pervaporation and vapor permeation since they are very selective and not limited by vapor-liquid equilibrium Rautenbach [57]. The choice of membrane type to be used in membrane reactor depends on parameters such as the productivity, separation selectivity, membrane life time, mechanical and chemical integrity at the operating conditions and, particularly, the cost. Consequently, a hybrid process consisting of membrane-assisted reactive distillation contributes to sustainable process improvement due to arising synergy effects and allows for reduction of investment and operational costs. A review of hybrid processes combining pervaporation with one or more other separation technologies was given by Lipnitzki et al. [58]. The analysis of hybrid separation processes combining membrane separation with conventional distillation was described in Kreis and Gorak [59]. They have presented various configurations corresponding to this hybridization which offers significant advantages like reduced energy requirement, lower production cost, etc.

Membrane hybrid processes can achieve separations which are impractical to achieve with either conventional process. An example for the investigation of a reactive hybrid process concept is the transesterification of methyl acetate and butanol to butyl acetate and methanol by the combination of reactive distillation and pervaporation, as studied by Steinigeweg and Gmehling [60]. They carried out experiment using various catalytic packing is like Katapak S, Katapak SP and Sulzer. However, experiment was not presented for the hybrid technique but only simulation results have been reported. The industrially operated hybrid process for the continuous production of fatty acid esters by reactive distillation and pervaporation was presented by Scala et al. [61]. Ozdemir et al. [62] presented an overview of the commercial polymers used as membranes as well as of other polymers having high potentially for application as a membrane material. However, many industrial processes involve operations at high temperatures. Luo et al. [63] showed that integration of a membrane unit for a side stream withdrawal from the section of reactive distillation where the azeoptrope is liable to form which further improves the product yield.

Polymeric membranes are not generally useful in hybrid reactor and therefore inorganic membranes are preferred. Gorak et al. [64] have shown higher efficiency and capacity of membrane assisted reactive distillation with special focus on Pervaporation unit. Author has deeply identified the current challenges and future predicted trends for implementation of this hybrid technique in the field of chemical and biochemical industry. Holtbruegge et al. [65] represented synthesis of dimethyl carbonate using this hybrid technique to overcome the limitation of chemical equilibrium and azeotrope formation. Replacing membrane at various location in reactive distillation yield different efficiency, which is rigorously studied by Bida et al. [66]. They proved that placing pervaporation membrane at the bottom shows remarkably improved performance with effective economy and energy efficiency. Thus, we can say that combination of pervaporation and reactive distillation exploits the advantages of minimization of cost by reducing energy expenditures and making higher degree of separation.

2.5 Biodiesel in reactive divided wall column: design and control

Due to gradual depletion of world petroleum reserves and impact of environment pollution there is a need for alternative fuels for use in diesel engine. Biodiesel has emerged as a promising alternative because it is renewable and environment friendly and leads to reduction of exhaust emission. Masjuki et al. [67] first researched on the various aspects of use of biodiesel as future fuel by considering the rising cost and increased pollution from conventional carbon containing fuels. They proposed that biodiesel can be manufactured from easy to available raw material like animal oil or used vegetable oil which is generally discarded as a waste. This mishandling also serves as a matter for pollution in water or soil. Thus, biodiesel also minimizes the waste in one way or other. We [68] worked on the effect of molecular weight of fatty acid on the octane rating of biodiesel. There are many processes to convert vegetable oil into biodiesel, but transesterification reaction was found to be most viable process of oil modification. Biodiesel can be produced from animal as well vegetable oil, which is reported by Cr et al. [69]. Today economic factors along with environmental concerns are playing a key role in increase in thermal efficiency. Studies show that biodiesel is much better fuel than fossil fuel-based diesel in term of engine performance, emission reduction, lubricity, and environmental benefits. The production of biodiesel by transesterification in existing conventional processes requires excess alcohol. This excess alcohol must be recovered and purified for reusing by rectification and distillation, which involves additional capital and operating cost. Kiss et al. [70] first reported reactive divided wall distillation column consisting of one condenser, one reboiler, reactive zone a pre fractionators and main column in a single shell leads to process integration and intensification, leading to cost saving and increased purity of final product and side streams product. Later on, several research works have been carried out in the field of reactive divided wall distillation column for biodiesel production. Kiss et al. [71] described the increase in purity and energy reduction of 30% in a divided wall column in which at the bottom of diving wall section, the vapor flow was split proportionally to the cross sectional area of each side. They focused on enhanced methanol recovery from the DWC unit. Delgado et al. [72] recommended the use of petluyk distillation column when the molar fraction of middle component is low. **Figure 4** shows divided wall distillation column.

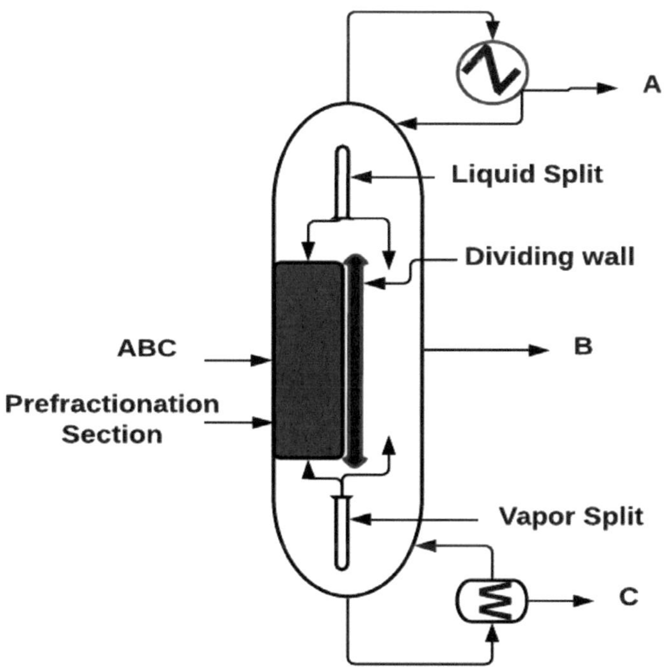

Figure 4.
Divided wall distillation column.

Since reactive divided wall distillation column is a case of process intensification, there is a complex interaction between vapor liquid equilibrium, vapor liquid mass transfer, intra catalyst diffusion and chemical kinetics. Such interactions and strong nonlinearity lead to multi steady states and complex dynamics. Bravo et al. [73] verified this nonlinearity of divided wall in laboratories as well as small pilot plants. The work has been carried out to study the various control system for reactive divided wall distillation column which can lead to process optimization. R-DWC was designed for a quaternary reactive system – two reactants (one in excess) and two products – more difficulties concerning the process control may be expected due to consideration of the high degree of integration of the process. Study was carried to tackle the optimal design, dynamics, and control of such an integrated unit and proposes an efficient control structure for a biodiesel process based on reactive DWC technology. Chongkhong et al. [74] proposed the novel distillation technologies for enhanced bioethanol dehydration, by extending the use of dividing-wall columns (DWC) to energy efficient Extractive Distillation (ED) and Azeotropic Distillation (AD). This technology is beneficial because the industrial production of anhydrous bioethanol requires energy demanding distillation steps to overcome the azeotropic behavior of the ethanol-water mixtures. A recently proposed process by Gomez- Castro et al. [75] depicts control of divided wall technology involving the use of short chain alcohols at supercritical condition that avoids the use of catalyst and this condition was applied to a reactive petlyuk column that results in thermal coupling and more of vapor liquid interactions. Aspen Plus and Aspen Dynamics were used as computer aided process engineering (CAPE) tools to perform the rigorous steady-state and dynamic simulations, as well as the optimization of the new R-DWC based biodiesel process. These control structure described the excellent performance of RDWC for biodiesel production.

2.6 Membrane assisted reactive divided wall column

The production of biodiesel by transesterification in existing conventional processes requires excess alcohol. This excess alcohol must be recovered and purified for reusing by rectification and distillation, which involves additional capital and operating costs. For production of biodiesel various techniques have been proposed such as Reactive Divided Wall. However, a major question remains that which material should be used to make that wall in middle of the column. For this Atadashi et al. [76] have reported membrane biodiesel production technique to provide high quality biodiesel fuel. In this technique the membrane system exploits the characteristic of high selectivity, high surface area and their potential for controlling mixing between two phases. The author has taken canola oil as the base oil for biodiesel synthesis and has studied the effect of membrane pore size and catalyst on the performance of membranes. The results show that membrane reactor restrict the passage of unreacted oils to the biodiesel product mixture and the use of alkaline catalyst result into soap formation while acid catalyst avoids the same.

There are various types of membrane separation processes available like ultrafiltration, microfiltration, pervaporation, etc. Vapor permeation along with RD is a novel technique in which the volatile components are separated by nonporous membrane. One possible process alternative using vapor permeation was suggested by Buchaly et al. [77] for n-propyl propionate synthesis, in which Amberlyst 46 was used to compete the side product formation. The author has used online data reconciliation by satisfying mass, component, and reaction rates as boundary conditions. In another work by Buchaly et al. [78] on same case, a comparison of most common modeling depths such as Maxwell Stefan's equation, equilibrium model with and without considering reaction kinetics was presented. Thermal coupling

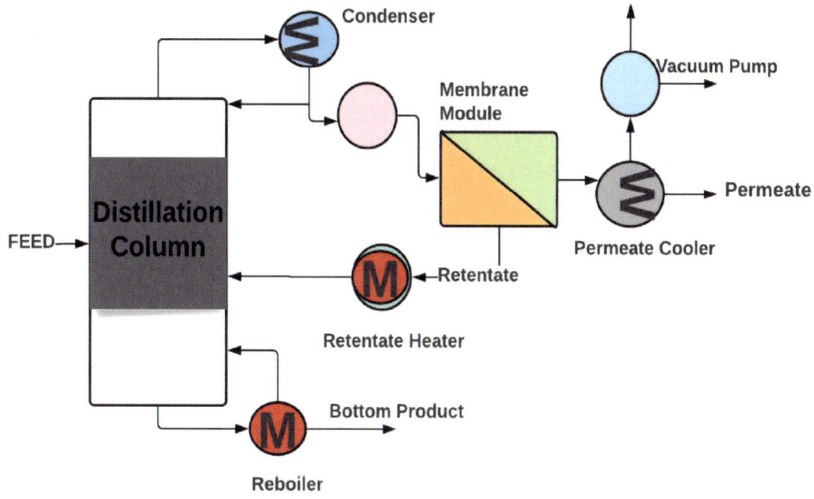

Figure 5.
Membrane reactive distillation column.

between two columns in a sequence has proven to be very successful in providing energy savings with respect to conventional trades. The control of thermally coupled membrane RDWC was presented by Wang et al. [79]. It was reported that controlled stage temperature shows no multiplicity and proper temperature control can maintain reactant inventory. These thermally coupled systems show higher thermodynamic efficiency. Some improvement for dividing wall distillation column likes catalytic packing and non-welded walls have also been proposed by Aspiron et al. [80]. A rigorous study of ASTM standards for biodiesel was carried out by researchers and reported that Membrane biodiesel separation processes provides high-quality biodiesel fuel. He et al. [81] and Saleh et al. [82] have shown that a refining step is necessary to be accompanied with transesterification of biodiesel. Also, the membrane separation processes for biodiesel were carried out under moderate temperature and pressure conditions and their scale-up process found less cumbersome. Sarmento et al. [83] critically examined the production and refining of biodiesel using membrane technology.

Some application and future aspects of dividing wall column with bioethanol production was investigated by Delgado [84]. He has also reviewed practical application of dividing wall column by performing simulation studies. The major emphasis was given on environmental impact of fuel burning and prices of oil and biofuels and the results obtained were compared and summarized by taking the difference between them. The schematic diagram of membrane reactive distillation is presented in **Figure 5**.

3. Conclusion

Reactive distillation as highly nonlinear and practically difficult to control process is challenging study. The current chapter is focused on giving deep review of work carried out by many researchers in the field of reactive distillation design, control and advances in conventional processes. The control of such highly interactive and nonlinear process is possible with several models predictive control techniques including neural network control, fuzzy models, hybrid control structures, adaptive model predictive control and optimization, reactive divided wall concept and membrane divided wall techniques. The focus is given to have wide range application and use of reactive distillation column, its control using advanced techniques and advanced synthesis methods. Many processes and reactions including catalytic reactions are equilibrium limited and reactive distillation is best techniques to get higher conversion. Many advanced techniques of controlling such highly nonlinear processes are proven to be cost effective. The increased use of membrane and synthesis in membrane assisted reactive distillation is also proven to get higher product purity and higher solvent recovery rate.

Acknowledgements

The author is gratefully acknowledge the assistance provided by AICTE, New Delhi and administrative support provided by Madhav Institute of Science & Technology, Gwalior, India and Manipal Academy of Higher Education, Dubai campuses to carry out extensive research work.

Author details

Vandana Sakhre
Manipal Academy of Higher Education, Dubai, UAE

*Address all correspondence to: vandana.sakhre@manipaldubai.com

References

[1] Amy R. Ciric, Deyao Gu, Synthesis of non-equilibrium reactive distillation processes by MINLP optimization, AIChE Journal, Process system engineering, Vol. 40, 1994, pp. 1479-1487.

[2] Vinay Amte, H. Nistala, S. M. Mahajani, R. K. Malik, Optimization based conceptual design of reactive distillation for selectivity engineering, European Symposium on Computer Aided Process Engineering, Vol. 20, 2010, pp. 1-10.

[3] M. Miniotti, M.F. Doherty, Design for simultaneous reaction and liquid-liquid extraction, Ind. Eng. Chem. Res., Vol. 37, 1998, pp. 4748-4755.

[4] Hoshang Subawalla, James R. Fair, Design Guidelines for Solid-Catalyzed Reactive Distillation Systems Ind. Eng. Chem. Res., Vol. 38, 1999, pp. 3696-3709.

[5] M.A. Schenk, Design of operable reactive distillation columns, PhD Thesis. University of London, 1999.

[6] M. F. Cardoso, R. L. Salcedo, Optimization of reactive distillation processes with simulated annealing, Chemical Engineering Science, Vol. 55, 2000, pp. 5059-5078.

[7] N. Vora, & P. Daoutidis, Dynamics and control of an ethyl acetate reactive distillation column, Industrial and Engineering Chemistry Research, Vol. 40, 2001, pp. 833-845.

[8] R. Taylor and R. Krishna, Reactive Distillation: Status and Future Directions, Modeling of Homogeneous and Heterogeneous Reactive Distillation Processes, Vol. 55, 2000, pp. 5183-5229.

[9] Amit M. Katariya, R. S. Kamath, K.M. Moudgalaya, S. M. Mahajani, Non equilibrium stage modeling and nonlinear dynamic effects in the synthesis of TAME by reactive distillation, Computers and Chemical Engineering, Vol. 32, 2008, pp. 2243-2255.

[10] Kamyar Mehran, Takagi-Sugeno Fuzzy Modeling for Process Control, Industrial Automation, Robotics and Artificial Intelligence (EEE8005), Thesis, 2000.

[11] S. Gruner, K.D. Mohl, A. Kienle, E.D. Gilles, G. Fernholz and M. Friedrich, Non-linear control of a reactive distillation column, Control Engineering Practice, Vol. 11, 2003, pp. 915-925.

[12] J.A. Heath, I. Kookos and J.D. Perkins, Process control structure selection based on Economics, American Institute of Chemical Engineers, Vol. 46, 2000, pp. 1998-2016.

[13] M. Schenk, R. Gani, D. Bogle and E. N. Pistikopoulos, A Hybrid Modelling Approach for Separation Systems Involving Distillation, Chemical Engineering Research and Design, Vol. 77, 1999, pp. 519-534.

[14] M.C. Georgiadis, M. Schenk, E.N. Pistikopoulosand, R. Gani, The interactions of design, control and operability in reactive distillation systems, Computers and Chemical Engineering, Vol. 26, 2002, pp. 735-746.

[15] Monroy-Lopera, E. Perez-Cisneros and J. Alvarez-Ramirez, A robust PI control configuration for a high purity ethylene glycol reactive distillation column, Chemical Engineering Science, Vol. 55, 2000, pp. 4925-4937.

[16] M.G. Sneesby, M.O. Tade, T.N. Smith, A multi-objective control scheme for an ETBE reactive distillation column", Chemical Engineering

Research and Design, Vol. 78, 2000, pp. 283-292.

[17] Al-Arfaj and W.L. Luyben, Comparative control study of ideal and methyl acetate reactive distillation. Chemical Engineering Science, Vol. 57, 2002, pp. 5039-5050.

[18] PVS Ravi Chandra, Ch. Venkateshwarlu, Multistep model predictive control of ethyl acetate reactive distillation column, Indian journal of Chemical Technology, Vol. 14, 2007, pp. 333-340.

[19] Sonal Goyal, *A. rani*, V. Singh, SVR tuned PID controller design for reactive distillation process, International Journal of Applied Engineering Research,Vol.7, 2012, pp. 1-6.

[20] Muhammad Shoaib Nizami, Development of a Fuzzy Logic Controller for a Distillation Column Using Rockwell Software, Thesis, 2007.

[21] Zhao Lei , Chunhai Yi , Bolun Yang, Design, optimization, and control of reactive distillation column for the synthesis of tert-amyl ethyl ether, Chemical Engineering Research and Design, Vol. 91, 2013, pp. 819-830.

[22] A.C. Dimian, C.S. Bildea, F. Omata, A.A. Kiss, Innovative Process for fatty acid esters by dual reactive distillation, Computers and Chemical Engineering, Vol.33, 2009, pp. 743-750.

[23] Kwantlp Konakom, A. Saengcham, P. Kittisupakorn, High purity ethyl acetate production with a batch reactive distillation column using dynamic optimization strategy, Proceedings of the World Congress on Engineering and Computer Science , Vol. 2, 2010 , pp. 978-988.

[24] M.G. Sneesby, O.M. Tade and T.N. Smith, Two-point control of a reactive distillation column for composition and conversion, Journal of Process Control, Vol. 9, 1997, pp. 19-31.

[25] F. Zhao, J. Ou, and W. Du, Pattern based fuzzy predictive control for a chemical process with dead time, Eng. Appl. of Artificial Intelligence, Vol. 13, 2000, pp. 37-45.

[26] J.E. Seem, A new pattern recognition adaptive controller with application to HVAC systems, Automatica, Vol. 34, 1998, pp. 969-982.

[27] M.J. Jang and C.L. Chen, Fuzzy successive modelling and control for time-delay system, Int. J. System Science, Vol. 27, 1996, pp. 1483-1490.

[28] Y.C. Tian, Inference of conversion and purity for ETBE reactive distillation, Brazilian J. of Chem. Eng., Vol. 17, 2000, pp. 617-625.

[29] Vikrant Bansal, V. Sakizlis, R. Ross, J. D. Perkins, New algorithm for mixed integer dynamic optimization, Computers and Chemical Engineering, Vol. 21, 2003, pp. 647-668.

[30] S. C. Patwardhan, S. Narasimhan, J. Prakash, R.B. Gopaluni, S. L. Shah, Nonlinear Bayesian State Estimation, Review and Recent Trends, Control Engineering Practice, Vol. 20, 2012, pp. 933—953.

[31] V.M. Becerra, P.D. Roberts, G.W Grifiths, Applying the extended Kalman filter to systems described by nonlinear differential-algebraic equations, Control Engineering Practice, Vol. 9, 2001, pp. 267-281.

[32] R. K. Mandela, R. Rengaswamy and S. Narasimhan, Recursive state estimation techniques for nonlinear differential algebraic systems, Chemical Engineering Science, Vol. 65, 2010, pp. 4548-4556.

[33] B. Huang, and R. Kadali, Dynamic Modeling, Predictive Control and

Performance Monitoring. Springer publication, 2009, pp.1-10.

[34] Brent M Akesson, H.T. Toivonen, A neural network model predictive controller, Journal of Process Control, 2006, Vol. 16, pp. 937-946.

[35] N. Sharma, Control of Reactive Distillation- A review, International Journal of Chemical Reactor Engineering, 2010, pp. 4-10.

[36] Robert Babusk, Fuzzy Systems, Modeling and Identification, IEEE Transactions on systems, man, and cybernetics, Vol.15, 1985, pp. 116-132.

[37] Fatima Barcelo Rico, Jose M. Gozalvez-Zefrilla, Jose Luis Diez, and Asuncion Santafe-Moros, Modeling and Control of a Continuous Distillation Tower through Fuzzy Techniques, Chemical Engineering Research and Design, 2011, Vol. 89, pp.107-115.

[38] M. Pekkanen, A local optimization method for the design of reactive distillation, Computers & Chemical Engineering, Vol. 19, 1995, pp. 235-240.

[39] D.O. Araromi, J. O. Emuoyibofarhe, J. A. Sonibare , Fuzzy Hybrid Modeling of a Reactive Distillation Column for Ethyl Acetate Process, International Journal of Engineering and Technology, Vol.2, 2012, pp. 888-899.

[40] C. Sumana, Ch. Venkateshwarlu, Optimal selection of sensors for state estimation in a reactive distillation process, Journal of Process Control, 2009, Vol. 19, pp. 1024-1035.

[41] Xudong Wang, Rongfu Luo, Huihe Shao, Designing a Soft Sensor for a Distillation Column with Fuzzy Distributed Radial Basis Function Neural Network, Decision and Control, Vol. 2, 1996, pp. 1714-1719.

[42] Jaroslava Zilkova, Nonlinear System Control Using Neural Networks, Acta Polytechnica, Vol. 3, 2006, pp. 85-90.

[43] S.R. Vijaya Raghavan, T.R. Radhakrishnan, K.Srinivasan, Soft Sensor Based Composition Estimation and Controller Design for an Ideal Reactive Distillation Column, ISA transactions, 2011, Vol. 50, pp. 61-70.

[44] K.J. Jithin Prakash, Neuro estimator based GMC Control of Batch Reactive Distillation, ISA Transaction, Vol. 50, 2011, pp. 537-539.

[45] Almila Bahar, and Canan Ozgen, State Estimation and Inferential Control for a Reactive Batch Distillation Column, Engineering Applications of Artificial Intelligence, Vol. 23, 2010, pp. 260-270.

[46] S. Nithya, N.S ivakumaran, T. Balasubramanian, Controllers implementation based on soft-computing for non-linear process, Proceedings of the World Congress on Engineering and Computer Science, Vol. 2, 2010, pp. 978-988.

[47] Elrashid Idris, S. Engell, Real Time Optimization Nonlinear Control Applied to a Continuous Reactive Distillation Process, International Federation of Automatic Control, Vol. 18, 2011, pp. 4892-4897.

[48] V. Sujatha, R. C. Panda, Control configuration selection for multi input multi output processes, Journal of Process Control, Vol.23, 2013, pp. 1567-1574.

[49] Bidyadhar Subudhi, Nonlinear system identification using memetic differential evolution trained neural networks, Neurocomputing, Vol. 74, 2011, pp. 1696-1709.

[50] Hung-Ching Lu, Ming-Hung Chang, Cheng-Hung Tsai, Parameter estimation of fuzzy neural network controller based on a modified differential evolution Neurocomputing, Vol. 89, 2012, pp.178-192.

[51] Maciej Lawrynczuk, Explicit nonlinear predictive control algorithms with neural approximation, Neurocomputing, 2014, Vol. 129, pp. 570-584.

[52] K. K. C. W. Kandanapitiya, Modeling of Reactive Distillation for Acetic Acid Esterification, Journal of the Institution of Engineers, Vol. 48, 2015, pp. 17-23.

[53] Matthias Wierschem, Continuous Enzymatic Reactive Distillation with Immobilized Enzyme Beads, AICHe, Annual meeting, 2015.

[54] Tianping Zhang, Xiaonan Xia, Decentralized adaptive fuzzy output feedback control of stochastic nonlinear large scale systems with dynamic un certainties, Information Sciences, Vol. 315, 2015, pp: 17-18.

[55] Zhouhua Peng, Dan Wang, Hongwei Zhang, Yejin Lin, Cooperative output feedback adaptive control of uncertain nonlinear multi-agent systems with a dynamic leader" Neurocomputing, Vol. 149, 2015, pp. 132-141.

[56] Guozeng Cui, Zhen Wang, Guangming Zhuang, Yuming, Chu, Adaptive Centralized NN control of large scale stochastic nonlinear time delay systems with unknown dead zone inputs, Neurocomputing, Vol. 158, 2015, pp. 194-203.

[57] Rautenbach, Separation potential of pervaporation, Journal of Membrane Science, Vol. 25, 1997, pp. 25-31.

[58] F. Lipnitzki, R. W. Field and P. K. Ten, Pervaporation-based hybrid process: a review of process design, applications and economics, Journal of Membrane Science, 1999, Vol. 153, pp. 183-210.

[59] P. Kries, A. Gorak, Process analysis of hybrid separation processes combination of distillation and Pervaporation, Chem. Eng. Res. Des. 2006, Vol. 84, pp. 595-600.

[60] Sven Steinigeweg, Jurgen Gmehling, Transesterification processes by combination of reactive distillation and pervaporation, Chemical Engineering and Processing, 2004, Vol. 43, pp. 447-456.

[61] Von Scala, J. Fassler, E. Gerla, 2005, Maus, Kontinuierliche Herstellung von kosmetischen Fetts aureestern mittels Reaktivrektifikation and Pervaporation, Chem. Ing. Tech., Vol. 77, pp. 1809-1813.

[62] S.S. Ozdemir, Catalytic Polymeric Membrane: Preparation & Application, Appl. Catal A. Gen, Vol. 307, 2006, pp. 167-183.

[63] G.S. Luo, M. Niang, P. Schaetzel, Separation of ethyl tert-butyl ether-ethanol by combined pervaporation and distillation, Chemical Engineering Journal, Vol. 68, 1997, pp. 139-143.

[64] A. Gorak, Reactive and membrane assisted distillation: recent developments and perspective, Chemical Engineering Research, and design, 2013, Vol. 91, pp. 1978-1991.

[65] J. Holtbruegge, M. Wierschem, Hybrid configuration of reactive distillation and vapor permeation for the production of dimethyl carbonate and propylene glycol, A Thesis, 2013.

[66] L.V. Bida, Gongping Liu, Xueliang Dong, Wang Wei, and Wanqin Jin, Novel Reactive Distillation–Pervaporation Coupled Process for Ethyl Acetate Production with Water Removal from Reboiler and Acetic Acid Recycle, *Ind. Eng. Chem. Res.*, 2012, Vol. 51, pp. 8079-8086.

[67] H. Masjuki, Biofuels as diesel fuel alternative: An overview, Journal of

energy heat mass transfer, 1993, Vol. 15, pp. 293-304.

[68] Kloptenstem We, Effect of molecular weight of fatty acid esters on cetane numbers as diesel fuels, J Amer oil chemical society, 1988, Vol 65, pp. 1029-1031.

[69] Engler Cr, Le Pori We, Animal fats as alternative diesel fuels from renewable resources, Proc Altern Energy conference Amer Society, 1992, pp. 89-96.

[70] Anton A. Kiss, J. J. Pragt, C. J. G. van Strien, Reactive dividing-wall columns: towards enhanced process integration, Distillation Absorption, 2010, pp. 253-258.

[71] Anton A. Kiss, 2012, Enhanced bioethanol dehydration by extractive and azeotropic distillation in dividing wall columns, Separation and Purification technology, Vol. 86, pp. 146-153.

[72] Raul Delgado, S. Hernandez, F. Omar, B. Munoz, J. Gabriel, S. Hernandez, A.J.C. Montoya, From simulation studies to experimental tests in a reactive dividing wall distillation column, Chemical Engineering Research and Design, 2012, Vol. 90, pp. 855-862.

[73] J.L. Bravo, A. Pyhalathi and H. Jaervelin, Investigation in a catalytic distillation pilot plant: vapor/ liquid equilibrium, kinetics and mass transfer issues, Ind. Eng. Chem. Res., 1993, vol. 32, pp. 2220-2225.

[74] S. Chongkhong, Biodiesel production by esterification of fatty acid distillate, Biomass and Bioenergy, 2007, Vol. 31, pp. 563-568.

[75] Fernando Israel Gomez-castro, V. R. Ramirez, J.G.S. Hernandez, S. Hernandez, Feasibility study of a thermally coupled reactive distillation process for biodiesel production, Chemical Engineering and processing, Vol. 49, 2010, pp. 262-269.

[76] I. M. Atadashi, M.K. Aroua, A.R. Abdul, N.M.N Sulaiman, Membrane biodiesel production and refining technology: A critical review, Renewable and Sustainable Reviews, 2011, pp. 1-12.

[77] Carsten Buchaly, Peter Kreis, Andrej Gorak, Experimental Investigation of reactive Distillation in Combination with Membrane Separation, Chair of Fluid Separation processes, Vol. 152, 2006, pp. 373-383.

[78] Carsten Buchaly, Peter Kreis, Andrej Gorak, Hybrid separation Processes- Combination of Reactive Distillation with Membrane Separation, European Congress of Chemical Engineering, Vol. 6, 2007, pp. 1-17.

[79] San Jang Wang, David S.H. Wong, Shuh-Woei Yu, Design and control of transesterification reactive distillation with thermal coupling, Computers and Chemical Engineering, Vol.32, 2008, pp. 3030-3037.

[80] Asprion N, Kaibel G. Dividing wall columns: Fundamentals and recent advances, Chem Eng Process, 2010, Vol. 49, pp. 139-146.

[81] H.Y. He, X. Guo, Comparison of membrane extraction with traditional extraction method for biodiesel production, Journal of American Oil Chemist Society, Vol. 83 2006, pp. 457-460.

[82] Saleh, Effect of soap, methanol and water on glycerol particle size in biodiesel purification, Energy Fuels, Vol. 24, 2010, pp. 6179-6186.

[83] L. A. Sarmento, C.B. Spiricigo, Performance of reverse osmosis membrane in separation of CO_2 and

essential oil, Journal of membrane Science, 2004, Vol. 237, pp. 71-76.

[84] Raul Delgado-Delgado, Some operational aspects and applications of dividing wall columns: energy requirements and carbon dioxide emissions, Clean Technologies and Environmental Policy, 2015, Vol. 17, pp. 657-665.

Chapter 6

Treatment of Wastewater by Nanofiltration

M. Amine Didi

Abstract

In recent years, some countries have implemented regulations governing aqueous discharges. With a view to sustainable development, manufacturers are looking for wastewater treatment technologies to control their discharges. Nanofiltration seems particularly suitable for the separation characteristics that it allows with regard to the size of the target molecules. Pollution by rare earths and heavy metals affects groundwater and surface water. This changed the quality of the water and made it unsafe to use. Water pollution is a big problem, given the diversity of sources and characteristics of polluting species, the main ones being industrial, urban and agricultural discharges, generated by human activity. The great difficulty being that heavy metals are not biodegradable and tend to accumulate in living organisms (fish, mollusks, vegetables, etc.) consumed by humans. For these concerns, environmental laws have become more severe. For this, the treatment of aqueous effluents has become important. It can be concluded that separation and purification chemistry is an area of topical research. The discharges coming from the industry contain heavy metals (chromium, copper, zinc, nickel, iron, cobalt, cadmium, lead, ...) which are harmful for the human health, the fauna and flora. It is necessary to be well controlled. This chapter presents a study of nanofiltration for industrial wastewater treatment.

Keywords: nanofiltration, industrial wastewater treatment, pollution, environment

1. Introduction

Increased demands by limited water resources have triggered in worldwide for innovative water practices. Many processes are used to purify water contaminated with rare earths and heavy metals, such as solvent extraction, precipitation, ion exchange, absorption and liquid–liquid extraction. The underground water resources can be polluted by wastewaters [1].

There is a lot of proceed of separation which can be applied to the treatment of discharges, among these techniques we mention: chemical precipitation, coagulation–flocculation, flotation, ion exchange, and the membrane processes as supported membrane liquid [2–4], nanofiltration (NF) [5], ultrafiltration (UF) [6, 7], reverse osmosis (RO) [8] and microfiltration recently show by [9].

Nanofiltration has a lot of applications in industry [10]. Among membrane technologies, nanofiltration is the best opportunity to solve environmental problems, such as: desalination recently shown by [11], wastewater and ground water treatment [12, 13], and heavy metals elimination [14].

These methods are valid but have many drawbacks which require the use of organic solvents which are harmful to the environment and time consuming.

With technological development in industries and the differentiation in quantity and type of waste, the development of new, more efficient techniques has become necessary. Among these techniques, membrane processes.

Currently, membrane extraction is the most widely used method for treating industrial waste and purifying wastewater. It is a very active field in separation sciences and many companies are currently producing and developing new membrane extraction techniques.

Nanofiltration is ecological technique; this advantage is particularly linked to this operation without the need for organic solvents and also because the extraction requires little time to perform.

Currently, membrane techniques play a very important role in water purification and open up new possibilities for beneficial use for water sources. They were difficult to use before for technical and economic reasons.

Membrane methods are commonly used in the purification of water and the purification of wastewater [15, 16].

Using these techniques, every five years the creation of water purification plants is multiplied by ten [17].

The "naturalists" are the first who approached the selective transport of substances through membranes [18].

In order to explain the transfer of the solvent, Abbé Nollet supposes the presence of forces between the parts of the membrane, through the membranes the passage of substances is influenced by their molecular mass [19].

The use of membrane processes in industry began in the 1960s. There is a wide variety of membranes due to the existence of several fields of application for membrane processes.

The membranes are used in the separation and concentration of ionic species or molecules in the solution and/or to separate microorganisms (bacteria, viruses, etc.) or suspended matter.

Perm selective membranes are used in membrane processes [19, 20].

The selective displacement of certain components through a membrane, under the application of a force, is the concept of membrane separation [20–22].

All membrane processes use tangential filtration to limit the accumulation of material [20, 23].

The solution of the particles which do not pass through the membrane is called "retentat", and "permeat", the solution of the components which pass through the membrane [24].

Various advantages, among which:

- Reliability and quality of the final product [20].
- Low energy consumption [20].
- The ability to treat water containing several metals.
- Good selectivity [25].
- It works without adding chemicals [26], and the ease and use of industrial integration [20].
- It works without the need for secondary chemicals [27–29].

For these reasons, the water treatment sectors use these technologies [30].

2. Membrane processes by nanofiltration

The nanofiltration technique is located between ultrafiltration (UF) and reverse osmosis (OI), it is used to separate components of size close to the nanometer order.

This type of membrane does not retain non-ionized organic compounds with a molar mass of less than 200 g/mol and monovalent salts. On the other hand, non-ionized organic compounds with a molar mass greater than 250 g/mol and multivalent ionized salts (calcium, magnesium, aluminum, sulfates, etc.) are retained [31].

Nanofiltration has lower electrical energy consumption, because it requires lower pressures than reverse osmosis [32].

The nanofiltration technique (**Figure 1**) is based on two separation mechanisms, separation under the effect of electrical repulsion for charged species, and separation under the effect of size for uncharged solutes [33].

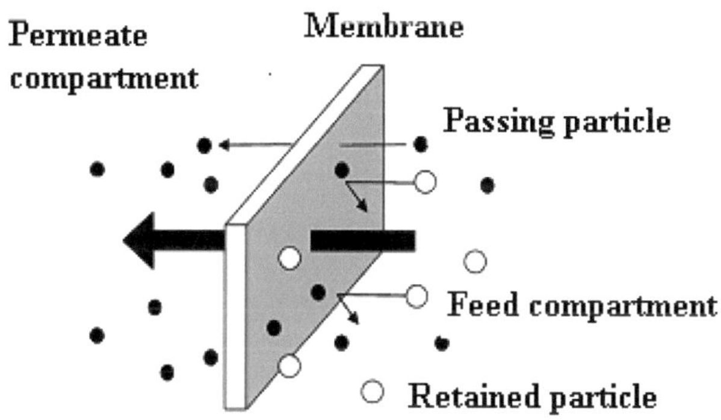

Figure 1.
Selective membrane [15].

3. Classification of membranes

The comparison between membranes of the same type requires the evaluation of the performance of these membranes under real conditions.

Using a few criteria such as: selectivity, permeability and lifetime, one can consider that one membrane is more adequate than another to be separated [27]. The information provided by the membrane manufacturers is insufficient to compare the membranes. To understand these membranes, the acquisition of this information is essential in order to master the transfer mechanisms and improve their performance.

Two categories of parameters are often sought:

- performance-related parameters:
    - Permeability.
    - Retention.

The concentration factor.

- Parameters related to morphology:
    - The membrane thickness.

- o Pore sizes.
- o Distribution of pores.
- o The charge density.
- o Hydrophobicity.
- o The adsorption and absorption capacities.

4. Principle of membrane separation

4.1 Tangential filtration

Under the action of a pump, the fluid circulates parallel to the membrane from a reservoir (**Figure 2**) [29].

4.2 Front filtration

The flow of the feed solution is perpendicular to the membrane (**Figure 3**) [17, 28].
Classification according to the separation mechanism
The processes responsible for membrane filtration are:

- Sieving.
- The friction on the walls of the pores.
- The diffusion in the membrane material and in the pores of the membranes.
- Repulsive or attractive surface forces.
- Electrostatic repulsion.

According to their separation mechanisms, the membranes are classified:

- Non-porous membranes: these membranes can be considered as dense media, the species are diffused in the volumes located between the molecular chains of the material.
- Porous membranes: the effects of friction, sieving and surface forces are dominant in these membranes.

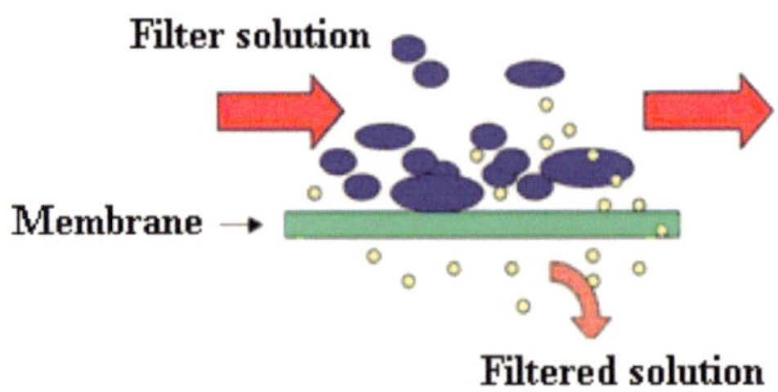

Figure 2.
Tangential filtration [34].

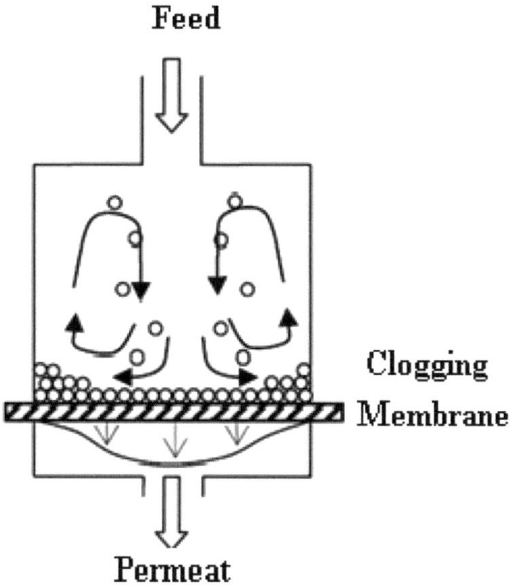

Figure 3.
Front filtration [35].

- Ion-exchange membranes: consist of dense gels, it is a special type of non-porous membrane, they can be anion-exchange (with a positive charge) or cation-exchange (with a negative charge).

Classification according to the geometry of the membranes

According to the geometry under which the membranes are manufactured, they are classified as:

- Flates modules (**Figure 4**):
- Spiral membranes (**Figure 5**):
- Tubular modules (**Figure 6**): are membranes with an internal diameter greater than 3 mm.
- Hollow fiber modules (**Figure 7**):

Classification according to chemical nature

The membranes are synthesized from inorganic materials and organic polymers. There are also mixed membranes made from inorganic materials and polymers [38].

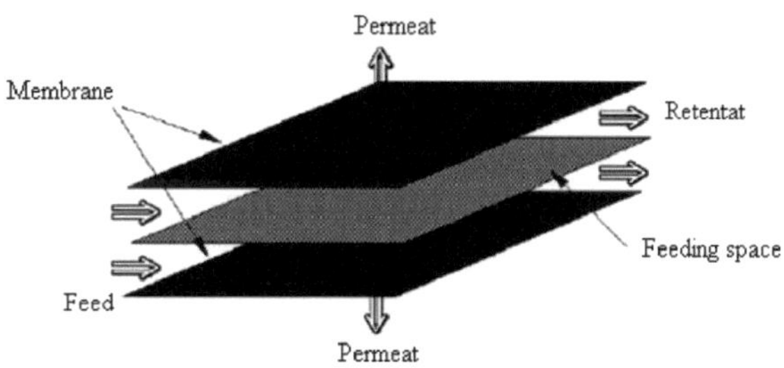

Figure 4.
Flat module [36].

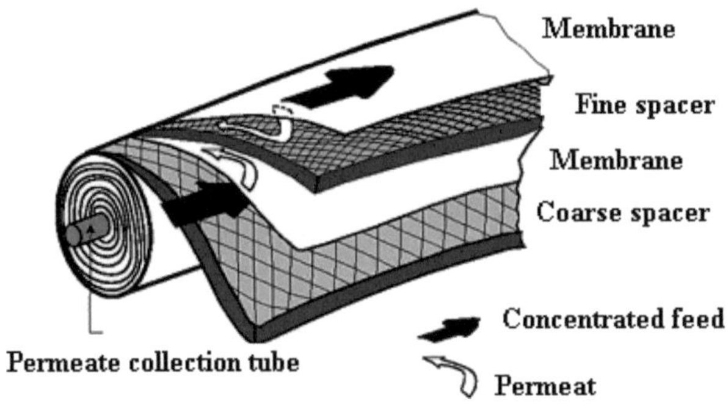

Figure 5.
Spiral module [37].

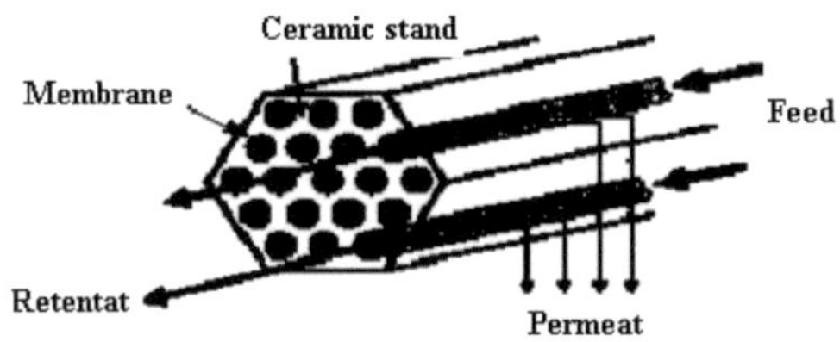

Figure 6.
Tubular module [36].

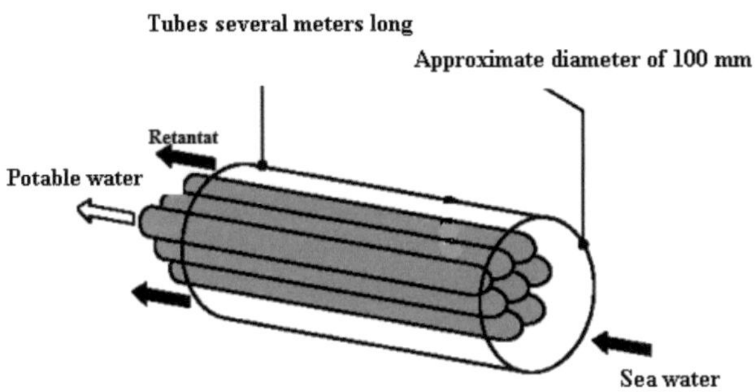

Figure 7.
Hollow fiber module [36].

The majority of usable membranes are manufactured either from inorganic materials or from polymers.

The main polymers used for the manufacture of membranes are:

- Cellulose derivatives: are inexpensive, considered more hydrophilic, this type of polymer has a low tendency to adsorb.
- Polyamides: these polymers are very sensitive to chlorine, polyamides having chemical and thermal properties superior to those of cellulose derivatives.
- Polysulfone and polyethersulfone, these polymers have good mechanical, and thermal.

Inorganic membranes are made of ceramic (zirconium, titanium, aluminum oxides). These membranes are expensive, have a mechanical, chemical and thermal stability superior to polymeric membranes, they are brittle [38].

Classification according to morphology

According to their structure, the membranes can be classified:

- Membranes with symmetrical structure (isotropic membranes): have the same structure over their entire thickness.
- Membranes with asymmetric structure (anisotropic membranes): The structure of the membrane is different from one layer to another.

There are two subtypes of asymmetric membranes:

- Membranes made from the same material.
- Composite membranes: are mainly composed of two layers:

Skin: a layer with a very low thickness, this layer is responsible for the selectivity of the membrane.

Support layer: it is a layer with a much greater permeability than that of the skin layer, which is thicker and which retains the mechanical resistance.

In order to increase the permeability of the membrane, the majority of commercial membranes are manufactured with an asymmetrical structure [39].

4.3 Size characteristics of membranes

Physical characteristics

Permeate retention and flow

The performance of filtration is characterized by the observed retention and permeates flow. Retention is expressed as follows [40]:

$$Y(\%) = \left(1 - \frac{C_p}{C_0}\right).100 \quad (1)$$

C_p: Concentration in permeat and C_0: Concentration in feed.

This parameter characterizes the effectiveness of the treatment. It quantifies the retention of a target compound.

- **Permeability of a membrane (Lp)**
  - It represents the volume or mass flow crossing the membrane per unit of membrane area.
  - Permeability is a function of temperature and pressure.
  - According to Darcy's law, the flow of solvent (Jv) is proportional to the transmembrane pressure.
  - The hydraulic permeability (Lp) is given by the following equation, this equation is valid for all membranes.

The permeate flow (Jv) characterizes the productivity and is expressed as follows:

$$J_v = L_p\left(\Delta P - \sigma\Delta\pi\right) \quad (2)$$

The permeability of the membrane is deduced from the slope to the right of the permeate flow Jv as a function of the transmembrane pressure ΔP.

- **Hydraulic resistance (Rm)**

This is the resistance of the membrane to the flow of the fluid to be filtered; the resistance is related to the permeability by Eq. (3).

$$R_m = \frac{1}{\mu L_p} \tag{3}$$

μ is the viscosity of the permeate (water).

The resistance can be calculated from the permeate flow through the membrane and the transmembrane pressure.

- **Cut-off threshold**

The SC of a membrane is the molecular weight above which the membrane retains at least 90% of the molecules. It is expressed in g/mol or in Dalton [40].

The cutoff is used in the characterization of membranes, but from a scientific point of view it is not rigorous, as it depends on the operating conditions and the characteristics of the solute.

- **Lifespan**

This is a characteristic of the membrane, because beyond which the membrane will no longer be effective [41].

- **Conversion rate**

The conversion rate is the flow rate fraction of the liquid passing through the membrane [41]:

μ is the viscosity of the permeate.

The resistance can be calculated from the permeate flow through the membrane and the transmembrane pressure.

$$Y = \frac{Q_p}{Q_f} \tag{4}$$

Chemical characteristics

Depending on the chemical nature of the membrane, there are interactions between the membrane and the solutes to be filtered, in particular at the level of clogging [20].

- Hydrophobicity and hydrophilicity:

Hydrophilicity depends on the ionized or polar groups of the polymers used, by nature organic membranes are hydrophobic [20].

Due to their hydrophilic properties, regenerated cellulosic membranes are widely used in ultrafiltration [42, 43].

- Surface electrical charge:

Organic membranes have two functional groups one is amine (basic) and the other is carboxylic (acid) having positive or negative charges respectively.

Due to the partial hydrolysis of the amide functions, NF membranes are negatively charged, inorganic membranes have amphoteric surfaces.

4.4 Phenomenon limiting the transfer of matter

Clogging

The clogging phenomenon (**Figure 8**) occurs when dissolved or suspended matter is deposited on the outer surface or inside the pores [43]. Due to clogging, the performance of the membrane decreases [45].

Upon contact between the membrane and the particles of the solution, a reversible or irreversible modification of the membrane was caused [40]. After this modification, the membrane needs to be replaced or cleaned [45].

The clogging mechanisms are linked to:

- The adsorption of particles through the pores (partial blockage).
- Complete blockage of the membrane pores.
- Internal blockage of the membrane pores.
- Formation of a deposit of particles on the membrane surface (forming a cake) [40, 45].

In view of the nature of the sealant, we can distinguish the type of plugging:

- Clogging by precipitation of soluble salts (scale)
- Bio-clogging (biofilm formation)
- Clogging by colloidal substances [43].

The clogging reduces the flow of water at constant pressure, so a cleaning of the clogged membrane must be applied in order to recover its initial characteristics.

Clogging is a function of:

- Type of membrane,
- The nature and concentration of solutes and solvents,

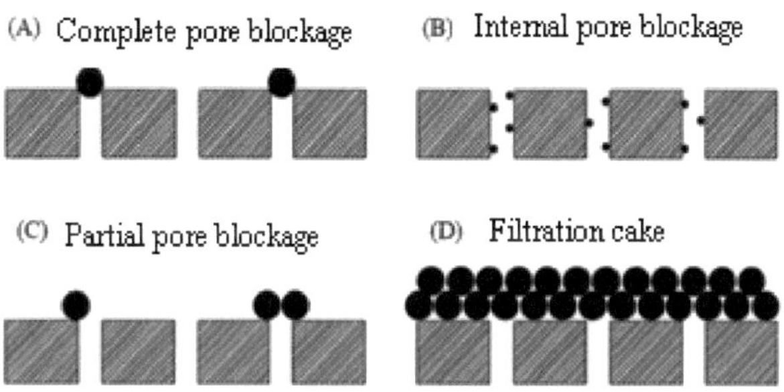

Figure 8.
Clogging mechanisms [44].

- The characteristics of the membrane surface,
- Membrane materials. The hydrodynamics of the module.
- Distribution of pore size.

Polarization of concentration

The accumulation of solutes retained on the surface of the membrane leads to the phenomenon of concentration polarization [43], which causes an increase in the concentration of solute at this surface [41].

When the driving force is interrupted and the permeate flow quenched, the concentration polarization is invisible [45].

Assuming that the true concentration is unidirectional along an axis perpendicular to the membrane, the total permeation flux is the sum of the retro-diffusive flux and the convective flux [44].

Cleaning

To remove the clogging elements, cleaning with chemicals (acids and bases) and/or physical (or mechanical) cleaning and/or by using a specific cleaning solution containing appropriate detergents [46].

Physical (mechanical) cleaning

Physical cleaning is used to remove and loosen the material accumulated on the membrane.

Backwashing is the most common procedure: part of the permeate passes through the membrane in the opposite direction of flow.

This requires a membrane that must physically support the inverted pressure gradient, other practices use pulsed flows.

Air/gas scrubbing: used to inject air or gas into the membrane [47].

If the membrane is not fully restored, chemical cleaning is necessary.

Chemical cleaning

Chemical cleaning contains two parts of acid and basic cleaning as well as rinsing.

Acid washing is used to dissolve the scale layers of metal oxides and thus prevent the formation of insoluble hydroxides that are difficult to remove [47].

The purpose of alkaline washing is to hydrolyze silica, inorganic colloids and organic and biological matter.

Surfactants are also used as cleaning solutions, they are used to:

1. Move the clogging elements in the surface.
2. Emulsified oils.
3. The solubilization of hydrophobic elements [47].

For example for polyether sulfonated membranes Tween 20 is used for cleaning [7].

5. Material

Pilot equipment

Nanofiltration experiments were carried out with the separation unit illustrated in **Figure 9**. On an industrial scale, this unit is multiplied according to needs.

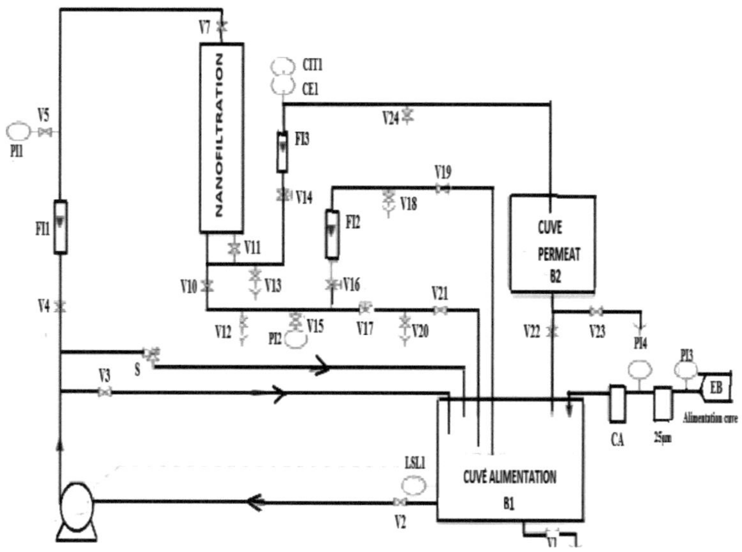

Figure 9.
Experimental setup [48].

CA is the cartridge filter with activated carbon and 25 μm of wound cartridge filter. S is the safety valve (14 bars). B1 is the feed tank (100 L). B2 is the permeate tank (20 L). C2 is the nanofiltration membrane. FI1 is the upstream flow meter (100–1000 l/h). FI2 is the downstream flow meter of retentive. FI3 is the downstream flow meter of permeate. PI1 & PI2 are the manometers at upstream and downstream of module (0–16 bars). PI3 & PI4 are the monitoring manometers of filters state (0–2.5 bars). LSL1 is the low level sensor (pump safety). CE1 is the sensor of permeate conductivity measuring. Y is the emptying, CIT1 to the electrical display cabinet. V1–5, 7, 10, 11, 14–16, 19 & 22 are the pressure regulation valves for nanofiltration process. P is the multistage centrifugal pump (high pressure).

5.1 Materials

Analyzes of water are carried out on turbidimeter, pH meter, conductimeter, atomic absorption spectrophotometer, UV–visible, apparatus for measuring DBO5 and DCO, and ICP-MS.

6. Applications

- With membrane techniques, gas separation is more economical.
- Membrane processes are better for the environment because they work without chemicals.
- Are easy and effective compared to other techniques.
- A low investment cost for their systems.
- The membranes do not require frequent regeneration.
- The determination of various compounds [49].

During nanofiltration, the electromigration, diffusion and convection are the major transport mechanisms. The atoms ions can pass through the channels between the molecular groups in the molecular structure of the membrane. The selectivity of nanofiltration membrane was defined by nature and size of these passages, ionic size, shape, pore density, pore diameter, and membrane surface charge [50, 51].

The suitability of a thin film polyamide nanofiltration composite membrane (SNTE NF270–2540) to extract heavy metal ions (Zn (II), Cr (III), Cu (II), Fe (III), Cd (II), Pb (II), Co (II) & Ni (II)) from industrial wastewater was examined [52]. The operating conditions including feed pH value, feed metal concentration and pressure were optimized. The retention of iron at pH = 5.3 and nickel at pH = 7.2 for pressure from 6.0 to 13.5 bars was quantitative (100%). For lead at pH = 4.4 for pressure from 6.0 to 10 bars, the retention was quantitative (100%).

The influence of divalent cations on the rejection of trivalent cations was examined [52].

The membrane was characterized by using the ions transport model and the consumption energy was calculated. The effect of ionic concentration and the diffusion fluxes were studied [52].

7. Conclusion

The water purification by nanofiltration membrane techniques is presented, as well as the different classes of membranes. The influence of physico-chemical parameters is discussed.

The effect of pH on the adsorption capacity depends on the surface charge or zeta potential of the sorbent at different pH and presence of different electrolyte.

Purification by membrane filtration is highly dependent on the species present in water and properties of the membrane.

Membrane nanofiltration in combination with others techniques showed a high rejection.

NF membrane can be used in any types of water and it can treat organic or inorganic effluents.

For a given module, we observe that there is interest in working at a high conversion rate to limit heating of the solution (energy consumption). However, the solution to be treated is concentrated very quickly. Scaling of the cartridge will occur very quickly. We must find a compromise for the three parameters C_0, Y and Q_p.

Acknowledgements

We gratefully acknowledge the DGRSDT - Algeria for the financial support.

Author details

M. Amine Didi
Laboratory of Separation and Purification Technologies, Chemistry Department, Faculty of Sciences, Tlemcen University, Algeria

*Address all correspondence to: madidi13@yahoo.fr; ma_didi@mail.univ-tlemcen.dz

References

[1] Maher A, Sadeghi M, Moheb A. Heavy metal elimination from drinking water using nanofiltration membrane technology and process optimization using response surface methodology. Desalination. 2014;**352**:166-173

[2] Didi MA, Sekkal AR, Villemin D. Procédés de dépollution par extractions sur membranes liquides supportées et liquide-liquide du Chrome(III) en milieu chlorure avec l'acide di-2-éthylhexyl phosphorique et l'oxyde de trioctylphosphine - Basé sur leur mélange synergique. Scientific Study & research, VII. 2006;**2**:349-361

[3] Medjahed B, Didi MA. Removal of Copper Ions Using Aliquat 336/TBP Based Supported Liquid Membrane. XIV: Scientific Study & Research; 2013. pp. 163-172

[4] Medjahed B, Didi MA, Villemin D. Factorial design in optimization of extraction procedure for copper using Aliquat 336/TBP based supported liquid membrane. Desalination and Water Treatment. 2014;**52**:3237-3245

[5] Tansel B, Sager J, Rector T, Garland J, Strayer RF, Levine L, et al. Significance of hydrated radius and hydration shells on ionic permeability during nanofiltration in dead end and cross flow modes. Separation and Purification Technology. 2006;**51**:40-47

[6] Tham HM, Wang KY, Hua D, Japip S, Chung TS. *From ultrafiltration to nanofiltration: Hydrazine cross-linked polyacrylonitrile hollow fiber membranes for organic solvent nanofiltration*. Journal of Membrane Science. 2017;**542**:289-299

[7] Amiri Largani M, Saljoughi E, Mohammadi T. Improvement of permeation performance of Polyethersulfone (PES) ultrafiltration membranes via addition of Tween-20. Journal of Applied Polymer Science. 2010;**115**:504-513. DOI: 10.1002/app.30814

[8] Qi S, Fang W, Siti W, Widjajanti W, Wang R. *Polymersomes-based high-performance reverses osmosis membrane for desalination*. Journal of Membrane Science. 2018;**555**:177-184

[9] He C, Wang X, Liu W, Barbot E, Vidic RD. *Microfiltration* in recycling of Marcellus Shale flow back water: Solids removal and potential fouling of polymeric *microfiltration* membranes. Journal of Membrane Science. 2014;**462**:88-95

[10] Pruksasri S, Lanner B, Novalin S. Nanofiltration as a potential process for the reduction of sugar in apple juices on an industrial scale. LWT. 2020 https://doi.org/10.1016/j.lwt.2020.110118

[11] Kaya C, Sert G, Kabay N, Arda M, Yüksel M, Egemen O. Pre-treatment with nanofiltration (NF) in seawater desalination - preliminary integrated membrane tests in Urla, Turkey. Desalination. 2015;**369**:10-17

[12] Van Der Bruggen B, Vandecasteele C. Removal of pollutants from surface water and groundwater by nanofiltration: Overview of possible applications in the drinking water industry. Environmental Pollution. 2003;**122**:435-445

[13] Hussain AA, Al-Rawajfeh AE. Recent patents of nanofiltration applications in oil processing, desalination, wastewater and food industries. Recent Patents Chemical Engineering. 2009;**2**:51-66

[14] Murthy ZVP, Chaudhari LB. Application of nanofiltration for the rejection of nickel ions from aqueous solutions and estimation of membrane transport parameters. Journal of Hazardous Materials. 2008;**160**:70-77

[15] Mallevialle J, Odendaal PE, Wiesner MR. The emergence of membranes in water and waste water treatment, In: Water treatment membranes process, Chapitre 1, McGraw-hill. In: p10. 1996

[16] Mika KA, Kallioinen MM. Two-stage nanofiltration for purification of membrane bioreactor treated municipal wastewater - minimization of concentrate volume and simultaneous recovery of phosphorus. Separation and Purification Technology. 2020. DOI: https://doi.org/10.1016/j.seppur.2020.117255

[17] P. Bacchin. Principes de base des Technologies à Membranes. 2eme Ecole d'Eté Franco-Maghrébine" Sciences et Technologies a Membranes ", Sep 2005, p9.

[18] Audinos R. Les membranes artificielles. Paris: Presses universitaires de France; 1983

[19] Aptel P, Moulin P, Quemeneur F. Microfiltration et ultrafiltration: conduite des Essais pilotes. Lavoisier; 2002

[20] Bimbenet JJ, Albert D, Gilles T. Génie des procédés alimentaires. Ed: Des bases aux applications; 2002

[21] Schäfer AI, Fane AG, Waite TD, editors. Nanofiltration: Principes and Applications. Oxford, UK: Elsevier; 2005

[22] Bouroche A, Le bras M. Techniques de séparation par membranes. Vocabulaires Français – Anglais – Allemand: Ed; 1994

[23] P. Marty. traitement des effluents par filtration membranaires; industries alimentaires &Agricoles. In: Ed. 1999

[24] R. D. Noble, S.A. Stern. Membrane séparations technologie: principales and applications, Elsevier science B.V, 1995.

[25] Aimar P, Daufin G, Rene F. Les séparations à membranes dans les procédés de l'industrie alimentaire, technique et documentation. Lavoisier; 1998

[26] Juang LC, Tseng DH, Lin HY. Membrane processes for water reuse from the effluent of industrial park waste water treatment plant: A study on flux and fouling of membrane. Desalination. 2007;**202**:302-309

[27] Bouranene S, Fievet P, Szymczyk A, El-Hadi Samar M, Vidonne A. Influence of operating conditions on the rejection of cobalt and lead ions in aqueous solutions by a nanofiltration polyamide membrane. Journal of Membrane Science. 2008;**325**:150-157

[28] Al-Rashdi B, Somerfield C, Hilal N. Heavy metals removal using adsorption and nanofiltration techniques. Separation and Purification Reviews. 2011;**40**:209-259

[29] Lin SW, Sicairos SP, Navarro RMF. Preparation, characterization and salt rejection of negatively charged polyamide nanofiltration membranes. Journal of the Mexican Chemical Society. 2007;**51**:129-135

[30] Snyder AS, Adham S, Redding AM, Cannon FS, Decarolis J, Oppenheiner J, et al. Role of membranes and activated carbon in the removal of endocrine disruptors and pharmaceuticals. Desalination. 2007;**202**:156-181

[31] Berland JM, Juery C. Les procédés membranaires pour le traitement de l'eau. In: Ed. 2002

[32] Mehiguene K, Garba Y, Taha S, Gondrexon N, Dorange G. Influence of operating conditions on the retention of copper and cadmium in aqueous solutions by nanofiltration: Experimental results and modelling. Separation and Purification Technology. 1999;**15**:181-187

[33] Brun JP. Procédés de séparation par membrane; transport, techniques

membranaires, application. Masson Paris Milan Barcelone Mexico. 1996;**1989**:88-136

[34] http://www.viticulture-oenologie-formation.fr/vitioenoformlycee/info/info-tk-tc1-10-11/avril-2011/filtre-tangentiel/principe-filtre-tangentiel.html

[35] G. Belford, R. H. Davis, A. L. Zydney. The behavior of suspensions and Macromolecular solutions in cross flow microfiltration, Ed. 1994.

[36] Metaiche. M. Technology membranaire, Ed. 2014.

[37] Bouchard. C, Kouadio. P, Ellis. D, Rahni. M, Lebrun. R. Les procédés à membranes et leurs applications en production d'eau potable, Vecteur Environnement, Ed. 2000.

[38] Aptel P, Buckley CA. Categories of membrane operations, In: Water treatment membrane process, Chapitre 2. McGraw-Hill.

[39] Wentao Shang, Feiyun Sun, Wei Jia, Jiaxin Guo, Shengming Yin, Pak Wai Wong, Alicia Kyoungjin An (2020). High-performance nanofiltration membrane structured with enhanced stripe nano-morphology. Journal of Membrane Science. 600. https://doi.org/10.1016/j.memsci.2020.117852

[40] Charcosset C. Principles on Membrane and Membrane Processes, Membrane Processes in Biotechnologies and Pharmaceutics, p335, ed. Amsterdam: Elsevier; 2012. pp. 1-41

[41] Jaffrin M. Procédés de filtration membranaire, 1er édition. Book. 2014:p75

[42] Jokinen JN, Nystrom M. Comparison of membrane separation processes in the internal purification of paper mill water. Journal of Membrane Science. 1996;**119**:99-115

[43] Babu PR, Gaikar VG. Membrane characteristics as determinant in fouling of ultrafiltration membranes. Separation and Purification Technology. 2001;**24**:23-34

[44] Cui ZF, Jiang Y, Field RW. Fundamentals of pressure-driven membrane separation processes. Membrane technology: A practical guide to membrane technology and applications in food and bioprocessing, Butterworth–Heinemann, Elsevier. UK. 2010;**12**:p18

[45] Cui ZF, Jiang Y, Field RW. Fundamentals of pressure-driven membrane separation processes. Membrane Technology: A Practical Guide to Membrane Technology and Applications in Food and Bioprocessing, Butterworth–Heinemann, Elsevier, UK. 2010;**12**:p18

[46] Fane AG, Tang C, Wang R. Membrane technology for water: Microfiltration, ultrafiltration, nanofiltration, and reverse osmosis. Treatise on Water Science. 2011:301-335

[47] Bhattacharya A, Ghosh P. Nanofiltration and reverse osmosis membranes: Theory and application in separation of electrolytes. Reviews in Chemical Engineering. 2004;**20**(1-2)

[48] Nanofiltration pilote technical sheet. MP72/N°24. Delta Lab.

[49] Huang R, Chen G, Sun M, Gao C. Preparation and characterization of quaterinized chitosan/poly (acrylonitrile) composite nanofiltration membrane from anhydride mixture cross-linking. Separation and Purification Technology. 2008;**58**(3):393-399

[50] Chitry F, Pellet-Rostaing S, Gozzi C, Lemaire M. Separation of lanthanides (III) by nanofiltration-complexation in aqueous medium. Separation Science and Technology. 2001;**36**(4):605-618

[51] Benko K, Pellegrino J, Mason LW, Price K. Measurement of water

permeation kinetics across reverse osmosis and nanofiltration membranes: Apparatus development. Journal of Membrane Science. 2006;**270**(1-2):187-195

[52] Aoufi B, Didi MA, Azzouz A. Influence of operating conditions on the retention of severe industrial wastewater by nanofiltration. International Journal of Environmental Analytical Chemistry. 2020. DOI: 10.1080/03067319.2020.1736057

Chapter 7

Principles of Membrane Surface Modification for Water Applications

Yilmaz Yurekli

Abstract

Membrane technologies offer efficient and reliable solutions to separate components from aqueous media. Among them, pressure driven membrane separation processes namely microfiltration (MF), ultrafiltration (UF), nanofiltration (NF) and reverse osmosis (RO) have been preferred in many industrial operations (food, pharmaceutical, chemical, drinking water, wastewater) due to the intrinsic advantages such as high selectivity, stability, ecocompatibility, scalability, flexibility, small footprint and low operational cost. This chapter will focus on the latest developments of surface modified polymeric membranes via the Layer-by-layer self-assembly approach and incorporation/decoration of nanomaterials. Variable parameters including size and charge of polyelectrolyte, ionic strength of the media, number of bilayers, and different types of nanomaterials on the bulk and surface property, water permeability, selectivity, antifouling, antibacterial, and adsorptive properties of the resultant composite membranes will be reviewed by comparison with the neat membranes. Membrane stability in terms of throughput and rejection characteristics during long-term filtrations will be addressed in this chapter.

Keywords: fouling, layer-by-layer self-assembly, membrane, nanometaloxide, surface modification

1. Introduction

The global water scarcity is one of the critical issues faced by human beings. Sustainability of the available water resources is very important for society's development which renders the transformation of wastewater into clean water is mandatory. One of the most challenges in the treatment of industrial and municipal wastewater is the quality and the corresponding cost of the treated water. Recent improvements in membrane technology have emerged as the most important and reliable treatment method for wastewater separation and recycling by the unique features including no need for chemical additives, thermal inputs, and spent media regeneration. The fact that the membrane market in the water and wastewater segment around the world is projected to reach USD 39.2 billion with a compound annual growth rate (CAGR) of 10.8% from 2013 to 2019 is a true indicator of this appeal [1]. Among different kinds of membrane materials, polymer-based membranes have the most common use owing to their relatively cheap manufacturing costs and simple fabrication processes [2, 3]. Polymeric UF, NF, and RO membranes have been successfully used for the production of clean water and recent improvements have been summarized by Deng and Yin [4].

The hydrophobic nature of the polymeric membranes with their inherent permeability/selectivity trade-off is the most prominent problem that causes membrane fouling and lower throughput [5]. Applying one of the surface modification strategies (coating, grafting, blending, etc.) to convert surface non-polar groups into strong polar groups by the introduction of -OH, -COOH, $-NH_2$ has been accepted as a facile and robust way for the manufacturing of the membranes with desired hydrophilicity, leading to improved performance in terms of permeability, selectivity, and antifouling properties [6, 7]. It must be pointed out that the number of modification steps during the membrane fabrication process makes it difficult for large-scale production and the bulk structure of the membrane can be worse affected by the complex technological process, which will result in impairing the separation performance and mechanical strength of the membrane.

The LbL self-assembled surface modification via polyelectrolytes provides a defect-free ultra-thin surface accomplished on any negatively or positively charged surface by a single-step process. In addition, it is an environmentally benign process involving aqueous solution as the media at moderate temperatures. Another approach adopted to improve membrane performance is the impregnation/decoration of inorganic nanomaterials in/on the membrane. According to the literature, TiO_2, ZnO and Ag NPs [8–10] provide antibacterial, SiO_2 NPs [11] electrical conductance, carbon nanotubes (CNTs) such as single-walled CNTs (SWCNTs) and multi-walled CNTs (MWCNTs) [12] and graphene oxide (GO) [13] new water pathways, Fe catalytic property, and FeO NPs [14] magnetic property to the membrane. The hydrophilic nature of the nanomaterials with their high surface area to volume ratio, photocatalytic, antibacterial, and adsorptive capabilities have been widely utilized to modify the conventional polymeric membranes, aiming to overcome their limitations. For example, GO and TiO_2 nanoparticles have been attracted considerable attention, in which the former has abundant of oxygen-containing functional groups (e.g., carboxyl, carbonyl, epoxy groups, and hydroxyl), making them hydrophilic, hence improves membrane permeability [15], while the latter can contribute to continuous oxidation reactions, result in destruction or lethal effect on bacteria, virus, fungi, and algae [16, 17]. Zeolite nanoparticles with high ion exchange capability, on the other hand, add new functionality to the above-mentioned nanoparticles. Zeolites have well-defined porous structures and offer mobility of alkali and alkaline earth metals, in order to compensate net negative charge between Si4+ and Al3+ in the framework makes zeolites excellent adsorber for the removal of many target solutes [18–20]. The nanomaterials can be incorporated into polymer dope by physical blending [21] or deposited as a thin layer on the active layer of the membrane via layer-by-layer self-assembly [22], interfacial polymerization [23], surface grafting [24], or filtration [25, 26] methods.

In the following sections, recent developments in the fabrications and applications of membranes that meet the required throughput, selectivity, mechanical integrity, resistance to fouling, and low manufacturing cost will be discussed. Throughout the chapter, membrane modification techniques via layer-by-layer self-assembly and decoration/incorporation of inorganic nanoparticles (hybrid membranes) will be focused. The effect of variable parameters including size and charge of polyelectrolyte, ionic strength of the media, number of bilayers, and different types of nanomaterials on the bulk and surface property, water permeability, selectivity, antifouling, antibacterial, and adsorptive properties of the resultant composite membranes will be highlighted. Benefits and drawbacks of blending and coating methods will be discussed.

2. Surface modification strategies

This section focuses on the two methods commonly used for the modification of the surface properties of polymeric membranes. They are LbL self-assembly approach with and without inorganic nanoparticles, and surface decorated polyamid (PA) skin layer of thin film nanocomposite (TFN) membranes.

2.1 Surface modification based on LbL self-assembly

LbL self-assembly is a noninvasive method that does not impact the bulk properties of the supporting membranes. The superiority of this technique lies in the well-controlled of thickness, roughness, and charge of the layer. LbL uses polyelectrolytes which are normally hydrophilic and exhibit a charge. Chitosan, Polyethyleneimine (PEI), Poly(diallyldimethylammonium chloride) (PDADMAC), Poly(allylamine hydrochloride) (PAH) are extensively preferred cationic polyelectrolytes, while alginate, Poly(sodium 4-styrenesulfonate) (PSS), Sodium carboxymethyl cellulose are preferred as anionic polyelectrolytes [27, 28]. Deposition can be accomplished by alternating oppositely charged polyelectrolytes on support followed by rinsing after each step, which is used to remove excess and weakly adsorbed polyelectrolytes, hence defect-free ultra-thin layer around 2 nm is formed. A binary layer formed by two opposite polyelectrolyte deposits can be reproduced up to 60–100 multilayers [29, 30]. One of the coating processes including dip coating, spin coating or spray coating, which are schematically represented in **Figure 1** can be applied for LbL assembly. The main drawback of dip coating is the long process time due to diffusional resistance, and rinsing for the elimination of polyelectrolyte complex formation and hence flocculation on the surface. Successive deposition of highly ordered polyelectrolyte multilayers as a result of rapid rearrangement of polymer chains on the substrate is performed by a spin coating within a short time, however, the surface area of the material to be coated is limited. In the case of spray coating process, the polyelectrolyte solution is sprayed over a support membrane and the excess solution is drained by gravity. The processing time to finish deposition is almost two orders of magnitude lower than the dip-coating process and large surfaces can be easily coated with an automated spray coater.

The thickness and the morphology of the layers are influenced by the polyelectrolyte type, number of bilayers and deposition conditions (pH, salt concentration, polyelectrolyte concentration, and deposition temperature and time). For example, linear growth in the multilayer thicknesses was achieved for the PSS/PDADMAC system at 25 °C, while, the increase was reported exponential at 55 °C [32]. The number of sequential layered pairs is commonly known as to produce the thicker film, which corresponds to a lower permeability. However, according to Lajimi's results, the maximum flux, charge density, and hydrophilicity were observed when the number of CHI/ALG bilayers attained 15 [33]. This was explained by the transition of polyelectrolyte layers from a loose stratified structure to a tightened interpenetrating granular structure. In the following subsection, various parameters affecting the structure of the deposited layers will be discussed.

2.1.1 Factors affecting the structure of the LbL modified membranes

One of the most important parameters controlling the thickness, stability, and structure of layers is the salt concentration (ionic strength). Increasing ionic strength leads to thicker layers with a rougher surface [34]. The polyelectrolytes at high salt concentrations turn to coiled and loopy structures (instead of a flat

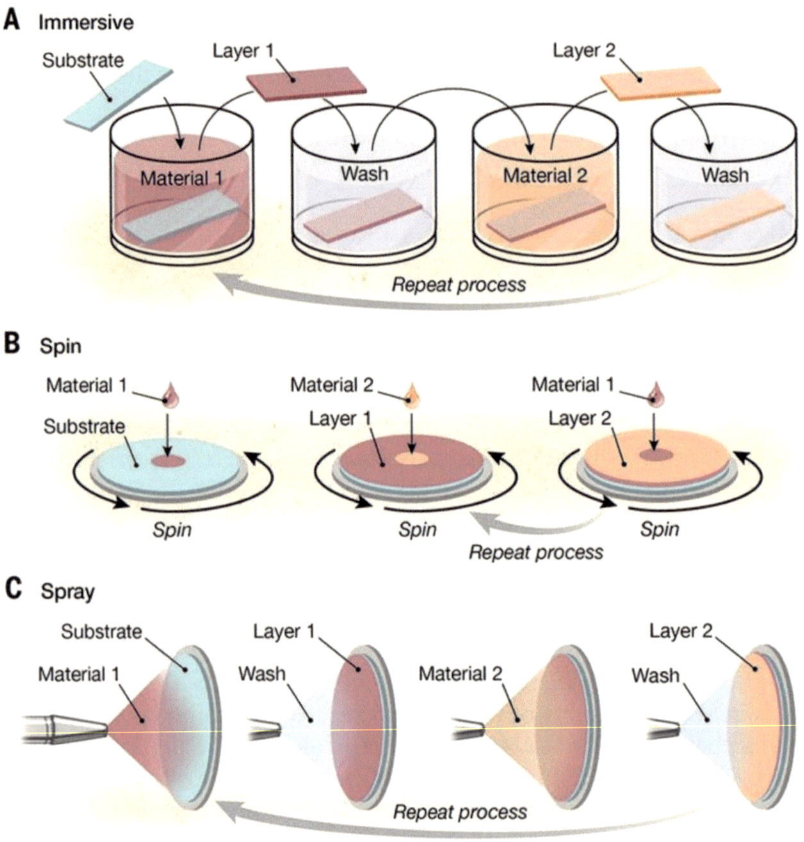

Figure 1.
Schematic representation of different processes used for LbL assembly: (A) dip coating, (B) spin coating and (C) spray coating. [31].

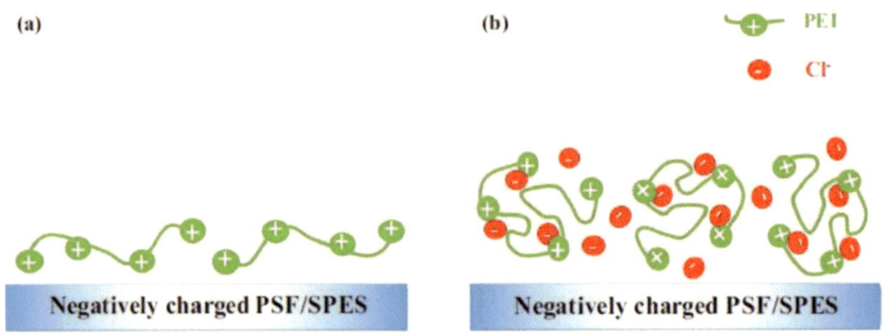

Figure 2.
Variation of the PEI deposition layer with respect to ionic strength of PEI solution. NaCl concentration in PEI solution (a) 0 M and (b) 0.5 M [35].

surface) with more charged chain segments, due to the screening effect, which prevents the electrostatic interaction between polyelectrolyte charges. This behavior is represented schematically in **Figure 2** [35]. In the presence of salt, more PEI adsorption occurs as a result of reduced segment/segment repulsions and increased surface/segment attractions. The reduction in repulsive forces between polyelectrolyte segments makes them small coils covering lower surface area per chain, causing to a larger area density of segments. The former is connected to the radius of gyration (Rg), directly related to selectivity, while the latter is related to surface charge density. Tekinalp showed that the surface charge groups of PEI deposited layer in the presence of salt was higher compared to the salt free case. It has been also proposed that the first polyelectrolyte solution should contain a high concentration of salt, to improve surface segment interactions and thus higher polyelectrolyte adsorption, where a higher unbound charge can be used to form a stable and thin second layer. On the other hand, there is no need to add salt to the second polyelectrolyte solution in order to prevent the formation of thick and rough layers.

The strength of the interactions between polyelectrolyte layers, hence stability of the deposition is depended on the salt counterions which participate in charge neutralization called extrinsic compensation. The following ion exchange reaction can be used to explain this phenomenon [36]:

$$Pol^{+}Pol^{-}_{m}Na^{+}_{aq} + Cl^{-}_{aq} \leftrightarrow Pol^{+}Cl^{-}_{m} + Pol^{-}Na^{+}_{m} \tag{1}$$

$$K = \frac{y^{2}}{(1-y)[NaCl]} = \left(\frac{y^{2}}{[NaCl]^{2}_{aq}}\right)_{y\to 0} \tag{2}$$

where, m, K, and y represent deposited layers, equilibrium constant, and extrinsically compensated polyelectrolyte multilayers, respectively. High salt concentration results in an increased charge screening as the extrinsic charge overcompensation ($Pol^{+}Cl^{-}_{m}Pol^{-}Na^{+}_{m}$) is dominant over the intrinsic charge compensation ($Pol^{+}Pol^{-}_{m}$). At this condition, chains are more mobile due to the poor interactions between polyelectrolytes, which will enhance the possibility of layers detachment [37]. In fact, multilayer formation is thermodynamically favorable as the polyelectrolyte complexation has a small enthalpy change and increase in entropy [38]. The displacement of counterions when polyelectrolytes adsorb on an oppositely charged polyelectrolyte creates an increase in entropy. The sum of three terms determine the Gibbs free energy change of the LbL process. The first one, which is known as intrinsic compensation is the attraction energy carried out between surface charge groups and opposite charges on the polyelectrolyte. The second one, which is not desirable through LbL process is the conformational change caused by entropy loss. The third term is another penalty for deposition and is related to the segment/segment repulsive energy. It is clear that detachment of the deposited layers is prone when the sum of the repulsive and conformation energies exceeds the attraction energy. Therefore, charge density of polyelectrolytes and the ionic strength of the solution predominantly control the stability of the deposited multilayer. However, in some cases, where long-term usage under harsh condition of water purification, stability would be required to be increased, alternatively by using crosslinker between each polyelectrolyte layer [39].

Solution pH is another parameter influencing the morphology of the deposited layer. Thicker layer forms during polyelectrolyte assembly carried out at a high pH. Thin layers with flat chain conformations are attained in the case of highly charged polyelectrolytes. Polyelectrolytes at high pH are partially ionized and incorporating more nonionized chain segments is prone to swell. Therefore, the solution pH determines the charge density of polyelectrolytes, hence surface charge density. At the deposition pH corresponding to the average of the pKa values of the polycation and polyanion, maximum density of ionic cross-links in the assembly is achieved. Unless the charge density is below a minimum value the charge reversal is possible for the formation of multilayers [40].

In order to overcompensate surface charges, charge density and polyelectrolyte concentration in dipping solution should be high. Temperature selected during polyelectrolyte assembly influences the final structure of the film. Higher temperature results in thicker layer due to the chain mobility leading to an increase in the number of loops and tails adsorbed to the membrane surface. The molecular weight of polyelectrolyte influences the stability of the deposited layer. Wang et al. studied the impact of molecular weight of cationic PEI on the permeability and rejection performance of the hydrolyzed PAN membrane [41]. 3.5 bilayers were established by keeping the molecular weight of sulfonated poly (ether ether ketone) constant, which was selected as an anionic polyelectrolyte. Water fluxes with the increase in layers were decreased for both high (25,000) and low (800) molecular weight PEI, however, no salt rejected in the case of low molecular weight PEI was observed. This was explained by the lower structure of the selective layer. Consequently, when using low molecular weight PEI, a higher number of layers could be necessary to achieve salt rejection. The support membrane is, on the other hand, limited to the first few bilayer depositions. Surface charge density and the relative dielectric permittivity of the support may alter the morphology of the multilayer assembly up to a thickness in the micrometer range [42]. In general, a good support membrane for the LbL assembly is expected to have a low surface roughness and a high surface charge density.

2.2 Surface modification containing nanomaterials (TFN)

The purpose of the surface modification of membranes used in water treatments is to reduce or eliminate fouling, which is the main problem in any membrane separation process. In the NF processes, the starting material is usually selected as UF membrane. The polyamide thin film composite (TFC-PA) membranes have been successfully used in water treatment for the purpose of desalination and decolorization, however, membrane fouling and chlorine intolerance cause to decline the permeation flux, shorten the service life, and increase the operating cost, and hence reduce the long term process efficiency. Therefore, researches based on nanoparticle decorations in a skin layer of the asymmetric membrane for NF applications have been rapidly increased.

In TFC-PA approach, nanoparticles are introduced either in organic or aqueous phase. However, the hydrophilic nature of the mostly inorganic nanomaterials necessitates their use in aqueous amine solution. Interfacial polymerization (IP) occurs as soon as acyl and amine monomers interact with each other, in which nanoparticles are either embedded within the polymer matrix or dispersed on top of the polymer film depending on the approach of introducing nanoparticles. The so-called membrane is typically rinsed with hexane or water followed by heat treatment to complete polymerization.

Carbon nanotube (CNT) has intensively attracted attention due to high aspect ratio, low density, mechanical strength, and stiffness. However, its hydrophobic nature makes dispersion problems in various solvents (NMP, DMAc, DMSO, DMF) as well as within a polymer matrix. Therefore, many efforts have been focused on introducing hydrophilic/functional moieties or macromolecules on a CNT surface [43]. Various methods, including acid treatments, plasma oxidation, chemical grafting, in situ polymerization, amination, hydrothermal treatment, and TiCl4 precipitation on the acid treated multi-walled carbon nanotube (MWCNT) have been successfully investigated for the addition of functional groups such as carboxylic, amine, hydroxyl etc.

MWCNT-NH_2 has been embedded in PA layer to improve separation performance of the NF membrane. Dispersion of 0.001 to 0.01 w% MWCNT-NH_2 in piperazine monomer solution followed by the interfacial polymerization revealed that the MWCNT-NH_2 was successfully dispersed within the PA layer, and the modified NF membranes had high hydrophilicity, smoothness, enhanced separation performance, and antifouling property [44]. Similarly, Xue et al. studied the effect of MWCNTs with different functional groups (MWCNT–COOH, MWCNT–OH, or MWCNT–NH) on the NF performances [45]. Piperazine solution consisting of 1 w% functionalized CNT was cast on PSf UF membrane. The coated membranes were than immersed into an organic acyl solution to initiate interfacial polymerization. The remarkable results for thin film nanocomposite NF membranes obtained from different fabrication approaches are summarized in **Table 1**.

Wu et al. used MF membrane to fabricate NF by vacuum filtration of functionalized CNT suspension followed by IP process [55]. The thickness and roughness of the intermediate layer determined the PA active layer morphology. Authors observed that the thickness of the active layer was increased with an increase of CNT layer which associated with the absorbed monomer on the coated membrane. Remarkable results from CNT loaded UF and NF membranes have been summarized in literature [43].

GO is another carbon-based nanomaterial, which has charged oxygen-containing functional groups. Owing to its laminar structures with high surface area, GO nanosheet is mostly preferred for MF, UF membrane surface modification via vacuum filtration or LbL assembly methods. The number of deposition cycles can adjust the thickness of GO layer at a molecular level. Thin layer formation occurs based on alternatively depositing polyelectrolytes and GO nanosheets through

Substrate	Method	Nanofiller	Remarkable features	Ref.
PSf	IP (PIP-TMC)	Modified SiO_2 NPs (100 nm)	Enhanced fouling resistance, long term stability and proper pore size	[46]
PSf	IP (TMC-MPD)	Biocidal GO nanosheets	Good bacterial inactivation without alteration intrinsic transport properties of the membrane	[47]
PSf	IP (TMC-MPD)	GO nanosheets	Enhanced water permeability	[48]
PES	IP (PIP-TMC)	ZIF 8/GO hybrid nanosheets	High antibacterial activity and salt retention	[49]
PES	PVA coating layer	TiO_2	Enhanced water flux and NaCl salt rejection	[50]
PAN	PEI-g-GA coating layer		Impressive prospect for the dye reuse	[51]
PES	Chitosan incorporated GO coating layer		High antibacterial activity	[52]
PP (MF)	LbL (CMCNa- PEI) crosslinked with GA		Highly potential to the application of dye removal and partial desalination with high permeability.	[53]
PVDF	Vacuum filtration	LDH@g-C3N4@ PDA/GO	Superior dye rejections, water flux, and photocatalytic self-cleaning ability	[54]

Table 1.
Thin film nanocomposite NF membranes fabricated by using different approaches and their performance summary.

electrostatic attractions. Zhao et al. fabricated ultrathin hybrid membranes via LbL self-assembly using gelatin (GE) and GO on hydrolyzed PAN membrane [56]. The positively charged GE interacted with negatively charged GO in the self-assembly process, leading to efficient multilayers.

Song et al. functionalized and anchored GO nanosheets with polyelectrolyte to further enhance the separation performance of the GO membranes [57]. GO modification was carried out with ethylenediamine (EDA) molecules, followed by poly (allylamine hydrochloride) (PAH) anchoring to amplify the surface charge density. Amine reduced GO (ArGO) anchored by PAH (PAH@ArGO) nanosheets with positive charge and PSS@GO nanosheets with negative charge were alternately deposited on the polycarbonate support via LbL assemblyThe selective layer thickness of the PE@ArGO membrane was about 160 nm, possessed high density positive/negative charge gated ion transport nanochannels and superior salt rejection by means of Donnan charge exclusion.

However, instability of GO (disintegration or re-dispersion) in water is one biggest block that limits its practical usage. Covalent crosslinking is a promising strategy for the solution of this problem. The functional groups on the GO nanosheets such as hydroxyl and carboxyl groups are convenient sites for the cross-linking reaction with different crosslinker. Mie et al. fabricated a GO membrane covalently cross-linked by 1,3,5- benzenetricarbonyl trichloride between acyl chloride and carboxyl groups [58]. Results revealed that cross-linking effectively ensured the GO membrane with necessary stability to prevent its inherent dispensability in an aqueous environment.

MOFs are porous crystalline materials possess superior compatibility in polymer matrix, apart from other inorganic nanomaterials. By their unique features including size, shape, and polarity, MOFs provide preferential passage for certain molecules, simultaneously rejected undesired substances, when embedded in membrane phase. Gong et al. prepared positively charged NF membrane by incorporating NH2-MIL-125(Ti) porous titanium based MOFs material into PEI and trimesic acid (TMA) crosslinking system [59]. The structure and the experimental procedure are illustrated in **Figure 3(1a-c** and **2)**. The effect of MOFs loading amount on the NF

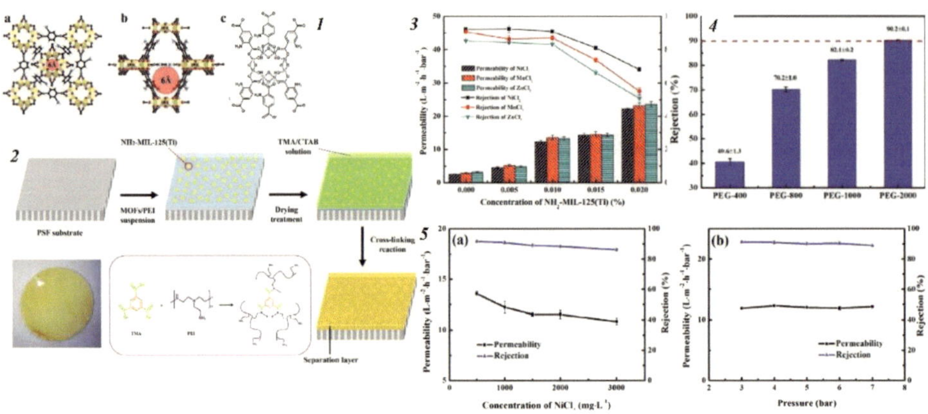

Figure 3.
(1) (a, b) Schematic of NH2-MIL-125(Ti) structure (H, white; C, black; O, red; N, blue; Ti, yellow polyhedral) and (c) the structural formula of NH2-MIL-125(Ti), (2) Schematic of the preparation of NH2-MIL-125(Ti)/PEI/TMA composite membranes, (3) The effect of NH2-MIL-125(Ti) loading on heavy metal removal performance of NH2-MIL-125(Ti)/PEI/TMA composite membranes, (4) Selectivity of different PEG solutions (pressure: 4 bar; solute concentration: 200 mg/L) by the MPT-0.010 composite membrane, (5) Heavy metal (Ni2+) removal performance of the MPT-0.010 composite membrane at different (a) feed concentrations (test pressure: 4 bar) and (b) test pressures (salt concentration: 1000 mg/L) [59].

performance, metal rejection, and stability were studied. The surface roughness of the membrane increased from 6.9 to 92.4 nm when the MOFs loading increased from 0.0 to 0.02 w%. The rejection of Ni^{2+}, Mn^{2+}, Zn^{2+} and permeability optima was obtained, at 0.01 w% of MOFs (**Figure 3 (3)**). Beyond this value, metal rejections reduced seriously. The MWCO of the membrane based on 90% or above PEG rejection (**Figure 3 (4)**) showed that the composite membrane could be considered as a loose NF membrane having an effective pore radius and PEG rejection of 1.5–2.2 nm and 1000–2000 Da, respectively. Furthermore, the resultant composite membrane was found to be positively charged over a large pH interval (3–11) that could be ascribed to the protonation of amine groups. The increase in hydraulic permeability, while maintaining with similar rejection by the introduction of MOFs was attributed to the preferential water channels and suitable window size, that can selectively cut off heavy metal ions allowing water molecules to pass through. In addition, the positive surface charge density of the nanocomposite membrane contributed to the rejection of heavy metal cations by electrostatic repulsive forces.

3. Blending method

The introduction of inorganic nanomaterials (eg. TiO_2, ZnO, GO, Al_2O_3, and SiO_2) into polymer dope to make mixed matrix membrane (MMM) has been favored over the other methods including coating and grafting due to its simplicity. These nanoparticles are attractive in wastewater treatment, because of their porous textures, high surface area to volume ratio, high pore volume and their surface functional groups (-OH) which impart enhanced hydrophilicity and surface properties of the resultant membranes. They form new water pathways, increase solute rejection and control the degree of fouling.

MMM can be configured in both flat sheet and hollow fiber forms based on phase inversion process. Homogenous solution is prepared by dispersing the nanomaterials in a suitable organic solvent (N-methyl-2-pyrrolidone, dimethylformamide, dimethylsulfoxide, dimethylacetamide). A certain amount of polyethyleneglycol or polyvinylpyrrolidone as pore former is introduced to the solution. After sonication, the suspension is slowly added into the main membrane forming material (PSf, PAN, PWDF etc.). The dope solution is continuously stirred overnight, followed by casting process with the required thickness. An asymmetric membrane is then produced by placing the composite film into coagulation bath for nonsolvent induced phase inversion. The incorporation of the inorganic nanoparticles alter the bulk structure of the membrane as they hinder the interaction between polymer and solvent molecules [60]. Their hydrophilic character changes the solvent exchange rate during phase inversion process leading to a thin and dense selective layer with a finger like support layer. The hydrophilic nature of the nanoparticles causes them to move towards the surface during phase inversion process. This enhances the surface properties of the resultant membrane.

3.1 Nanocomposite membranes prepared by blending method

The research in improving the properties of nanofiltration membranes is recently increased tremendously. Various nanomaterials including GO, CNT, metal organic frameworks (MOF) (ZIF-8), TiO_2, SiO_2, zeolite, have been incorporated to form a nanocomposite membrane with high performance. In the following part, a review will be provided to highlight the performance of these composite membranes.

3.1.1 Metal organic frameworks

MOFs consist of metal ions or clusters coordinated to organic ligands to form 1D, 2D or 3D structures. High crystallinity, porosity (up to 90%), internal surface areas (over 6000 m^2/g), and stability make MOF ideal candidate for the enhancement of the membrane performance. The possibility of synthesizing different structures having various sizes and functionalities for a specific application is another advantageous of the MOFs [61]. This is important, since the main problem encountered during incorporation of inorganic nanomaterials into polymeric matrix is their incompatibilities [62]. Filler-polymer compatibility can be improved by the organic constituents of MOFs.

Ze-Xian Low studied the effect of 2D ZIF-L nanoflakes on the performance of the PES UF membrane [63]. Incorporation of ZIF-L was significantly improved water flux without greatly altering the MWCO of the modified UF membrane. Similarly, the combined effect of lower surface roughness, zeta potential, and higher hydrophilicity, caused to a lower bovine serum albumin (BSA) attachment onto the surface of the composite membrane. With those outstanding features, fouling resistance of the so-called membrane against BSA enhanced almost twice with more than 80% flux recovery.

In literature, MOF type materials have been extensively studied in heavy metals adsorption as they provide tunable pores and high specific area [64, 65]. The adsorptive characteristics of UiO-66-NH2 MOF has been tested by incorporating into PAN/CHI dope solution to make composite nanofiber [66]. The adsorptive membrane incorporated with 10 w% of UiO-66-NH2 MOF showed maximum monolayer coverage of 441.2, 415.6, and 372.6 mg/g for Pb^{2+}, Cd^{2+}, and Cr^{6+}, respectively, in static condition. In crossflow filtration, carried out 20 mg/L initial metal concentration at 1 bar TMP, the permeation flux and metal removal were observed as for 452, 463, and 479 $L/m^2.h.$, and 94, 89, and 86%, respectively for Pb^{2+}, Cd^{2+}, and Cr^{6+}. During long term filtration, slightly reduced permeation flux and rejection were obtained up to 18 h., beyond this point, a significant reduction in both flux and rejection revealed that the nanofibrous membrane pores were saturated.

3.1.2 Zeolite NPs

Zeolite nanoparticles with their unique properties such as high ion exchange capacity and fast adsorption rate make it excellent choice for the separation of heavy metals in wastewater treatment [67] and desalination process [68]. In the study of Yurekli, variations in the morphologies and the filtration performances of the zeolite NPs blended PSf hybrid membranes have been investigated with respect to the loading amount of zeolite NPs [67]. **Figure 1** indicates surface and cross-sectional SEM microphotographs of the native and zeolite NPs incorporated PSf membranes. Formation of the new pores with larger diameters in **Figure 4** has been attributed to the phase separation occurred quickly and to the aggregation of the NPs. Compared to the native PSf membrane, zeolite loaded PSf membrane has more uniformly distributed finger-like pores which are extended through the thickness of the membrane that shortens the pathway (tortuosity) of the solutes, hence improve the hydraulic permeability. An increase in the hydraulic permeability value of 94% for the PSf10–30 membrane has been attained compared to the pristine PSf (23.2 L/m2.h.bar). It was observed that the retention of heavy metals through the PSf10–30 membrane was more pronounced at lower transmembrane pressures and heavy metals concentrations.

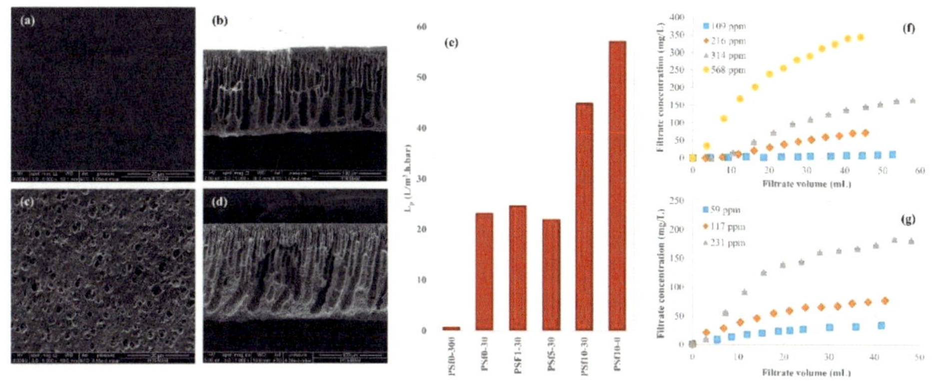

Figure 4.
SEM microphotographs of the native and zeolite incorporated PSf membranes; (a, c) top surface and (b, d) cross-sectional images of the native and 10% zeolite added PSf membranes, (e) hydraulic permeabilities of the PSf membranes prepared with different amount of zeolite NPs, (f) Pb2+ and (g) Ni2+ concentrations in permeate during filtrations of Pb2+ or Ni2+ aqueous solutions, respectively in different initial concentrations through PSf 10–30 membrane at 1 bar. [67].

3.1.3 GO NPs

Similarly, Mukherjee et al. studied the impact of GO NPs on the removal of heavy metals using mixed matrix membrane (MMM) [69]. GO NPs were synthesized based on the modified Hummer method, blended in different fractions with PSf dope solution and prepared the MMM by phase inversion method. The main results extracted from the study of Mukherjee is depicted in **Figure 5**. From **Figure 5(a)-(c)** addition of GO NPs into PSf matrix increased MWCO, porosity, negative charge density, and permeability of the MMM. Based on the preliminary cross-flow tests 414 kPa was selected as optimum TMP considering both rejection and permeabilities of the membrane. In order to investigate reusability of the MMM, the fouled membrane with 50 mg/L chromium aqueous solution

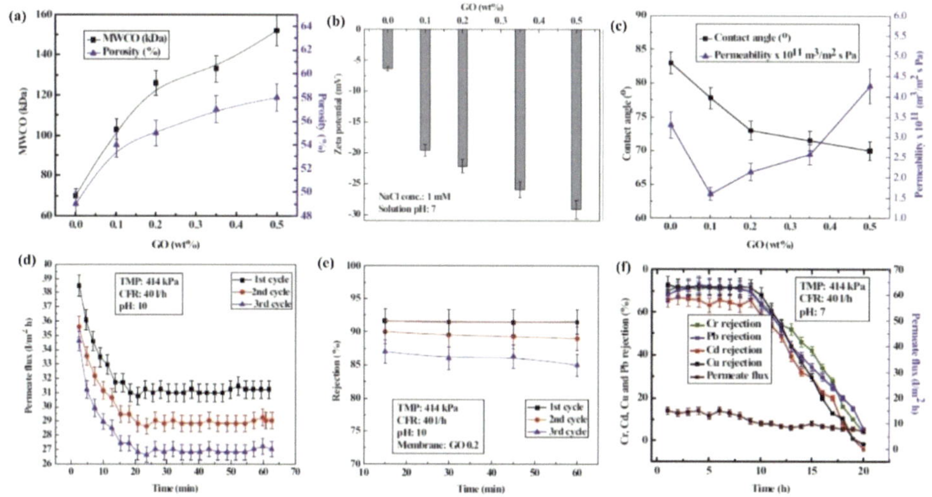

Figure 5.
Effect of GO concentration on (a) MWCO and porosity; (b) zeta potential at neutral pH (c) contact angle and permeability. Effect of regeneration on (d) permeate flux and (e) rejection by GO 0.2 membrane, with 414 kPa TMP, 40 l/h CFR and operated at pH 10. (f) Long duration filtration performance of GO 0.2 membrane in case of filtration of mixed metal solution at 414 kPa TMP and 40 L/h CFR. [69].

in cross-flow mode of 414 kPa and 20 L/h of retentate flow rate for 1 hour was washed first with water for 10 min and then with acidic solution at pH 5.5 for 30 min and again with water to remove all the acidic residuals. All the steps in the regeneration cycle were carried out in dynamic conditions similar to the case in fouling. The reduction in solution flux at each cycle was explained as the accumulation of chromium ion at the GO surface, which was also observed by the reduction in rejection values in the consecutive cycles (**Figure 5(a)** and **(b)**). The results from long term filtration performance of the MMM with respect to the mixed metal solution (Cr, Cu, Cd, and Pb) each having concentrations of 50 mg/L, as illustrated in **Figure 5(f)** demonstrated the permeate flux decreased continuously. The authors concluded that the simultaneous adsorption of different heavy metals, on the membrane increased the resistance of adsorption, which resulted in permeate flux reduction rapidly with time. Finally, the concentrations of all heavy metals in permeate remained almost constant up to breakthrough time (8 h) and increased thereafter till feed concentration.

3.1.4 TNT nanotubes

Subramaniam et al. fabricated PVDF hollow fiber UF membrane incorporated with titanate nanotubes (TNTs) for decolourization of aerobically-treated palm oil mill effluent [70]. TNTs which were synthesized based on the alkaline hydrothermal process were dispersed in NMP under sonication for 30 min. Hollow fiber membranes were fabricated by means of a dry-jet wet spinning method by changing the amount of TNTs in the bore solution of TNT/PVP/PVDF in between 0–1%. The spherical TiO_2 nanoparticles were reported to be converted into completely TNTs with an average diameter of 24 nm at the end of the hydrothermal process. The variation in the pore size was more pronounced by the addition of pore former than the addition of TNT into the dope solution meaning that the addition of TNT had no considerable effect on the pore size and finger-like structure of the membrane. The result was evidenced by the similar porosity values obtained for all the membrane formulations. Addition of TNT increased the membrane roughness, BSA rejection, color rejection and water flux simultaneously (**Figure 6a**). However, beyond a certain amount of TNT loading, a reduction in water flux has been observed, which was ascribed to the aggregation of NPs. The authors finally investigated the flux recovery and antifouling performances of the resultant membranes during 5 regeneration cycles as depicted in **Figure 6**. Regeneration was accomplished by first fouled the membrane with AT-POME at 1 bar for 240 min using cross-flow filtration, then the fouled membrane was washed with water for 30 min. In **Figure 6**, the fluxes in all membrane configurations decreased over time but the flux recoveries for all the membranes after 5 cycles were observed above 80% except the pristine PVDF, which exhibited gradual declination of flux throughout the test. Similarly, all the membranes comprised of TNT regardless of loading were able to recover rejection performance after water washing

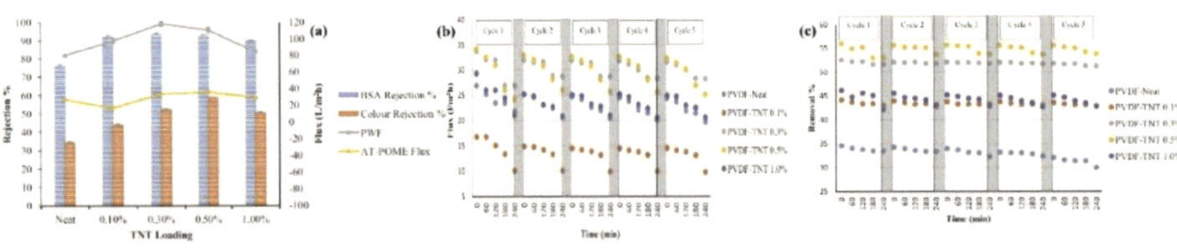

Figure 6.
(a) Pure water fluxes, AT-POME flux, BSA and color removal (b) AT-POME flux and (c) AT-POME color removal for five cycles of AT-POME filtration [70].

but pristine PVDF showed declining rejection through 5 cycles. The results has been related to the variation of the hydrophilic natures of the membranes.

3.1.5 Alumina NPs

Alumina NPs loaded PVDF membrane fabricated according to the phase inversion method has been demonstrated that the hydrophilicity, permeability, antifouling capacity, and mechanical stability were increased with no considerable change in pore density and size [71]. In another study, fouling resistive, reduced flux decline and less hydrophobic interaction between foulant and membrane surface are some of the outstanding features of the alumina NPs added PES UF membrane [72]. Different types of NPs have been incorporated into variable polymeric matrix to improve separation performances and fouling resistance of the hybrid membranes and their characteristics compared to the pristine ones collected from recently published studies have been listed in **Table 2**.

Nanoparticles used to fabricate a hybrid membrane can also enhance the mechanical stability compared to the pristine membrane by decreasing the impact of the membrane compaction. Compaction is known as the mechanical deformation of the polymeric membrane matrix and observed at the initial stage of the pressure driven membrane operations. Structure densification leads to a flux decline. Blending NPs with polymer matrix allows well distribution of NPs through the thickness of the membrane during phase inversion process. The stability of those NPs in the macrovoid region reduces the loss of membrane structure because compaction is known to occur predominantly in bulk macrovoid region. For instance, Pendergast et al. reported that stability of PSf membranes was improved when silica and zeolite NPs were included in the membrane structure, resulting in less compaction than pure PSf membrane [86].

3.1.6 Surface modified NPs

Although, the addition of the NPs into polymer matrix by simple blending method offers superior advantageous in terms of rejection, antifouling, and permeability, the tendency of NPs to agglomeration due to high surface free energy and high affinity to water molecules during phase separation on one hand, and the possibly release of NPs into filtrate due to weak interactions between NPs and polymer chains on the other hand, makes surface modification of NPs mandatory to obtain a hybrid membrane with well dispersed, stable, and compatable NPs. Modifications are based on either chemical treatment of NPs or grafting of functional polymers on hydroxyl groups available in NPs. Oxidation, for example, could be the desired pathway through acid treatment to create carboxyl and hydroxyl functional groups on carbon nanotubes. Different surfactants have also been proposed to be grafted on NPs surfaces to mitigate the agglomeration and their mechanisms have been reviewed detailed in the literature [87]. Besides the improvement of homogenous dispersivity, stability, and compatibility, the functionalization of NPs increases the membrane antifouling property as they offer higher surface charge density, which is useful for the rejection of the foulants. Ayyaru and Ahn studied with the PES nanocomposite membranes blended with surface modified TiO_2 NPs (anatase, 20–25 nm in size) for fouling mitigation [88]. The TiO_2 nanoparticles were sulfonated by replacing the surface hydroxyl groups with $-SO_3H$ group and the loading effect of $sTiO_2$ NPs was investigated. Surface roughness, porosity, and pore size of the modified membranes exhibited notable enhancement compared to the PES membrane. Increasing porosity revealed a good distribution of $sTiO_2$ NPs in the dope solution. The improved properties of $sTiO_2$ blended membranes such as high hydrophilicity permeability, anti-fouling performance, and improved BSA rejection were attributed

Polymer	Filler	Method	CA (°)	Foulant	The main contributions of nanoparticles to the resultant membrane	Ref.
PES	TMU-5 MOF	NIPS	67.2–51.6	8000 ppm powder milk solution	Hydraulic permeability increased from 44.4 to 60.7 L/m2.h.bar. Surface become smoother (roughness reduced from 43.9 to 5.2 nm.) and flux recovery ratio (FRR) increased to 25.5 to 98.7%. After 5 cycles 99% of water flux recovery was obtained.	[73]
PVDF	CNT/GO	NIPS	79.6–62.1	1000 ppm BSA solution	Hydraulic permeability increased from 75.5 to 125.6 L/m2.h.bar. Pore size after nanoparticle addition increased from 14.5 to 18.5 nm. Roughness exceptionally increased from 12.9 to 30.9 nm. FRR increased from 52 to 60.6%.	[74]
PVDF	Amine functionalized silica	Dip coating	66–32	Mixture of organic and inorganic foulants	The addition of nanoparticles improved pore size (30.4 to 50.3 nm.) and water permeability (648 to1512 L/m2.h.bar) of the resultant membrane. No significant change in roughness was observed and FRR improved from 53 to 80%.	[75]
PSf	Cuprous iodide nanosheets	NIPS	71.6–60.4	500 ppm BSA solution (pH 7.3)	The increase in pore size (from 58.2 to 92 nm.) and hydrophilicity resulted in an increase in hydraulic permeability (556 to 1473 L/m2.h.bar). FRR value was slightly improved (31–36%). At the end of 6 regeneration cycles 36% water flux recovered.	[76]
PES (MF)	CuO NPs	NIPS	69.8–64	500 ppm BSA solution	Hydraulic permeability increased from 266 to 435 L/m^2.h.bar. Pore size after nanoparticle addition increased from 13.9 to 17.8 nm. Surface roughness reduced from 3.2 to 2.5 nm. FRR was slightly improved from 45.5 to 48.2%.	[77]
CA (UF)	GO/MOF@GO	NIPS	73.2–49.5	1000 ppm BSA solution	Both rejection (91.4–95.4%) and hydraulic permeability (66–122 L/m^2.h.bar) enhanced. The reduction in surface roughness from 25.2 to 7.2 nm. caused to increase FRR from 49.8 to 88.1%.	[78]
PVDF (UF)	AC/TiO_2	NIPS	80.4–66.4	20 ppm HA solution	Pore size (18 nm.), surface roughness (22.4 nm.) values were unchanged, but hydraulic permeability improved from 90 to 172 L/m^2.h.bar. FRR was reduced from 90.6 to 83.3%, due to strong interaction of the foulants with AC. After 2 regeneration cycles 87% of water flux was recovered.	[79]
PSf (UF)	GO/CGO	NIPS	90.6–73.1	1000 ppm BSA solution	Hydraulic permeability increased from 60 to 123 L/m^2.h.bar with a rejection value of BSA larger than 99.5%. Smoother surface (reduction in surface roughness from60.7 to 23.1 nm.) and lower FRR value (from 68.6 to 52.7%) were reported.	[80]

Polymer	Filler	Method	CA (°)	Foulant	The main contributions of nanoparticles to the resultant membrane	Ref.
PES (UF)	HANTs	NIPS	83.6–67.1	1000 ppm BSA solution pH 7.4	Addition of nanoparticles caused all pore size (34.1–43.2 nm.), surface roughness (10–19.4 nm.), hydraulic permeability (169–439 L/m^2.h.bar), and FRR (54.4–69.3%) to increase. After three cycles, regeneration capability was enhanced from 45.1 to 57.9%.	[81]
PVC (UF)	Boehmite (30 nm)	NIPS	71.3–57.2	500 ppm BSA solution	Pore size (from 11.1 to 14.9 nm.), hydraulic permeability (211.3–350.7 L/m^2.h.bar), and FRR (from 44.3 to 60.4%) were improved by the addition of nanoparticles. At the end of the 4th regeneration cycles, water flux recovery improved from 20.6 to 47.7%.	[82]
PVDF (UF)	TiO_2/GO	NIPS	66–53	1000 ppm BSA solution	With the increased in pore size (50–57.6 nm.) and hydrophilicity, hydraulic permeability from37.9 to 200 L/m^2.h.bar. and FRR from 51.1 to 89.2% were increased.	[83]
PVDF (UF)	Vermiculate 82 nm CuO <50 nm Al_2O_3 < 50 nm SiO_2 < 200 nm	NIPS	82.6–57.3 66.4 62.5 64.8	1000 ppm BSA solution pH 7	Humic acid (HA) rejection (94.6, 88.3, 91.7, and 89%) and hydraulic permeability (444, 372.9, 425.7, and 405.7 L/m^2.h.bar) and stability at the end of 4th regeneration cycles (81.5, 61.4, 73.8, and 68.3%) were improved when vermiculate, cuprous oxide, alumina, and silica was added respectively. The properties of the neat membrane were reported as 80.9% HA rejection, 367.3 L/m^2.h.bar. Hydraulic permeability, and 45.5% water flux recovery at the end of 4th cycles.,	[84]
PSf (UF)	GO nanosheets	NIPS	94–83.3	200 ppm BSA solution	Hydraulic permeability and pore size increased from 98 to 294 L/m^2.h.bar. and from 19 to 31 nm., while rejection was unchanged. After two cycles, water flux recovery enhanced from 73.4 to 82.4%.	[85]

Table 2.
Filtration performances and physical characteristics of the hybrid membranes prepared by blending method.

to the hydrogen bonding force and more electrostatic repulsion properties of $sTiO_2$ NPs. The same group has also investigated the effect of the addition of sulfonated graphene oxide (sGO) NPs into PVDF membranes fabricated by the phase inversion method [89]. A gradual increase in water fluxes was obtained up to the sGO loading concentration of 0.8% meaning to the elimination of aggregation. Maximum water permeability attained at the 0.8 wt% of sGO addition was reported as 146% higher than the neat PVDF. The enhancement of the water flux has been explained by the improved charge density due to the availability of extra sulfonic groups on sGO supports that can attract more water layer. In addition, the attached $-SO_3H$ group in sGO provides stronger hydrogen-bonding with respect to –OH/-COOH groups available in native GO. The performances of the CNT and sulfonated CNT (sCNT) NPs blended PVDF UF membranes were compared for the objective of conserving the bacterial population and providing antifouling property [90]. The porosity, pore size, water flux, fouling recovery ratio values of PVDF-CNT and PVDF-sCNT were obtained as 81 and 84%, 50 and 60 nm, 360 and 680 $L/m_2.h$ and 72.7 and 83.5%, respectively. In addition, the BSA (bovine serum albumin) rejection was 90% in the PVDF-sCNT. Authors demonstrated that the fabricated composite membranes were nontoxic to the bacterial population, hence the proposed membrane architecture can be a promising approach for membrane bioreactor systems in wastewater treatment plants.

4. Conclusion

Membranes containing nanomaterials in the form of MMNMs or TFN, ensure outstanding features in terms of permeability, selectivity, antifouling, and self-cleaning ability for water applications due to their synergistic effects. However, there are still major challenges to developing high-performance membranes for large scale water treatments. Difficulties in dispersion of inorganic nanoparticles in a polymer matrix, the release of nanoparticles and associated environmental toxicity, long term stability, and the production cost of the nanocomposite membrane, needs to be further explored. Care should be taken that the nanoparticles selected do not sacrifice an ability of the membranes to improve another ability.

Conflict of interest

The author declares no conflict of interest.

Author details

Yilmaz Yurekli
Engineering Faculty, Bioengineering Department, Manisa Celal Bayar University, Manisa, Turkey

*Address all correspondence to: yilmazyurekli@gmail.com

References

[1] XueMei Tan, Denis Rodrigue, A Review on Porous Polymeric Membrane Preparation. Part I: Production Techniques with Polysulfone and Poly (Vinylidene Fluoride), Polymers 11 (2019) 1160.

[2] Z. Xu, T. Wu, J. Shi, W. Wang, K. Teng, X. Qian, M. Shan, H. Deng, X. Tian, C. Li, F. Li, Manipulating Migration Behavior of Magnetic Graphene Oxide via Magnetic Field Induced Casting and Phase Separation toward High Performance Hybrid Ultrafiltration Membranes, ACS Applied Materials & Interfaces, 8 (2016) 18418-18429.

[3] Z. Xu, T. Wu, J. Shi, K. Teng, W. Wang, M. Ma, J. Li, X. Qian, C. Li, J. Fan, Photocatalytic antifouling PVDF ultrafiltration membranes based on synergy of graphene oxide and TiO2 for water treatment, J. Membr. Sci., 520 (2016) 281-293.

[4] Jun Yin, Baolin Deng , Polymer-matrix nanocomposite membranes for water treatment, Journal of Membrane Science 479 (2015) 256-275.

[5] Meenakshi Sundaram Sri Abirami Saraswathi, Alagumalai Nagendran, Dipak Rana, Tailored polymer nanocomposite membranes based on carbon, metal oxide and silicon nanomaterials: a review, J. Mater. Chem. A 7 (2019) 8723-8745.

[6] Li-guang Wu, Xue-yang Zhang, Ting Wang, Chun-hui Du, Cai-hong Yang, Enhanced performance of polyvinylidene fluoride ultrafiltration membranes by incorporating TiO_2/graphene oxide, Chemical Engineering Research and Design 141 (2019) 492-501.

[7] Shujuan Yang, Qinfeng Zou, Tianhao Wang, Liping Zhang, Effects of GO and MOF@GO on the permeation and antifouling properties of cellulose acetate ultrafiltration membrane, Journal of Membrane Science 569 (2019) 48-59.

[8] Luo M.L., Zhao J.Q., Tang W., Pu C.S., Hydrophilic modification of poly (ether sulfone) ultrafiltration membrane surface by self-assembly of TiO2 nanoparticles, Appl. Surf. Sci.1 (2005) 76-84.

[9] Li X., Li J., Van der Bruggen B., Sun X., Shen J., Han W., Wang L., Fouling behavior of polyethersulfone ultrafiltration membranes functionalized with sol–gel formed ZnO nanoparticles, RSC Adv. 63 (2015) 50711-50719.

[10] Zhu X., Bai R., Wee K.H., Liu C., Tang S.L., Membrane surfaces immobilized with ionic or reduced silver and their anti-biofouling performances, J. Membr. Sci. 1 (2010) 278-286.

[11] Yin J., Kim E.S., Yang J., Deng B., Fabrication of a novel thin-film nanocomposite (TFN) membrane containing MCM-41 silica nanoparticles (NPs) for water purification, J. Membr. Sci. 423 (2012) 238-246.

[12] Mendez R., Constant B., Garzon C., Nisar M., Nachtigall S.M.B., Quijada R., Barrier, mechanical and conductive properties of polycaprolactam nanocomposites containing carbon-based particles: Effect of the kind of particle, Polymer 130 (2017) 10-16.

[13] Hassan M., Reddy K.R., Haque E., Faisal S.N., Ghasemi S., Minett A.I., Gomes V.G., Hierarchical assembly of graphene/polyaniline nanostructures to synthesize freestanding supercapacitor electrode, Compos. Sci. Technol. 98 (2014) 1-8.

[14] Homayoonfal M., Mehrnia M.R., Niassar M.S., Akbari A.,

Sarrafzadeh M.H., Ismail A.F., Fabrication of magnetic nanocomposite membrane for separation of organic contaminant from water, Desalin. Water Treat. 54 (2015) 3603-3609.

[15] C. Zhao, X. Xu, J. Chen, G. Wang, F. Yang, Highly effective antifouling performance of PVDF/graphene oxide composite membrane in membrane bioreactor (MBR) system, Desalination 340 (2014) 59-66.

[16] Robertson, J.M.C., Robertson, P.K.J., Lawton, L.A., 2005. A comparison of the effectiveness of TiO2 photocatalysis and UVA photolysis for the destruction of three pathogenic microorganisms. J. Photochem. Photobiol. A: Chem. 175, 51-56.

[17] Skorb, E.V., Antonouskaya, L.I., Belyasova, N.A., Shchukin, D.G., Mohwald, H., Sviridov, D.V., 2008. Antibacterial activity of thin-film photocatalysts based on metal-modified TiO2 and TiO2:In2O3 nanocomposite. Appl. Catal. B: Environ. 84, 94-99.

[18] A. Charkhia, M. Kazemeinia, S.J. Ahmadib, H. Kazemian, Fabrication ofgranulated NaY zeolite nanoparticles using a new method and study theadsorption properties, Powder Technol. 231 (2012) 1-6.

[19] Y. Wang, Z. Lei, B. Chen, Q. Guo, N. Liu, Adsorption of NO and N2O on Fe-BEAand H-BEA zeolites, Appl. Surf. Sci. 256 (2010) 4042-4047.

[20] S. Malamisa, E. Katsoua, A review on zinc and nickel adsorption on naturaland modified zeolite, bentonite and vermiculite: examination of processparameters, kinetics and isotherms, J. Hazard. Mater. 252-253 (252) (2013)428-461.

[21] Fang X., Li J., Ren B., Huang Y., Wang D., Liao Z., Li Q., Wang L., Dionysiou D.D., 2019, Polymeric ultrafiltration membrane with in situ formed nano-silver within the inner pores for simultaneous separation and catalysis, J. Membr. Sci., 579, 190-198

[22] Choi W., Choi J., Bang J., Lee J.H., 2013, Layer-by-Layer Assembly of Graphene Oxide Nanosheets on Polyamide Membranes for Durable Reverse-Osmosis Applications, ACS Appl. Mater. Interfaces, 52, 3, 12510-12519

[23] Lau W., Gray S., Matsuura T., Emadzadeh D., Chen J.P., Ismail A, 2015, A review on polyamide thin film nanocomposite (TFN) membranes: history, applications, challenges and approaches, Water Res., 80, 306-324.

[24] Liang S., Xiao K., Zhang S., Ma Z., Lu P., Wang H., Huang X., 2018, A facile approach to fabrication of superhydrophilic ultrafiltration membranes with surface-tailored nanoparticles, Sep. Purif. Technol., 203, 251-259.

[25] Ko K., Yu Y.J., Kim M.J., Kweon J., Chun H., 2018, Improvement in fouling resistance of silver-graphene oxide coated polyvinylidene fluoride membrane prepared by pressurized filtration, Sep. Purif. Technol., 194, 161-169.

[26] Cheng X., Zhou W., Li P., Ren Z., Wu D., Luo C., Tang X., Wang J., Liang H., 2019, Improving ultrafiltration membrane performance with pre-deposited carbon nanotubes/ nanofibers layers for drinking water treatment, Chemosphere, 234, 545-557.

[27] Mou Paul, Steven D. Jons, Chemistry and fabrication of polymeric nanofiltration membranes: A review, Polymer 103 (2016) 417-456.

[28] Filiz Yasar Mahlicli, Sacide Alsoy Altinkaya, Yilmaz Yurekli, Preparation and characterization of polyacrylonitrile membranes modified

with polyelectrolyte deposition for separating similar sized proteins, Journal of Membrane Science 415-416 (2012) 383-390.

[29] Nithya Joseph, Pejman Ahmadiannamini, Richard Hoogenboom, Ivo. F. J. Vankelecom, Layer-by-layer preparation of polyelectrolyte multilayer membranes for separation, *Polym. Chem.* 5 (2014) 1817-1831

[30] Bartosz Czerwieniec, Marcin Strawski, Ludomira H. Granicka, Marek Szklarczyk, AFM study of adhesion and interactions between polyelectrolyte bilayers assembly, Colloids and Surfaces A: Physicochemical and Engineering Aspects 555 (2018) 465-472

[31] Joseph J. Richardson, Mattias Björnmalm, Frank Caruso, Technology-driven layer-by-layer assembly of nanofilms, Science, 348 (2015) 2491.

[32] Salomäki, M., Vinokurov, I.A. & Kankare, J. (2005) Effect of temperature on the buildup of polyelectrolyte multilayers. Langmuir, 21, 11,232-11,240.

[33] R.H. Lajimi, et al., Change of the performance properties of nanofiltration cellulose acetate membranes by surface adsorption of polyelectrolyte multilayers, Desalination 163 (1-3) (2004) 193-202.

[34] Kolarik, L., Furlong, D.F., Joy, H., Struijk, C. & Rowe, R. (1999) Building assemblies from high molecular weight polyelectrolytes. Langmuir, 15, 8265-8275.

[35] Önder Tekinalp, Sacide Alsoy Altinkaya, Development of high flux nanofiltration membranes through single bilayer polyethyleneimine/alginate deposition, Journal of Colloid and Interface Science, 537 (2019) 215-227

[36] S.T. Dubas, J.B. Schlenoff, Swelling and smoothing of polyelectrolyte multilayers by salt, Langmuir 17 (2001) 7725-7727.

[37] N.G. Hoogeveen, M.A.C. Stuart, G.J. Fleer, Polyelectrolyte adsorption on oxides.2. Reversibility and exchange, J. Colloid Interface Sci. 182 (1996) 146-157.

[38] Schlenoff, J.B., Rmaile, A.H. & Bucur, C.B. (2008) Hydration contributions to association in polyelectrolyte multilayers and complexes: visualizing hydrophobicity. Journal of the American Chemical Society, 130, 13589-13597.

[39] Qing Chen, Pingping Yu, Wenqiang Huang, Sanchuan Yu, Meihong Liu, Congjie Gao, High-flux composite hollow fiber nanofiltration membranes fabricated through layer-by-layer deposition of oppositely charged crosslinked polyelectrolytes for dye removal, Journal of Membrane Science, 492, (2015), 312 321.

[40] Glinel, K., Moussa, A., Jonas, A.M. & Laschewsky, A. (2002) Influence of polyelectrolyte charge density on the formation of multilayers of strong polyelectrolytes at low ionic strength. Langmuir, 18, 1408-1412.

[41] Wang, J., Yao, Y., Yue, Z. & Economy, J. (2009) Preparation of polyelectrolyte multilayer films consisting of sulfonated poly (ether ether ketone) alternating with selected anionic layers. Journal of Membrane Science, 337, 200-207.

[42] Guillaume-Gentil, O., Zahn, R., Lindhoud, S., Graf, N., Voros, J. & Zambelli, T. (2011) From nanodroplets to continuous films: how the morphology of polyelectrolyte multilayers depends on the dielectric permittivity and the surface charge of the supporting substrate. Soft Matter, 7, 3861-3871.

[43] Merry Sianipar, Seung Hyun Kim, Khoiruddin, Ferry Iskandar, Gede Wenten, Functionalized carbon nanotube (CNT) membrane: progress and challenges, RSC Adv., 2017, 7, 51175.

[44] H. Zarrabi, M. E. Yekavalangi, V. Vatanpour, A. Shockravi and M. Safarpour, Desalination, 2016, 394, 83-90.

[45] S.-M. Xue, Z.-L. Xu, Y.-J. Tang and C.-H. Ji, ACS Appl. Mater. Interfaces, 2016, 8, 19135-19144.

[46] H. Wu, B. Tang, P. Wu, Optimizing polyamide thin film composite membrane covalently bonded with modified mesoporous silica nanoparticles, J. Membr. Sci. 428 (2013) 341-348.

[47] F. Perreault, M.E. Tousley, M. Elimelech, Thin-film composite polyamide membranes functionalized with biocidal graphene oxide nanosheets, Environ. Sci.Technol. Lett. 1 (2013) 71-76

[48] J. Yin, G. Zhu, B. Deng, Graphene oxide (GO) enhanced polyamide (PA) thin-film nanocomposite (TFN) membrane for water purification, Desalination 379 (2016) 93-101.

[49] J. Wang, Y. Wang, Y. Zhang, A. Uliana, J. Zhu, J. Liu, B. Van der Bruggen, Zeolitic imidazolate framework/graphene oxide hybrid nanosheets functionalized thin film nanocomposite membrane for enhanced antimicrobial performance, ACS Appl. Mater. Interfaces 8 (2016) 25508-25519.

[50] S. Pourjafar, A. Rahimpour, M. Jahanshahi, Synthesis and characterization of PVA/PES thin film composite nanofiltration membrane modified with TiO2 nanoparticles for better performance and surface properties, J. Ind. Eng. Chem. 18 (2012) 1398-1405.

[51] S. Zhao, Z. Wang, A loose nano-filtration membrane prepared by coating HPAN UF membrane with modified PEI for dye reuse and desalination, J. Membr. Sci. 524 (2017) 214-224.

[52] S.F. Seyedpour, A. Rahimpour, H. Mohsenian, M.J. Taherzadeh, Low fouling ultrathin nanocomposite membranes for efficient removal of manganese, J. Membr. Sci. (2017).

[53] Qing Chen, Pingping Yu, Wenqiang Huang, Sanchuan Yu, Meihong Liu, Congjie Gao, High-flux composite hollow fiber nanofiltration membranes fabricated through layer-by-layer deposition of oppositely charged crosslinked polyelectrolytes for dye removal, Journal of Membrane Science 492 (2015) 312-321.

[54] Yuchuan Liu, Zongxue Yu, Xiuhui Li, Liangyan Shao, Haojie Zeng, Super hydrophilic composite membrane with photocatalytic degradation and self-cleaning ability based on LDH and g-C3N4, Journal of Membrane Science 617 (2021) 118504.

[55] M.-B. Wu, Y. Lv, H.-C. Yang, L.-F. Liu, X. Zhang and Z.-K. Xu, J. Membr. Sci., 2016, 515, 238-244.

[56] J. Zhao, Y. Zhu, F. Pan, G. He, C. Fang, K. Cao, R. Xing and Z. Jiang, J. Membr. Sci., 2015, 487, 162-172.

[57] X. Song, R. S. Zambare, S. Qi, B. N. I. L. Sowrirajalu, A.P. James Selvaraj, C. Y. Tang and C. Gao, ACS Appl.Mater. Interfaces, 2017, 9, 41482-41495.

[58] M. Hu and B. Mi, Environ. Sci. Technol., 2013, 47, 3715-3723.

[59] Xin-Yu Gong, Zhi-Hao Huang, Hao Zhang, Wei-Liang Liu, Xiao-Hua Ma, Zhen-Liang Xu, Novel high-flux positively charged composite membrane incorporating titanium-based MOFs

for heavy metal removal, Chemical Engineering Journal 398 (2020) 125706.

[60] A. Sotto, A. Boromand, R. Zhang, P. Luis, J.M. Arsuaga, J. Kim, B. Van der Bruggen, Effect of nanoparticle aggregation at low concentrations of TiO2 on the hydrophilicity, morphology, and fouling resistance of PES–TiO2 membranes, J. Colloid Interface Sci. 363 (2011) 540-550.

[61] N.Abdullah, N.Yusof, A.F.Ismail, W.J.Lau, Insights into metal-organic frameworks-integrated membranes for desalination process: A review, Desalination 500 (2021) 114867.

[62] B. Zornoza, C. Tellez, J. Coronas, J. Gascon, F. Kapteijn, Metal organic framework based mixed matrix membranes: an increasingly important field of research with a large application potential, Microporous Mesoporous Mater. 166 (2013) 67-78.

[63] Ze Xian Low, Amir Razmjou, Kun Wang, Stephen Gray, Mikel Duke, Huanting Wang, Effect of addition of two-dimensional ZIF-L nanoflakes on the properties of polyethersulfone ultrafiltration membrane, Journal of Membrane Science, 460, (2014) 9-17.

[64] C. Wang, X. Liu, J.P. Chen, K. Li, Superior removal of arsenic from water with zirconium metal-organic framework UiO-66, Sci. Rep. 5 (2015) 16613.

[65] A. Abbasi, T. Moradpour, K. Van Hecke, A new 3D cobalt (II) metal–organic framework nanostructure for heavy metal adsorption, Inorganica Chim. Acta 430 (2015) 261-267.

[66] Sana Jamshidifard, Shahnaz Koushkbaghi, Seyedehgolshan Hosseini, Sina Rezaei, Alireza Karamipour, Azadeh Jafari rad, Mohammad Iran, Incorporation of UiO-66-NH2 MOF into the PAN/chitosan nanofibers for adsorption and membrane filtration of Pb(II), Cd(II) and Cr(VI) ions from aqueous solutions, Journal of Hazardous Materials, 368 (2019) 10-20.

[67] Y. Yurekli, Removal of heavy metals in wastewater by using zeolite nano-particles impregnated polysulfone membranes, J. Hazard. Mater. 309(2016) 53-64.

[68] S.M. Hosseini, S. Rafiei, A.R. Hamidi, A.R. Moghadassi, S.S. Madaeni, Preparation and electrochemical characterization of mixed matrix heterogeneous cation exchange membranes filled with zeolite nanoparticles: ionic transport property in desalination, Desalination 351 (2014) 138-144.

[69] Mukherjee R, Bhunia P, De S (2016) Impact of graphene oxide on removal of heavy metals using mixed matrix membrane. Chem Eng J 292:284-297.

[70] Subramaniam M.N., Goh P.S., Lau W.J., Tan Y.H., Ng B.C., Ismail A.F., 2017, Hydrophilic hollow fiber PVDF ultrafiltration membrane incorporated with titanate nanotubes for decolourization of aerobically-treated palm oil mill effluent, Chem. Eng. J. 316, 101-110.

[71] Yan L., Li Y.S., Xiang C.B., 2005, Preparation of poly (vinylidene fluoride)(pvdf) ultrafiltration membrane modified by nano-sized alumina (Al2O3) and its antifouling research, Polymer, 46, 7701-7706.

[72] Maximous N., Nakhla G., Wan W., Wong K., 2009, Preparation, characterization and performance of Al2O3/PES membrane for wastewater filtration, J. Membr. Sci., 341, 67-75.

[73] Gholami F., Zinadini S., Zinatizadeh A. A., Abbasi A. R., 2018, TMU-5 Metal-Organic frameworks (MOFs) as a novel nanofiller for flux increment and fouling mitigation in PES

ultrafiltration membrane, Sep. Purif. Technol., 194, 272-280.

[74] Xiao Y.T., Xu C.X., Geng H.Z., Ji Q., Wang L., He B., Jiang Y., Kong J., Lia J., 2020, Multifunctional PVDF/CNT/GO Mixed Matrix Membranes for Ultrafiltration 1and Fouling Detection, J. Hazard. Mater., 384, 120978.

[75] Liang S., Xiao K., Zhang S., Ma Z., Lu P., Wang H., Huang X., 2018, A facile approach to fabrication of superhydrophilic ultrafiltration membranes with surface-tailored nanoparticles, Sep. Purif. Technol., 203, 251-259.

[76] Zhou J., Sun D., Wang L., Guo L., Chen W., Yu F., Wang Y., Yang Y., 2019, Two-dimensional superstructures filled into polysulfone membranes for highly improved ultrafiltration: The case of cuprous iodide nanosheets, J. Membr. Sci., 576, 142-149.

[77] Nasrollahi N., Aber S., Vatanpour V., Mahmoodi N.M., 2019, Development of hydrophilic microporous PES ultrafiltration membrane containing CuO nanoparticles with improved antifouling and separation performance, Mater. Chem. Phys., 222, 338-350.

[78] Yang S., Zou Q., Wang T., Zhang L., 2019, Effects of GO and MOF@GO on the permeation and antifouling properties of cellulose acetate ultrafiltration membrane, J. Membr. Sci., 569, 48-59.

[79] Liu Q., Huang S., Zhang Y., Zhao S., 2018, Comparing the antifouling effects of activated carbon and TiO_2 in ultrafiltration membrane development, J. Colloid Interf. Sci., 515, 109-118.

[80] Jiang Y., Zeng Q., Biswas P., Fortner J.D., 2019, Graphene oxides as nanofillers in polysulfone ultrafiltration membranes: Shape matters, J. Membr. Sci., 581, 453-461.

[81] Mu Y., Zhu K., Luan J., Zhang S., Zhang C., Na R., Yang Y., Zhang X., Wang G., 2019, Fabrication of hybrid ultrafiltration membranes with improved water separation properties by incorporating environmentally friendly taurine modified hydroxyapatite nanotubes, J. Membr. Sci., 577, 274-284.

[82] Farjami M., Moghadassi A., Vatanpour V., Hosseini S.M., Parvizian F., 2019, Preparation and characterization of a novel high-flux emulsion polyvinyl chloride (EPVC) ultrafiltration membrane incorporated with boehmite nanoparticles, J. Ind. Eng. Chem., 72, 144-156.

[83] Wu L.G., Zhang X.Y., Wang T., Du C.H, Yang C.H., 2019, Enhanced performance of polyvinylidene fluoride ultrafiltration membranes by incorporating TiO2/graphene oxide, Chem. Eng. Res. Des., 141, 492-501.

[84] Isawi H., 2019, Evaluating the performance of different nano-enhanced ultrafiltration membranes for the removal of organic pollutants from wastewater, J.Water Process Eng., 31, 100833.

[85] Zhang G., Zhou M., Xu Z., Jiang C., Shen C., Meng Q., 2019, Guanidyl-functionalized graphene/polysulfone mixed matrix ultrafiltration membrane with superior permselective, antifouling and antibacterial properties for water treatment, J. Colloid Interf. Sci., 540, 295-305.

[86] Pendergast M.T.M., Nygaard J.M., Ghosh A.K., Hoek E.M., 2010, Using nanocomposite materials technology to understand and control reverse osmosis membrane compaction, Desalination, 261, 255-263.

[87] Heinz H., Pramanik C., Heinz O., Ding Y., Mishra R.K., Marchon D., Flatt R.J., Lopis I.E., Llop J., Moya S., Ziolo R.F., 2017, Nanoparticle decoration with surfactants: Molecular interactions,

assembly, and applications, Surf. Sci. Rep., 72, 1-58.

[88] Ayyaru S., Ahn Y.H., 2018, Fabrication and separation performance of polyethersulfone/ sulfonated TiO2 (PES–STiO2) ultrafiltration membranes for fouling mitigation, J. Ind. Eng. Chem., 67, 199-209.

[89] Ayyaru S., Ahn Y.H., 2017, Application of sulfonic acid group functionalized graphene oxide to improve hydrophilicity, permeability, and antifouling of PVDF nanocomposite ultrafiltration membrane, J. Membr. Sci., 525, 210-219.

[90] Ayyaru S., Pandiyan R., Ahn Y.H., 2019, Fabrication and characterization of anti-foulingand non-toxic polyvinylidene fluoride-Sulphonated carbon nanotube ultrafiltration membranes for membrane bioreactors applications, Chem. Eng. Res. Des., 142, 176-188.

Section 3

Ions Exchange

Chapter 8

A Comprehensive Method of Ion Exchange Resins Regeneration and Its Optimization for Water Treatment

Sameer Al-Asheh and Ahmad Aidan

Abstract

Ion exchange membranes, specifically resin technology, lie at the heart of electrolytically conductive systems used in the treatment of wastewater. This chapter deals with ion exchange deionization and the effect of resin amount as well as the concentration of acid and base on the product conductivity. The strong acidic cation polymeric exchanger resin is commercially called MERCK 104765 cation exchanger IV with capacity greater than 3.2 mmol/ml, while the strong basic anion polymeric exchanger resin is commercially called MERCK 104767 anion exchanger III with capacities greater than 1.0 mmol/ml. Water conductivity, as an indicator of regeneration efficiency, was monitored over time at the different conditions and scenario. In general, it was observed that the conductivity decreases with time until one point is reached and then starts to increase as a result of resin saturation. It was also noticed that the lowest conductivity is achieved when using 1-vol% NaOH and 5-vol% HCl in the cathodic and anodic resin tubes, respectively, and that water conductivity increases with the increase in the amount of water being used. The amount of resin significantly impacts the deionization efficiency; more ions are removed as the amount of resin increases.

Keywords: ionic resin, cationic resin, NaOH, HCl

1. Introduction and literature review

It is well known that such mineral as calcium and magnesium, in particular, along with iron and manganese cause water harness, the existence of which may cause scaling problems and serious failures in pipelines of boilers and heat- transfer equipment. Also, these divalent ions can react with soap anions and thus decreasing the cleaning efficiency resulting in high consumption of detergents. Temporary hardness is a type of water hardness caused by the presence of dissolved bicarbonate minerals (calcium bicarbonate and magnesium bicarbonate). When dissolved, these type of minerals yield calcium and magnesium cations (Ca^{2+}, Mg^{2+}) and carbonate and bicarbonate anions ($CO^{2-}{}_3$ and $HCO^-{}_3$). The presence of the metal cations makes the water hard. However, unlike the permanent hardness caused by sulfate and chloride compounds, this "temporary" hardness can be reduced either by boiling the water, or by the addition of lime (calcium hydroxide) through the process of lime softening. Boiling promotes the formation of carbonate from the bicarbonate and precipitates calcium carbonate out of solution, leaving water that is softer upon cooling. Permanent hardness (mineral content) is generally difficult to remove by boiling. If this occurs, it is usually caused by the presence of calcium sulfate/calcium chloride and/or magnesium sulfate/magnesium chloride in the water, which do not precipitate out as the temperature increases. Ions causing permanent hardness of water can be removed using a water softener, or ion exchange column.

Calcium is the most abundant mineral in the human body. Calcium plays vital roles in the structure and function of the human body [1]. It is substantial for intracellular metabolism, bone growth, blood clotting, nerve conduction, muscle contraction and cardiac function [2]. However, there is a significant association between calcium level in drinking water and colorectal, gastric and breast cancer [3]. Magnesium is a naturally occurring mineral that is found in food and other medical products. It is an essential component in the in bones and in muscles and other tissues. However, the too much supplement of magnesium may result in symptoms of toxicity, such as a fall in blood pressure, confusion, abnormal cardiac rhythm, muscle weakness, difficulty breathing and deterioration of kidney function [4].

Therefore, it is important to control the level of hardness in water. Several technologies are available for removal of water hardness (Ca^{2+}, Mg^{2+}), such as electrochemical processes [5], Enzyme catalyzed [6], Nanofiltration [7], electro-dialysis [8], ultrasound [9], ultra-filtration [10], ion exchange [11, 12], membranes [13] and pulsed spark discharge [14]. The two major methods which are typically used to remove hardness form water are lime soda softening and ion exchange softening. The high operating cost of lime soda softening, which is mostly used for municipal purposes [15], is attributed to the production of a large volume of sludge that requires post-treatment, excessive use of chemicals (e.g., lime soda ash, and caustic soda) and the addition of acids for pH adjustment [16]. The ion exchange process is primarily employed for residential water softening. Experimental studies have found the sodium level in softened water was 2.5 times higher than municipal water [17]. Adsorption is one of the few promising alternatives for this purpose, especially using low-cost sorbets. A three-dimensional super absorption polymer were used as a absorbents for heavy metal ions from water and other aqueous solution susficialy. Polymer like polyacrylic acid, polyacryliamide and its derivatives which have functional groups (such as carboxylic, hydroxyl and amide) can be used as absorbents for metal ions removal via the interaction between the metal ions and these groups [18].

Ion exchange resins are classified as cation exchangers, which have positively charged mobile ions available for exchange, and anion exchangers, whose exchangeable ions are negatively charged. Both anion and cation resins are produced from the same basic organic polymers. Classification of ion exchange resins based on their sources is illustrated in **Figure 1**.

In the early 1800s, several scientists discovered the ion exchange process. The ion exchange industry in the United States was born in the early 1900s when cation exchangers were first synthesized. Ion exchange resins are used for many water treatment applications. Of these applications, in terms of the volume of resins used, water softening and demineralization of water are the most significant. In 1905, the German chemist Gans used sodium aluminosilicate materials (*zeolites*) to soften water. These materials can absorb calcium or magnesium cations and liberate sodium ions, the basis of water softening. The zeolite could be regenerated by running concentrated NaCl solution through it to liberate $CaCl_2$ and $MgCl_2$. While aluminosilicates are seldom used anymore, the term zeolite softener is occasionally used for any cationic exchange resin [19]. In 1944, cationic exchange resins were

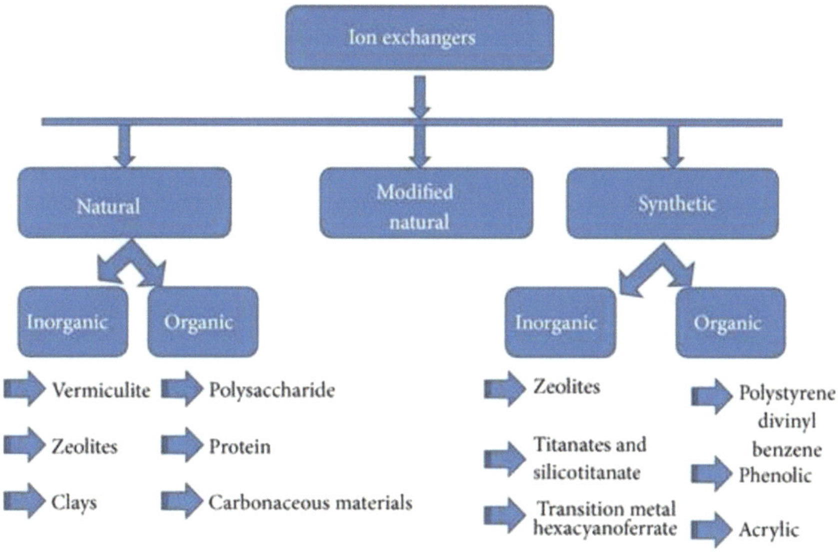

Figure 1.
Classifications of ion exchange resins.

developed based on polystyrene cross-linked with about 6–8% divinylbenzene. This resin had a much higher capacity than previous resins, meaning that the same volume of resin could exchange many times more ions than previous materials. In 1948, the anion exchange analog was developed. Using these two resins in sequence allows the complete demineralization of water, instead of merely exchanging cations for sodium ions. The macroreticular anion resins that have discrete pores are used first in the sequence to remove organic materials from the raw water that would otherwise cause fouling of the standard gelular resins. Many modifications of the original polystyrene divinylbenzene have been made to accomplish a wide variety of separations [19].

Water softening has been practiced commercially for a century or more, making use of a wide range of natural and synthetic products. As the variety of uses for purified water has increased, so has the need to soften and demineralize water. Demineralization has only been practiced since the discovery of synthetic anion exchange resins in the 1920s. Their usefulness increased greatly with the invention of strongly basic anion exchange resins, which can remove weakly acidic compounds such as silica and carbon dioxide, as well as mineral acids. This process of ion exchange can be used as a simple method to produce water of very high purity. In general, as industrial and domestic requirements have grown, specifications for water quality have become progressively more stringent, and regulations to enforce these have become more strict. Hence the choice of resin types for a particular application becomes increasingly complex [20].

The ion exchange (IX) technology, an already existing technology, is considered an economical and and environmentally sustainable method for removing hardness (Ca^{2+} and Mg^{2+} ions removal) [21]. The theory behind the ion exchange (electro-chemical process) is to exchange positive and negative ions in water with hydronium (H^{+}) and hydroxide ions (OH^{-}). In order for this process to work, a 1-mm porous bed compromising of insoluble polymers and countless ion exchange sites must be used. When water passes through the bed, positive metallic ions (such as sodium, calcium and aluminum) are exchanged with H^{+}, while others such as chloride, nitrate and sulfate will be replaced by OH^{-} [22]. It can also be used to remove nitrates [23], arsenic [24], dissolved organic carbon (DOC) [25] and other heavy metal like cobalt [26], nickel, silver zinc and copper [27]. In general, the ion exchange method is vastly adequate in its ability to purify and separate a wide range of chemical compounds in the water treatment field. The medium used in most cases are IX polymer resin that is made of organic polymer structures [21]. Ion exchange process is considered a cost-effective solution and can remove microscale ion particles. However, it does not remove bacteria or pyrogens, and once all exchange sites are used up no ion exchange will occur. Furthermore, any ion exchange unit consists of two main types of resins: Cation exchange resins (CER) and Anion exchange resins (AER). CERs are used to remove cations from water and wastewater, particularly calcium and magnesium ions which are the two main contributors to water hardness [28].

The water from the main water supply enters the softening system and essentially flows through the bed of resin; the working principle of the system is such that the positively charged Ca and Mg ions are attracted to the resin beads [8]. An ion exchange then takes place wherein the Ca and Mg ions are replaced by weaker Na ions. Once the "hard" minerals are fully extracted, the water is termed as "soft" water [29]. As time goes by, the resin beads become saturated with hard ions and need to be regenerated to continue the ion exchange process [29]. Conventionally, the resins are made up of cross-linked polymer chains with ionically active sites and are usually in a bead-like form [30]. The cationic membranes are charged with a bulk of negative ions and available positive counter ions for the ion exchange process [30]. Characterization of these ionic exchange membranes and relating this to their different structures, including their physiochemical properties are reported elsewhere [31]; this involved procuring different types of ion-exchange membranes from different companies namely, Russia, Japan, USA and China. In addition, laboratory samples were also prepared from aromatic polymers and several perflourinated compounds [32]. The chemical characteristics of the ion-exchange resins are demonstrated by studying the kinetics of a sodium-hydrogen ion exchanger on sulphonated cross-linked polystyrenes [33]. The rate of exchange was measured using an indicator and the results showed a higher rate of ion exchange at a Na^+ concentration greater than 1 N; however, at very high concentrations, the rate of exchange was found to be independent of the sodium concentration and inversely proportional to the particle radius [33].

Ion-exchange provides several advantages over other water treatment methods. Ion exchange process is environmentally friendly, is low-cost since the resin can be fervently regenerated and produces high output [34]. However, it has some disadvantages as well, adsorption of organic matter and calcium sulfate fouling are the most important ones [34]. Calcium sulfate precipitate blocks the resin beads and may also cause blocking of pipes in the system; adsorption of organic matter may reduce resin efficiency, and thus reducing quality of water. Therefore, eventually at one point the resin becomes saturated and ineffective in exchanging ions, and thus must be recharged or regenerated.

Regeneration is the process at which ion exchanging capabilities of saturated resins are recovered. Available regenerators may include salts like NaCl and KCl, acids such as acetic and citric acids, and alkalis including NaOH and KOH. One of the methods to maximize the efficiency of the regeneration process is the flow pattern, in which the regenerator can be introduced from the bottom to the upper side of the ion exchange bed. As a result, the resins at the bottom are cleaner as they undergo more regeneration than the resins in the upper side of the ion exchange unit. Thus, the treated water leaving the exchanger (downward flow) will be in contact with cleaner less saturated resin, which increases the ion exchange efficiency and results in purer water [35]. Ion regeneration is the removal of the ions that block or plug the internal exchange sites on resins; it includes the following three main steps: backwashing, regeneration and rinsing. The regeneration process is capable of restoring only 60–80% of the resin capacity, as some ions and hardness are retained on the resins. When service starts, these ions might leach off the resin and leave the bed with the treated water effluent, this case is known as leakage [36]. Thus, the regeneration cycle continues until the ions that were removed from the feed water during the service process are recovered from the resins or meet the allowable limits. This can be monitored by electrical conductivity measurements, where the change of electrical conductivity is measured at different volumes of rinsed water during succeeding resin washing cycles. It is expected that after each washing cycle, the electrical conductivity of water will decrease until it reaches distilled water conductivity [37].

Brining process referred to the one when a sodium cycle is used to soften water, where concentrated sodium chloride (NaCl) is passed through the bed to remove calcium and magnesium ions. It can be done in up flow or down flow mode. If sodium cycle is employed to soften water containing the iron ions, this may lead to lowering the resin capacity. This is because iron will be oxidized to the ferric insoluble form in which it blocks the exchange sites on the resin. This can be minimized by performing more frequent and quick regeneration so that iron would not have enough time to oxidize and blocks the sites. Moreover, the process can be enhanced by mixing sodium bisulfate with sodium chloride. This will reduce the insoluble ferric ion to its soluble form [38].

Deionized water has many applications: it can be used as a cooling medium, an agent for lab testing, in car engines and many more other applications [39]. In this work, tap water was deionized using HCl and NaOH as regeneration agents for the cathodic resin and the anodic resin, respectively. The effect of acid and concentrations, amount of water treated and mass of resins on the deionization process will be considered in this work.

2. Material and methods

2.1 Apparatus

The CE 300 Ion exchange demonstration unit (Gunt Hamburg, Germany) is used in this work. The apparatus facilitates tests relating to water softening and demineralization; it is equipped with both cation and anion exchangers with strong and weak basic or acidic contents. The unit layout is illustrated in **Figure 2**. CE 300 enables water deionization with the aid of cation and anion exchangers. The raw water is pumped from the tank into the top of the cation exchanger. In the softening process the water flows from there back into the collecting tank. To desalinate the raw water, it is then additionally routed through the anion exchanger. From there the treated water passes into the collecting tank. In the regeneration process, acid or caustic is fed into the ion exchangers from below using the same pump. The acid and caustic used is collected in the collecting tank. The flow rate of the pump is adjustable, and can be read from a flow meter before it enters the first ion exchanger. For continuous evaluation of the process, a conductivity sensor is installed upstream of the inlet into the collecting tank. The measured values can be read from a conductivity meter. Samples can be taken at all relevant points. Tap water can be used as raw water.

Commercial cation and anion exchangers were provided by Gunt Hamburg (Germany). The strong acidic cation polymeric exchanger resin is commercially called

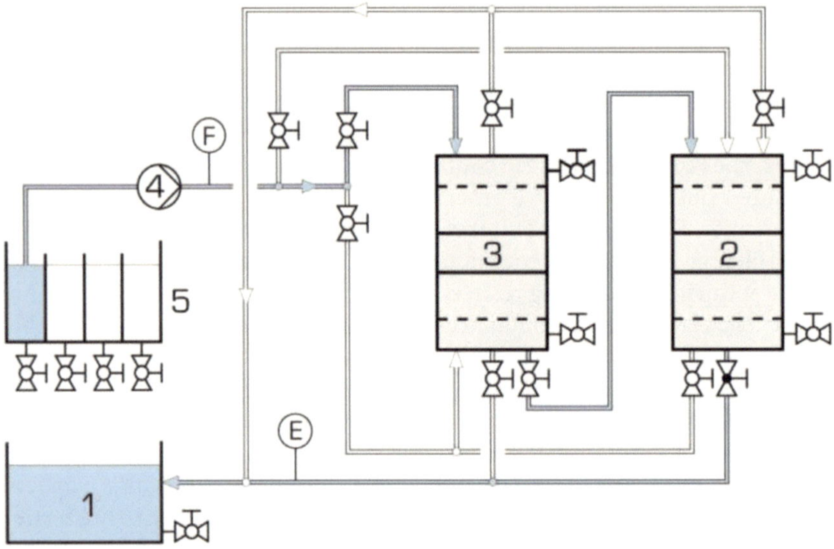

Figure 2.
CE300 unit diagram: Flow path with the two ion exchangers configured in series (desalination): 1 collecting tank, 2 anion exchanger, 3 cation exchanger, 4 pump, 5 raw water tank; E conductivity, F flow rate.

MERCK 104765 cation exchanger IV with capacity greater than 3.2 mmol/ml; while the strong basic anion polymeric exchanger resin is commercially called MERCK 104767 anion exchanger III with capacities greater than 1.0 mmol/ml.

2.2 Methodology

The set-up was prepared by opening/closing valves as described. Solutions of 5 vol% of HCl and 0.1 vol% of NaOH each in 100 mL distilled water were prepared. The cation tube was filled with the 5% HCl solution and the anion tube with the 0.1% NaOH solution; each containing 20 g polymeric resin. The process started by pumping the hard water through the column, and water was allowed to pass through the outlet tubes. Once a steady flow is passing through the outlet, the conductivity of the water at the outlet was recorded at time intervals of 10 seconds. The pump turned off when the conductivity values start to increase until reaching a steady state. The experiment was repeated with other concentrations of HCl and NaOH with 0.5 and 0.1 increments, respectively.

For the purpose of studying the effect of amount of water treated, the process was repeated at certain concentrations of HCl and NaOH with the tubes filled with different amounts of hard water. Also, the effect of amount of resin used was studied by pacing different amounts of resin at the different runs for given concentrations of HCl and NaOH.

3. Results and discussion

3.1 Effects of acid and base concentrations on deionazation

Resin generation is important from cost point of view as well as minimizing solid waste. In this work, resin regeneration using different combinations of acid in the cationic resin and base in the anionic resin, each at different concentrations, was accomplished. Each bed was fed with 20 g of polymeric resin and each filled with 2 L of hard water. The hard water is municipality tab water with a hardness conductivity of 100,000 μS and a TDS of 600 mg/L. The experiment was started off by putting 0.5 vol% HCl and 0.1 vol% NaOH in the cathodic and anodic resin tubes, respectively; and in every different run the vol% was increased by 0.5 and 0.1 for each of the aforementioned resins. The results for water conductivity at different combinations of acid-base are shown in **Figure 2**, having in mind that the conductivity of purely deionized water is 1.1 micro Siemens (μS).

The experiment was started off by using 1-vol% NaOH and 5-vol% HCl in the cathodic and anodic resin tubes, respectively. It is seen (**Figure 3**) that conductivity starts to decrease indicating that the resin is deionizing the water until a point is reached where the conductivity starts to increase again, indicating that the resins have reached accumulation point. Accumulation point was reached after 5 seconds of running the experiment, and the maximum value of regeneration was found to be 1550 μS. When 0.9 vol% and 4.5 vol% of NaOH and HCl, respectively, were used the conductivity kept decreasing until a value of 1076 μS was reached in 45 seconds. The same trend was noticed when the NaOH and HCl vol% were decreased to 0.8 and 4% respectively; however, in this run a higher conductivity of 2270 was obtained in 45 seconds. This shows that as the amount of resin decreases, the conductivity of water increases indicating that fewer ions were removed. In **Figure 3**, a conductivity of 2140 μS was obtained at 45 seconds, when 0.9 and 4.5% NaOH and HCl, respectively, are used. More time was required to deionize the water for the case of 0.9 and 4.5% NaOH and HCl, respectively; at 45 seconds the conductivity was found to be 2360 μS. Furthermore, the conductivity was found to be 2260 μS at 45 seconds for the case 0.6 and 3.0% NaOH and HCl, respectively. Furthermore,

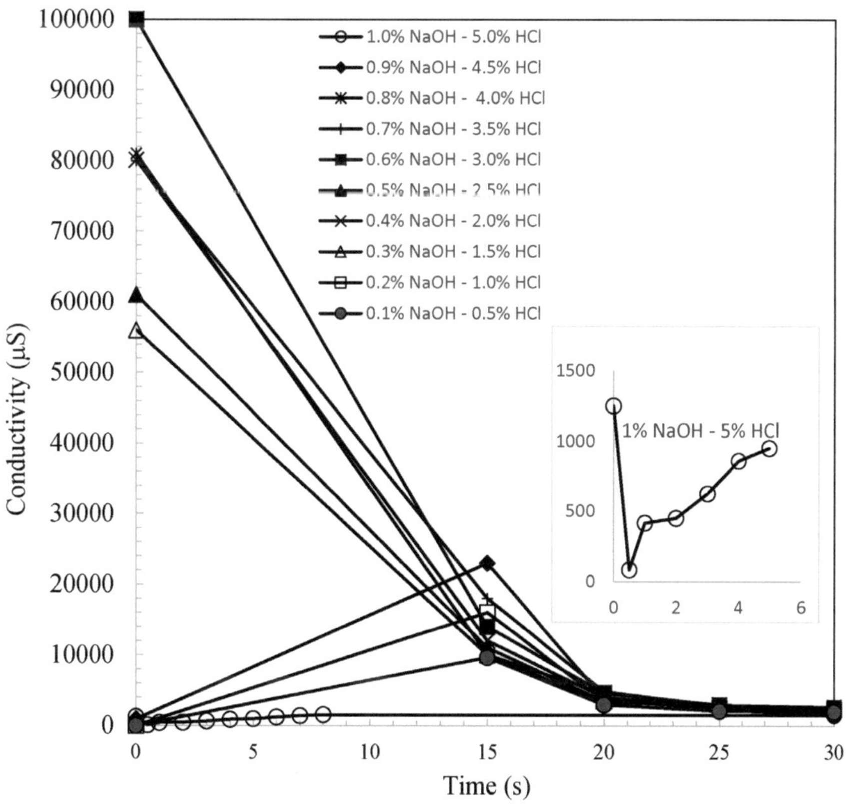

Figure 3.
Effect of acid-base concentrations on resin regeneration; resin amount: 20 g in both tubes; volume treated: 2 L.

as the vol% decreased, the conductivity started giving similar results at 45 seconds. For example, for the cases of 0.3% NaOH—1.5% HCl, 0.2% NaOH—1.0% HCl, and 0.1% NaOH—0.5% HCl, the conductivity was found to be 2300, 2100, and 1867 μS, respectively, which are close to each other. In conclusion, to have best results, the resin vol% should be high.

3.2 Effect of treated volume on deionization

Effect of treated voume of water was also investigated using same amount of resins in each bed (20 g) and same eluants concentrations in the cationic and anionic resins (5 vol% HCl and and 1 vol% NaOH, respectively). The results for water conductivity at different amounts of treated water are shown in **Figure 4**. As shown in **Figure 3** for the case of 1 L, the conductivity started to decrease indicating that the resins are deionizing the passing water. It started off with a conductivity of 864 μS which is the conductivity of tap water and ended with a conductivity of 73.19 μS. When the volume increased to 2 L, the lowest conductivity was found to be 70 μS after 140 seconds have passed. On the other hand, when 3 L of water was added, 1.14 L of water was treated and the lowest conductivity was found to be around 77.2 μS; after this point, the conductivity increased again and that happened at 70 seconds (**Figure 4**). For the case of 4 L of water being added, an amount of 1.63 L of water was treated. The lowest conductivity was found to be 242 μS and then it started to increase at 70 seconds. This shows that as more water is being treated, the conductivity increases.

3.3 Effect of amount of resin

The effect of amount of resin used in cationic anionic resin tubes on deionixation is also studied by putting different amounts of resins in each tube with

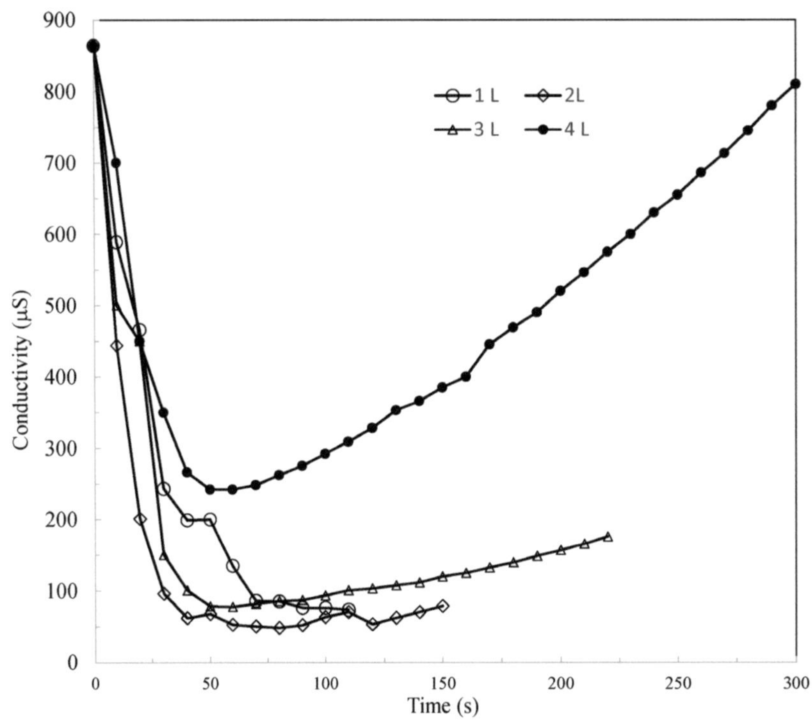

Figure 4.
Effect treated water on resin regeneration; resin amount: 20 g in both tubes; 5 vol% HCl—1 vol% NaOH.

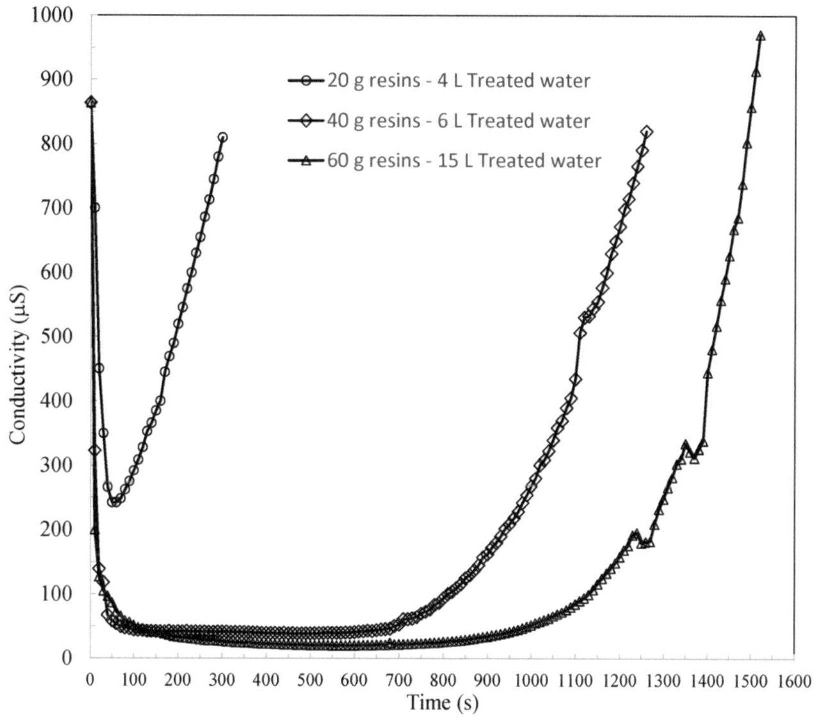

Figure 5.
Effect amount of resins on resin regeneration using 5-vol% HCl in cationic resin tube and 1-vol% NaOH in anionic resin tube.

different volumes of hard water as 3 using 5 vol% HCl in the cationic resin tube and 1 vol% NaOH in anionic resin tube. The results are shown in **Figure 5**. The case of 20 g resin is the same as the described in the previous section. When the amount of resin was increased to 40 g, the same trend was obtained as that of the 20 g resin. However, more water was deionized in this case because more resin was used. In other words, it required around 480 seconds to saturate the resin and to stop the deionization process. This shows that increasing the resin amount, helps in increasing the deionization efficiency. The resin amount was again increased to 60 g. The water was being deionized until 560 seconds were reached and a conductivity of 19.8 μS was obtained. After this point, the conductivity started to increase again indicating that the resin has been saturated. As the resin amount increased, the apparatus was able to give better results in terms of removing more ions from the water. For example, at 500 seconds the conductivity was found to be 38.6 and 20.4 μS for the 40 and 60 g, respectively. This shows that the amount of resin is related to the deionization efficiency. As the amount of resin increases, more ions are removed.

4. Conclusions

Water conductivity decreases with the increase in resins concentrations; the lowest conductivity is achieved when using 1-vol% NaOH and 5-vol% HCl in the cathodic and anodic resin tubes, respectively. The results of this work show that water conductivity increases with the increase in the amount of water being used. The amount of resin significantly impacts the deionization efficiency; more ions are removed as the amount of resin increases. The optimization implemented in this work is considered superior compared to other deionization techniques due to life time and efficiency of the reused resins.

Acknowledgments and Disclaimer

The work in this book chapter was supported, in part, by the Open Access Program from the American University of Sharjah.

This book chapter represents the opinions of the author(s) and does not mean to represent the position or opinions of the American University of Sharjah.

Author details

Sameer Al-Asheh* and Ahmad Aidan
Department of Chemical Engineering, American University of Sharjah, Sharjah, UAE

*Address all correspondence to: sslasheh@aus.edu

References

[1] Schroeder HA, Nason AP, Tipton IH. Essential metals in man: Magnesium. Journal of Chronic Diseases. 1969;**21**:815-841

[2] Akyilmaz E, Kozgus O. Determination of calcium in milk and water samples by using catalase enzyme electrode. Food Chemistry. 2009;**115**:347-351

[3] Yang CY, Chiu HF. Calcium and magnesium in drinking water and risk of death from rectal cancer. International Journal of Cancer. 1998;**77**:528-537

[4] Graber TW, Yee AS, Baker FJ. Magnesium: Physiology, clinical disorders and therapy. Annals of Emergency Medicine. 1981;**10**:49-57

[5] Zeppenfeld K. Electrochemical removal of calcium and magnesium ions from aqueous solutions. Desalination. 2011;**277**:99-105

[6] Mary AA, Kristen SB, Minteer SD, He Z. Enzyme catalyzed electricity-driven water softening system. Enzyme and Microbial Technology. 2012;**51**:396-401

[7] Charis MG, Fountoulis G, Gekas V. Nanofiltration of brackish groundwater by using a polypiperazine membrane. Desalination. 2012;**286**:277-284

[8] Kabay N, Demicioglu M, Ersöz E, Kurucaovali I. Removal of calcium and magnesium hardness by electrodialysis. Desalination. 2002;**149**:343-349

[9] Entezari MH, Tahmasbi M. Water softening by combination of ultrasound and ion exchange. Ultrasonics Sonochemistry. 2009;**16**:356-360

[10] Abrahamse AJ, Lipreau C, Heijman SGJ. Removal of divalent cation reduces fouling of ultrafiltration membranes. Journal of Membrane Science. 2008;**323**:153-158

[11] Muraviev D, Noguerol J, Valiente M. Separation and concentration of calcium and magnesium from sea water by carboxylic resins with temperature-influence selectivity. Reactive & Functional Polymers. 1996;**28**:111-126

[12] Şahin M, Görçay H, Kir E, Şahin Y. Removal of calcium and magnesium using polyaniline and derivative modified PVDF cation-exchange membranes by Donnan dialysis. Reactive & Functional Polymers. 2009;**69**:673-680

[13] Park J-S, Song J-H, Yeon K-H, Moon S-H. Removal of hardness ions from tap water using electromembrane processes. Desalination. 2007;**202**:1-8

[14] Yang Y, Kim H, Starikovskiy A, Fridman A, Cho YI. Application of pulsed spark discharge for calcium carbonate precipitation in hard water. Water Research. 2010;**44**:3659-3668

[15] Bergman R. Membrane softening versus lime-softening in Florida: A cost comparison update. Desalination. 1995;**102**:11-24

[16] Randtke S, Hoeha R. Removal of DBP Precursors by Enhanced Coagulation and Lime Softening. Denver, Co: American Water Work Association Research Foundation and American Water Work Association; 1999. Available from: http://www.waterrf.org/projects reports/Public Report Library/RFR 90783-1999-814.pdf [Accessed: 30 April 2012]

[17] Yarows S, Fusilier W, Wider A. Sodium concentration of water from softeners. Archives of Internal Medicine. 1997;**157**(2):218-222

[18] Snakeskin J, Gefeniene A. Co-sorption of metal cations and

nonionic surfactant in polyacrylic acid functionalized cation-exchanger. Reactive and Functional Polymers. 2000;**46**:109-115

[19] Bajpai P, Biermann's Handbook of Pulp and Paper. 3rd ed. London, United Kingdom: Elsevier; 2018

[20] Irving J. Encyclopedia of Separation Science. New Jersey, USA: Wiley; 2000

[21] Sata T, Jones GN, Sata T. Ion Exchange Membranes: Preparation, Characterization, Modification and Application. Cambridge: Royal Society of Chemistry; 2007

[22] ELGA LabWater—Water Purification Technologies, Ion Exchange, *Elgalabwater.com*. [Online]. Available from: http://www.elgalabwater.com/en-gb/technologies-en/ion-exchange-en [Accessed: 05 March 2017]

[23] Amini A, Kim Y, Zhang J, Boyer T, Zhang Q. Environmental and economic sustainability of ion exchange drinking water treatment for organics removal. Journal of Cleaner Production. 2015;**104**:413-421

[24] Burge S, Halden R. Nitrate and perchlorate removal from groundwater by ion exchange. American Water Works Association. 1999:85-96

[25] Song H, Li A, Zhou Y. Selective removal of DOM on anion-exchange resin from water. In: Functions of Natural Organic Matter in Changing Environment. New York, USA: Springer; 2012. pp. 921-924

[26] Rengaraj S, Moon S-H. Kinetics of adsorption of Co(II) removal from water and wastewater by ion exchange resins. Water Research. 2002;**36**(7):1783-1793

[27] Hubicki Z, Koodynsk D. Selective Removal of Heavy Metal Ions from Waters and Waste Waters Using Ion Exchange Methods. Hubicki and Kolodynska, licensee InTech: Ion Exchange Technologies; 2012

[28] Comstock SEH, Boyer TH. Combined magnetic ion exchange and cation exchange for removal of DOC and hardness. Chemical Engineering Journal. 2014;**241**:366-375

[29] How Does an Ion Exchange Water Softener Work? | Ecowater, *Ecowater-softeners.co.uk* [Online]. 2019. Available from: https://www.ecowater-softeners.co.uk/how-does-ion-exchange-water-softener-work [Accessed: 10 May 2019]

[30] Wheaton RM, Lefevre LJ. DOWEX: Ion exchange resins. In: Fundamentals of Ion Exchange. USA: Dow Chemical, Lenntech; 2000

[31] Berezina NP, Kononenku NA, Dyomina OA, Gnusin NP. Characterization of ion-exchange membrane materials: Properties vs structure. Advances in Colloid and Interface Science. 2008;**139**(1-2):3-28

[32] Reichenberg D. Properties of ion exchange resins in relation to their properties. III. Kinetics of exchange. Journal of the American Chemical Society. 1953;**75**(3):589-597

[33] What Is the Difference Between Cation and Anion Exchange Resins? [Online] *www.samcotech.com*, 2019. Available from: https://www.samcotech.com/difference-cation-anion-exchange-resins/ [Accessed: 08 May 2019]

[34] The Disadvantages of Ion Exchange, *Sciencing.com* [Online]. 2019. Available from: https://sciencing.com/disadvantages-ion-exchange-8092882.html [Accessed: 08 May 2019]

[35] Alchin D. Ion Exchange Resins, Research Gate [Online] 2016. Available from: https://www.researchgate.net/publication/305658612_ION_EXCHANGE_RESINS [Accessed: 05 March 2019]

[36] What Is Regeneration, *Purolite.com* [Online]. 2019. Available from: https://www.purolite.com/about-us/what-is-regeneration [Accessed: 05 March 2019]

[37] Chandrasekara N, Pashley R. Study of a new process for the efficient regeneration of ion exchange resins. Desalination. 2015;**357**:131-139. DOI: 10.1016/j.desal.2014.11.024 [Accessed: 10 May 2019]

[38] Wachinski A. Environmental Ion Exchange: Principles and Design. 2nd ed. Milt: CRC Press LLC; 2016

[39] Thedistilledwatercompany.com. 6 Uses for Deionised Water [Online]. 2019. Available from: https://www.thedistilledwatercompany.com/6-uses-deionised-water

Section 4

Modeling

Chapter 9

CFD Simulations in Mechanically Stirred Tank and Flow Field Analysis: Application to the Wastewater (Sugarcane Vinasse) Anaerobic Digestion

Hélène Caillet, Alain Bastide and Laetitia Adelard

Abstract

Anaerobic digestion is a widely used process for waste treatment and energy production. This natural process takes place in a controlled environment, anaerobic digesters. Mixing is one of the main operating parameters. The understanding of the flows during the agitation of the medium is crucial for the optimization of the process yield. In fact, the mass and heat transfers are enhanced by the agitation. However, the complex biochemical reactions can be inhibited with overly vigorous agitation. A detailed and in-depth understanding of the phenomena occurring during agitation requires modeling studies. In this chapter, we propose a general approach, based on computational fluid mechanics (CFD), to analyze the mechanical mixing of an anaerobic reactor. We apply this work to the anaerobic digestion of the sugarcane vinasse, which is a liquid waste generated during the production of alcohol. The single-phase Reynolds-averaged Navier-Stokes (RANS) simulations of mechanical agitation of Newtonian fluids for different rotational speeds are presented. The equations system is closed with the standard k-epsilon turbulence model. The flow field is analyzed with the velocity profiles, the Q and Lambda2 fields, the pressure and the vorticity.

Keywords: anaerobic digestion, computational fluid dynamics (CFD), Newtonian, mechanical mixing, vinasse, RANS

1. Introduction

Anaerobic digestion is a widely used process for organic waste treatment, such as manure, sludge or biowaste. This natural process is based on the digestion of the material in five biological reactions. Each reaction involves a specific group of micro-organisms. First, the complex polymers are hydrolyzed into water-soluble monomers and oligomers by the hydrolytic and fermentative bacteria. Then, the acidogenesis consists in the transformation of the soluble organic molecules into volatile fatty acids, organic acids, hydrogen, carbon dioxide, alcohols and acetate by acidogenic bacteria [1, 2]. This reaction is followed by the acetogenesis, during which the products of hydrolysis and acidogenesis are converted into acetate and carbon dioxide by the action of acetogenic bacteria [1, 2]. Finally, the methanogenesis, the last step of the degradation process, forms methane through two metabolisms. Acetotrophic methanogens transform acetate into methane and carbon dioxide, while hydrogen methanogens combine hydrogen and carbon dioxide to form methane and water [1, 2]. On a laboratory or industrial scale, this natural process takes place in a controlled environment in anaerobic digesters. The main operating parameters are temperature, agitation, organic load and hydraulic retention time.

Anaerobic digestion is based on a set of biological reactions under the action of various groups of micro-organisms. Therefore, the contact between the organic matter and each group of microorganisms must be guaranteed to provide biogas production. This condition is met in the case of adequate digester agitation. The stirring system is used in order to enhance mass and heat transfers. The stirring can be done by recirculation of gas or leachate or mechanical system. The understanding of the flows during the agitation of the medium is crucial for (1) the comprehension of the impact on biochemical reactions and (2) the optimization of the mixing system. According to the choice of agitation mode, the medium can be well-agitated, homogenous or there might be presence of dead zones or isolated turbulent zones. In this context, the flow analysis by numeric approach is a promising method to identify these zones and two mains issues occur. First, to promote the anaerobic digestion process, it is necessary to limit the volume of dead zones without stirring too vigorously. In fact, in the dead zones, the organic matter could be not digested due to the lack of contact between the substrate and each group of micro-organisms. In this sense studies are being conducted to evaluate the volume and the position of dead zones with the aim of reducing their volumes [3–5]. Second, in the case of highly mixed zones could imply the destruction of the methanogenic centers and inhibit the biochemical reactions. As a consequence, it is evident that anaerobic digestion yields can be improved by the agitation optimization.

The qualitative description of the effect of flow velocity on the biogas yield is illustrated in **Figure 1**.

As a matter of fact, the properties of the medium are impacted by the mixture type. Consequently, the fluid characterization is an important factor. In this work, we propose first to study the mechanical parameter of the sugarcane vinasse, with the consideration of the rheology of the digestion medium. The sugarcane vinasse is a liquid waste generated from the alcohol production. In the case of liquid digestion, defined by a solids content of less than 15%, the medium may both have a Newtonian or non-Newtonian behavior. The behavior of the fluid has a direct consequence on the flows within the digester. Depending on the waste type, the digestion medium will also have a different viscosity. Viscosity is the term to express the resistance of a fluid to flow and motion. It is assumed that the sugarcane vinasse is Newtonian [6].

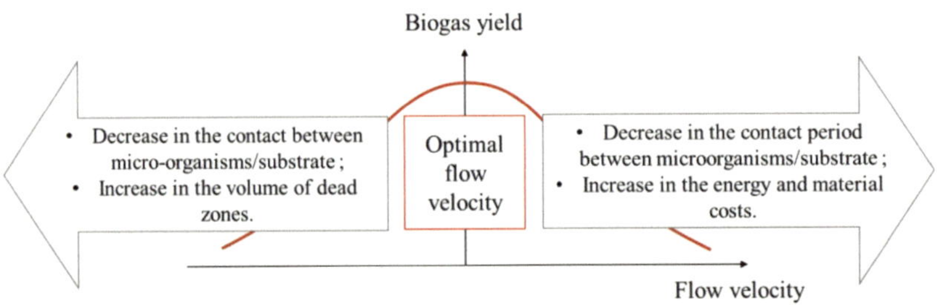

Figure 1.
Qualitative description of the effect of flow velocity on the biogas yield.

In order to model the flows within a digester, it is therefore necessary to know both its technical characteristics (geometry, stirring mode) and the fluid properties. All these choices will have an incidence on the mechanical energy and thus on the expended electrical energy to produce methane.

Consequently, the CFD simulations allow to analyze the flows within a digester of a specific waste mixture, considering its rheological properties. In the literature, the CFD models developed are based on the Reynolds-averaged Navier-Stokes (RANS) equations, the large eddy simulation (LES) as well as direct numerical simulation (DNS). DNS method provides the more precise results by solving directly the Navier-Stokes equations but it is also the most time-consuming method. RANS and LES are specific methods developed for turbulence studies. They are both used by many authors. LES method is more time-consuming than RANS method but provides more accurate outcomes. For example, Fan and al. used RANS method for simulating the hydrodynamic behaviors in a stirred tank [7], and Foukrach et al. used RANS method for studying mixing performance in a gas-mixed anaerobic digester [8].

In the current work, we propose a case study on the vinasse. The RANS approach is used, closed with the standard k-epsilon turbulence model. Simulations for different rotational speeds are carried out in order to analyze the flow field in mechanically stirred digester in function of the rotational speed.

2. Theoretical model

2.1 Assumptions

The anaerobic digestion occurs at the mesophilic regime at 37°C. As previously said, it is assumed that the sugarcane vinasse is Newtonian [6] and the fluid is isothermal and incompressible. The rheological properties of the medium are related to the total solid (TS) content of the substrates. The properties of the substrate are detailed in the following part in the **Table 1**.

In the present study, we consider single-phase turbulent flow.

The sugarcane vinasse from the Rivière du Mât distillery is a recalcitrant waste with a chemical oxygen demand (COD) of 86.70 $g_{O2}.L^{-1}$, 6.64% TS content, 4.04% volatile solid content and a pH of 4.84 [10]. The biochemical methane potential is 185.59 $NL_{CH4}.kg_{COD}{}^{-1}$ [10].

In the next section, we present the geometry used in this work and the mesh generation suitable to the mechanical agitation study.

2.2 Geometry and mesh

This study is carried out on an existing geometry in literature. Based on the study carried out by Wu [11], the geometry consists in a 260 mm height tank (liquid

Fluid	Density ($kg.m^{-3}$)	Dynamic viscosity (Pa.s)	Sources
Vinasse	1044.69	0.0010097	[6]
Water	1000	0.001	
Screened and diluted dairy cow manure	1000	0.0009	[9]

Table 1.
Rheological properties of Newtonian fluids.

height) with a diameter of 152 mm. The digester bottom is conical with a height of 25.88 mm and a diameter of 41 mm. The tank geometry is shown in **Figure 2**. Wu [11] validated his model from computer automated radioactive particle tracking (CARPT) experiment developed by Hoffmann et al. [9]. Then, Wu [11] also confronted his results with particle image velocimetry (PIV) experiment carried out by Bugay et al. [12]. The geometry of the impeller is shown in **Figure 2**.

The Lightnin A310 impeller, which is a hydrofoil impeller, is used. Hydrofoil impellers were developed for applications where axial flow is important and low shear is desired [13]. The impeller diameter is 62 mm and the axis diameter is 8 mm. The impeller axis is positioned at a height of 50 mm from the bottom of the digester.

The mixing is modeled using the sliding mesh method. The sliding mesh model is a time-dependent solution approach in which the grid surrounding the rotating component(s) physically moves during the solution [13]. Two zones are defined: the stationary zone and the moving zone. The impeller is located in the moving zone. The interface between the two zones is a regular surface mesh size. The interface meshes of the two zones must be identical.

2.3 Governing equations

The sliding mesh (SM) method is suitable for unsteady flow [14]. In our case study, we are in the case of unsteady flow. Indeed, initially, the mechanical agitation is zero and increases gradually until the desired agitation. The flow is steady only when the rotational speed of the stirrer is reached. Therefore, the SM method is used in this work. In order to model the impeller rotation, the digester tank is thus divided in two domains, the rotating zone which contains the impeller and the stationary zone. The arbitrary mesh interface (AMI) is used to link the two domains. The AMI interface is a pair of detached surfaces, giving the AMI1 and AMI2 boundaries. One belongs to the mobile zone and the other to the

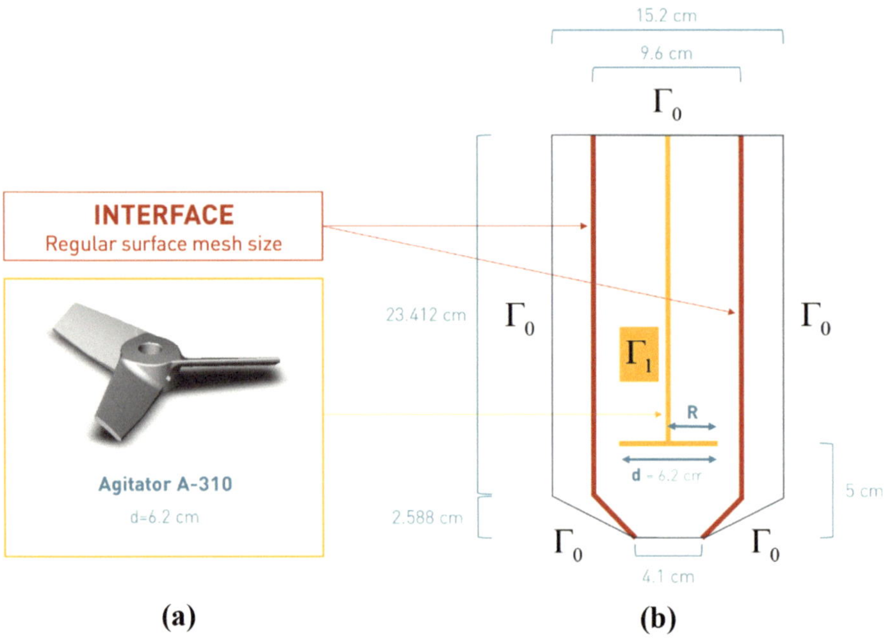

Figure 2.
Cross section of the pilot geometry with the moving zone and the stationary zone (a) and the impeller geometry (b).

stationary zone. These two boundaries are identical and have the same initial and boundary conditions. The rotational speed is assigned directly to the moving zone.

The mathematical model is based on RANS equations associated with the forces linked to the impeller rotation: the Coriolis and centrifugal forces. The governing equations for the rotating zone and the stationary zone are respectively:

$$\begin{cases} \dfrac{\partial u_R}{\partial t} + \nabla \cdot (u_R \otimes u_I) = -\nabla p + \nabla \cdot \left(\nu_{eff}\left(\nabla u_I + (\nabla u_I)^T\right)\right) \\ \nabla \cdot u_R = 0 \end{cases} \quad (1)$$

$$\begin{cases} \dfrac{\partial u_I}{\partial t} + \nabla \cdot (u_I \otimes u_I) = -\nabla p + \nabla \cdot \left(\nu_{eff}\left(\nabla u_I + (\nabla u_I)^T\right)\right) \\ \nabla \cdot u_I = 0 \end{cases} \quad (2)$$

where u_I and u_R are the absolute velocities viewed respectively from the stationary and the rotating frames (m.s^{-1}), t is the time (s), ν_{eff} is the effective kinematic viscosity (m^2.s^{-1}) and p is the pressure (Pa).

The incompressible solvers on OpenFoam use the modified pressure (kinematic pressure) P (m^2.s^{-2}), calculated by the following equation:

$$P = \frac{p}{\rho} \quad (3)$$

where p is the pressure (kg.m^{-1}.s^{-2}) in and ρ the fluid density (kg.m^{-3}).

The system is closed with the standard $k - \varepsilon$ turbulence model, where k is the turbulent kinetic energy (m^2.s^{-2}) and ε is the turbulent dissipation (m^2.s^{-3}). The model implemented in OpenFOAM does not include the buoyancy contribution. The two equations of this model are:

$$\frac{\partial}{\partial t}(k) = \nabla \cdot (D_k \nabla k) + G_k - \frac{2}{3}(\nabla \cdot u)k - \varepsilon + S_k \quad (4)$$

$$\frac{\partial}{\partial t}(\varepsilon) = \nabla \cdot (D_\varepsilon \nabla_\varepsilon) + \frac{C_1 G_k \varepsilon}{k} - \left(\frac{2}{3}C_1 - C_{3,RDT}\right)(\nabla \cdot u)\varepsilon - C_2\frac{\varepsilon^2}{k} + S_\varepsilon \quad (5)$$

where u is equal to u_I or u_R, S_k is the internal source term for k, S_ε is the internal source term for ε, G_k is the production of k, C_1 and C_2 are model constants, D_k the effective diffusivity for k and D_ε is the effective diffusivity for ε. The values of the model constants used in OpenFoam, as in the standard model are: $C_\mu = 0.09,\ C_1 = 1.44,\ C_2 = 1.92,\ \sigma_k = 1$ and $\sigma_\varepsilon = 1.3$ [15].

The production term G_k is:

$$G_k = \nu_t S^2 \quad (6)$$

The assumption of the standard $k - \varepsilon$ turbulence model turbulence is the following expression of the turbulent viscosity ν_t (m^2.s^{-1}):

$$\nu_t = C_\mu \frac{k^2}{\varepsilon} \quad (7)$$

For the initialization of the numerical simulations, the turbulent kinetic energy for isotropic turbulence k_0 and the turbulent dissipation rate ε_0 are calculated as follows:

$$k_0 = \frac{3}{2}\left(I|u_{ref}|\right)^2 \tag{8}$$

$$\varepsilon_0 = \frac{C_\mu{}^{0.75} k_0^{1.5}}{L} \tag{9}$$

where I (%) is the turbulence intensity with a default value of 0.05, u_{ref} (m.s^{-1}) is a reference velocity (the stirring velocity) and L (m) is the reference length scale (the digester radius).

The mesh must be refined at the walls. In fact, the standard turbulent model is derived under the assumption of a high local turbulent Reynolds number [16]. This low turbulent Reynolds number region is called the viscous sub layer. Launder and Spalding suggested the following wall function equation [15] to reduce cell number in these zones:

$$u^+ = \frac{1}{\kappa} \ln y^+ + C \tag{10}$$

where $u^+ = |u|/u_T$ is the dimensionless velocity where u_T is the friction velocity (or shear velocity). $y^+ = u_T y/\nu$ is the dimensionless local Reynolds number where y is the width of the boundary layer (normal distance from the wall). The range of the local Reynolds number is $11.06 \le y^+ \le 300$ [17]. Platteeuw et al. recommend that the local Reynolds number of the first cell should be in the range of $20 \le y^+ \le 100$ [16]. κ is the von Karman constant and C is a parameter related to the wall roughness. In the case of smooth walls, the values are respectively 0.419 and 5.24.

The boundary conditions are:

$$\begin{aligned} &\text{No slip: } u = 0, && \frac{\partial}{\partial n} p = 0, \quad \text{on } \Gamma_0 \\ &\text{Imposed velocity field: } v_r = \omega R, && \frac{\partial}{\partial n} p = 0, \quad \text{on } \Gamma_1 \end{aligned} \tag{11}$$

where the index r is the center of the elementary surfaces of each cell the agitator, ω is the angular velocity, R is the impeller radius, Γ_0 is the digester walls and Γ_1 is the impeller walls.

The rotational speed of the agitator is defined by the speed of rotation of the mobile zone of the mesh. A sudden increase in the stirring speed at the first time step causes a sharp increase in the CFL. Thus, a velocity ramp is necessary for the simulations starting to maintain the Courant number lower than 0.5.

2.4 Rheological expression

We refer to the pseudo-plastic model of Ostwald. It allows to express the shear stress τ (Pa) as a function of the shear rate $\dot{\gamma}$ (s^{-1}). It is calibrated by two parameters: the consistency index k (Pa.s^n) and the flow index n. It is expressed as follows:

$$\tau = k\dot{\gamma}^n \tag{12}$$

The viscosity is expressed as follows:

$$\eta = k\dot{\gamma}^{n-1} \tag{13}$$

The components of the tensor of the shear rate are:

$$\dot{\gamma}_{ij} = \left(\frac{\partial u_i}{\partial x_j} + \frac{\partial u_j}{\partial x_i}\right) \tag{14}$$

$$S_{ij} = \eta\left(\frac{\partial u_i}{\partial x_j} + \frac{\partial u_j}{\partial x_i}\right) = k\dot{\gamma}^{n-1}\left(\frac{\partial u_i}{\partial x_j} + \frac{\partial u_j}{\partial x_i}\right) \tag{15}$$

The rheological properties of the vinasse, the water and diluted dairy cow manure are shown in **Table 1**. It can be seen that the properties of the vinasse are closed to the properties of water. The difference between these two substrates is mainly the presence of suspensions. In anaerobic digestion process, the particles settle, accumulate at the bottom of the digester and result in the formation of sludge. In the present study, as a monophasic model is considered, the suspensions are not taken in account.

2.5 Flow field analysis

The Reynolds number for mechanical agitation of Newtonian fluids is [13, 18, 19]:

$$\text{Re} = \frac{\rho N d^2}{\mu} \tag{16}$$

where N is the impeller speed (rev.s^{-1} or rps), ρ is the fluid density (kg.m^{-3}), d is the impeller diameter (m) and μ is the dynamic viscosity (Pa.s).

The generalized Reynolds number valid for non-Newtonian fluids is [5]:

$$\text{Re}_g = \frac{\rho U_\infty^{2-n} d^n}{K\left(0.75 + \frac{0.25}{n}\right)^n 8^{n-1}} \tag{17}$$

where U_∞ is the average velocity of the fluid (m.s^{-1}), k is the consistency index (Pa.s^n), n is the flow index and d is the reference length (m).

The evaluation of the energy consumption and consequently its economic impact is possible with the calculation of the power calculation. The power consumption P (W) calculation based on the torque of the impeller is:

$$P = 2\pi N T \tag{18}$$

where N is the impeller rotational speed and T is the impeller torque (N.m). The torque is defined as [20]:

$$T = \left(\int_S \left(\vec{r} * \left(\bar{\bar{\tau}} \cdot \hat{n}\right)\right) dS\right) \cdot \hat{\alpha} \tag{19}$$

where S represents the surfaces including the rotating parts, $\bar{\bar{\tau}}$ Is the total stress tensor, $\hat{n}$ is a unit vector normal to the surface, $\vec{r}$ is the position vector and $\hat{\alpha}$ is a unit vector parallel to the rotation axis.

The mixing energy level (MEL) (W.m^{-3}) is calculated by [21]:

$$MEL = \frac{P}{V} \tag{20}$$

where P is the power consumption (W) and V is the working volume (m^3).

The flow rate through the moving zone Q (m^3.s^{-1}) is used for the evaluation of the fluid circulation through the digester. We create a surface for the discharge zone for computing the flow rate. The surface consists of two discs above and below the stirrer for the axial flow and a cylinder (AMI section) for the radial flow.

The power and flow (also called pumping number) numbers are two dimensionless numbers commonly used to characterize the stirred tank flows and mixing processes. The flow number is a measure of the pumping capacity of an impeller [13]. The power number is a dimensionless parameter that provides a measure of the power requirements for the operation of an impeller [13]. These two mixing parameters can be calculated from CFD results and allow to compare the results of simulations with the experimental results. In our case, the experimental values for hydrofoil impeller are N_P equal to 0.3 and N_Q equal to 0.56 [11, 22]. The power and flow numbers are respectively expressed by:

$$N_P = \frac{P}{\rho N^3 d^5} \tag{21}$$

$$N_Q = \frac{Q}{N d^3} \tag{22}$$

The spatially average characteristic velocity gradient G (s^{-1}) is used for describing the rotational speed through the digester [23], as well as characterizing the turbulent shear rate [22]. It is defined with the following expression [22, 23]:

$$G = \left(\frac{\overline{\varepsilon}}{\nu}\right)^{1/2} \tag{23}$$

where ν is the kinematic viscosity and $\overline{\varepsilon}$ ($m^2.s^{-3}$) is the global average turbulent energy dissipation rate calculated as follow [22, 23]:

$$\overline{\varepsilon} = \left(\frac{N_P N^3 d^5}{V}\right) \tag{24}$$

The circulation time t_c is expressed by [22, 24]:

$$t_c = \frac{V}{N_Q N d^3} = \frac{V}{Q} \tag{25}$$

The vorticity ξ (s^{-1}) is a measure of the local rotation in the fluid. More specifically, it is a vector field that gives a microscopic measure of the rotation at any point in the fluid. The vorticity describes the local spinning motion. The helicity H provides indications on the alignment of the vorticity vector and the velocity vector U. It allows to illustrate the longitudinal vortices, or spiral motion, as is often found in vortex cores.

The vorticity is defined as the curl of the velocity vector:

$$\xi = \nabla \times U \tag{26}$$

The helicity is expressed as:

$$H = U \cdot \xi = U \cdot (\nabla \times U) \tag{27}$$

The angle between the vorticity vector and the velocity vector (which is 0° or 180° in a vortex core) is given by:

$$\alpha = \cos^{-1}\left(\frac{H}{|\xi||U|}\right) \tag{28}$$

2.6 CFD simulations

The numerical simulations are performed on OpenFOAM software. The PIMPLE algorithm is used, which is a combination of the PISO (pressure-implicit with splitting of operators) [25, 26] and SIMPLE (semi-implicit method for the pressure linked equations) [27] algorithms. The PISO algorithm is pressure implicit with splitting of operators [25] and allows the use of the sliding mesh method. The SIMPLE algorithm is the semi-implicit method for pressure linked equations [28]. The pimpleFoam solver assumes incompressible, unsteady and viscous flows. The simulations are launched in parallel on 16 processors. The simulation time, for which we get a steady state, is a few seconds. The calculation time is about a few days.

3. Results and discussions

3.1 Mesh

The tank geometry and the mesh are done on OpenFOAM and the impeller geometry on Sketch Up. Then, the OpenFOAM snappyHexMesh command is used to obtain the final mesh. The mesh is refined to meet the quality criteria of mesh such as cell skewness and mesh non-orthogonality. To meet these criteria, we reach a very large number of cells and thus long computing times. Consequently, we do not perform grid independence tests.

The surface mesh of the digester tank and the cross section of the mesh at the impeller level are shown in **Figures 3** and **4**.

The characteristics of the mesh are shown in **Table 2**. In total, this mesh is composed of 2,841,535 cells with mostly hexahedra. The total mesh faces is 9,150,640. The impeller edge has the larger amount of faces with a value of 465,732 due to its geometry. Concerning the interface between the moving zone and the stationary zone (AMI1 and AMI2), the two surface meshes must be identical with the same faces number in order to avoid numerical errors. Globally, the maximum cell skewness is 3.4 and the maximum mesh non-orthogonality is 65.0, which reflect an acceptable mesh quality. The dimensionless local Reynolds number y^+ is below 1.

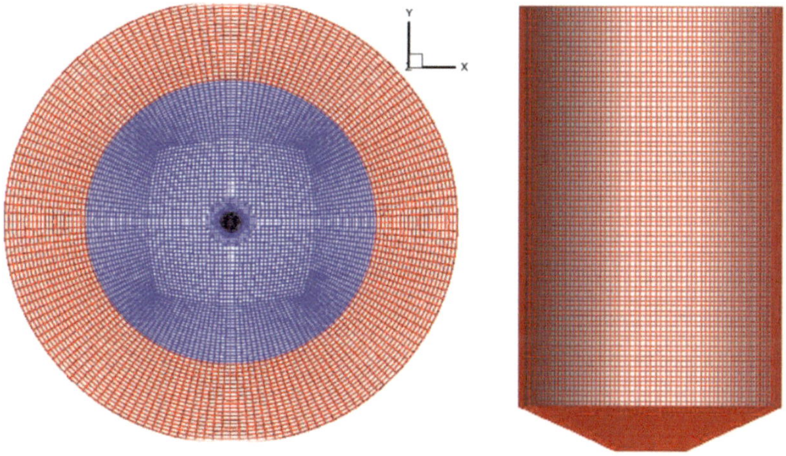

Figure 3.
Surface mesh of the digester: top (left) and wall (right).

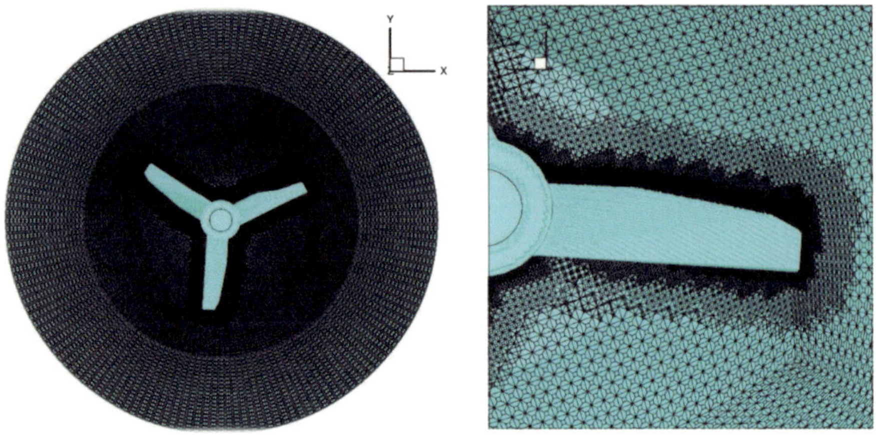

Figure 4.
Cross section of the mesh at the impeller level.

Characteristics		
Total mesh cells	2,841,535	
Total mesh points	3,485,073	
Total mesh faces	9,150,640	
Total mesh internal faces	8,591,189	
Number of hexahedra	2,578,249	
Number of prisms	10,298	
Number of wedges	3	
Number of polyhedra	252,985	
Maximum cell skewness	3.40025	
Maximum mesh non-orthogonality	65.0182	
Average mesh non-orthogonality	21.3783	
Breakdown of polyhedra by number of faces		
Faces	Number of cells	
<10	216,780	
≥10	36,205	
Patch topology		
Patch	Faces	Points
Upper edge (moving zone)	5719	5952
Bottom edge (moving zone)	4800	4881
Impeller edge (moving zone)	465,732	483,277
Upper edge (stationary zone)	3200	3360
Bottom edge (stationary zone)	3200	3360
Lateral wall (stationary zone)	25,760	25,760
AMI1	25,760	25,760
AMI2	25,760	25,760

Table 2.
Mesh characteristics.

3.2 Velocity profiles and shear stress

In this part, we present the simulation results for four rotational speeds: 20, 40, 60 and 100 rpm. **Figure 5** shows the cross section of the velocity field within the digester at the agitator height and **Figure 6** extends the longitudinal section of the velocity flied.

The results obtained are consistent. In fact, by increasing the rotation velocity of the stirrer, the maximum velocity increases, the volume of dead zones decreases and the volume of continuously agitated zones increases. The maximum velocity value obtained for each case is the agitator velocity. The maximum velocity is obtained at the blade extremities. The velocity decreases gradually with the distance from the agitator.

For each case, we find that the velocity at the edges of the digester is zero. Likewise, the velocity at the axis of the agitator increases with the increase of the rotational speed. The interest of the study of the velocity field at this level relates in particular to the influence of the agitation on the biofilm which tends to develop on

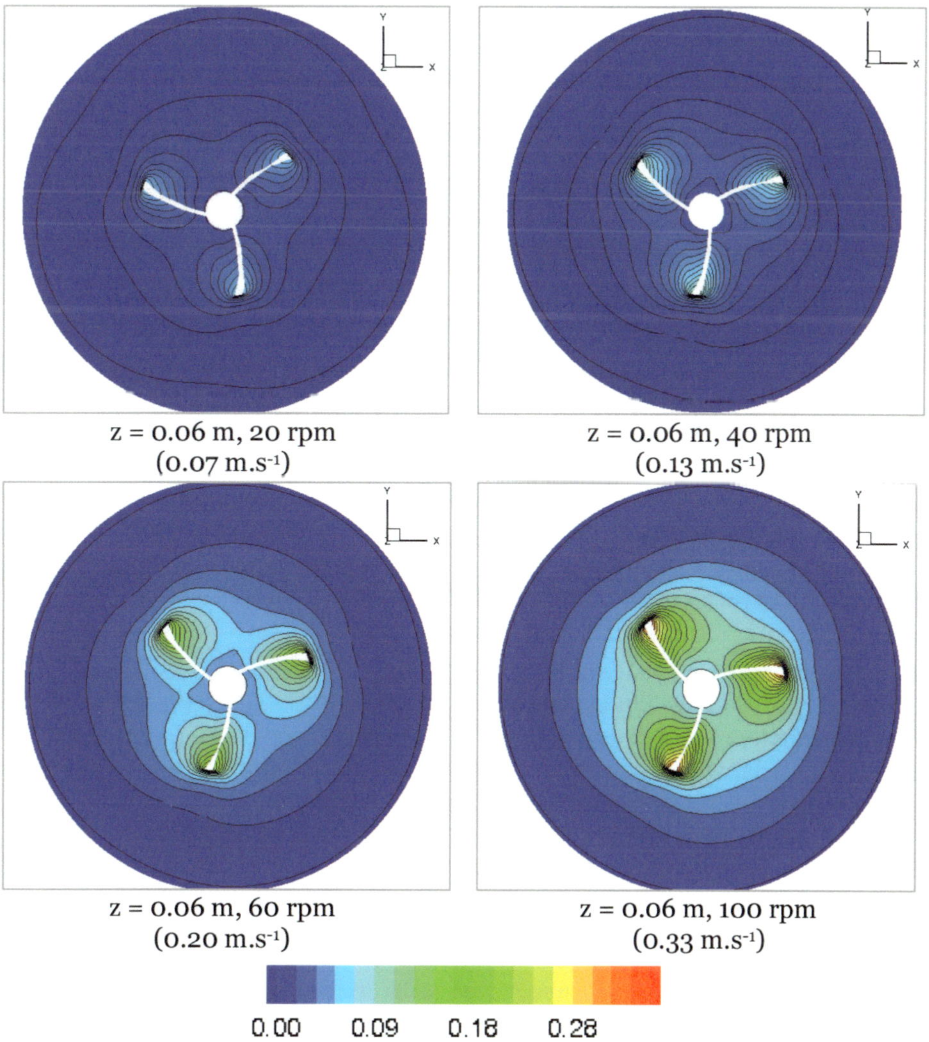

Figure 5.
Axial cross-section of the velocity flied (m.s^{-1}) within the digester at the agitator level (z = 0.06 m) for four rotational speeds: 20, 40, 60 and 100 rpm (the maximum velocity in m.s^{-1} at the impeller extremity is mentioned below the figures).

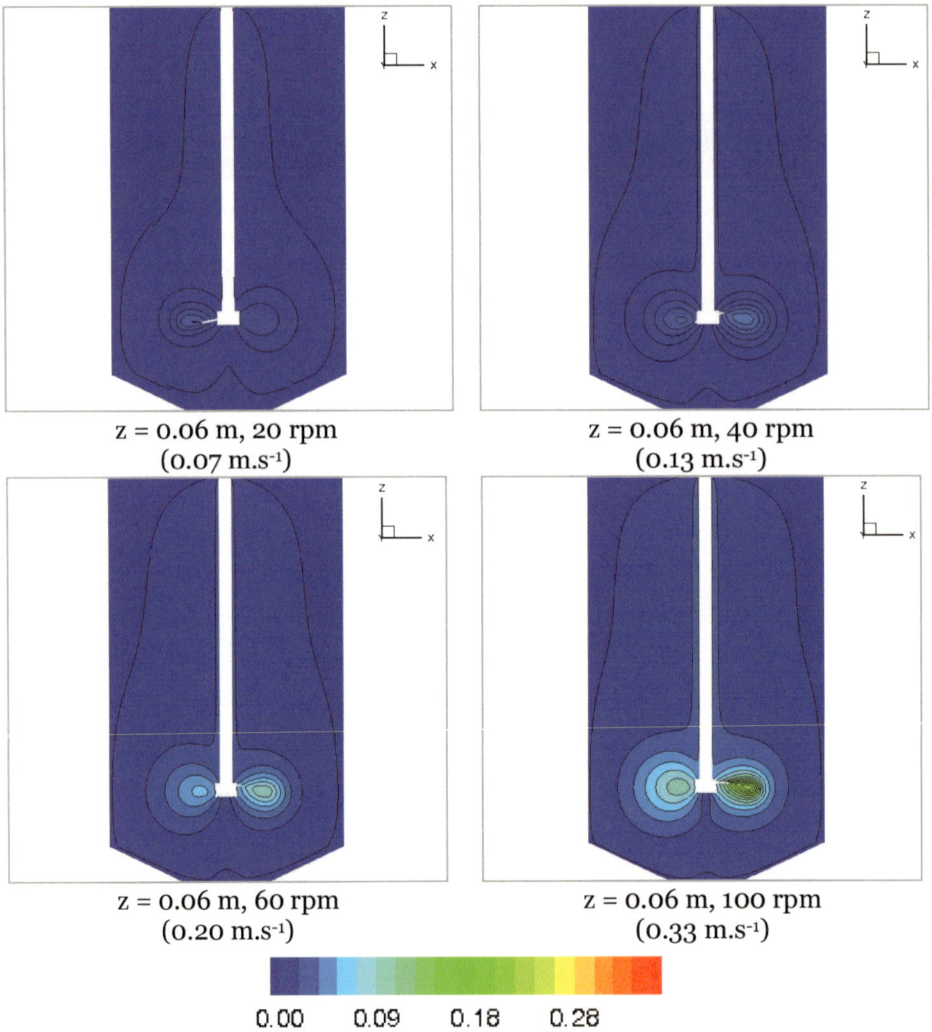

Figure 6.
Longitudinal cross-section of the velocity field ($m.s^{-1}$) within the digester for four rotational speeds: 20, 40, 60 and 100 rpm (the maximum velocity in $m.s^{-1}$ at the impeller extremity is mentioned below the figures).

the axis of the agitator. The interest is even greater in the case of intermittent mixing. This mode of agitation is particularly interesting because it leads to the reduction of energy consumption. Overall, the optimization of the mechanical agitation must allow to properly stir the medium, so as to improve the biogas yields, while reducing the rotational speed and frequency of agitation.

From both the axial and longitudinal sections of the flow field, we observe that the maximum velocity field is mainly concentrated in the area close to the agitator and quickly becomes zero away from the influence zone of the impeller blades.

We assume that the dead zones are defined by a velocity magnitude less than 0.001 $m.s^{-1}$ as [5]. Considering this hypothesis, we find that, in our case study, almost the whole area is dead zones except in the zone of influence of the agitator, which is more important with the increase of the rotation speed. Regarding the contours of the longitudinal cross-section, it can be seen that low-speed velocities are observed near the impeller axis. Furthermore, the area at the upper edges of the digester as well as the lower area of the conical section remains dead zones despite a rotational speed of 100 rpm. CFD simulations are useful for defining the volume of dead zones. Indeed, the dead zones within a digester lead to the heterogeneity of the medium. There will be areas where the substrate will not be digested or acidic areas. This can eventually lead to intoxication of the digester generating significant costs on an industrial scale.

Experimentally it would result in an accumulation of solid and undigested matters, and thus the formation of sludge. It is therefore necessary to adapt the agitation according to the digester technology (with or without sludge bed). In addition, in the case of anaerobic digestion of a recalcitrant waste, such as vinasse, it is important to apply sufficient agitation to re-suspend the undigested matter in order to improve yields.

Similarly, since the upper part of the digester is not agitated, the digestion medium would not be homogenized. Overall, this mode of agitation (mechanical, a three-bladed agitator) would result in a significant stratification of the reaction medium with the presence of dead zones.

In view of the obtained results, it would be recommended to use a stirring system with several rows of blades to agitate the lower and upper areas of the digester to avoid stratification of the digestion medium. In addition, it would be desirable that the diameter of the stirrer be close to that of the digester in order to limit the development of biofilm on the walls. In addition, theses simulations provide information on the velocity field within the digester as a function of agitator rotational speed; however, we have no information on their impact on the biochemical reactions and the biogas production and quality.

A study has been carried out on the impact of shear stress and impeller design on the production of biogas in anaerobic digesters [29]. An impact of the shear stress on the biogas production was highlighted [29] and the abrasion of the anaerobic sludge granule due to the shear rate increase above 5 s^{-1} [29, 30]. Therefore, it was suggested to estimate the shear stress value for the impeller type and mixing rotation [29]. In the present study, the shear stress is negligible in the digester expected at the wall and impeller levels. We obtained 0.31 Pa at 20 rpm and 1.24 Pa at 100 rpm at the tank wall. At the impeller wall, we reported 453 Pa at 20 rpm and 1883 Pa at 100 rpm. Therefore, the shear stress was multiplied by 4.15 at the impeller wall and by 3.9 at the digester wall, with the increase of the stirring speed from 20 to 100 rpm.

CFD simulations accompanied by experimental studies allow to relate the stirring velocity, the volume of dead zones, the homogenization of the medium (physicochemical properties) and the biogas yields. Thus, a better understanding of the physical phenomena involved in the digesters is necessary for the optimization of anaerobic digestion on an industrial scale. An experimental study on the anaerobic digestion of vinasse on this same geometry (digester and agitation system) would provide knowledge on the yields obtained with the velocity fields obtained in modeling. This would make it possible to relate the yields obtained experimentally to the flows.

3.3 Pressure

During the anaerobic digestion process, the pressure applied on the medium has an impact. Indeed, there are processes involving free microorganisms that are in the form of flocs or biofilm [31]. There are two types of biofilms, those formed on a mineral or organic support (fixed or mobile) and granules (natural agglomeration of microorganisms from a few tens of microns to several millimeters in diameter). In fact, the flow study through the digester provide information on the pressure imposed on the digestion medium. The interest is to evaluate the impact of pressure on the flocs and biofilms.

Figures 7 and **8** show the kinematic pressure in continuous flow regime at 40 and 100 rpm respectively. The pressure is due to the impeller rotation. At 20 rpm, it varies from 0 to $5.86.10^{-5}$ $m^2.s^{-2}$ (0 to $5.86.10^{-2}$ Pa). The pressure range at 40 rpm is −0.50 to 0.40 $m^2.s^{-2}$ which corresponds to −500 to 400 Pa. At 60 rpm, the pressure varies from −0.73 to 0.61 $m^2.s^{-2}$ (−730 to 610 Pa). At 100 rpm, it varies from −1.16 to 1.05 $m^2.s^{-2}$ (−1160 to 1050 Pa).

3.4 Turbulence phenomena analysis

The turbulence phenomena can be studied from different parameters. In the present study, we propose the description of the Reynolds number, the isotropic turbulence, the Q field, the Lambda2 fields and the vorticity.

Considering the impeller diameter for the characteristic length, the Reynolds number is 1326 at 20 rpm, 2651 at 40 rpm, 3977 at 60 rpm and 6629 at 100 rpm.

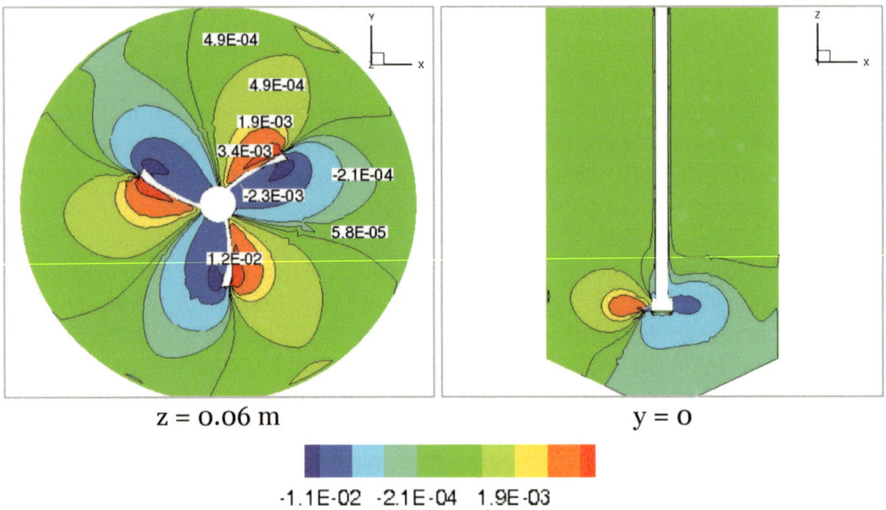

Figure 7.
Kinematic pressure ($m^2.s^{-2}$) at 40 rpm (z = 0.06 m, y = 0).

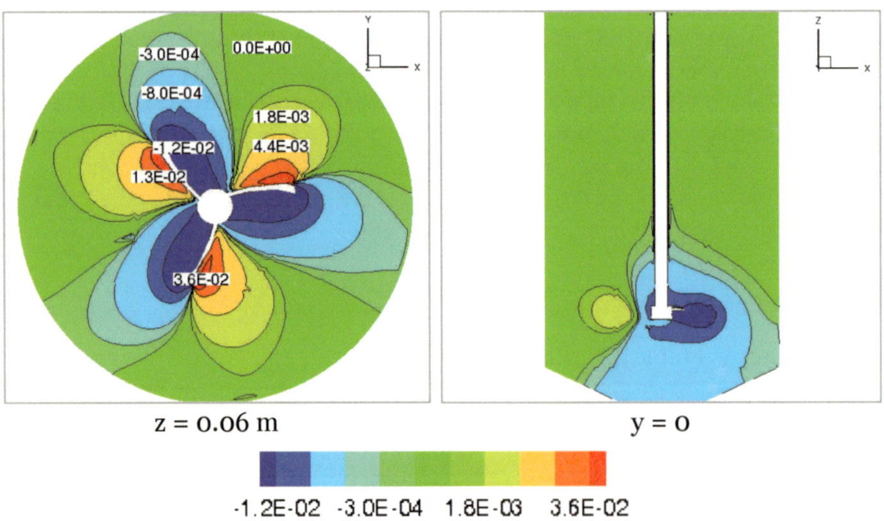

Figure 8.
Kinematic pressure ($m^2.s^{-2}$) at 100 rpm (z = 0.06 m, y = 0).

Therefore, local turbulence phenomena can occur around the impeller, however, the turbulence phenomena dissipate rapidly away from the impeller. If the isotropic turbulence k is very low, we consider that we are in a laminar regime. The maximum k value is 2.2×10^{-16} $m^2.s^{-2}$ at 20 and 40 rpm, 3.0×10^{-11} $m^2.s^{-2}$ at 60 rpm and 3.9×10^{-8} $m^2.s^{-2}$ at 100 rpm. Therefore, we observe these local important values at the impeller level.

Q and Lambda2 fields provide a precise description of the turbulence and local rotation. The Lambda2 (s^{-2}) function object computes the second largest eigenvalue of the sum of the square of the symmetrical and anti-symmetrical parts of the velocity gradient tensor. Q iso-surfaces are good indicators of turbulent flow structures. The Q function object computes the second invariant of the velocity gradient tensor (s^{-2}):

$$Q = \frac{1}{2}\left[(tr(\nabla u))^2 - tr(\nabla u \cdot \nabla u)\right] \tag{29}$$

Figure 9 presents the velocity flood with the Lambda2 contours. The Lambda2 and velocity profiles are similar in zones between the blades but different at the impeller extremities.

Figure 10 shows the Q field at the impeller level at 40 and 100 rpm. The extremum values of the Q field are obtained at the impeller extremities. The Q field varies from – 5171 to 39,405 s^{-2} at 40 rpm and from −27,000 to 289,000 s^{-2} at 100 rpm. The value is close to zero in areas away from the impeller. The Q field varies from −1383 to 8924 s^{-2} at 20 rpm and from −9314 to 94,137 s^{-2} at 60 rpm.

The Lambda2 field at the impeller level at 40 rpm is shown in **Figure 11**. It varies from – 3677 to 31,177 s^{-2} at 40 rpm and from – 15,561 to 226,008 s^{-2} at 100 rpm. Both the Q and Lambda2 field maximum values are about seven times higher at 100 rpm than at 40 rpm. The Lambda2 field varies from −1125 to 8449 s^{-2} at 20 rpm and from −5234 to 73,482 s^{-2} at 60 rpm.

Figure 12 presents the vorticity module (s^{-1}) at different digester heights in continuous flow regime at 40 rpm (z = 0.06 m, z = 0.08 m, z = 0.259 m and y = 0). The vorticity is a pseudo vector field that describes the local spinning motion (the curl of the velocity). The maximum vorticity module is 314.8 s^{-1} at 20 rpm, 621.7 s^{-1} at 40 rpm, 959.4 s^{-1} at 60 rpm and 1680.7 s^{-1} at 100 rpm.

The extremum values are observed at the impeller level. We notice that the maximum value of vorticity is multiplied by 2.7 from 40 to 100 rpm. We can

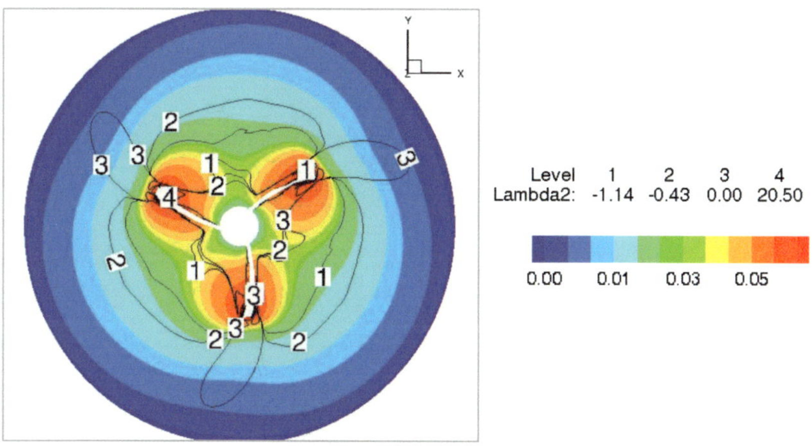

Figure 9.
Velocity ($m.s^{-1}$) flood and Lambda2 (s^{-2}) contour at 40 rpm (z = 0.06 m).

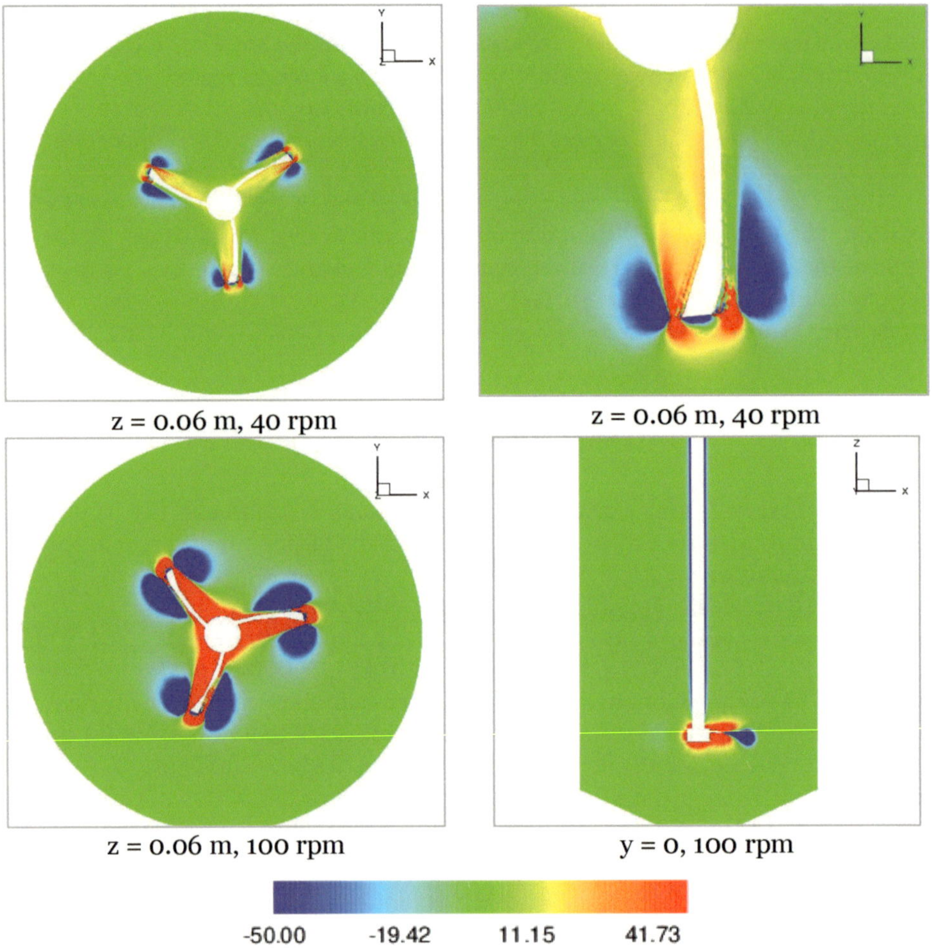

Figure 10.
Q field (s^{-2}) at the impeller level at 40 and 100 rpm.

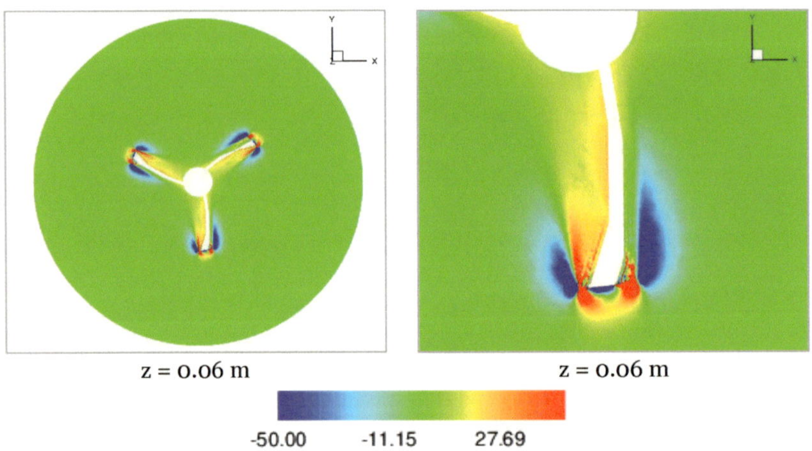

Figure 11.
Lambda2 field (s^{-2}) at the impeller level at 40 rpm.

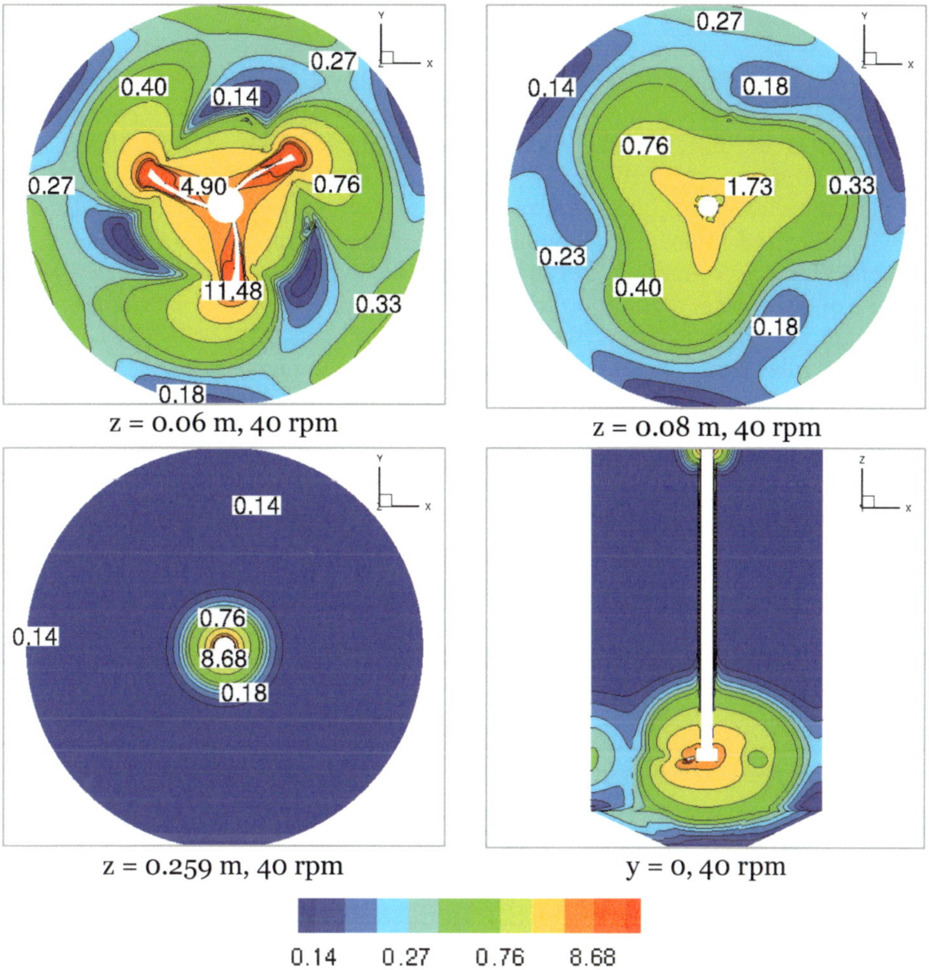

Figure 12.
Vorticity module (s^{-1}) in continuous flow regime at 40 rpm (z = 0.06 m, z = 0.08 m, z = 0.259 m and y = 0).

consequently note that the variations of the Q and Lambda2 fields are more significant than the variation of the vorticity with the increase of the impeller mixing speed.

Figure 13 summarizes the variation of maximum vorticity, pressure, Q and Lambda2 values in function of mixing speed. The Q maximum value variation is the most pronounced among the parameters observed.

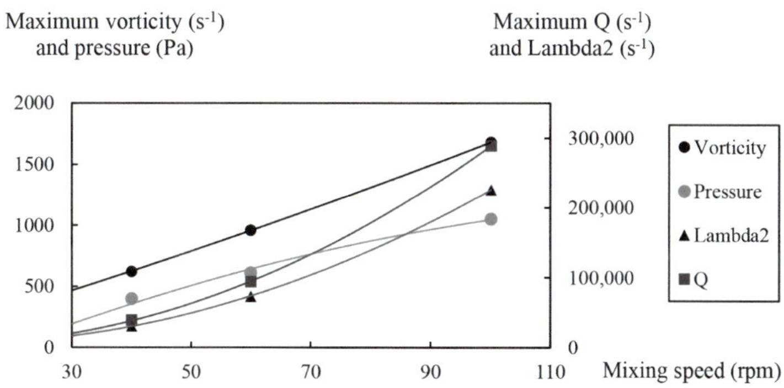

Figure 13.
Variation of maximum vorticity, pressure, Q and Lambda2 values in function of mixing speed.

4. Conclusion

Three-dimensional CFD simulations of the mechanically stirred digester were carried out. The model was based on RANS equations and standard k-epsilon turbulence model and the fluid was Newtonian. The sliding mesh method was conducted to characterize the impeller rotation. The outcomes were the velocity profiles, the shear stress, the pressure and the turbulence phenomena analysis with the Q and Lambda2 fields, the Reynolds number, the isotropic intensity and the vorticity. The area of influence of the agitator was in the zone of the blades. The volume of dead zones was important for the four rotational velocities studied. The maximum velocity was observed at the blades extremities. In conclusion, several rows of paddles would be required to reduce the dead zone volume.

In perspective of this work, the flow study will be carried out for different tank geometry, agitation configuration and fluid rheology (non-Newtonian).

Acknowledgements

This work was supported by the Region Reunion (France) as part of the funding of a research thesis in the PIMENT laboratory at the Reunion Island University.

Author details

Hélène Caillet*, Alain Bastide and Laetitia Adelard
PIMENT Laboratory, Reunion Island University, Saint-Denis, France

*Address all correspondence to: helene.caillet@univ-reunion.fr

References

[1] Bollon J. Etude des mécanismes physiques et de leur influence sur la cinétique de méthanisation en voie sèche: essais expérimentaux et modélisation. Lyon, INSA, 2012.

[2] Kerroum D, Mossaab B-L, Hassen MA. Production of biogas from sludge waste and organic fraction of municipal solid waste. INTECH Open Access Publisher; 2012.

[3] López-Jiménez PA, Escudero-González J, Montoya Martínez T, Fajardo Montañana V, Gualtieri C. Application of CFD methods to an anaerobic digester: The case of Ontinyent WWTP, Valencia, Spain. Journal of Water Process Engineering 2015;7:131–40. https://doi.org/10.1016/j.jwpe.2015.05.006.

[4] Vesvikar MS, Al-Dahhan M. Flow pattern visualization in a mimic anaerobic digester using CFD. Biotechnology and Bioengineering 2005;89:719–32. https://doi.org/10.1002/bit.20388.

[5] Wu B, Chen S. CFD simulation of non-Newtonian fluid flow in anaerobic digesters. Biotechnology and Bioengineering 2007;99:700–11. https://doi.org/10.1002/bit.21613.

[6] Koerich DM, Rosa LM. Optimization of bioreactor operating conditions using computational fluid dynamics techniques. The Canadian Journal of Chemical Engineering 2017;95:199–204. https://doi.org/10.1002/cjce.22635.

[7] Fan W, Yuan L, Qu X. CFD simulation of hydrodynamic behaviors and aerobic sludge granulation in a stirred tank with lower ratio of height to diameter. Biochemical Engineering Journal 2018;137:78–94. https://doi.org/10.1016/j.bej.2018.05.012.

[8] Foukrach M, Bouzit M, Ameur H, Kamla Y. Influence of the vessel shape on the performance of a mechanically agitated system. Chemical Papers 2019;73:469–80. https://doi.org/10.1007/s11696-018-0606-4.

[9] Hoffmann RA, Garcia ML, Veskivar M, Karim K, Al-Dahhan MH, Angenent LT. Effect of shear on performance and microbial ecology of continuously stirred anaerobic digesters treating animal manure. Biotechnology and Bioengineering 2008;100:38–48. https://doi.org/10.1002/bit.21730.

[10] Caillet H, Adelard L. Start-Up Strategy and Process Performance of Semi-Continuous Anaerobic Digestion of Raw Sugarcane Vinasse. Waste Biomass Valor 2020. https://doi.org/10.1007/s12649-020-00964-z.

[11] Wu B. Large eddy simulation of mechanical mixing in anaerobic digesters. Biotechnology and Bioengineering 2012;109:804–12. https://doi.org/10.1002/bit.24345.

[12] Bugay S, Escudié R, Liné A. Experimental analysis of hydrodynamics in axially agitated tank. AIChE Journal 2002;48:463–475.

[13] Paul EL, Atiemo-Obeng VA, Kresta SM, American Institute of Chemical Engineers, editors. Handbook of industrial mixing: science and practice. Hoboken, NJ: Wiley-Interscience; 2004.

[14] Cortada-Garcia M, Dore V, Mazzei L, Angeli P. Experimental and CFD studies of power consumption in the agitation of highly viscous shear thinning fluids. Chemical Engineering Research and Design 2017;119:171–82. https://doi.org/10.1016/j.cherd.2017.01.018.

[15] Launder BE, Spalding DB. The numerical computation of turbulent flows 1973:21.

[16] Platteeuw PDA, Loeven GJA, Bijl H. Uncertainty Quantification Applied to the k-epsilon Model of Turbulence Using the Probabilistic Collocation Method. 49th AIAA/ASME/ASCE/AHS/ASC Structures, Structural Dynamics, and Materials Conference
16th AIAA/ASME/AHS Adaptive Structures Conference 10t, Schaumburg, IL: American Institute of Aeronautics and Astronautics; 2008. https://doi.org/10.2514/6.2008-2150.

[17] Kuzmin D, Mierka O, Turek S. On the implementation of the $k - \varepsilon$ turbulence model in incompressible flow solvers based on a finite element discretization 2007:8.

[18] Wu B. Computational Fluid Dynamics Study of Large-Scale Mixing Systems with Side-Entering Impellers. Engineering Applications of Computational Fluid Mechanics 2012;6:123–33. https://doi.org/10.1080/19942060.2012.11015408.

[19] Marshall EM, Bakker A. Computational Fluid Mixing. In: Paul EL, Atiemo-Obeng VA, Kresta SM, editors. Handbook of Industrial Mixing, Hoboken, NJ, USA: John Wiley & Sons, Inc.; 2003, p. 257–343. https://doi.org/10.1002/0471451452.ch5.

[20] Wu B. CFD analysis of mechanical mixing in anaerobic digesters. Transactions of the ASABE 2009;52:1371–1382.

[21] Wu B. CFD simulation of mixing in egg-shaped anaerobic digesters. Water Research 2010;44:1507–19. https://doi.org/10.1016/j.watres.2009.10.040.

[22] Spicer PT, Keller W, Pratsinis SE. The effect of impeller type on floc size and structure during shear-induced flocculation. Journal of Colloid and Interface Science 1996;184:112–122.

[23] Clark MM, Flora JR. Floc restructuring in varied turbulent mixing. Journal of Colloid and Interface Science 1991;147:407–421.

[24] Alliet-Gaubert M, Sardeing R, Xuereb C, Hobbes P, Letellier B, Swaels P. CFD analysis of industrial multi-staged stirred vessels. Chemical Engineering and Processing: Process Intensification 2006;45:415–427.

[25] Issa RI. Solution of the implicitly discretised fluid flow equations by operator-splitting. Journal of Computational Physics 1986;62:40–65.

[26] Issa RI, Gosman AD, Watkins AP. The computation of compressible and incompressible recirculating flows by a non-iterative implicit scheme. Journal of Computational Physics 1986;62:66–82. https://doi.org/10.1016/0021-9991(86)90100-2.

[27] Patankar SV, Spalding DB. A calculation procedure for heat, mass and momentum transfer in three-dimensional parabolic flows. Numerical Prediction of Flow, Heat Transfer, Turbulence and Combustion, Elsevier; 1983, p. 54–73. https://doi.org/10.1016/B978-0-08-030937-8.50013-1.

[28] Caretto LS, Gosman AD, Patankar SV, Spalding DB. Two calculation procedures for steady, three-dimensional flows with recirculation. Proceedings of the third international conference on numerical methods in fluid mechanics, Springer; 1973, p. 60–68.

[29] Aline L, Stéphane D, Philippe M, Fabrice B, Stéphane P, Michel F, et al. Impact of shear stress and impeller design on the production of biogas in anaerobic digesters.pdf. Bioresource Technology 2017.

[30] Jiang J, Wu J, Poncin S, Li HZ. Effect of hydrodynamic shear on biogas production and granule characteristics in a continuous stirred tank reactor. Process Biochemistry 2016;51:345–51.

https://doi.org/10.1016/j.procbio.2015.12.014.

[31] Moletta R. Technologies du traitement des effluents par méthanisation 2002:20.

Section 5

Wastewater Treatment Techniques

Chapter 10

Evaluation of the Use of Advanced Ozone Oxidative Process in Reducing the Danger of Environmental Toxicity by Endocrine Interferences of Magistral Pharmacy

Thais Francinne, Suellen Zucco Bez, Julia Carolina Soares, Sabrina Martins da Rosa, Aline Mirian Paszuck, Luciana Ferreira Karsten and Luciano Henrique Pinto

Abstract

The presence of emerging pollutants in the waters has been observed worldwide, resulting from improper domestic disposal, non-recommended veterinarian use, and product waste from pharmaceutical industrics and magistral pharmacies. The contamination provoked, besides causing damage to the environment, remains in potable water even after passing through the treatment plants. The objective of this work was to verify the existence of environmental toxicity of raw effluents from gross pharmacy laboratories, as well as the same effluent treated with POA via ozone in the time of 1 hour, having as a risk parameter the changes that they cause in *Euglena gracilis* algae. Photosynthetic efficiency tests were conducted via PAM, and chlorophyll concentration and behavioral evaluation were checked via NGTOX. The results demonstrate that the hormone laboratory was considered the most impacted effluent treated, with lower production and significant chlorophyll reduction. It presented reduction in photosynthetic post-ozonation activity, due to the hormone decomposition, oxidative potential and ethylene formation. Effluents from psychotropic and solid laboratories presented different production demand, but similar follow-up, with impact on the behavior and algae's photosynthetic activity, due to the presence of active substances on cellular action potentials. The treated effluent from dermocosmetics laboratory influenced the chlorophyll concentration, as well as the general speed and velocity of surface ascent. The behavioral differences between the laboratories and the pre and post-ozonation conditions demonstrate that the effluent treatment should be distinguished, according to the characteristics of the manipulated substances in each laboratory.

Keywords: *Euglena gracilis*, oxidative process, ozonation, biomonitoring

1. Introduction

All over the world, the presence of medicines and cleaning and health product waste has been identified in the waters, and nowadays they are classified as emergent pollutants. This contamination is the result of many factors, like improper domestic disposal; non-recommended veterinarian use, which is exacerbated, making the excretion of medicine active metabolites reaches the groundwater in higher-than-expected quantities; and product waste from pharmaceutical industries and magistral pharmacies (also called manipulation pharmacy), that dispose their compounds to the effluents. Although the wastewater treatment plants treat this water, many medicines still remain in the drinking water [1].

In general, it has been seen that traditional wastewater treatment processes are not very efficient in removing this kind of emergent pollutants. In biological processes, for example, the degradation efficiency is highly influenced by the presence of other macrocompounds, what makes the drug degradation, besides rarely, only partially [2]. Systems based on absorption processes have been recently proposed, which use standard (active carbon) and modern (pre-absorbed micelles in montmorillonite) sorbents. However, their efficacy is questionable [3].

In this perspective, investigations indicating the environmental risk of these pollutants and the methods of removing these contaminants are increasingly more needed, since neither the treatment approaches nor the awareness of this issue are enough. The legislation can be cited here. It must be updated when it comes to emergent pollutants.

The Brazilian Water Resources Management Policy aims to assure the proper water availability to human consumption [4]. The Order no. 2.914/11, of the Ministry of Health, defines the potability patterns to water consumption. In this document, the drugs with potential risk to human health are not mentioned [5]. This condition makes the compounds neither be identified nor even treated on the wastewater treatment plants [6].

Differently from the Brazilian reality, organizations like European Union, the United States Environmental Protection Agency and the World Health Organization have already published guidelines and rules that warn about the risks of the presence of medicines in water and require studies that lead to their removal, in order to stablish acceptable limits for drinking water [7].

In this scenario, there are the magistral pharmacies, which, in the last decade, manipulated 8% of all the prescriptions in Brazil [8]. Nowadays, there are more than 7000 facilities like them over the country, and they are responsible for the small-scale and personalized medicine production, besides all the precautions required by the current legislation. In Joinvilly city specifically, the study site, at present 25 magistral pharmacies are registered with the Pharmacy Board and Sanitary Surveillance – which also attend the traditional segment –, dealing with drug and cosmetic demands, beyound the specialized pharmacies, that work with veterinarian products and hormone manipulation.

When reaching the environment, the hormones are then called endocrine disruptors (ED). These ED are defined as natural or synthetic exogenous chemical substances that, when in the environment, are capable of modifying the endocrine system, since they simulate the actions of natural hormones. These compounds might cause disorders that affect human and animal health [6], provoking, for example: breast and uterine cancer, increase in the incidence of polycystic ovarian, reduction in male fertility and prostatic neoplasm [9]. Considering the disorders the ED may cause in health and the environment, their chemical removal has been largely studied.

A technique used to remove ED is based on the development of advanced oxidation processes (POA), that correspond to a type of water treatment. POA promotes the composition of highly reactive and little selective hydroxyl radicals, being able to act on chemical oxidation of a wide range of organic substances, like medicines, converting them in substances that do not present, *a priori*, the same biological interactions than the original molecule does. In ED situation, there is the estradiol, whose oxidant action leads to the decomposition of pharmacophores and cessation of estrogenic activity [10].

A way to obtain the advanced oxidation is through the ozone, which is highly used along with other oxidizing agents, like hydrogen peroxide, titanium dioxide and ultraviolet. This process has been showing efficiency in emerging environmental decontamination [11].

The use of POA in this case is justified by the previously presented points related to the need of reducing the risks. However, it is important to say that compounds originated from degradation (COD) will be formed, and their evaluation will be relevant as well. The evaluation of the environmental impact provoked by these COD and the toxicity hazards in different trophic levels may clarify the use of this decontamination procedure and the results on the suppression of a certain environmental risk.

In this investigation, the results found out by Pinto et al. [8] with *Euglena gracilis* algae will be taken into account, in relation to raw effluents from a hormone manipulation laboratory after the ozone/ultraviolet POA, being analyzed the alterations and solutions COD may cause, comparing these results to other ones observed in other (psychotropics, dermo-cosmetic and solids) laboratories.

2. Materials and methods

2.1 Study design

This experimental study was performed in the Laboratory of Photochemistry and Photobiology and in the Environmental Laboratory, at the University of the Region of Joinville. It involved the *Euglena gracilis* KLEBS algae from the University of Göttingen's collection, Germany. Behavioral changes, photosynthetic activity and chlorophyll level alterations were analyzed, when the algae were submitted to chlorinated water from the pharmacy, as well as the raw effluents and the post-ozone/UV-POA effluents.

2.2 Sample collection

Three types of samples were analyzed:

- Water from access: chlorinated, to be used as control;
- Raw effluent, collected from the pharmacy;
- Post-POA-treated effluent.

All of them were collected from the four pharmacy laboratories.

In order to conduct the study, there was the collaboration of a magistral pharmacy from Joinville city, northeast of Santa Catarina, the same facility that was part of Pinto et al.'s investigation [8]. The pharmacy made possible the samples collection from the four production environments of the pharmacy:

- psychotropics laboratory, responsible for the manipulation of controlled-sale medicine prescriptions, accordingly to the Order 344/98;

- hormone laboratory, responsible for the manipulation of strictly hormonal prescriptions;
- solids laboratory, responsible for tablets and other solids formulation production;
- dermo-cosmetic laboratory, responsible for the solids and semisolids formulation production of dermatological properties.

The samples considered as water of access were collected directly from the faucets of the washing sinks in each laboratory, through previously sterilized borosilicate glass jars. They were used here as control samples (water from the wastewater treatment plant). The raw effluents samples were collected from the syphons connected to the washing sinks using a peristaltic pump, and here sterilized borosilicate glass jars were also used. Samples were taken up to 12 L^{-1} from each laboratory. Afterwards, the samples were kept in polystyrene boxes with ice and away from light up to their packaging in the freezer.

2.3 Production estimate of the magistral pharmacy laboratories

In order to conduct the study, there was the collaboration of a magistral pharmacy from Joinville city, northeast of Santa Catarina, that made possible the samples collection from the four production environments of the pharmacy.

An important factor to be considered here is the quantity of actives and other substances disposed through the sinks and that compound the laboratories' raw effluent. For this purpose, the pharmacy's average monthly production of six months was taken into account, and the monthly average production was calculated, in order to verify the laboratories' activity average, accordingly Eq. (1):

$$MP = \left(\frac{\sum Monthlyquantityofproducts}{30} \right) x \left(\frac{\sum mgofrequestsmade}{quantityofrequests} \right) \quad (1)$$

2.4 Fractions tests preparation

2.4.1 Removal process

Removal reaction occurred in a 500-mL^{-1} reactor, which contained the raw effluent samples from all the studied laboratories. The other samples removals occurred after 1-hour ozonation, through a Trump TCB ozone generator, that injected the ozone in a 10-mg L^{-1} flow.

The total time was of 2 hours, in accordance with Ferreira [9]. The volume removed was up to 10% of the total volume (50 mL), following the recommendation, to avoid interferences related to a larger oxidizing agents' exposure to a smaller contaminant volume [12]. Afterwards, the samples were kept in freezer to be later analyzed.

2.5 Tests for environmental toxicity risks

For the purpose of tests for environmental toxicity risks, four sample categories were investigated:

- Algae pure culture;
- Water of access;
- Raw effluent;
- Treated-post-POA effluents.

2.5.1 Tests with Euglena gracilis algae

From each one of the four sample categories, a 5-mL aliquot was removed, and it was added in a 40-mL *Euglena gracilis* algae culture. The collections were performed for photosynthetic efficiency tests, chlorophyll concentration and behavioral evaluation via NGTOX after a period of at least 48 hours, according to Ekelund [13].

2.5.1.1 Algae photosynthetic efficiency test via PAM

For testing the photosynthetic efficiency via PAM, the photosynthetic parameters were measured through a modulated pulse-width PAM 2000 fluorimeter (Walz, Effeltrich, Germany). The PAM measurement principle is based on changes on the chlorophyll fluorescence level, after the application of saturated light pulses. The photosynthesis performance (Yeld) was, then, calculated accordingly Eq. (2), on Yeld photosynthetic efficiency.

$$Yeld = \frac{fm - f0}{f0} \tag{2}$$

Approximately 5 mL^{-1} of the tested cultures were taken and transferred to the cuvette of the PAM equipment. They were then submitted to saturating light pulse emission, for photosynthetic activity evaluation. The saturating light pulse emission made possible the detection of the maximum fluorescence Fm, indicating total reduction of the electrons FSII receptor. The light-curve response was determined for all the treated samples. The algae were exposed to an increasing luminous intensity (generated by an internal halogen bulb) in 10 steps, from 0 to 3111 molm-2/s. After 10 s of each luminous step, a saturating pulse was applied, and the photosynthetic performance and the electron transport rate were measured automatically.

Afterwards, the average performance on test situation was calculated, considering all the values obtained on the saturation process. The global photosynthetic efficiency was measured according with Eq. (2):

$$EFG = \frac{\sum Yeld\ during\ saturation}{Quantity\ of\ submitted\ irradiating\ pulses} x\,100 \tag{3}$$

This way, it was intended to analyze the interference the compound of raw effluents and the waste will promote in the algae culture, when compared to the control one.

2.5.1.2 Evaluation of algae behavioral changes through biomonitoring via NGTOX

The behavioral tests with *Euglena gracilis* and the samples from the laboratories were conducted using a real-time biomonitoring tool called NGTOX, developed and homologated by Ecobabitonga Tecnologia Ltda. The instrument has monitored, through the analysis of real-time images, the algae behavior, considering different movement parameters of the photosynthetic single-celled flagellate [12].

The equipment is a system of connections that involve four silicon tubes responsible for the suction of *Euglena gracilis* cells culture, water samples with hormones for the test, water for sample dilution and the analyzed material disposal.

Three pumps activated by peristaltic motors transported the cells, the diluent and the sample up to a glass cuvette of 22 mm of internal diameter and 0.2 mm of thickness. The trial bodies in contact with the control were blended and transferred to an observation cuvette, connected to a microscope, which captured the images of the cells in movement. The images were recorded by a charged coupled device camera and digitalized by a board connected to a microcomputer, in which they were presented in a monitor. Then, the software calculated the movement parameters, the movement speed, the ascent rate, the average cell size, etc. After that, the samples were added separately, and the analyses were performed by the software one more time 10 minutes later. Any alteration on the movements, average speeds, ascent rates and cell size were calculated and compared with the previous results [14].

2.5.1.3 Test for alterations on concentration of chlorophyll present in algae: chlorophyll removal from the algae and UV analysis (160 SHIMADZU)

This test had the objective of verifying if the parameters previously analyzed affected the chlorophyll concentration. After the time of exposure to contaminants, 5-mL of the culture media submitted to the presence of the test samples and the control were taken. These aliquots were treated according with the procedures conducted by [15]. Aliquots were vacuum-filtered through Whatman® 47-mm filter paper.

The papers containing the filtered (precipitated cells) were transferred to a Falcon tube, received 5 mL of ethanol and were afterwards kept at 4°C for 60 minutes, for the pigments extraction. Then, the mixtures were centrifugated at 6000 g for 10 min at 4°C in order to aggregate on the waste cells.

The absorption spectrum of the supernatant was measured accordingly Lorenz's equation for the calculation of chlorophyll concentration.

2.6 Data statistical analysis

The data were evaluated through ANOVA, a univariate technique which deals with the quantitative data relating them to three-level independent category variables.

For the groups' analysis (tests and control), comparing all the effects, the used technique was an ANOVA extension, called ANOVA for repeated measures, that consists on a better developed approach for paired data. Then, comparisons of results and averages based on the samples' quantitative items were performed.

Following, the other variables were described, since formally there is not a statistical hypothesis test, although it works on confirming or not the a-priori expectations about the results.

The statistical analysis on algae behavior evaluated via NGOTX were conducted by ImagingTox®, a software especially developed and written for Microsoft platform, with multilingual Net 64-bit and MS SQL Server database. It has seven threads (the main one, three for video 1 and three for video 2), two functions (one for controlling the PC and NGTOX connection and the other one for database connection and validation), making possible the storage of bioassays performed for forensic analysis and real-time results exhibition screen. ImagingTox® conducted the 5-PL integrated statistical analysis.

3. Results and discussion

The environmental toxicity related to emergent pollutants has been increasingly causing concern among the scientific community, especially because a lot of studies have been pointed out to clear health and environmental risks [6]. Recent investigations have found essential results for other researches, as well as for the ones which study the development of decontamination processes of these products that are emergent and dangerous to the environment. Then, nowadays there is a scenery that leads to the revision and reorientation of conducts and legislations that involve environmental issues, both in national and international context [2].

In this perspective, magistral pharmacies are potent candidates to generate emergent pollutions potentially harmful to health and to the environment. Studies performed by the National Health Surveillance Agency (ANVISA) have showed that in Brazil around 120 thousand tons of garbage is generated every day, and between 1 and 3% of this total is produced by health facilities, comprising the magistral pharmacies (manipulation of magistral formula). From 10 to 25% of health waste represents risks to the environment and population's health, including the medicines [16].

Among the magistral pharmacy aspects that worked as effluent collection, an important factor to be considered was the quantity of actives and other substances disposed though the sinks and which compound the raw effluents from the laboratories. In the present investigation, the idea was not only to know the characteristics of hormone laboratory, but also the other ones of the other laboratories, in order to compare the found results and to evaluate the LMH risk before the production and other points. To know how many items the pharmacy averagely produced, the average monthly production of six months was examined, accordingly Eq. (1), previously presented.

The found results were collected and placed in a worksheet (**Table 1**):

On the results, it was seen a more prominent production of solids formulae, which include orally used tablets. They represent more than half of the formulae produced in the verified period. On second place, there are the dermato-cosmetic formulae, and among them there are creams in which a multiple of actives are incorporated to. In smaller quantity, there are the Order 344/98 formulae, that, in accordance with the current legislation, require an extra laboratory. The same occurs with the hormones – main study object here. Regarding the manipulation of hormones, almost all the formulae (99%) corresponded to the manipulation of estradiol valerate, with a similar structure of the 17 β estradiol used for many clinical conditions.

As the purpose of this investigation was to evaluate the impact of ED in the effluents and of the formed COD, a CGMS analysis of the LMH sample was

Laboratory	Quantity of formula	Representation (%)
Psychotropics	3150	18.6
Hormones	728	4.3
Solids	9388	55.5
Dermato-cosmetics	3638	21.5
Total	16,904	99.9

Source: pharmacy's registers.

Table 1.
Quantity of formulae produced between January and July of 2014.

conducted, in order to identify the presence of estradiol valerate ED, and another analysis was performed after ozone-based POA and UV for 120 minutes. The acquired chromatograms are showed in **Figure 1**.

3.1 Influence of COD in raw effluents on the behavior and chlorophyll concentration in algae

The presence of hormones in the effluent has significantly affected the velocity of surface ascent rate, since it was inhibited (**Figure 2**). The same was observed in the algae's r-value. The phenomenon was similar to the one observed by Pinto et al [8].

However, it is important to say the COD, unlike the original ED, presents a certain potential of chlorophyll degradation.

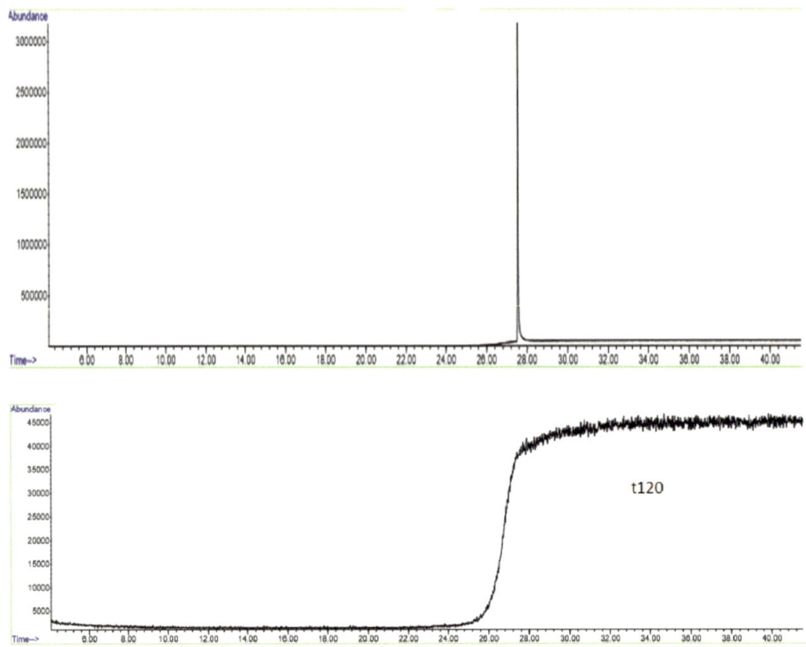

Figure 1.
CGMS chromatogram showing ED absence after POA/UA for 120 minutes.

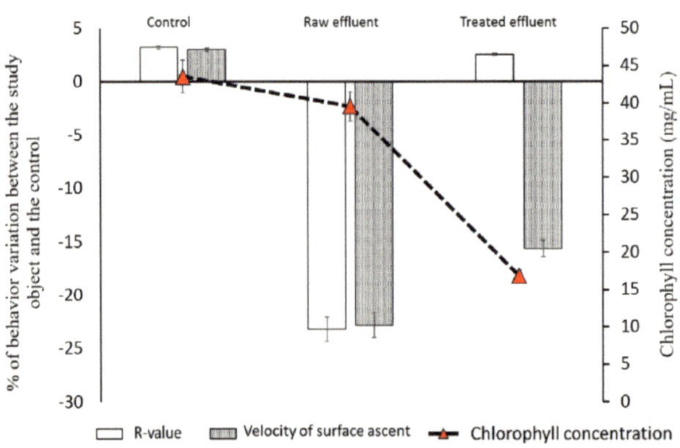

Figure 2.
Changes in behavioral parameters of Euglenas gracilis exposed to post-ozonating effluents at the Laboratory of Hormones, compared to the raw effluent. Pearson correlation test: There are significant connections between the pairs of variables (P > 0,050).

To better comprehend this effect on chlorophyll degradation by COD, a comparison between the concentration before and after the ozonating process was performed (**Figure 3**). Considering the influence over the chlorophyll concentration, it was seen a clear difference between the laboratories and the pre- and post-ozonating, as it is possible to seen in **Figure 3**.

In this comparison, it was seen that in the solids and psychotropics laboratories there was reduction in *a*-chlorophyll in the presence of raw effluent. The result was very similar when compared with the same waste treated using ozonating. However, in the dermato-cosmetics laboratory, as well as in the LMH, it was difficult to see influence of raw effluent, but is was observed influence of the treated effluent, which now has a concentration similar to the ones from the other laboratories treated with ozone/UV. It is important to emphasize the ozone in excess was not present, because in reaction with potassium permanganate the result was negative for the ozone. The justification for the LHM was the antioxidating activity, while a probable explanation for the dermato-cosmetics was the small quantity of highly lipophilic products manipulated there, besides the little influence in the algae physiology [17].

Another relevant variable is the quantity of actives and substances disposed through the sinks and that compound the raw effluent of the laboratories.

For this purpose, the pharmacy's average monthly production of six months was taken into account, and the monthly average production was calculated, in order to verify the laboratories' pre-ozonating activity average. The results obtained accordingly Eq. (1) are presented in **Figure 4**.

The most impacting treated effluent that has COD, in an independent way, on *a*-chlorophyll concentration came from the laboratory of hormones, since it has the lowest production and significant reduction of *a*-chlorophyll concentration. At the same time, in the presence of raw effluent, its concentration was similar to the control one.

Then, it is clear that the removal process via ozone/UV, in the total estrogenic activity removal, interferes on *a*-chlorophyll degradation with COD. Both the raw effluent and the treated one influence on *a*-chlorophyll concentration, reducing it, except at LMH. The reduction of chlorophyll concentration might affect the algae's photosynthetic efficiency and cause ecotoxicity hazards, when there is not compensatory mechanism to assure the algae's survival. Anyway, the condition does not dispense the monitoring of effluents [18].

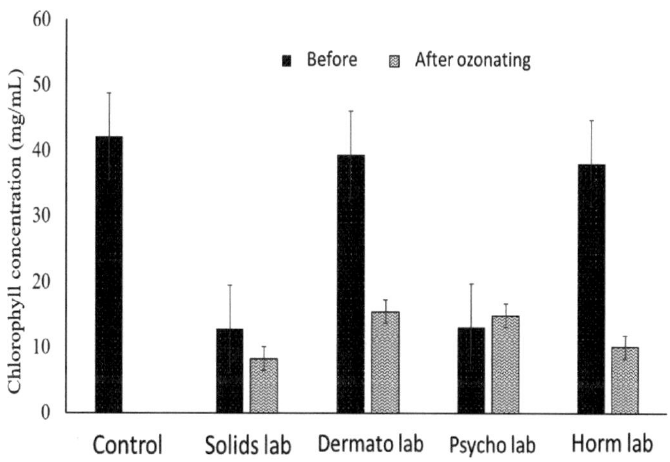

Figure 3.
Influence on chlorophyll concentration. Variance analysis on Kruskal-Wallis variables. H = 7.812 with 2 liberty degrees. P (est.) = 0.020 P (exact) = 0.011. The difference among the average value between the treatment groups are greater than the expected. There is a statistically significant difference (P = 0.011).

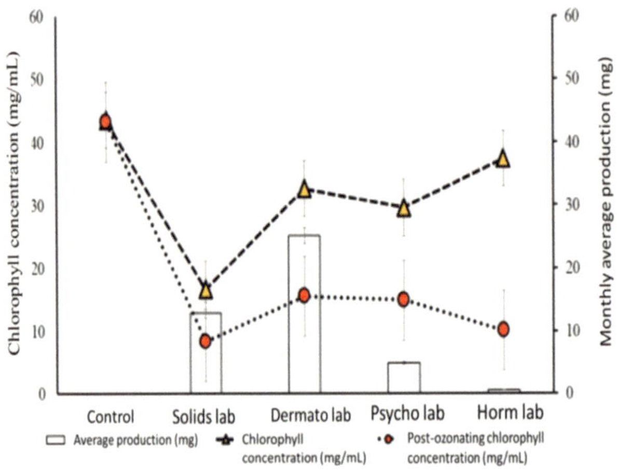

Figure 4.
Influence of the produced quantity versus pre-ozonating chlorophyll concentration. Pearson correlation test: Alteration on chlorophyll concentration at the laboratory (P > 0,050).

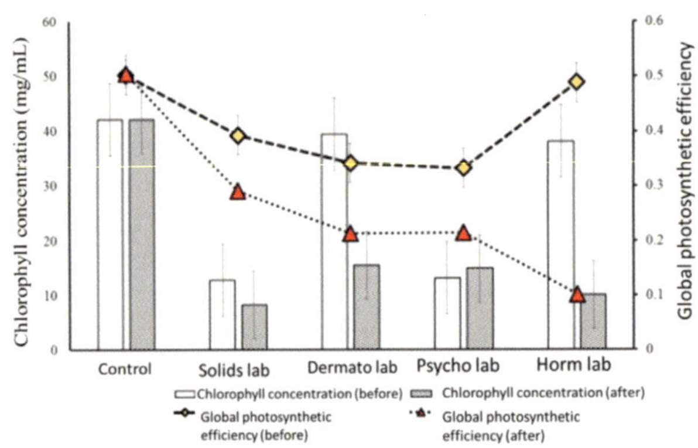

Figure 5.
Comparison between chlorophyll concentration and global photosynthetic efficiency. Variance analysis on Kruskal-Wallis variables. H = 7.812 with 2 liberty degrees. P (est.) = 0.020 P (exact) = 0.011. The difference among the average value between the treatment groups are greater than the expected. There is a statistically significant difference (P = 0.011).

Thus, considering the reduction of *a*-chlorophyll concentration was strongly influenced by the COD, it was also pursued if these reductions would interfere on the global photosynthetic efficiencies of the post-ozonating samples, in order to confirm if the chlorophyll concentration influence anyhow this physiologic parameter, denoting one more risk/environmental hazards (**Figure 5**).

The global photosynthetic efficiency has presented variations among the raw effluent samples, revealing a direct correlation between the chlorophyll concentration and the global photosynthetic efficiency in all the laboratories, except in the hormones one, in which the probable antioxidating action attributed to the steroidal hormone structure helps the electrons transference and contributes to the global photosynthetic efficiency improvement [8, 19].

The global photosynthetic efficiency was influenced by the COD in all laboratories, and the laboratory of hormones has showed the lowest value in the presence of COD and the highest difference between the efficiencies found in the exposure of raw effluents and the treated ones. The hormones' antioxidating activity has assured a good global photosynthetic efficiency performance. The reduction of global photosynthetic efficiency after the ozonation occurred because the hormone decomposition and its oxidizing potential, as well as the possible ethylene formation, which affects the chlorophyll activity [20].

Regarding the solids and psychotropics laboratories, the global photosynthetic efficiency reduction was due to the presence of substances that act on action potentials, listed in the production of both laboratories. These changes may influence the algae's movement ability, being an extra factor on the global photosynthetic efficiency, or the reduction is only attributed to the decline on chlorophyll concentration.

Maybe, the potential cellular action attributed to the manipulated medicines in these laboratories prevent these mechanisms, which affect the algae's flagellar mobility [21].

4. Conclusions

The present investigation made possible an analysis on how the ozone-based oxidative processes influence the reduction of the ecotoxicity risk caused by emergent pollutants. The obtained results have showed algae's behavioral changes among the four examined laboratories–hormones, solids, dermo-cosmetics and psychotropics–, comparing raw effluent samples, treated effluent samples and control.

Significant alterations on *Euglena gracillis* behavior were observed for the effluents from the laboratories of hormones and dermo-cosmetics, with modifications in general speed and velocity of surface ascent. The psychotropics and solids laboratories have not presented significant statistically difference over the algae's behavior. However, complementary studies are appropriate, in order to confirm the long-term toxicity, since the algae find compensatory mechanisms to fit the adversities.

The variations on algae's behavior before the exposure to different pollutants have suggested it is important to distinguish the effluent treatment, according with the characteristics of the substances manipulated in each laboratory, to reduce the environmental toxicity risks.

Therefore, the found biomonitoring data were relevant for better knowing and to be more aware of the issue, indicating the environmental toxicity caused by the effluents from magistral pharmacies may provoke great impact to the environment if revisions of actions and legislation are not performed in a way that in long term the environmental points related to this type of emergent pollutants are reduced.

For future studies, the evaluation of fish's behavior, before similar conditions, may point out better comprehension over the influence of pharmaceutical ecosystem risks.

Acknowledgements

The researchers thank Univille and the Research Support Funds, that make possible the Environmental Impacts Integrated Project (ECOSAM) development.

Author details

Thais Francinne[1], Suellen Zucco Bez[1], Julia Carolina Soares[1], Sabrina Martins da Rosa[2], Aline Mirian Paszuck[2], Luciana Ferreira Karsten[3] and Luciano Henrique Pinto[4*]

1 Pharmacy Course, Health and Environmental Research Laboratory, University of the Region of Joinville (UNIVILLE), Joinville, SC, Brazil

2 Nursing Course, Health and Environmental Research Laboratory, University of the Region of Joinville (UNIVILLE), Joinville, SC, Brazil

3 Department of Nursing, University of the Region of Joinville (UNIVILLE), Joinville, SC, Brazil

4 Department of Pharmacy/Nursing/Medicine, Health and Environmental Research Laboratory, University of the Region of Joinville (UNIVILLE), Joinville, SC, Brazil

*Address all correspondence to: lucianoefar@gmail.com

References

[1] VERLICCHI, P., AL AUKIDY, M., ZAMBELLO, E.; Occurrence of pharmaceutical compounds in urban wastewater: Removal, mass load and environmental risk after a secondary treatment—A review. *Sci. Total Environ.* 429

[2] KOCK-SCHULMEYER, M; GINEBREDA, A.; POSTIGO, C; LOPEZ-SERNA, R. PEREZ, S.; BRIX, R.; LLORCA, M., LOPEZ DE ALDA, M., PETROVIC, M., MUNNÉ, A., TIRAPU, L. Wastewater reuse in Mediterranean semi-arid areas: The impact of discharges of tertiary treated sewage on the load of polar micro pollutants in the Llobregat river (NE Spain). Chemosphere, 82 P. 670 – 678, 2011.

[3] CRUZ, L. H. Degradação fotocatalítica de Sulfametoxazol, Trimetropina e Diclofenaco em solução aquosa. Química Nova, vol. 33, No. 6, 1270-1274, 2010.

[4] BRASIL. Lei N° 9.433, de 8 de Janeiro de 1997. Institui a Política Nacional de Recursos Hídricos, cria o Sistema Nacional de Gerenciamento de Recursos Hídricos, regulamenta o inciso XIX do art. 21 da Constituição Federal, e altera o art. 1° da Lei n° 8.001, de 13 de março de 1990, que modificou a Lei n° 7.990, de 28 de dezembro de 1989. Brasília, DF, 1997. Available at: <http://www.planalto.gov.br/ccivil_03/leis/l9433.htm> Accessed on: July 10, 2014.

[5] CORDEIRO, D. Uso de bioindicador de efeito endócrino e validação do método para determinação de hormônios na água da represa municipal de São José. Mastering dissertation. Instituto de Química de São Carlos. Universidade de São Paulo. São Carlos, 2007.

[6] KUNKEL, U., RADKE, M.; Fate of pharmaceuticals in rivers: Deriving a benchmark dataset at favorable attenuation conditions. Water Res. 46(17), 5551-5565,2012.

[7] ESPUGLAS, S., BILA, D. M., KRAUSE, L. G. T., DEZOTTI, M. Ozonation and advanced oxidation technologies to remove endocrine disrupting chemicals (EDCs) and pharmaceuticals and personal care products (PPCPs) in water effluents. Journal of Hazardous Materials, vol. 149, 631-642, 2007.

[8] PINTO L. H., CARDOZO G., SOARES J.C.; ERZINGER, G. S.; Toxicidade ambiental de efluentes advindo de diferentes laboratórios de uma farmácia magistral. Ambiente & Água - An Interdisciplinary Journal of Applied Science 1761, 09 Jul. 2016.

[9] FERREIRA, M. G M; Remoção da Atividade Estrogênica de 17β-Estradiol e de 17α-Etinilestradiol pelos Processos de Ozonização e O_3/H_2O_2. Universidade Federal do Rio de Janeiro, Chemical Engineering Doctoring, Rio de Janeiro; 2008.

[10] LOPEZ-SERNA, R., PETROVIC, M., BARCELO, D.; Occurrence and distribution of multi-class pharmaceuticals and their active metabolites and transformation products in the Ebro river basin (NE Spain). *Sci. Total Environ.* 440, 280-289, 2012.

[11] SHI, W., WANG, L., ROUSSEAU, D.P., LENS P.N.; Removal of estrone, 17alpha-ethinylestradiol, and 17beta-estradiol in algae and duckweed-based wastewater treatment systems. Environ Sci Pollut Res Int. 2010 May;17(4):824-33. doi: 10.1007/s11356-010-0301-7. Mar 7, 2010.

[12] ERZINGER, G. S., DEL CIAMPO & HÄDER, D. P. Equipamento e Processo para Análise de Toxicidade em Sistemas Aquáticos. Instituto Nacional de Propriedade Industrial – INPI, N°.0000221105523696. 2011.

[13] EKELUND, N. G. A., NILSSON, L. Effects of estrogenic substances on the movement of *Euglena Gracilis* Verh. Internat. Verein. Limnol. 2008, vol. 30, Part. 2, Stuttgart, April, 2008.

[14] HÄDER, D.; AZIZULLAH, A., RICHTER, P., JAMIL, Chronic toxicity of a laundry detergent to the freshwater flagellate *Euglena gracilis*. Ecotoxicology, v. 21, n. 7, p. 1957-1964, 2012.

[15] SUMIDA, S.; LYMAN, H., KIYOHARA, N.; OSAFUNE, T.; Mechanism of Conversion from Heterotrophy to Autotrophy in *Euglena gracilis*. Cytologia 72(4): 447-457, 2007

[16] BRASIL. Agência Nacional de Vigilância Sanitária. Anvisa e ABDI discutem descarte de resíduos de medicamentos. Available at: <. Accessed on: July 10, 2014.

[17] ARONSSON, K. A.; ECKELUND, N. G. A.; Effects on motile factors and cell growth of *Euglena gracilis* after exposure to wood ash solution: assentment of toxicity, nutrient, availability and pH-dependency. Water, Air and Soil Pollution, v.162, p.353-368, 2005.

[18] HÄDER, D-P., LEBERT, M. Real time computer controlled tracking of motile microorganisms. Photochemistry and Photobiology, vol. 42, 509-514, 1985.

[19] STACEY, A.; MAOYUN, T.; RAVI, S; Estradiol-17β as an antioxidant: Some distinct features when compared with common fat-soluble antioxidants. The Journal of Laboratory and Clinical Medicine. Volume 128, Issue 4, Page A1, October, 1996.

[20] STREIT, N. et al. As clorofilas. Cienc. Rural, Santa Maria, v. 35, n. 3, p. 748-755, June 2005.

[21] GOODMAN & GILMAN: As Bases Farmacológicas da Terapêutica. 12ª ed. Rio de Janeiro: McGraw-Hill, 2012.

Chapter 11

Water Quality Parameters and Monitoring Soft Surface Water Quality Using Statistical Approaches

Romana Drasovean and Gabriel Murariu

Abstract

Water is the matrix of life and is indispensable on Earth. Water has a multitude of applications and all known life forms depend on it. Therefore, water quality is important for all of us. Water quality can be represented by a set of physical, chemical, biological and bacteriological characteristics. These parameters allow water to be classified in multiple categories leading to its use for a specific purpose. This chapter establishes the connections between external causes and their effect on water quality parameters. In order to provide information on water quality, different Water Quality Index (WQI) models can be used. In order to study the association between water quality parameters, several correlation coefficients have been developed. For a coherent statistical approach, we have used Pearson and Spearman correlations. In order to exemplify the manner in which WQI can be calculated and interpreted, we used a series of data from our previous work, consisting of 13 parameters measured for water samples taken from the Danube River, from Galati City area, Romania.

Keywords: water quality, correlation coefficient, quality index

1. Introduction

Water is a common "good" of the whole society and is essential for human, animal and plant life and has a multitude of uses.

1.1 Water classification

Water can be classified according to its source into [1, 2]:

1. Surface water (water located at the surface of the soil) - Surface water sources are represented by running waters, water seas, oceans, rivers, lakes, icebergs;

2. Groundwater water below the surface in the saturated area and in direct contact with soil or subsoil. Groundwater sources are represented by groundwater aquifers, deep aquifers, springs; and,

3. Atmospheric water.

According to the field of use, the water is classified as:

1. Drinking water: for domestic consumption and for agriculture; and,

2. Industrial water: auxiliary in manufacturing processes, raw material for various industries, power generator, coolant agent, heating agent, etc.

Wastewater is water that has changed its original properties through use, in other words has been contaminated by human beings [2].

1.2 Water quality

Water quality is represented by a set of physical, chemical, biological and bacteriological characteristics. These characteristics are also called parameters or indicators. Physical, chemical, biological and bacteriological parameters allow water to be classified in some categories, leading to its utilization for a specific use.

Water quality requirements depend on the purposes for which the water will be used. Thus, drinking water must not contain chemicals or micro-organisms which can affect the human health. Water used in agriculture must not contain large amounts of sodium ions, high concentrations of nitrates or high concentrations of other contaminants. Requirements for water use in industry are less rigorous than drinking water [3].

Water quality also depends on the type of water source and changes with geological, meteorological and land use conditions. The World Health Organization (WHO) has established regulations and standards for water safety in support of public health [3]. The European Union has, also, established a legal framework for water protection [4]. Water quality criteria in all countries have been established in accordance with the WHO guidelines [3]. In European countries, the framework directives of the European Union are closely followed [4].

The EU Framework Directive requires that operational monitoring should be specific and based on monitoring relevant biological, hydro-morphological and physic-chemical parameters. These world environmental monitoring systems provide for water quality measurements in three categories of parameters:

1. Basic parameters: Temperature, pH, Conductivity, Dissolved Oxygen (DO), Total Dissolved Solids (TDS), Coli bacilli, Biochemical Oxygen Consumption (CBO_5), Turbidity, Nitrites and Phosphates concentration;

2. Indicators of persistent pollution: Cadmium, Mercury, Organo-halogen compounds and Mineral oils; and,

3. Optional parameters: Total Organic Carbon (TOC), Anionic detergents, Heavy metals, Arsenic, Boron, Sodium, Cyanides, Total oils, Streptococci [3].

Water quality monitoring indicators established by the European Union rules are grouped according to their common characteristics as it follows [4, 5]:

1. Indicators that give information about oxygen condition: Dissolved Oxygen, Biochemical Oxygen Consumption (CBO_5), Chemical Oxygen Consumption with chromium, Chemical Oxygen Consumption with manganese;

2. Indicators that give information about the presence of nutrients that contribute to eutrophication: Ammonium (NH_4^+), Nitrates (NO_2^-), Nitrites (NO_3^-), Total nitrogen (N), Orthophosphates (PO_4^{3-}), Total phosphorus (P);

3. Salinity indicators: Chlorides (Cl^-), Sulphates (SO_4^{2-}), Ca^{2+}, Mg^{2+}, Na^+;

4. Heavy metals (e.g. Fe, Cu, Cd, Pb, Hg); and,

5. Other relevant indicators: Phenols, Anionic surfactants, Absorbed organic halides [4, 5].

To understand the overall health of an ecosystem and the condition of water, a number of water quality parameters or indicators must be analyzed and monitored. In 1998, Sene and Farquharson [6] stated that monitoring of the surface water quality is necessary to assess spatial and temporal regional variations. The process of monitoring the quality of ambient water has led to the development of water standards and the periodic assessments of the environment.

The monitoring program and the parameters to be measured for the study of water quality should be chosen specifically for each locality and each type of water. Although many parameters of water are important for human health or the health of an ecosystem, the analysis of all parameters is not feasible. The standards recommend the analysis of specific parameters for both drinking water and non-drinking water [6, 7].

Chemical and physical parameters are important in the rapid determination of water quality while biological parameters provide a detailed and complex analysis of the environment [8].

1.2.1 General physico-chemical parameters

Temperature is an important parameter that influences the chemical properties of water. Temperature affects the density and stratification of water, the density and viscosity of transported sediments, solubility of dissolved gases, vapor pressure [9].

pH is determined only at the place where the sample was collected, directly from the water source while also determining the air temperature. Due to the presence of carbon dioxide, bicarbonates, and carbonates, the pH of the water varies very little from the neutral pH. The pH of natural water is usually in the range 6.5–8. The pH of the wastewater can be alkaline if pH is higher than 7 or acidic if pH is lower than 7 [10].

The conductivity of water is given by the presence of ions in the solution, ions that have the property of transmitting electric current. The higher the ionic concentration of the solution is, the higher the conductivity gets. The conductivity value depends on the amount of substances dissolved in the water. As a rule, high turbidity value also implies a high conductivity [10].

Turbidity expresses the amount of light reflected or absorbed by particles suspended in a water sample and is a measure of its relative clarity. Turbidity is due to solid particles in the form of suspensions or in a colloidal state [10].

1.2.2 Chemical parameters

1.2.2.1 Oxygen regime indicators

Dissolved Oxygen is an indicator of water quality whose values are dependent on the type of the water. The amount of oxygen dissolved in water depends on water temperature, air pressure and the quantity of acidic substances and microorganisms.

Oxygen is necessary for aquatic life. A series of aerobic chemical processes take place through dissolved oxygen: the oxidation processes of organic matter, oxidation of mineral substances, and bio-chemical decomposition of the dead bodies in water [9]. With the decrease of oxygen, the self-purification capacity of natural water is reduced, favoring the persistence of pollution with its undesirable consequences. Other indicators of oxygen regime are Biochemical Oxygen Consumption (CBO_5) and Chemical Oxygen Consumption (CCO). *Biochemical Oxygen Consumption* (CBO_5) is the amount of oxygen consumed by microorganisms, during a 5 day period, for the biochemical decomposition of organic substances contained in water, at a temperature of 20°C. *Chemical Oxygen Consumption with chromium* (CCOCr) is an integral index of the existence of difficult degradable organic substances. *Chemical Oxygen Consumption with manganese* (CCOMn) is a comprehensive index of the existence of easily degradable organic substances [11].

1.2.2.2 Biogenic indicators

Nitrites in water represent the incomplete oxidation of organic nitrogen. Their presence in the water indicates an old pollution, because the transformation of organic substances containing nitrogen under the action of microorganisms first convert into ammonia then ammonia converts into nitrites. Therefore, the concentration of nitrites in the water may indicate an old pollution because all these transformations take time. Under normal oxygenation of natural water, nitrogen appears in the form of nitrates. The chemical forms of nitrite and ammonium are present when water pollution occurs and are toxic to living organisms [10].

Nitrates represent the final stage of oxidation of organic nitrogen. If ammonia, nitrites and nitrates are present simultaneously in the water, this indicates a continuous pollution. The simultaneous presence of ammonia and nitrates in the water indicates an intermittent pollution [11].

The phosphate content in natural water is relatively low. High amounts of phosphorus in water can come from excessive use of nitrogen and phosphorus fertilizers. Higher concentrations of phosphorus in surface water can result in eutrophication.

1.2.2.3 Salinity indicators

Salinity is the content of mineral salts in water, mainly metal salts such as sodium, magnesium and calcium. The salts present in natural water are formed by the following cations Ca^{2+}, Mg^{2+}, Na^+, K^+ and anions HCO_3^-, SO_4^{2-}, Cl^-.

The chlorides in the water come either from natural soil layers, pollution or animal origin. The amount of chlorides that are released from the soil is relatively constant and varies slightly over time. A significant increase in chloride content is usually an index of organic pollution [11].

Hardness is an indirect indicator of the degree of mineralization of water.

1.2.2.4 Heavy metals

Heavy metals are those metals that have a high density (i.e. 5 g/cm^3) [10]. In low concentrations, heavy metal ions are essential for the development of metabolic processes in plants and animals. These metals (e.g., cadmium, chromium, cobalt, lead, nickel, mercury, selenium) can come from natural or anthropogenic processes. If certain concentrations are exceeded, then they become toxic substances for the living organisms.

1.2.3 Biological and bacteriological indicators

Water quality and its changes due to various forms of pollution may influence the composition of aquatic biocenoses. Biological analysis consists of an inventory of phytoplankton, zooplankton, benthic organisms or periphyton from water samples.

The microbial flora found in the water can be classified into two categories: water-specific microbial flora and microbial impurity flora. Water-specific microbial flora consists of microorganisms that commonly inhabit water and soil: cocci bacilli, different fungi and bacterial species which play a role in the natural degradation processes of organic substances. Microbial impurity flora consists of species of microorganisms of human or animal origin. This category can include pathogenic saprophytes. These microbes are generally accompanied by high concentrations of organic matter which provide their nutritional support [11].

In bacteriological analysis of water, the total number of germs and the determination of the bacillus coli have been adopted as bacteriological indicators.

2. Statistical analyses for assessing the surface water quality parameters

Water quality is determined by the biological, chemical and physical parameters of the water. Most often, it is not enough to measure these water quality indicators. In order to draw some solid conclusions, it is necessary to apply adequate statistical method to the measurements. These statistical methods can provide useful information that can lead to actionable advice regarding water management. There are a large number of statistical methods for examining water quality.

The main differences between these methods are the statistical techniques used and the significance of the values determined for each parameter. Statistical indices developed using water quality parameters can be linear, non-linear, segmented linear or segmented non-linear [12]. In order to have a global vision of the changes of the water quality in space and in time, various indices have been developed [13].

The water quality index (WQI) is represented by a number that expresses the general water quality in a particular location, over time, based on several water quality parameters. The aim of this index is to transform a large number of complex water quality measurements into information that is easy for water managers and the public to understand and to use. Are a multitude of methods for calculating water quality indices (WQI). In the following, we present the weighted average method. This method was proposed by Horton in 1965 and developed by Brown et al. in the year 1970 [14].

For the calculation of the WQI, the following expression was used [15–19]:

$$WQI = \frac{\sum q_n W_n}{\sum W_n} \tag{1}$$

where:

n is the number of the water quality parameters.

W_n is the Unit Weight:

$$W_n = \frac{K}{S_n} \tag{2}$$

q_n is the Quality rating:

$$q_n = \frac{V_n - V_{id}}{S_n - V_{id}} \times 100 \quad (3)$$

where:

V_n represents the measured value for the nth parameter of the water corresponding to a given sample;

V_{id} is the ideal value for the nth parameter corresponding to pure water; V_{id} values are zero for most parameters except for pH and dissolved oxygen. V_{id} for the pH is 7 and for dissolved oxygen (DO) is 14.6 mg/l [20].

S_n represents the standard value allowed for the nth parameter;

K it is a constant of proportionality calculated with the formula:

$$K = \frac{1}{\sum \frac{1}{S_n}}. \quad (4)$$

Water quality is Excellent if the WQI index score is between 0 and 25; Good for values of 26–50; Poor for WQI = 51–75; and, Very Poor for values between 76 and 100. If the value of the WQI index exceeds the value of 100, then the water is unsuitable for drinking and cannot be transformed into drinking water by any process [19, 21].

To study the relationship between two parameters of water samples, several correlation coefficients can be used. The statistics used most often are Pearson and Spearman coefficients. Linear correlation can be determined using the Pearson correlation coefficient while non-linear correlation can be determined using the Spearman coefficient. The Pearson correlation coefficient is a statistical technique that measures and describes the degree of linear association between two normally distributed continuous quantitative variables [21]. Let x and y be two variables, in our case two indicators of water quality. The Pearson coefficient, r, is calculated using the expression:

$$r = \frac{S_{xy}}{S_x \bullet S_y}; r = \frac{\sum(x-\overline{x})(y-\overline{y})}{\sqrt{\sum(x-\overline{x})^2(y-\overline{y})^2}} \quad (5)$$

where S_{xy} represents the covariance, S_x and S_y are the standard deviations of the two variables x and y; $\overline{x}$ and $\overline{y}$ and are the mean values of the two variables x and y [21]. The Pearson coefficient takes values between −1 and + 1. The value of the coefficient indicates the strength of the relationship between parameters while the sign of the coefficient indicates the direction of the linear association. If the sign is positive, the two variables are directly correlated and, if the sign is negative, the two variables are inversely correlated. The closer of the Pearson correlation coefficient is to the value of 1, the stronger the "intensity" of the linear relationship between the two variables [21]. The variables x and y are independent if r has the value 0 (**Figure 1**).

The minimum value of the Pearson coefficient (r = 0) is not an indicator of independence of the two characteristics (variables), but only of their non-correlation. The coefficient of determination (r^2) is the square of the Pearson coefficient. The coefficient of determination indicates the percentage of the total variation of the dependent variable (y) which is explained by the independent variable (x).

Spearman method is a non-parametric method used when the relationship between two variables is not linear (monotonic correlation) [23–25]. The Spearman coefficient addresses some limitations of the Pearson coefficient. It is denoted either

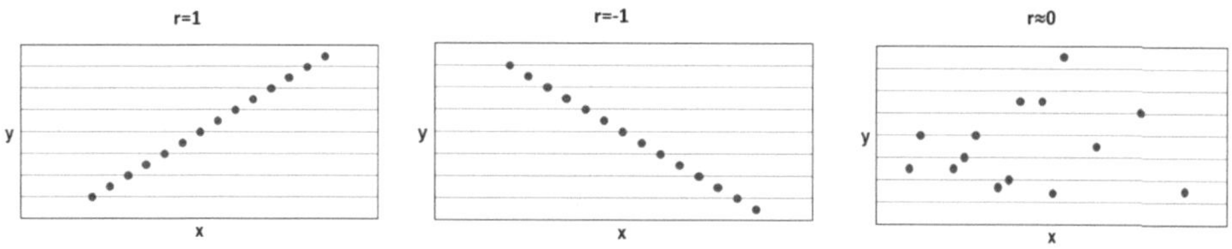

Figure 1.
Examples of the geometric significance of the Pearson coefficient.

with ρ or with r_S and represents an alternative to the Pearson coefficient. To calculate the coefficient, the data must have an order or rank. Coefficient can be calculated using the formula:

$$r_S = 1 - \frac{6\sum_i^n d_i^2}{n^3 - n} \tag{6}$$

where:
n is the number of pairs of values ordered in ascending order,
d_i is the difference between the orders of each pair(i) of values or the rank of the value:

$$d_i = r_{x_i} - r_{y_i} \tag{7}$$

where:
r_{x_i} is the rank of the value of x_i in the ascending ordered system and r_{y_i} is the rank of the value y_i in the ascending ordered system [21–25].

Eq. (6) is usually used when all n ranks are distinct integers or do not have tied ranks. When there are tied ranks, Eq. (6) is replaced by the following form:

$$r_S = \frac{S_{xy}}{S_x \bullet S_y} = \frac{\frac{1}{n}\sum_i^n \left(r_{x_i} - \overline{r_x}\right)\left(r_{y_i} - \overline{r_y}\right)}{\sqrt{\frac{1}{n}\sum_i^n \left(r_{x_i} - \overline{r_x}\right)^2 \frac{1}{n}\sum_i^n \left(r_{y_i} - \overline{r_y}\right)^2}} \tag{8}$$

where $\overline{r_x}$ and $\overline{r_y}$ are the mean ranks of value x and value y [23].

Spearman coefficient values are in the range $[-1, 1]$. The interpretation of these values is similar with that of the Pearson coefficient [21].

For a correct interpretation, the correlation coefficient must be accompanied by a significance test. The correlation coefficient has statistical significance if the value level of confidence factor $p < 0.05$. This significance coefficient p means the probability of making erroneous statements. If $p < 0.05$, we could reject the null hypothesis H0 and the computed results has certain statistical significance [24]. If the p result of the test is less than the significance threshold α (α = 0.05), hypothesis H1 is accepted: there is monotonic correlation. If p is greater than 0.05, then the H0 hypothesis is valid, which considers that there is no monotonic correlation [25].

3. Case study: monitoring water quality of the Danube River using the statistical approach

In this section, we provide an example of how to apply these methods in order to achieve a rapid assessment of water quality. The data set chosen for statistical analysis comes from our previous work [15, 16] and consists of 13 water quality parameters that were determined from samples taken from the Danube River. Sampling points were located along the river in the neighborhood of Galati. Galati is a Danube port city in the south-eastern part of Romania. Water samples were collected from November 2016 to December 2017. We will use data from 3 locations coded with D1, D4 and D7. All locations are along the Danube's left bank, D1 being located upstream and D7 downstream (**Figure 2**). The measured parameters were: potassium and calcium ions, nitrites, nitrates, total nitrogen, ammonium, chlorides, total phosphorus, sulphates, cadmium, chrome, copper, lead, iron, zinc, density, dissolved oxygen, chemical oxygen demand (CCO-Cr), biochemical oxygen demand (CBO_5), electrical conductivity, the density of the conductivity, resistivity, pH, salinity, total dissolved solids [15].

From our previous work [15, 16], the scatter plot diagrams and the box plot diagrams of the parameters indicated that quality class thresholds were exceeded during certain time periods. Correlations between the measured parameters could not provide a clear conclusion on the water quality condition.

For these reasons to provide clear information on the water quality condition, we calculated the Water Quality Index (WQI).

The Water Quality Index evaluation consisted of several stages. It is important to scale and weight the values of the monitored parameters according to the allowed limit values.

The water quality standards, S_n, were determined from Romanian legislation [26]. In accordance with the requirements of the "Normative on the classification of surface water quality in order to establish the ecological status of water bodies" [26], the limit values of the parameters are given for five water quality classes. In accordance with this legislation, we have used the water standards (S_n) for the third quality class. The surface water belonging to this class is considered moderately polluted.

Table 1 presents the intermediate results obtained from the application of the Water Quality Index method. The Unit Weights (W_n), the constant of proportionality (K), the ideal values (V_{id}) have the same values for all three locations – D1, D4 and D7. The Quality rating (q_n) was calculated with Eq. (3) for each parameter. The last stage of the method consists in calculating WQI using Eq. (1).

The obtained values for the water quality index corresponding to the three locations are presented in **Figure 3**.

According to the diagram from **Figure 3**, during the time interval November 2016–June 2017, the WQI values for the Danube River water were found in the

Figure 2.
Sampling points [15, 16].

Parameter	Standard (interval)	K from Eq. (4)	W_n from Eq. (1)	V_{id}
pH	6.5–8.5	0.01585	0.001864787	7
CCO-Cr (mg O_2/l)	50		0.000317014	0
CBO5 (mg O_2/l)	7		0.002264384	0
TDS (ppm)	200		7.92534E-05	0
Dissolved oxygen (mg/l)	5		0.003170138	14.6
Chlorides (mg/l)	250		6.34028E-05	0
Nitrites (mg/l)	0.06		0.264178133	0
Nitrates (mg/l)	5.6		0.00283048	0
Ammonium (mg N/l)	1.2		0.013208907	0
Total nitrogen (mg/l)	12		0.001320891	0
Sulphates(mg/l)	250		6.34028E-05	0
Lead(mg/l)	0.025		0.63402752	0
Cadmium(mg/l)	2		0.007925344	0
Total Phosphorus(mg/l)	0.75		0.021134251	0
Iron(mg/l)	1		0.015850688	0
Zinc(mg/l)	0.5		0.031701376	0

Table 1.
Intermediate results obtained from WQI method.

range of 100 to 2310. Between July and December 2017 the values decreased in the range 0–25. In the first-time interval (November 2016 to June 2017), the water quality index shows that the water was not suitable for consumption and cannot be transformed into drinking water by any process. However, by the end of the monitoring time interval (December 2017) the water quality was good or excellent.

According to our previous work [15, 16], during November 2016–June 2017, the following indicators had exceeded the limit permitted by Romanian law: all metals, Chlorides, Nitrates, Nitrites, Ammonium, Total Phosphorus, Sulphates, Solvent Extractable Substances and Anionic Surface Agents, Chemical Oxygen Consumption with chromium (CCOCr), Biochemical Oxygen Consumption (CBO_5). The high values obtained for these indicators were determined by the wastewater discharges into the Danube water. The high values of these indicators determined high values of WQI. At the end of the monitoring time interval the values of the studied

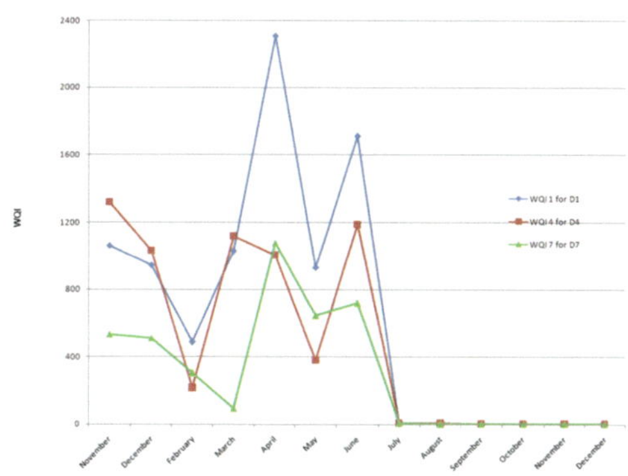

Figure 3.
Monthly values of WQI in the three locations during the monitoring.

indicators have improved. This improvement is found in the low values of WQI. The substantial improvement in water quality that occurred is due to the actions taken by the organizations responsible for environmental protection.

Figure 4 shows the boxplot diagram representing the main values of WQI in the 3 chosen locations. The average values of WQI are influenced by the extreme values. According to the 3rd quartile (Q_3) and the median of the upper half of the data set, 75% of the values in the data set lie below Q_3. The high average and median values, the values of the 3rd quartile frame depict waters as having severe pollution.

The value of the third quartile indicates that 75% of the determined values of WQI fall into the category of highly polluted waters. Based on **Figure 3**, only 25% (the 1st quartile - Q_1) of the values of the WQI lie below low values that would classify the studied water into the category of unpolluted waters. The information obtained from the WQI calculation was particularly useful in order to analyze how the overall water quality has evolved over time.

An easy method to identify possible sources of pollution is to calculate the correlations between the measured parameters. Using a Pearson Correlation Matrix [15] there was a strong positive linear correlation between TDS and Salinity ($r = 0.9394$) and TDS and Electrical Conductivity EC ($r = 0.9174$). Significant correlations also existed between the nitrites concentration and pH and between the nitrates concentration and pH there was a moderate negative corelation ($r = -0.65$ and $-$ 0.68 respectively).

To identify possible sources of pollution, the Pearson correlation matrix was computed between WQI and a series of measured parameters (**Table 2**).

In the absence of dedicated statistical software, the correlation coefficients can also be determined using free tabular software tool. We could exemplify quite easily this technique, for the Pearson coefficient between WQI and CCOCr, for the first location D1 (**Table 3**).

Table 3 shows the values of Pearson Correlation Coefficient (r) and coefficient of determination (r^2) for the water quality data set. The major influence of several parameters on the high values of WQI is due to strong positive correlation values. Therefore, excessive pollution was likely due to the presence of high concentrations of chlorides, nitrates, nitrites, ammonium, sulphates, lead, cadmium, iron, zinc.

The values of coefficient of determination (r^2) indicate that 89% of the variance of WQI is explained by the chlorides and cadmium concentrations while 87% is due to effect of iron and zinc. Nitrates concentration in the Danube River water explains 86% variation of the WQI. The high levels of covariance explained by the three groupings suggest significant co-linearity among the nutrient groups.

The strength and direction of monotonic association between water quality variables can be highlighted by the Spearman correlation. **Table 4** shows the

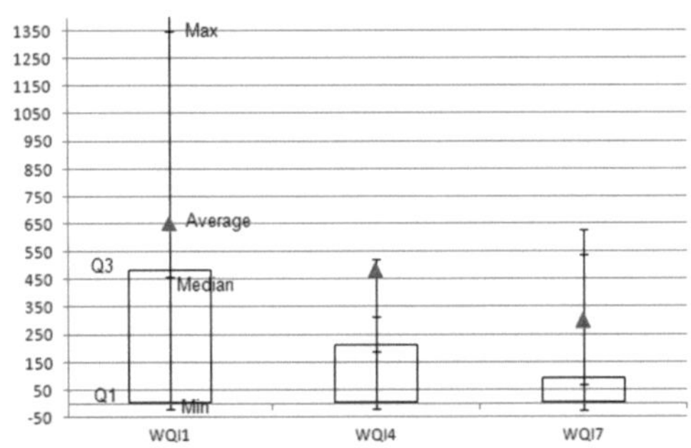

Figure 4.
Boxplot diagram of WQI in the locations D1, D3, D7.

WQI		
Parameter	**r**	**r^2**
pH	0.614	0.376996
CCO-Cr	0.28	0.0833
CBO5	−0.4297	0.1846
TDS	0.3092	0.0956
DO	−0.51102	0.2611
Chlorides	0.944033	0.891198
Nitrites	0.881528	0.777091
Nitrates	0.930999	0.86676
Ammonium	0.847447	0.718166
Total nitrogen	0.624669	0.390211
Sulphates	0.855873	0.732519
Lead	0.763807	0.583401
Cadmium	0.946214	0.895321
Total Phosphorus	−0.03909	0.001528
Iron	0.934618	0.873511
Zinc	0.934375	0.873057

Table 2.
Pearson correlation coefficients.

$x - \overline{x}$	$y - \overline{y}$	$(x - \overline{x})^2$	$(y - \overline{y})^2$	$(x - \overline{x})(y - \overline{y})$	S_x	S_y	r
406.854	−65.837	165530.052	4334.5	−26786.0468	2623.25	186.16	0.28858
288.854	−65.837	83436.544	4334.5	−19017.2808			
−166.696	−58.457	27787.608	3417.212	9744.548072			
375.374	34.163	140905.524	1167.116	12823.90196			
1656.854	52.663	2745164.668	2773.4	87254.9022			
276.854	52.163	76648.052	2720.987	14441.5352			
1055.854	56.163	1114827.344	3154.291	59299.9282			
−648.946	−65.837	421131.111	4334.5	42724.6578			
−648.036	−65.837	419950.857	4334.5	42664.74613			
−649.106	34.163	421338.799	1167.116	−22175.40828			
−649.146	34.163	421390.729	1167.116	−22176.7748			
−649.596	24.163	421975.163	583.854	−15696.18815			
−649.116	34.163	421351.781	1167.116	−22175.74991			
		$\sum$ = 6881438.232	**$\sum$ = 34656.209**	**$\sum$ = 140926.771**			

Table 3.
Calculation of the Pearson correlation coefficient between WQI (x) components and chemical oxygen consumption with chromium-CCOCr (y).

WQI		
Parameter	r_S	**p (2-tailed)**
pH	−0.7459	0.00342
CCO-Cr	0.24304	0.42366
CBO5	−0.3499	0.24126
TDS	−0.1658	0.58839.
DO	−0.674	0.01153
Chlorides	0.95461	0
Nitrites	0.93095	0
Nitrates	0.92562	0
Ammonium	0.86046,	0.00016
Total nitrogen	0.8138	0.00071
Sulphates	0.83252	0.00041
Lead	0.75674	0.00275
Cadmium	0.85436	0.0002
Total Phosphorus	0.59311	0.03263
Iron	0.82855	0.00047
Zinc	0.8856	0

Table 4.
Spearman rank correlation coefficients.

x	y	r_x	r_y	$r_{x_i} - \overline{r_x}$	$r_{y_i} - \overline{r_y}$	$(r_{x_i} - \overline{r_x}) \cdot (r_{y_i} - \overline{r_y})$	$(r_{x_i} - \overline{r_x})^2$	$(r_{x_i} - \overline{r_x})^2$	r_s
1060	10	11	2.5	4	−4.5	−18	16	20.25	0.24
942	10	9	2.5	2	−4.5	−9	4	20.25	
486.45	17.38	7	5	0	−2	0	0	4	
1028.5	110	10	8.5	3	1.5	4.5	9	2.25	
2310	128.5	13	12	6	5	30	36	25	
930	128	8	11	1	4	4	1	16	
1709	132	12	13	5	6	30	25	36	
4.2	10	5	2.5	−2	−4.5	9	4	20.25	
5.11	10	6	2.5	−1	−4.5	4.5	1	20.25	
4.04	110	4	8.5	−3	1.5	−4.5	9	2.25	
4	110	2	8.5	−5	1.5	−7.5	25	2.25	
3.55	100	1	6	−6	−1	6	36	1	
4.03	110	3	8.5	−4	1.5	−6	16	2.25	
						$\sum$=43	$\sum$=182	$\sum$=172	

Table 5.
Intermediate results for the Spearman correlation.

Spearman coefficients between the WQI and the water quality indicators that were measured.

Table 4 shows the high values obtained for WQI are associated with the high values obtained for chlorides, nitrates, nitrites, ammonium, total nitrogen, sulphates, lead, cadmium, iron, zinc. The association between these variables would be considered statistically significant.

Table 5 exemplifies such a calculation for the correlation coefficient between WQI and CCO-Cr.

For chlorides, nitrates, nitrites, ammonium, sulphates, lead, cadmium, iron, zinc, total nitrogen, the values of p coefficient are less than 0.001 (i.e., highly significant with confidence greater than 99.99%). For pH and DO, $p < 0.01$ means the statistical links are significant and the confidence is 99%.

Once the correlations between pollutants and WQI are identified, the sources of pollution can be established or the related process.

4. Conclusions

This chapter highlighted the importance of using statistical methods to display the water quality condition, using WQI evaluation and Pearson and Spearman correlations.

In order to exemplify the statistical methods, we have used a series of data from our previous work, consisting of 13 parameters measured for water samples taken from the Danube River, from Galati City area, Romania. Statistical correlations were made between quality parameters and Water Quality Index; thus, it was possible to identify which are the pollutants that determined an advanced degree of water pollution. The excessive pollution which occurred during the time interval November 2016–June 2017 is due to the presence of high concentrations of chlorides, nitrates, nitrites, ammonium, sulphates, lead, cadmium, iron and zinc. In recent times there are many statistical software for water quality analysis. If we do not have, for various reasons such programs, the statistical approach can be done classically. Water Quality Index (WQI) provides information on the overall quality of the water, while the correlation coefficients may indicate the parameters that influenced the changes in water quality.

Acknowledgements

This study was funded by the main author's personal resources.

Author details

Romana Drasovean* and Gabriel Murariu
Department of Chemistry, Physics and Environment, "Dunarea de Jos" University, Galati, Romania

*Address all correspondence to: rdrasov@ugal.ro

References

[1] Schneider S.H. (editor). Water Resources. Encyclopedia of Climate and Weather. Vol 2. New York: Oxford University Press 2011. 823 p. DOI: 10.1093/acref/9780199765324.001.0001

[2] Newman P.J. Classification of surface water quality management. Oxford: Heinemen Professional Publishing; 1988.189p.

[3] Water Quality Monitoring - A Practical Guide to the Design and Implementation of Freshwater Quality Studies and Monitoring Programs, United Nations Environment Program and the World Health Organization [Internet].1996. Available from:https://apps.who.int/iris/bitstream/handle/10665/41851/0419217304_eng.pdf?sequence=1&isAllowed=y

[4] Directive 2000/60/EC establishing a framework for Community action in the field of water policy [Internet].2000. Available from:https://eur-lex.europa.eu/legal-content/EN/TXT/?uri=celex%3A32000L0060

[5] Directive 2009/90/CE establishing the technical specifications for chemical analysis and monitoring of water status [Internet].2009. Available from: https://eur-lex.europa.eu/LexUriServ/LexUriServ.do?uri=OJ:L:2009:201:0036:0038:EN:PDF

[6] Sene K.J, Farquharson F.A.K, Sampling Errors for Water Resources Design: The Need for Improved Hydrometry in Developing Countries. Water Resources Management. 1998:12 (2): 121–138.

[7] Murariu G, Iticescu C, Murariu A, Rosu B, Munteanu D, Buruiana D.L, Assessment of Water Quality State Dynamics Using Adaptive Filtering Methods and Neural Networks Approaching Case study-Danube River in Galati area. Revista de Chimie. 2019; 70 (6):1914–1919. DOI :10.37358/RC.19.6.7246

[8] Savan B, Morgan A, Gore C, Volunteer environmental monitoring and the role of the universities: The case of Citizens' Environment Watch. Environmental Management. 2003; 31 (5): 561–568. DOI:org/10.1007/s00267-002-2897-y

[9] Quagliano J.V, Vallarino L.M. Chemistry. 3rd ed Florida State University: Prentice Hall, Inc., Englewood Cliffs; 1969. 844p.

[10] Petropol Serb G,D, Surse de apa si Ingineria apelor reziduale. Note de Curs. Craiova: Universitatea din Craiova; 2010. 173p.

[11] Bumbu I, Bumbu I, Virlan L. Controlul si Monitoringul Mediului. Curs de lucrari practice si laborator. Chisinau: Universitatea Tehnică a Moldovei; 2006. 55p.

[12] Abbasi T, Abbasi S.A. Water quality indices. Amsterdam, Netherland: Elsevier; 2012. 361p.

[13] United States Environmental Protection Agency (EPA), ENVIRONMENTAL QUALITY INDEX Overview Report [Internet]. 2014. Available from:https://edg.epa.gov/EPADataCommons/public/ORD/NHEERL/EQI/EQI%20Overview%20Report__Final.pdf

[14] Popa P, Murariu G, Timofti M, Georgescu L.P, Multivariate Statistical Analyses of Danube River Water Quality at Galati, Romania. Environmental Engineering and Management Journal. 2018; 17(5):1249–1266.

[15] Drasovean R, Murariu G, Constantinescu G, Circiumaru A, Assessment of surface water quality of Danube in terms of usual parameters

and correlation analyses. Revista de Chimie.2019; 70(2): 398–406. DOI: 10.37358/RC.19.2.6924

[16] Drasovean R, Murariu G, Condurache-Bota S, Constantinescu G, Studies on the water quality of the Siret river, near Galati city. In: Proceedings of 18th International Multidisciplinary Scientific Geoconference SGEM 2018; 2–8 July 2018; Albena; 18 (3.1) p.687–695. DOI: 10.5593/sgem2018/3.1/S12.

[17] Iticescu C, Georgescu L.P, Topa C, Assessing the Danube Water Quality Index in the City of Galati, Romania. Carpathian Journal of Earth and Environmental Sciences. 2013; 8(4): 155–164.

[18] Shweta T, Bhavtosh S, Prashant S, Rajendra D, Water Quality Assessment in Terms of Water Quality Index. American Journal of Water Resources. 2013; 1(3): 34–38. DOI:10.12691/ajwr-1-3-3

[19] Iticescu C, Georgescu L. P, Murariu G, Potential Effects of pH Variation Depending on the Temperature in the Drinking Water Supply System. Journal of Environmental Protection and Ecology. 2012;13(3) : 1324–1332.

[20] Malviya N, DeoS , Inam F, Determination of water quality index to assess water quality for drinking and agricultural purposes. International Journal of Basic and Applied Chemical Sciences. 2011; 1 (1) :79–88. DOI: http://www.cibtech.org/jcs.htm

[21] Iticescu C, Georgescu L. P, Murariu G, Topa C, Timofti M, Pintilie V, Arseni M, Lower Danube water quality quantified through WQI and multivariate analysis. Water, 2019; 11 (6):1305. DOI:10.3390/w11061305

[22] Scradeanu D. Modele Cantitative Statistice. Bucuresti: Editura Universitatii Bucuresti; 2013.103p.

[23] Clef T. Exploratory Data Analysis in Business and Economics. Springer International Publishing; 2014.215p. DOI: 10.1007/978-3-319-01517-0

[24] Mundry R, Fischer J, Use of statistical programs for nonparametric tests of small samples often leads to incorrect *P* values: examples from *Animal Behaviour.Animal Behaviour.*1998; Volume 56, Issue 1: 256–259. DOI: org/10.1006/anbe.1998.0756

[25] Opariuc C.D. Statistica Aplicata in Stiintele Socio-Umane. Notiuni de bazaă- Statistici univariate. Constanta: Editura Asociatia de Stinte; 2009. p330.

[26] Decree 161/2006 from Romanian Legislation on environmental protection.

Chapter 12

Reliability and Problems of Wastewater Treatment Processes in the Algerian Sahara

Rachid Zegait, Saber Kouadri, Samir Kateb and Mohamed Azlaoui

Abstract

This modest chapter deals more particularly with the reliability and the problems of the different processes used at the level of several treatment plants installed in the Algerian Sahara with the aim of eliminating the nuisances and the risks of contamination in the urbanized areas, protecting the receiving environment and water resources, the possibility of reusing treated effluents for irrigation. Through an evaluation of the performance of these stations after years of operation which confronted with climates such as the high temperature and evaporation and the impact of the sand winds on the efficiency of the basin, technical and anthropic problems such as the salinity and mismanagement of the waters on the other hand.

Keywords: westwater, purification, Sahara, Lagoon, problems

1. Introduction

In recent decades, humanity has become more and more aware of the danger threatening the planet as a result of the great demographic growth and the enormous technological advances which generate unsanitary conditions of the environment. Water is a big problem today affecting the whole of the earth. To do this, it must then be preserved by all possible means: reduction of waste; reuse of wastewater and its introduction into special recycling techniques. The reuse of wastewater is a widely used practice in regions of the world affected by water resource shortages. The Mediterranean Basin is one of the regions of the world where the reuse of urban effluents is practiced at a low rate.

In Algeria, this area is not very developed, and the system put in place does not allow the desired prospects to be achieved to deal with the problems emanating from wastewater. Several regions of the Algerian Sahara where wastewater discharged into nature without prior treatment with large and increasing volumes which are sources of pollution. They generate many water-borne diseases and the spread of epidemics [1]. The Algerian authorities have agreed to plan several purification stations, with a view to eliminating pollution and the risks of contamination in urban areas. Protecting the receiving ecosystem and the water supplies of these regions of the Algerian Sahara, in particular the water table, the possibility of reusing filtered effluent for irrigation, becoming an important biodiversity area and a breeding ground [2].

This chapter, aiming to shed light on the wastewater treatment component in the Algerian Sahara, through a description of the different systems used and to discuss their reliability and constraints.

2. The Saharan area in Algeria

The Sahara covers nearly 90% of Algerian territory, it is crossed ergs by sand dunes, regs of stony land, as well as volcanic massifs in the far south.

In the north, the grass of the steppe slowly becomes scarce as the species change to make way for the reg. The erg, the sand desert covers only a fifth of the Sahara. The great eastern erg borders the Wadi Righ, a succession of oases, stretches along the underground wadi. The Saoura valley limits the great western erg to the west. Between these two great ergs, is the M'Zab valley carved into a plateau.

This northern Sahara, still dotted with oases, is opposed to that of the South, dominated by the Hoggar massif at an altitude of more than three thousand meters. Vast monotonous hamadas like the Tademaït plateau between El Goléa and In Salah connect the large geographical areas of the Sahara [3]. **Figure 1** present the localizations of deferent WWTP concerning in this chapter.

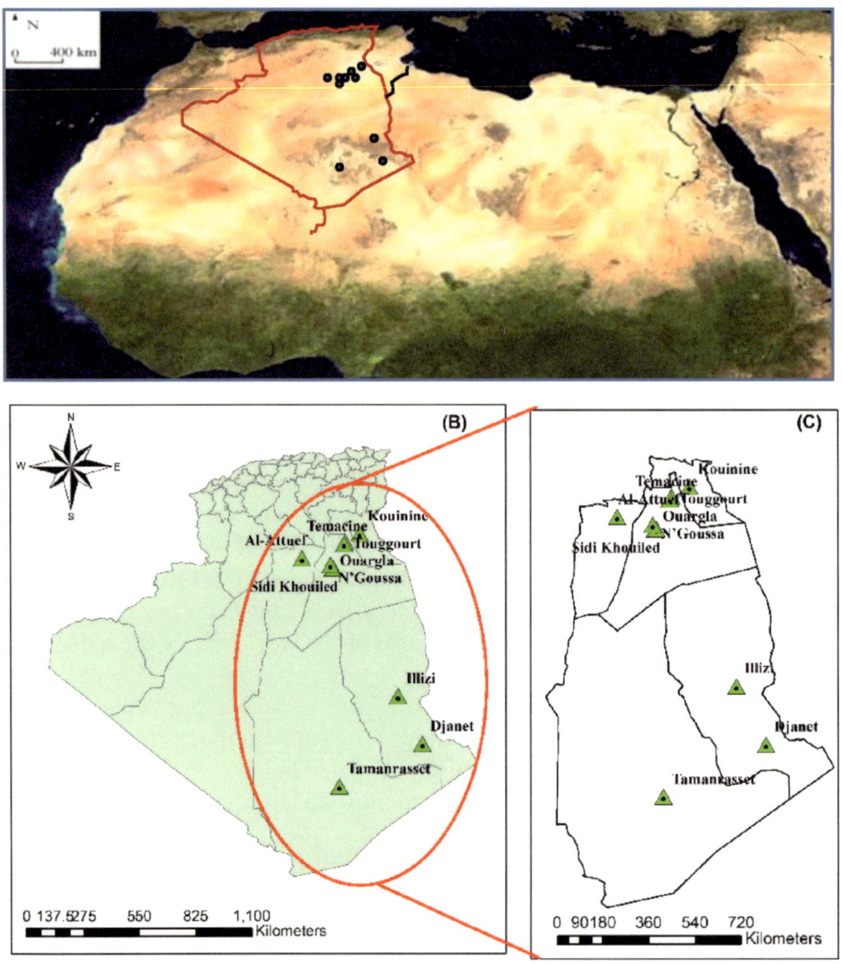

Figure 1.
Situation of the study area. (a) International location map of the study area. (b) Algerian location map of the study area. (c) Regional location map of the study area.

N°	Province	Number municipalities	Area Km^2	Population 2008
01	Ouargla	21	211 980	558 558
02	Ghardaia	13	86 105	363 598
03	El-Oued	25	54 573	647 548
04	Illizi	06	284 618	52 333
05	Tamanrasset	10	557 906	92 635
Total		75	983 202	1 714 672

Table 1.
Administrative breakdown [4].

The Sahara is characterized by low rainfall, recording less than 100 mm per year. It does happen, however, that it snows and floods sometimes revive the wadis that have dried up since prehistoric times. High temperatures can exceed 45 ° C especially in summer, winter is generally mild with average temperatures ranging from 8 to 12 ° C. The subsoil is full of water in the Albian tablecloth which extends under a large part of the Algerian Sahara, a vestige of the steppe climate that the region experienced 10,000 years ago. In 2018, it is home to a population of 3,600,000 inhabitants, or 10.5% of the Algerian population. Our investigation in this document concerns 10 wastewater treatment plants in 5 provinces in southern Algeria (**Table 1**).

3. Extensive processes in the Algerian Sahara

Groundwater is the main source of water in the Algerian Sahara; the preservation of this precious resource from all types of pollution is essential; to this end, the treatment of wastewater in the Algerian Sahara is a requirement and an inescapable social and environmental issue. The extensive process is used in most of the Saharan areas, it is ecological insofar as it does not use any chemical product to treat waste water and evacuate it safely to the receiving natural environment [5].

We noted more than 90% of the planned purification stations in the Algerian Sahara have extensive processes, the extensive solutions characterized with a large area and totally or partially dependence on natural purification processes in which the concentration of purifying organisms is low [6]. They do not involve recycling of bacterial liquor. Among these processes we note: lagooning and spreading. They bring into play complex natural self-purification phenomena dependent on climatic conditions. These extensive solutions require large footprint areas, implementation is essential and the quality of construction depends in part on the purification performance. They generally require little electromechanical equipment and are known for their hardiness and ability to adapt to variations in organic and hydraulic loads [7].

3.1 Natural lagoon

Natural lagooning is an interesting wastewater treatment process, particularly for small communities, and has been relatively developed since the 1970s [8].

Among the ten stations observed 3 stations (**Table 2**). planned using the natural lagooning process, we are talking about the stations of Djanet, M'Zab, and Illizi (**Figures 2–4**), these stations were abused in 2011; 2012 and 2019 respectively. The largest is that of Ghardaïa, which was built during the period 2008–2012

N°	Province	Station	Zone	UTM coordinates		
				X	Y	Z
01	Ghardaïa	Al-Attuef	31S	575518.34 m E	3589342.52 m N	435
02	Illizi	Illizi	32R	441295.27 m E	2930379.60 m N	554
		Djanet	32R	546961.85 m E	2703766.31 m N	995

Table 2.
Coordinates of natural lagoon stations.

Figure 2.
Ghardaia wastewater treatment plant.

Figure 3.
Illizi wastewater treatment plant.

Figure 4.
Djanet wastewater treatment plant.

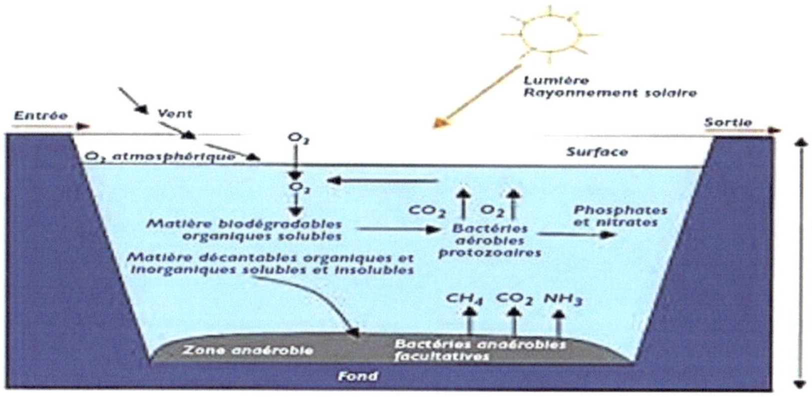

Figure 5.
Mechanisms involved in natural lagoon basins [10].

by the companies AMENHYD SPA as a production company and AQUATECH-AXOR (Canada): a control and monitoring design office; it was commissioned in November 2012 for a maximum capacity of 46,400 m^3/d, corresponding to 331,700 eq/inhab. by 2030. The station spread over 79 ha, has 16 basins divided into 02 floors finalized by 10 drying beds [9].

The purification is ensured through to a long residence time, in several watertight basins arranged in series. The most commonly encountered number of basins is 3. However, using a configuration with 4 or even 6 basins allows more thorough disinfection. The basic mechanism on which natural lagooning is based is photosynthesis. The upper water section of the basins is exposed to light. This allows the existence of algae which produce the oxygen necessary for the development and maintenance of aerobic bacteria (**Figure 5**). These bacteria are responsible for the degradation of organic matter [10].

The carbon dioxide formed by bacteria, as well as the mineral salts contained in wastewater, allow algae to multiply. There is thus a proliferation of two interdependent populations: bacteria and planktonic algae, also called "microphytes". This cycle is self-sustaining as long as the system receives solar energy and organic matter.

At the bottom of the pool, where light does not penetrate, it is anaerobic bacteria which degrade the sediments resulting from the settling of organic matter. A release of carbon dioxide and methane occurs at this level [10].

This type of process was chosen in these regions for various reasons, such as the low energy use or almost non-existent, especially if the difference in level is favorable; good elimination of nutrients phosphorus, nitrogen and pathogenic germs especially in summer when the high temperature. It adapts well to strong variations in hydraulic head which can favor the presence of the water table close to the surface. We also noted in these stations an absence of noise pollution with stable sludge and good integration into the landscape.

On the other hand; these stations encountered several obstacles during the design and operation, such as the large footprint (10 to 15 m^2/pe); investment cost very dependent on the nature of the subsoil. In the Saharan regions, where the terrain is sandy and unstable, this type of process is not recommended; which remains a contradiction with the location of these stations.

For the technical aspect, we noted that the purification performance is lower than intensive processes on organic matter. However, the release of organic matter takes place in the form of algae, which is less harmful than dissolved organic matter for the oxygenation of the environment downstream; and the control of the biological balance and the purification process remains limited. With a variable quality of

Figure 6.
Ouargla wastewater treatment plant.

Figure 7.
Tamanrasset wastewater treatment plant.

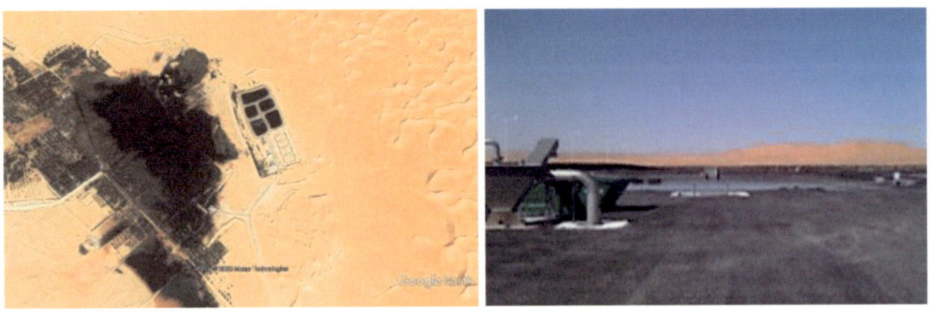

Figure 8.
Sidi khouild wastewater treatment plant.

Figure 9.
Kouinin wastewater treatment plant.

N°	Province	Station	Zone	UTM coordinates		
				X	Y	Z
01	Ouargla	Ouargla	31R	723557.44 m E	3542455.78 m N	129
		Sidi Khouiled	31R	729649.40 m E	3542586.63 m N	135
02	El-Oued	Kouinine	32S	300095.71 m E	3699943.49 m N	69
03	Tamanrasset	Tamanrasset	31Q	756780.31 m E	2519993.29 m N	1360

Table 3.
Coordinates of aerated lagoon stations.

Station	Year	Capacity Eq/inhab	Treated flow (m^3/d)	Receiving medium
Ouargla	2008	260.000	13986	Sebkhat Sefioune
Sidi khouiled	2010	7.156	995	Oum Raneb
Kouinine	2009	246.300	33251	Chott Haloufa
Tamanrasset	2011	228667	18000	Oued Tagrambait

Table 4.
Characteristics of aerated lagoon stations.

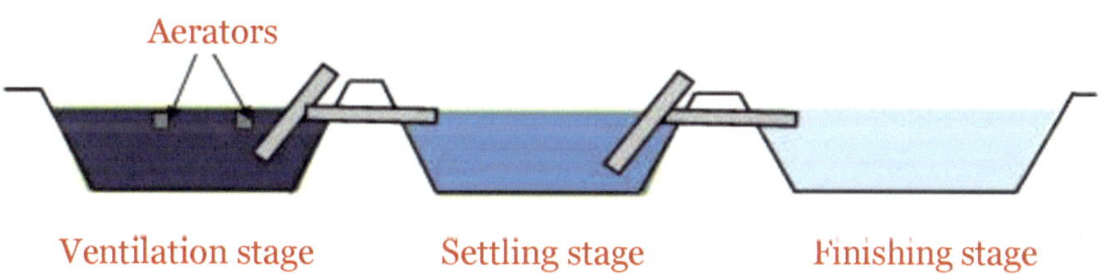

Figure 10.
Principle of aerated lagooning [12].

discharge depending on the season, especially in summer when the temperature is very high, which causes very strong evaporation, which influences the concentration of water in the lagoons and the formation of algae due to the long insolation period.

3.2 Aerated lagoon

In the Algerian Sahara, the aerated lagoon system is used frequently; we have identified four stations (**Figures 6–9**) among ten (Ouargla, Sidi Khouiled, Kouinin, Tamanrasset), where the largest nominal capacity is that of Kouinin with a flow of 33,251 m^3/d (**Tables 3** and **4**).

In this type of process, oxygenation is provided mechanically by a surface aerator or air blowing (**Figure 10**). This principle differs from activated sludge only by the absence of a sludge recycling system or continuous sludge extraction. The energy consumption of the two sectors is, at equivalent capacity, comparable (1.8 to 2 kW/kg BOD5 eliminated) [11].

The Algerian authorities often orient themselves towards this system since this system is very tolerant to variations in significant hydraulic and organic load as in the case of the regions mentioned above as well as their tolerance to effluents imbalanced in nutrients.

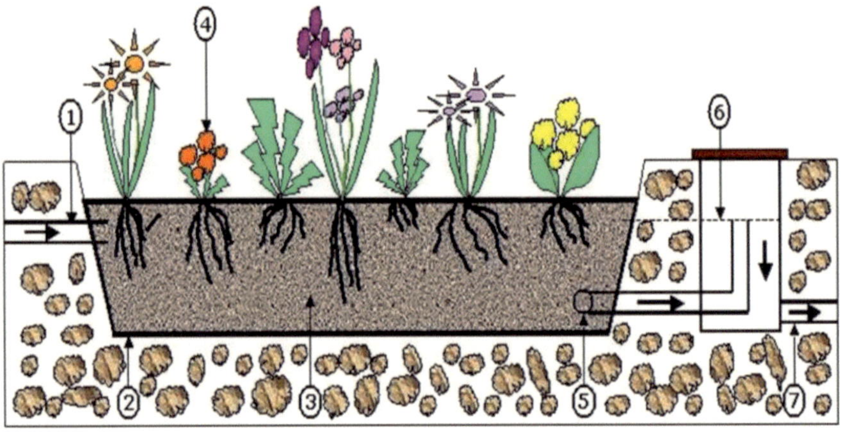

Figure 11.
Principle of wastewater gardens system [12].

According to our investigations, in addition to the average quality of rejection on all parameters and noise pollution, the station managers have some problems such as high energy consumption, and the presence of electromechanical equipment requiring maintenance by a specialized agent; which sometimes takes a considerable amount of time with a non-operating station.

Among the most apparent problems in the region is the inflow of water from the water table and the excess parasitic irrigation in the sewerage network which can influence the performance of the station in two due to the high salinity of the water who can wait 13000 μ/cm, and the rapid change in raw sewage concentration.

3.3 Wastewater gardens system (WWG)

The Wastewater Gardens unit is an eco-technology that uses ecologically based wastewater treatment principles [11].

Is a waterproofed basin, filled with gravel and plants whose roots are tolerant of water saturated conditions (**Figure 11**). There may be one or more compartments, depending on the size of the system and the area available for construction. The efficiency of a Waste Water Gardens pond is based on the time the wastewater residences within it before they flow into the drain area.

We have identified two stations in our region using this system.

3.3.1 Horizontal flow planted filters

The Temacine pilot station (**Figure 12**) is located in the city of Temacine next to old Ksar (**Table 5**), it was mainly created with the aim of treating 15 m^3/d of wastewater by the production of 100 people and at a reasonable rate. of 150 l/inhabitant/day. The surface of the WWG basin is 400 m2, and the total volume is 260 m^3. The water level in the basin is 0.5 m, covered by a layer of gravel ranging from 10 to 15 cm (gravel is a physical filter). The WWG basin is also filled with plants that can live in an environment saturated with wastewater (plants collect their nutrients and water through their roots).

This system offers low energy consumption; no need for advanced qualification for maintenance; and good reaction to load variations. On the other hand; this system has not been developed in the region because of their strong footprint, including the surrounding area, this is of the order of 10 m^2/pe (equivalent to the footprint of a natural lagoon).

Figure 12.
Temacine wastewater treatment plant.

N°	Province	Station	Zone	UTM coordinates		
				X	Y	Z
1	Ouargla	Temacine	32S	221565.67 m E	3657276.93 m N	80

Table 5.
Coordinates of horizontal flow planted filters stations.

3.3.2 Planted vertical flow filters

The wastewater treatment plant by vegetation located in the commune of N'goussa (**Figure 13**). The station is located at the lowest point of the sewage network (**Table 6**). Installed in 2010 and has been in operation since 2011. Use of solar energy pilot project of the WWTP. The characteristics of the releases are typically those of a domestic release. The treatment plant with filters planted with vertical flow reeds is made up of four parallel basins planted with reeds, each basin is divided into three equal parts operating alternately. Each basin is made up

Figure 13.
N'Goussa wastewater treatment plant.

N°	Province	Station	Zone	UTM coordinates		
				X	Y	Z
1	Ouargla	N'Goussa	31R	715432.22 m E	3558668.84 m N	118

Table 6.
Coordinates of vertical flow planted filters stations.

of three main inlets evenly distributed along the basin (one entry to each part), where each inlet fork tubes intended for supply by tarpaulins. The treated water collects in front of the second basin for reuse in watering the trees of the station and the rest is thrown towards the Sebkha of N'Goussa. The station has a capacity of 11000 Eq/hab and a nominal flow of 1743 m^3/d for a residence time of 3 days for each basin.

This system offers ease and low operating cost with reduced management to a minimum of organic deposits retained on the filters of the 1st stage; and good adaptation to seasonal population variations [13].

Besides the risk of the presence of insects or rodents, managers talk about the constraints of regular exploitation, annual mowing of the aerial part of the reeds, manual weeding before the predominance of reeds.

Using this sector for capacities greater than 2000 pe remains very delicate for questions of control of hydraulics and cost compared to conventional sectors.

4. Intensive processes in the Algerian Sahara

4.1 Activated sludge

The Touggourt wastewater treatment plant (**Figure 14**) is located in Ben Yassoued (**Table 7**), in the municipality of Tebesbest. It covers an area of 5 hectares. It was commissioned on 20/11/1993 and rehabilitated in 2004 currently managed by the national sanitation office (ONA), planned for 62,500 population equivalents and a daily flow of 9360 m^3/d.

Among the ten stations treated in this document, we noted that the only station with an intensive process is that of Touggourt by activated sludge, its principle resides in an intensification of the self-purification processes that are encountered in natural environments [14]. The bacteria float in flakes in the wastewater and the purification process takes place under intense aeration (**Figure 15**). Aerobic and anaerobic (oxygen poor) conditions can be altered in space and time so that nutrients too (such as nitrogen and phosphorus) can be removed [15].

Figure 14.
Touggourt wastewater treatment plant.

N°	Province	Station	Zone	UTM coordinates		
				X	Y	Z
1	Ouargla	Ben Yassoud	32S	228323.38 m E	3666543.82 m N	64

Table 7.
Coordinates of activated sludge station.

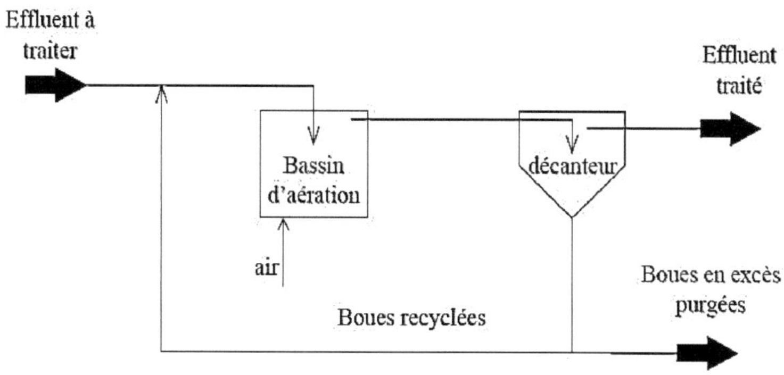

Figure 15.
Basic diagram of the activated sludge process [14].

We have noted that this system offers good elimination of all the pollution parameters (COD, BOD5, MES, N) by nitrification and denitrification; plus an easier implementation of dephosphorization. It is a system suitable for the protection of sensitive receiving environments such as that of Oued Righ.

On the other hand, in addition to the investment cost and energy consumption of this type of stations are quite high, the station encountered several technical and managerial problems such as the need for qualified personnel and regular monitoring; sensitivity to hydraulic overload and difficulty in controlling sludge production.

As the situation of the station in the middle of an area of palm groves, among the most apparent problems in the region is the inflow of water from the water table and the surplus of parasitic irrigation in the sanitation network which can affect the performance of the two stations by the high salinity of the water that can be expected, and the rapid variation in the concentration of raw sewage.

5. Conclusion

Through this chapter, we tried to shed light on the wastewater treatment component in the Algerian Sahara through a description of the different systems used and to discuss their reliability and constraints.

The purification techniques adopted in the Algerian Sahara are limited to three processes: lagooning process, plant filter and activated sludge. Furthermore, the cost of installing and maintaining these sectors requires significant human and financial resources which are difficult to bear in our region.

We noted that the main objective of these stations in the Sahara is the protection of the environment and the almost non-existent tertiary treatment, for this the practice of reuse of purified wastewater remains very limited to the time that these regions have a subsoil. of the largest freshwater table in the world (Intercalary Continental).

The document clearly explained that 70% of stations in the Algerian Sahara using lagooning where 40% are stations by aerated lagooning, which remains a choice which has been subject to constraints such as high energy consumption, and the presence of electromechanical equipment. Requiring periodic maintenance by specialists. Only two pilot stations using plant purification; this system has not been developed in the region because of their strong footprint, the use of this system remains very delicate for issues of hydraulic control and cost compared to conventional channels.

Due to investment costs there are almost no active mud stations and the energy consumption of this type of station is quite high.

The improvement of these wastewater treatment techniques has resulted in an increase in by-products, these by-products, and particularly the sludge which represents the largest volumes, must be conditioned and disposed of in the most appropriate way, this which adds additional equipment to the wastewater treatment plants.

Therefore, it is necessary to explore new wastewater treatment technologies, reliable loads and adapted to the realities of our country Algeria.

Author details

Rachid Zegait[1*], Saber Kouadri[2], Samir Kateb[2] and Mohamed Azlaoui[1]

1 Faculty of Science and Technology, Ziane Achour-Djelfa University, Algeria

2 Faculty of Applied Sciences, Kasdi Merbah-Ouargla University, Algeria

*Address all correspondence to: zegait.rachid@gmail.com

References

[1] Hannachi, A., R. Gharzouli, and Y. DJELLOULI TABET. Gestion et valorisation des eaux usées en Algérie. LARHYSS Journal P-ISSN 1112-3680/E-ISSN 2521-9782 19 (2014).

[2] Bouchaala, L., Charchar, N., & Gherib, A. (2017). Ressources hydriques: traitement et réutilisation des eaux usées en Algérie. Algerian journal of arid environment, 7(1), 84-95.

[3] Kouzmine, Y (2007); Dynamique et mutation territorial du Sahara algérien ; Thèse de Doctorat en Géographié à l'université de Franche-Comté

[4] ABHS ; (2014) ; Cadaster hydraulique du bassin hydrographique du Sahara septentrionnal

[5] Chabaca, N. M., Kettab, A., Nakib, M., Karef, S., Benziada, S., Benmamar, S., ... & Djillali, Y. (2017). The lagunage for the purification of waste water in the Sahara: an approach integrated into the environmental conditions. Algerian Journal of Environmental Science and Technology, 3(2).

[6] Chaoua, S., Boussaa, S., Khadra, A., & Boumezzough, A. (2018). Efficiency of two sewage treatment systems (activated sludge and natural lagoons) for helminth egg removal in Morocco. Journal of infection and public health, 11(2), 197-202.

[7] Degremont. (1989)-Mémento technique de l'eau : vol. 1, 9ème édition. Edition Technique et Documentation Lavoisier, pp 592

[8] Durot, M. A., & Molle, P. (2015). Amélioration du rejet des lagunes d'épuration. Synthèse bibliographique (Doctoral dissertation, irstea).

[9] Zegait, R. 2020. "Eau et Assainissement Dans Les Oasis Du M'ZAB: Rejets Urbains et Pollution de La Nappe (Cas de l'oasis d'El-Atteuf)." Thèse de doctorat à l'ENSH.Algerie

[10] OIE, (2001). Procédés extensifs d'épuration des eaux usées adaptés aux petites et moyennes collectivités. Office International de l'Eau, ISBN 92-894-1690-4, 42 pages.

[11] Kaymai, (2002) ; Conception et dimensionnement du lagunage aéré ; 7p

[12] Nair, J. (2008). Wastewater garden—a system to treat wastewater with environmental benefits to community. Water science and technology, 58(2), 413-418.

[13] Rahmani.A (20015) ; Épuration des eaux usées de la région de N'goussa (Ouargla) par des vegetaux. Thèse a l'université de Ouargal.

[14] Bernard BAUDOT ET Prudencio PERERA, GUIDE, 2001. Procédés extensifs d'épuration des eaux usées adaptés aux petites et moyennes collectivités. Office International de l'Eau.

[15] Dahou A, (2013). Lagunage aere en zone aride performance epuratoires cas de (region d'ouargla), these, université d'ouargla

Chapter 13

Photo-Processes as Effective and Low-Cost Methods for Laundry Wastewater Treatment

Endang Tri Wahyuni

Abstract

In this chapter, surfactants as cleansing agent in detergent used in laundry, are described. The negative effects of the laundry wastewater on the environment and human health are highlighted. Several methods examined for laundry wastewater treatment are also illustrated. Among the treatment methods, photo-process in the presence of TiO_2 photocatalyst and Fenton reagents are described in more detail. Furthermore, the factors influencing the effectiveness of photo-process including reagent dose, reaction time, and pH are discussed. Additionally, modifications of the photo-process to improve its performance that is associated with effectiveness and operational cost are also demonstrated. The photo-methods discussed in this chapter offered low-cost due to simplicity and effective technique for treating the laundry wastewater.

Keywords: laundry, wastewater, treatment, photo process, TiO_2, photo Fenton

1. Introduction

Laundry activity is intensively and routinely conducted in domestic activities including homes, hotels, hospitals, as well as public laundry services. In the laundry activity, large amount of detergent as cleansing agent must be used. Further, in general, washing machines can typically produce from 50 to 200 L of effluent per wash [1], implying that laundry activity always disposes large volume of wastewater. The active component with high content in the detergent is anionic surfactant prior to linear alkyl benzene sulfonate (LAS) [1–21]. It is reasonable therefore that high concentration of LAS is contained in the laundry wastewater, as reported [2], that was around 200 mg/L from the first rinse. The presence of LAS in water can cause damage to the ecosystem thereby affecting the environment, and consumption of LAS above 0.5 mg/L can be harmful to health [1]. Considering the negative effects, treatment of LAS from laundry wastewater before reaching the environment is urgent.

Various methods have been dedicated to remove LAS surfactant in water and wastewater, such as adsorption [3–4], coagulation [2, 5–6], and filtration [7–8]. By adsorption, coagulation, and filtration techniques, the surfactant of LAS is only replaced from water to the adsorbents, coagulants and membranes with the same toxicity [1], then they are collected as hazardous solid wastes. Further the hazardous solid wastes must create new environmental problems.

In recent years, various destructive methods including biological, chemical and combination of physical–chemical techniques have been employed for removal of the LAS surfactants from waters. The destructive techniques that have been developed for removal LAS are biodegradation [9–10], ozonation [1], photocatalytic degradation over TiO_2 [11–16], and Fenton and photo-Fenton [16–21]. Biodegradation of LAS in water was found to be less effective for high concentration of LAS, since the LAS is harmful for the bacteria [1]. Ozonation method for treatment of wastewater is believed to be uneconomical due to the use of the high dose of the ozone and pressurized and complicated equipment [22]. On the other hand, photo-degradation of LAS over TiO_2 photocatalyst under UV irradiation and by photo-Fenton process are intensively used as the effective methods to destroy the hazard LAS into smaller and saver molecules [11]. In addition, the methods only need light, and low cost and harmless chemicals, allowing them to be applied in large scale.

2. Surfactant in laundry waste water

Laundry activity always uses detergent that contains surfactant as the cleansing agent. The word surfactant is short for "Surface Active Agent." In general surfactants are constructed by hydrophobic long alkyl chain as tail, and a hydrophilic group as a head, as illustrated by **Figure 1**. In general, they are chemicals that, when dissolved in water or organic solvent, orient themselves at the interface (boundary) between the liquid and a solid (i.e. the dirt or grease that want to be removed), and modify the properties of the interface [23]. The cleansing dirt or grease occurs when the hydrophobic long chain is attracted to dirt, while the hydrophilic part of the molecule is attracted to water. When dirt or grease is present, the surfactants surround it then it is dislodged from the boundary. The dirt/grease removed from the fabric will come into water [23].

The hydrophobic long alkyls in the surfactants can refer as branched and linear chains. One of the branched long alkyl used in the detergent surfactant was dodecyl having molecular formula $C_{18}H_{30}$ or $(CH_3)3(CH_2)_{10}CH_2$ [23]. The branched alkyl offered superior tolerance to hard water and better foaming. Unfortunately,

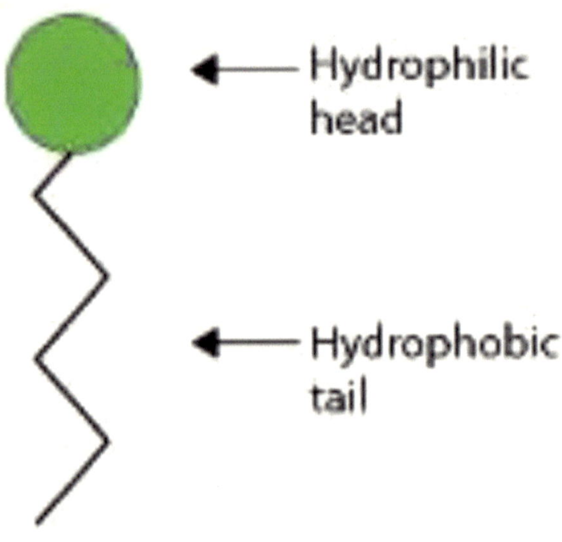

Figure 1.
The schema of surfactant structure [23].

highly branched tail made it difficult to biodegrade, that was widely blamed for the persistent foam in sewage treatment plants, streams, and rivers, and created environmental problems. Hence, the branched surfactants have been replaced by linier alkyl long chain, that is environmentally friendly and easily biodegrades to simpler substances [23]. For the linear alkyl long chain usually used are C10–15, such as hexadecyl ($C_{16}H_{33}$).

The hydrophilic part of the surfactant is found as non-ionic, cationic, and anionic forms as shown by **Figure 2**. These different groups refers the names of the surfactants as non-ionic, cationic, and anionic surfactants. Structurally, non-ionic surfactants combine uncharged hydrophilic and hydrophobic groups that make them effective in wetting and spreading and as emulsifiers and foaming agents [23]. One of the major types of nonionic surfactants includes alkyl phenol ethoxylate as seen as **Figure 2a** [23]. Nonionic surfactants represent a major component material for applications ranging from personal care to a wide range of industrial uses [23]. Concurrently, such surfactants have minimal skin and eye irritation effects and exhibit a wide range of critical secondary performance properties [23]. Cationic surfactants are positively charged in the hydrophilic part, as an example is hexadecyl trimethyl ammonium bromide or cetyl trimethyl ammonium bromide (CTAB) as seen in **Figure 2c** [23]. The cationic surfactants are much less used in laundry detergents, due to their tendency to rapidly adsorb to – and not desorb from – the fabric having negatively charged surfaces under normal conditions [23]. The surfactant is bounded strongly by the fabric, inhibiting in the removal of the dirt from the fabric [23].

One of the major groups of anionic surfactants are linear alkyl benzene sulfonates (LAS), that are characterized by an anion hydrophilic of sulfonate [1–21]. The commercially produced LAS comprises alkyl chains of 10–14 carbon atoms, such as dodecyl benzene sulfonate (DBS) as seen in **Figure 3** [11–12]. LAS type surfactants pose a lot of usage because of its high cleaning power and efficiency [1–21]. The superior property originates from the fact that the anionic sulfonate group is repulsed to attach strongly with fabric of cloths, that results in the maximal cleansing the dirt [12].

LAS surfactants are the most commonly used detergents, that is more than 1.8 9 10^6 tons/year) in the past 40 years [12]. In terms of quantity, LAS is the most prominent group of anionic surfactants which is mainly used in heavy-duty laundry powders, light-duty liquid dish detergents, heavy-duty laundry liquids, and specialty cleansers [12]. Consequently, a significant amount of LAS exists in municipal

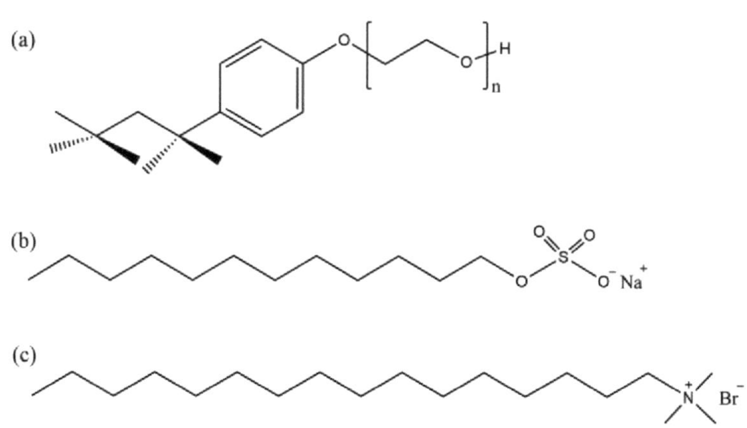

Figure 2.
The structures of (a) non-ionic, (b) anionic and (c) cationic surfactants [23].

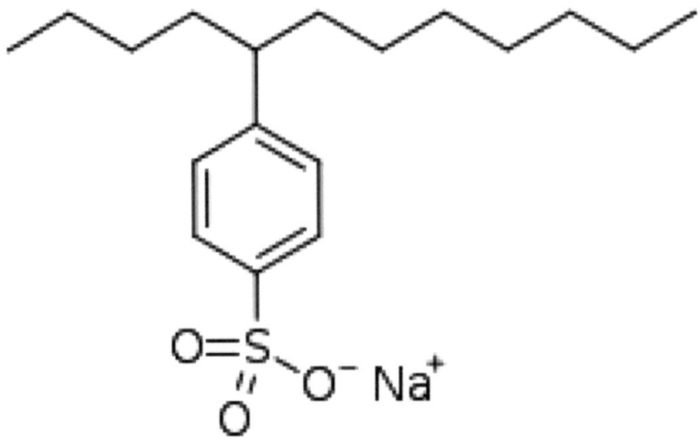

Figure 3.
The structure of LAS type [1].

disposed from homes, hotels, hospitals, and public laundry services, and industrial wastewater such as textile, leather, food, paint, cosmetics, polymer, oil recovery, mining and paper industries [12].

There was evident that less effective biodegradation of LAS in the water environment can occur [1]. Accordingly, LAS concentration of 3.5 mg/L has been measured in untreated sewage [12]. The values for concentrations of LAS in domestic effluents ranged from 3 to 21 mg/L [1] was also reported. From the laundry public services, 137 to 200 mg/L of the concentration of anionic surfactant in sewage was detected by other researchers, [2, 16].

Determination of the concentration of LAS in the solution or wastewater can be conducted by using visible spectrophotometry method. It is important hence to describe the analysis method. The method is based on the reaction between LAS with methylene blue to form a pair of methylene blue- linear alkyl sulfonate (MB-LAS), as shown by **Figure 4**, to form blue solution. The MB-LAS is insoluble in water, that has to be extracted into organic solvent such as chloroform. Then the MB-LAS as blue chloroform solution is measured by using spectrophotometer instrument at 650 nm of the wavelength, to obtain its absorbance [2, 12].

To calculate the LAS concentration in the solution or in the wastewater, the absorbance is plotted in the respective standard curve. The standard curve of LAS is displayed in **Figure 5**, showing that the straight linear line are obtained at 2–10 mg/L of the standard solution concentrations [14].

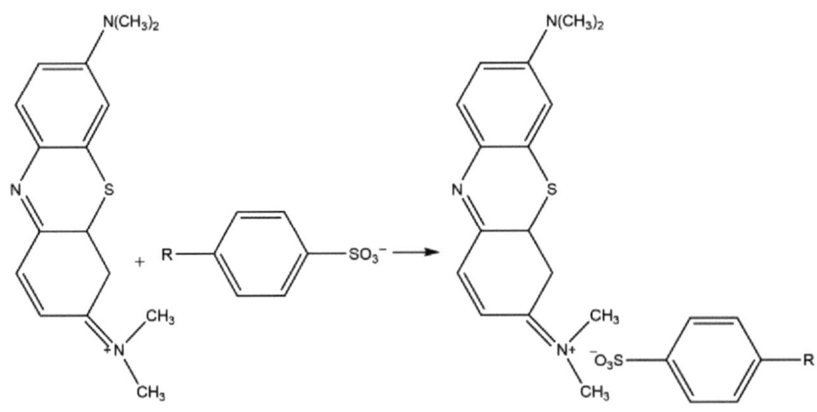

Figure 4.
The formation of MB-LAS pair giving blue color in chloroform solution [12].

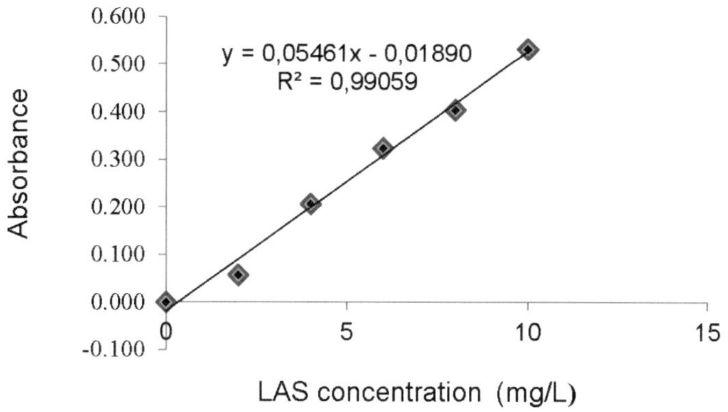

Figure 5.
The standard curve of LAS used for concentration determination [14].

LAS is reported to be toxic for aquatic organisms in higher than 0.1 mg/L [11]. Moreover, it can be accumulated in fish and consequently spread in the whole ecosystem. As a result it alters the natural balance of water which changes water into a harmful source for the aquatic organisms and human [11]. It also has some pathological, physiological, biological and other effects on aquatic animals [11]. For specific aquatic plants, LAS damages their cholorophyll protein and membrane leading to delay in growth and metabolism of cells [1]. With the same mechanism, LAS can also decrease the soil fertility [14]. Moreover, consumption of LAS above 0.5 mg/L can be harmful to people health [1].

Due to negative effects of anionic surfactant on organisms and environment, many environmental and public health regulatory authorities have considered restrictions. As an example, Indonesian Government regulates that anionic surfactants in laundry wastewater has to be lower than 1.0 mg/L, as MBAS (Methylene Blue-Active-Substance), allowed to be discharged to the environment [2].

3. The laundry wastewater treatment methods

Various methods have been developed to remove LAS surfactant in water and wastewater, such as adsorption [3–4], coagulation [2, 5–6], and filtration [7–8]. By adsorption, coagulation, and filtration techniques, the surfactant of LAS is only replaced from water to the adsorbents, coagulants and membranes with the same toxicity [1], then they are collected as hazardous solid wastes. Further the hazardous solid wastes must create new environmental problems.

The methods presented in this chapter are focused on photo-degradation process by photocatalysis over TiO_2, and by photo-Fenton.

3.1 TiO_2 photocatalyst and the photocatalysis process

TiO_2 is a semiconductor with electronic structure that is characterized by valence band filled with electrons and empty conduction band, separated by gap as much as 3.0–3.2 eV [24], as illustrated by **Figure 6**. The gap is known as band gap energy (Eg), that is equal to the light with wavelength lower than 387 nm emerging the UV light. This fact allows TiO2 to absorb the UV light, resulting in the excitation of electron in the valence band (e_{CD}^-) into conduction band while leaving a hole in the valence band (h_{VB}^+), as presented in Eq.1 [24–28]. The releasing electron and hole formation processes are demonstrated by **Figure 6**.

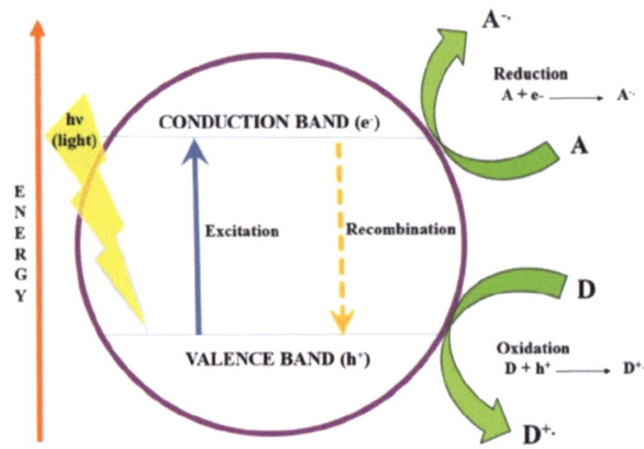

Figure 6.
The schema of semiconductor structure, and the electron excitation and hole formation [1].

In water medium, the holes formed can react with water and also with the surface of TiO_2, to form OH radicals, while the electrons released can react with O_2 dissolved in water to form superoxide radical anions (O_2–•), and further the anionic radicals will react with hydrogen ions from water to form hydroperoxy radicals (HO_2•). The reactions of various radical formations are shown by Eq. 2 up to Eq. 4. Hydroxyl radical (•OH) is a strong oxidizing agent with oxidation potential (E) as much 2.80 V, that is stronger than H_2O_2 with E as 1.23 V [24] and ozone with E of 2.07 V [24–25]. The two other radicals are also oxidizing agent, but the activity is weaker than the OH radicals strength [25].

$$TiO_2 + hv \rightarrow TiO_2\left(e_{CB} - + h_{VB}{}^{+}\right) \quad (1)$$

$$TiO_2\left(h_{VB}{}^{+}\right) + H_2O \rightarrow TiO_2 + H^{+} + \bullet OH \quad (2)$$

$$TiO_2\left(e_{CB} -\right) + O_2 \rightarrow TiO_2 + O_2 \bullet - \quad (3)$$

$$O_2 \bullet - + H^{+} \rightarrow HO_2 \bullet \quad (4)$$

The strong OH radical from TiO_2 has been proven to be able to degrade various organic pollutants such as amoxicillin [22, 27], dyes [24], and phenols [26] effectively. This process is called as photocatalysis degradation, that has also been intensively examined to remove LAS in water media as well as to treat laundry wastewater in the lab scale [11–16]. The reaction of the LAS degradation takes place by hydroxyl radicals (•OH), superoxide radical anions (O_2^{-}•), and hydroperoxy radicals (HO_2•). Under the photocatalysis degradation, the long chain hydrocarbons of LAS are primarily degraded into smaller organic compound then to form CO_2 and H_2O, and then the sulfonate group is oxidized into sulfate ions $SO4^{2-}$ [11]. The possible degradation pathway of alkyl-benzene sulfonate in photocatalytic oxidation as shown in **Figure 7** [11]. From the reaction, it is clear that the effective degradation of LAS yields smaller and harmless molecules.

The effectiveness of the degradation of LAS surfactant, whether in water and in the laundry wastewater under photocatalysis process is controlled by several factors, that are photocatalyst dose, irradiation time, pH of the water, and initial LAS concentration in the wastewater. The influence of these factors are describe below.

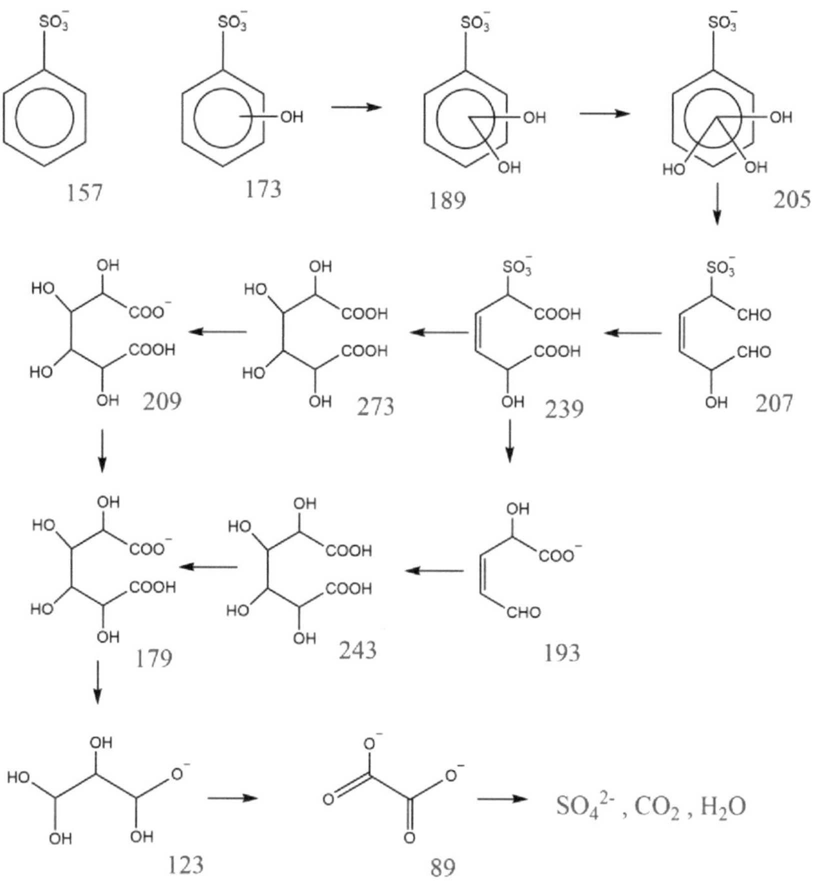

Figure 7.
The reaction mechanism of LAS degradation by OH radicals [11].

3.1.1 The influence of photocatalyst mass

The mass or dose of TiO_2 photocatalyst deals with the active surface in providing OH radicals, where the number of OH radicals provided will enriched as the photocatalyst mass is enlarged. The correlation of the photocatalyst dose with the LAS degradation has been reported by some studies [11–16], and as an example is displayed in **Figure 8** [16].

It is observable that increasing the photocatalyst dose gives raise the degradation, but for the further enlargement of the dose, the degradation effectiveness is found to decrease [2, 12–15]. Extending photocatalyst dose provides more OH radicals and so promotes more effective degradation. In contrast, higher dose than the optimal level, leads to an increase in the turbidity, causing filter effect of the light enters. As a consequence, the interaction of the light with TiO_2 is inhibited, resulting in the less of OH radicals, decreasing the degradation. The optimum level of the photocatalyst mass obtained was varied from one to other authors [11–16], ranges from 50 mg/L to 750 mg/L depending on the initial LAS concentration and the reaction time.

3.1.2 The influence of the irradiation time under UV light

For the industrial removal process of LAS, reaction time is a key factor. The irradiation time is associated to the time of contact between light with TiO_2 and

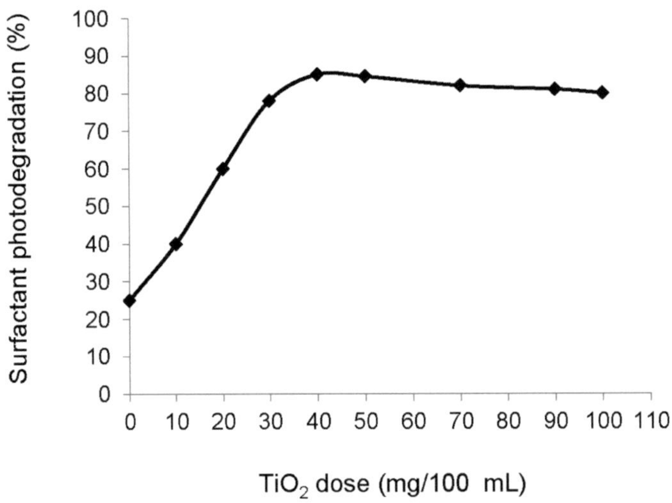

Figure 8.
The influence of the photocatalyst mass on the degradation of LAS in the wastewater [16].

between OH radicals with LAS molecules. Some studies [2, 11, 13, 15–16] have observed the effect of the UV irradiation time on the LAS degradation and they have similar trend, as seen in **Figure 9** [16]. The improvement of the LAS degradation appears with the expanding irradiation time but longer than the optimum time, the degradation effectiveness is independence on the time. In the beginning of the reaction, effective contact between light and TiO_2 and between OH radical with LAS proceed effectively. Prolong the irradiation time allows more effective contact and further results in higher effectiveness of the degradation up to reach the saturated condition. The optimum time reported were varied, one study found 60 min [11], while others reported of 50 min [12] and 100 min [15]. Very long irradiation optimum time was also possible, that was 24 h, due to high LAS initial concentration and photoreactor construction [2, 16].

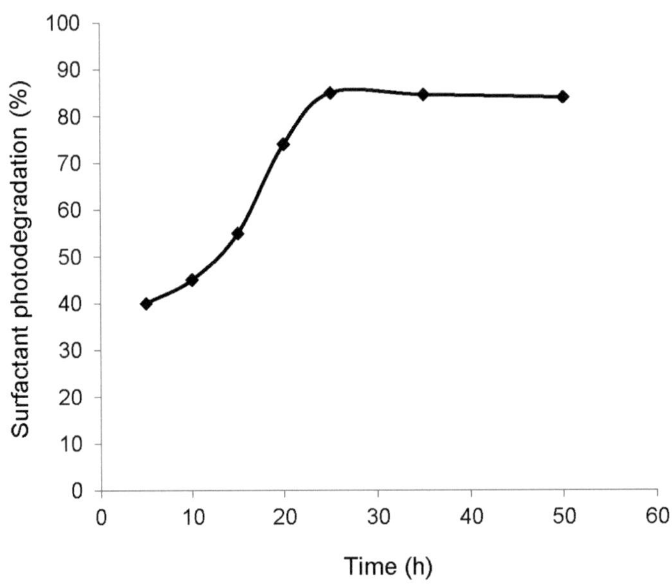

Figure 9.
The influence of the irradiation time on the degradation of LAS in the laundry wastewater [16].

3.1.3 The influence of the solution pH

The influence of the pH on the LAS degradation is one of the important factor, since pH determines the species of TiO_2 surface as well as LAS structure. **Figure 10** assigns [16] a trend of the degradation as function of the solution pH. It is observable an increase of the LAS degradation as the pH elevation up to 7, but further increase of pH causes a decline in the degradation. The trend can be explained based on the speciation of TiO_2 and LAS due the pH alteration.

At low pH, the surface of TiO_2 is protonated to form $TiOH_2^+$ that is difficult to provide OH radicals. With respect to LAS, at low pH, the LAS structure is also protonated that changes from negative to neutral surface. This condition can prevent the LAS to be adsorbed on the TiO_2 surface. Consequently, only little amount of LAS can interact with OH radicals, and further low degradation can occur. Increasing pH up to 7, most TiO_2 is found in neutral charge as TiOH [11, 13, 15]. It is important to takes a note that the zero point charge of TiO_2 is at pH 6.5 [11], referring uncharged TiO_2 surface, that can provide OH radicals maximally. At the same pH, LAS structure may form as anionic species, that allow them to be adsorbed on TiO_2 surface effectively. This high LAS adsorption can promote more effective LAS degradation.

At higher pH than the zero point charge, that is higher than 7, both TiO_2 and LAS are existed as negative species, that creates electrostatic repulsion. Hence, the LAS adsorption on the TiO_2 surface determents and further declines the LAS degradation. It is clear that pH strongly influences on the adsorption of the LAS on the TiO_2 surface, that effects the degradation effectiveness. The interactions at low and high pH are described as Eq. (5) and (6) [11, 13, 15].

$$TiOH + H^+ \rightarrow TiOH_2^+ \quad (5)$$

$$TiOH + OH^- \rightarrow TiO^- + H_2O \quad (6)$$

From the lab study for LAS in the artificial wastewater [11–15], it is demonstrated that the effective degradation is reached at low pH, that was 4 [11], while other study for real laundry wastewater obtained the most effective degradation at pH 5 [16]. However, in the application for real laundry wastewater having pH 5–6, adjusting pH is not required.

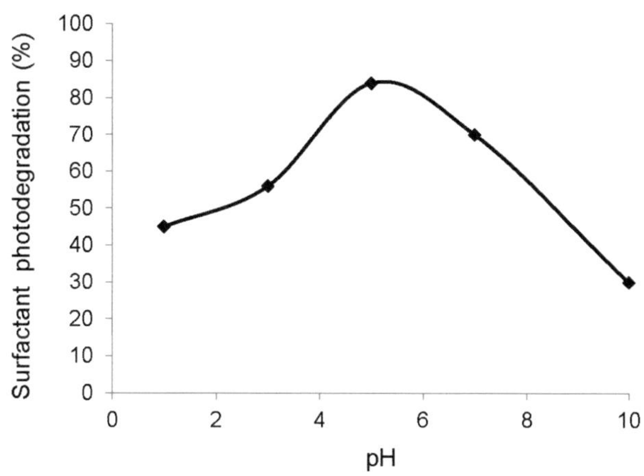

Figure 10.
Influence of pH on the degradation of LAS in the laundry wastewater [16].

3.1.4 The influence of the initial LAS concentration and the kinetic

The influence of the initial concentration of LAS in the real laundry wastewater is investigated by diluting the wastewater into the various desired concentrations. It was reported [11, 13] that increasing the initial LAS concentration leads to a decrease in the degradation. It can be explained that when the initial concentration of LAS is increased, more LAS adsorbed on the TiO_2 surface inhibiting the formation of OH radicals. Therefore less OH radicals are available, that decreases the LAS degradation [11, 13].

A kinetic study of the LAS photodegradation is desirable as it describes information about the rate of the degradation, which is important for efficiency of the process. The rate of a reaction is represented by rate constant (k), that depends on the concentrations. The relation between k and the concentration depends on the order of the reaction. The formulas of the first and second orders are given as Eq. 7 and Eq. 8, respectively. Ct represent the substrate concentration left in the media after t time of the reaction. Co is the initial substrate concentration.

$$\ln Ct = -kt + \ln Co \tag{7}$$

$$\frac{1}{Ct} = kt + \frac{1}{Co} \tag{8}$$

For determination of the reaction order, a curve is constructed generally by plotting time versus concentration. When a curve of ln Ct versus time gives a straight line, it is confirmed that the reaction agrees with first order reaction. Further, in order to confirm the second order reaction, a curve of 1/Ct versus time should be created, that results in the straight line.

In the LAS photodegradation by OH radicals, the rate of the LAS degradation reaction is determined by concentrations of LAS and OH radicals. When the reaction depends on both the concentrations of LAS and OH radicals, the LAS degradation should follow the second order model. The second order has been reported [11] with k value as much as 0.0031 L/mg. min. When the degradation is only controlled by the LAS concentration, the first order reaction must be followed, as obtained by Ghanbarian et al. [13], with k as much 0.020 1/min. The other possible condition is found as follow [15]. The reaction is dictated by both LAS and OH radical concentrations, but because the OH radicals are in the excessive amount that are assumed to be constant during the reaction. Accordingly the reaction rate is only influenced by the LAS concentration. Such condition allows the reaction rate to agree with the pseudo first-order. From the curve k as much 0.01–0.014 1/min is obtained [15].

3.2 Photo-Fenton process

Fenton is a process by using ferrous ion (Fe^{2+}) and hydrogen peroxide (H_2O_2), called as Fenton's reagent. In this process, hydrogen peroxide is decomposed catalytically by ferrous ions at acidic pH value, yielding hydroxyl radicals (•OH) and hydroxide anionic (^-OH), while ferrous ions are transformed to ferric ions. In general the accepted mechanism of Fenton reaction to form hydroxyl radicals is presented as Eq.(9) and Eq. (10) [16–21, 29–34]:

$$Fe^{2+} + H_2O_2 \rightarrow \bullet OH + Fe^{3+} + OH^- \tag{9}$$

$$Fe^{3+} + H_2O \rightarrow FeOH^{2+} + H^+ \quad (10)$$

Further, photo-Fenton is a process involving a combination of Fenton reagents (H_2O_2 and Fe^{2+}) with UV radiation (λ < 310 nm) that gives rise to extra OH radicals [18–19, 21, 30–31]. The major reactions in the photo-Fenton process for the formation of •OH radical include Fenton reaction, photolysis of H_2O_2 and photoreduction of Fe^{3+}, as shown in Eq. (11) and (12) [30–31].

$$H_2O_2 + h\surd(\lambda < 310\,nm) \rightarrow 2\bullet OH \quad (11)$$

$$FeOH^{2+} + h\surd(\lambda < 580\,nm) \rightarrow Fe^{2+} + \bullet OH + H^+ \quad (12)$$

The addition of UV or artificial light to Fenton's process is detected to accelerate Las degradation as it influences the direct formation of •OH radicals [18–19]. Consequently, the organic degradation rate of photo-Fenton is accelerated compared to Fenton process. The improvement is due to the continuous reduction of ferric ions (Fe^{3+}) to ferrous ions (Fe^{2+}) under illumination, and then the Fe^{2+} reacts back with H_2O_2 to result in Fe^{3+} and OH radicals. The Fenton reaction can be terminated when H_2O_2 is exhausted. The OH radicals from Fenton and photo-Fenton processes, as produced from photocatalys of TiO_2, also own strong ability as an oxidizing agent, that can destroy various organic pollutants in acid condition [29–31].

The primary benefits of Fenton type process are their ability to convert a broad range of pollutants to harmless or biodegradable products and the fact that their relatively cheap reagents are safe to handle and are environmentally acceptable. Fenton process because of high oxidation power, rapid oxidation kinetics, being relatively cheap with easy operation and maintenance is used for treating various industrial wastewaters, including phenol [29], dyes [30], and various organic pollutant in the wastewater [31].

Considering the reagent involved in the Fenton and photo-Fenton processes, the effectiveness of LAS degradation is controlled by H_2O_2 and Fe^{2+} (Fenton's reagent) concentrations. In addition, reaction time, solution pH, and initial concentration of the substrate also contribute in the degradation effectiveness. Following are discussion of the effect of the factors on the LAS photodegradation by Fenton and photo-Fenton processes.

3.2.1 Effect of H_2O_2 concentration

The concentration of H_2O_2 is a critical variable in the degradation through Fenton and photo-Fenton processes. Many researchers have observed the influence of H_2O_2 concentration on the LAS degradation by Fenton and photo-Fenton methods. One example data is taken and exhibited in **Figure 11** [16]. It is seen that the low concentration of H_2O_2 did not generate enough •OH in solution, giving less effective degradation. Increasing H_2O_2 concentration improved the LAS degradation due to more •OH available. Addition of H_2O_2 above the optimum level lead to a decrease in the LAS degradation, that is caused by the depletion of the •OH amount due to free radical scavenging by the excess H2O2 to produce hydroperoxy radicals (•O_2H). Then the hydroperoxy radical will further react with OH radical to form water and O_2 [17–20]. The reactions are exhibited by Eq. (13) and Eq. (14) below:

$$\bullet OH + H_2O_2 \rightarrow \bullet O_2H + H_2O \quad (13)$$

$$\bullet O_2H + \bullet OH \rightarrow H_2O + O_2 \quad (14)$$

It is obvious that there is an optimum H_2O_2 concentration to achieve the maximum percentage of LAS removal, although the values of the concentration range varies for different conditions. In a study the optimum photo-Fenton condition was mediated by a $[H_2O_2]/[Fe^{2+}]$ ratio = 40 [16]. The effect of mode of reagent addition was also studied giving ratio of 10 [17]. Similar results were obtained in other studies, that were 1.4 [18], 7.6 [20], and 11 [21].

3.2.2 Effect of Fe^{2+} concentration

The amount of ferrous ions is one of the primary parameters that influences the Fenton and photo-Fenton processes. In a study [16], it was observed that the extent of degradation increases with increasing initial Fe^{2+} concentration, promoted by more OH radicals, as presented by **Figure 12** [16].

Contrary, the excessive Fe^{2+} ion produced larger amount of Fe^{3+} ions (reaction in Eq. 7) that further allowed them to react with hydroxide ions to form $Fe(OH)_3$ precipitate, as also seen in Eq. (13) [17–21]. The precipitate formation created turbid solution that could inhibit the light entering into the solution. This situation depleted the number of OH radical formed, that further declined the degradation. This finding was in a good agreement with the other observations elsewhere [17–20]. However, the optimal values of Fe^{2+} concentration was varied among the reports, that were 5 mg/L [16], 56 mg/L [17], 40 mg/L [18], 130 mg/L [20], and 120 mg/L [21].

$$Fe^{3+} + 3^{-}OH \rightarrow Fe(OH)_{3(solid)} \quad (15)$$

3.2.3 Effect of initial pH

The solution pH plays an important role in the efficiency of the photo-Fenton reaction, since it greatly influences the speciation of Fe, H_2O_2 and LAS. The relationship between pH alteration and the effectiveness of LAS degradation as

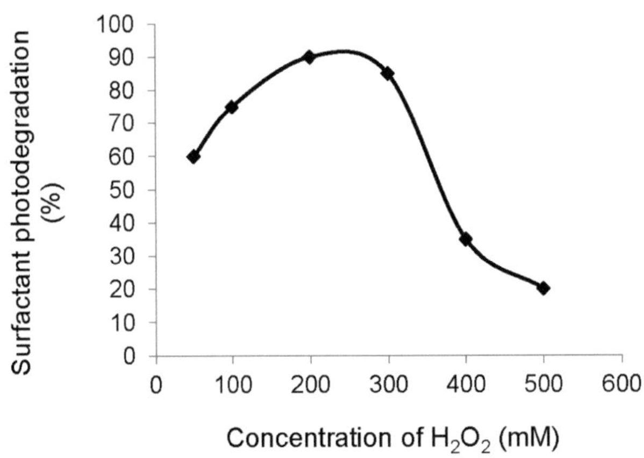

Figure 11.
Effect of H_2O_2 concentration on the LAS degradation effectiveness through photo-Fenton process [16].

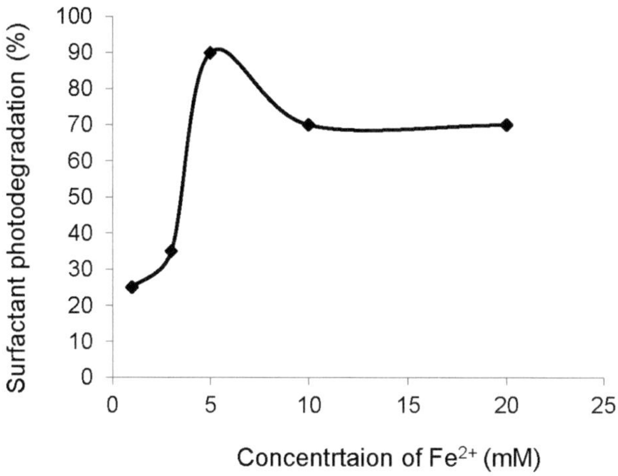

Figure 12.
Effect of Fe^{2+} concentration on the LAS degradation effectiveness [16].

reported by a study [16], is displayed as **Figure 13**. It can be observed a trend, that the LAS degradation is less efficient at very low pH, and the efficiency of the degradation improves considerably when the pH is increased up to 3. The higher pH than 3 causes a sharp decrease in the degradation.

At very low level of pH, hydrogen ions (H^+) were present in large amount, that could protonate H_2O_2 to form protonated hydrogen peroxide or $H_3O_2^+$ [17–20], as shown by Eq. (16).

$$H_2O_2 + H^+ \rightarrow H_3O_2^+ \tag{16}$$

The protonated hydrogen peroxide can inhibit the hydroxyl radical generation, resulting in small number of OH radicals, that further led to the lower photodegradation. An other reason proposed is that Fe^{2+}, found in abundant, may form a stable complex with H_2O_2, which neutralized the Fe^{2+} catalyst [16]. The neutral catalyst

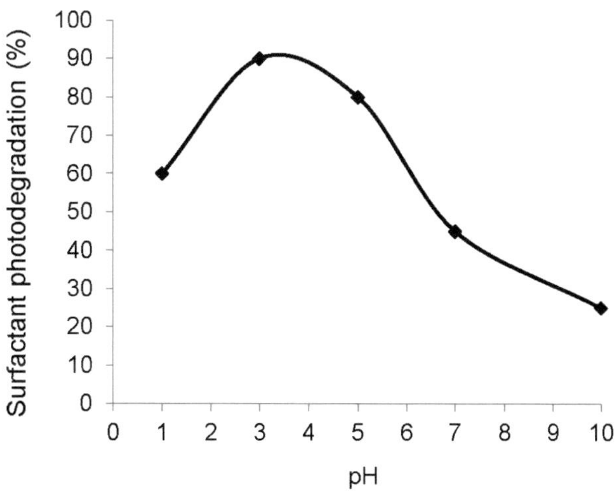

Figure 13.
Effect of pH on the LAS degradation effectiveness [16].

could only generate few amount of OH, that significantly declined of the photo-degradation. Further, increasing pH up to 3, provided smaller amount of H^+ than at pH 1, so that the protonation of H_2O_2 could be prevented, and further enhances the number of the OH radicals formed. In addition, at pH 3, the complex of Fe^{2+} with H_2O_2 should be decomposed allowing Fe^{2+} to catalyze H_2O_2 maximally, and much OH radicals could be provided [17–21]. These explained clearly the highest photo-degradation occurred at pH 3.

When the pH was increased up to 7, the number of hydroxide ion (^-OH) were enriched, allowing Fe^{3+} ions to deposit as $Fe(OH)_3$ (Eq. 15). As an effect, the sufficient Fe^{2+} catalyst did not remain in the solution. This caused lower decomposition of H_2O_2 and reduced the efficiency of the Fenton's process. Also, studies have shown that at higher pH, the oxidative potential of OH radical decreased and H_2O_2 was believed to be less stable [16, 18–19]. All the mentioned conditions obviously reduced the produced of OH radicals, and hence the amoxicillin degradation. The finding optimum pH (= 3) agreed with several other studies [17–21].

4. Modifications of photocatalysis and photo-Fenton processes

4.1 Photocatalysis method

Photocatalytic degradation using TiO_2 has recently received considerable attention for removal of the persistent organic pollutants (POPs) due to its cost-effective technology, non-toxicity, fast oxidation rate, and chemical stability [24–28]. However, the wide band gap of TiO_2, that is 3.2 eV for anatase, allows it only to be excited by photons with wavelengths shorter than 385 nm or UV region that limits its application under visible light [14, 22]. Therefore, an effort has been focused to overcome this deficiency, such as by doping TiO_2 structure with either non-metal, and metals elements.

Doping Ag metal on TiO_2 to increase the activity on the degradation of LAS in the laundry wastewater under visible light has been studied [14]. The results are seen as **Figure 14**. The increase of the TiO_2-Ag activity is promoted by the smaller Eg allowing TiO_2-Ag to be activated by visible light to provide OH radicals in adequate number. In contrast, TiO_2 with higher Eg (3.0–3.2 eV) is difficult to be excited by the visible light, that can only form fewer number of OH radicals. Moreover, the process with TiO_2-Ag under visible light takes place faster than TiO_2-Ag under UV

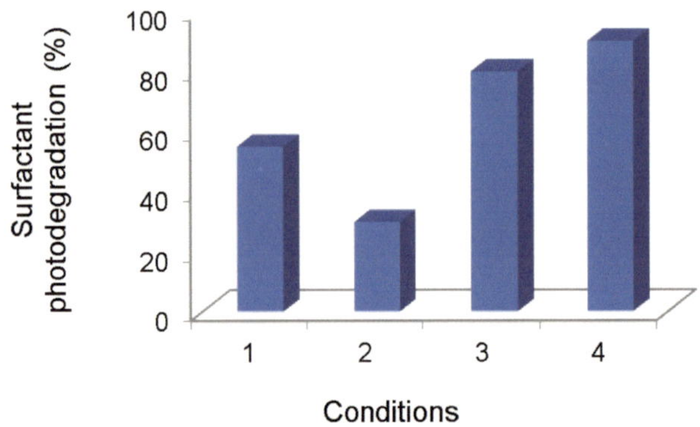

Figure 14.
The effectiveness of the LAS degradation with conditions of: (1) TiO_2/UV light, (2) TiO_2/visible light, (3) TiO_2-Ag/UV light, and (4) TiO_2-Ag/visible light [14].

light. In this case, the metal dopant can act as a separation center, where electron transfer from the TiO_2 conduction band to Ag particles at the interface is thermodynamically possible because the Fermi level of TiO_2 is higher than that of Ag metal [14, 22]. This doping resulted in the formation of a Schottky barrier at metal semiconductor contact region and improved the photocatalytic activity of TiO_2. Hence doping Ag atoms essentially reduced the band gap of TiO_2 for the photo-excitation or red shift, and simultaneously reduced the recombination rate of photogenerated electron–hole pairs [14, 22].

4.2 Photo-Fenton modification

The photo-Fenton process appears as an attractive alternative for removing emerging contaminants. Photo-Fenton processes are reported to be effective in removing several classes of contaminants, such as phenols [29], amoxicillin [30], and dyes [31]. On the other hand, the use of photo-Fenton process is restricted to acidic pH values, with associate high operating costs for industrial scale applications. To overcome these drawbacks, photo-Fenton processes modified by adding selected chelating agents such as polycarboxylates and amino polycarboxylates compounds, can be successfully performed at neutral pH. The chelating agent acting as a ligand is able to form strong complexes with Fe^{3+} that can prevent the precipitation of $Fe(OH)_3$ [32–33].

As pointed out in Eq. (17) and (18), such ligand (L) should be able to form stable complexes with Fe^{3+} which significantly absorb UV–vis light and then undergo photochemical reductions leading to Fe^{2+} ions [33].

$$Fe^{3+} + L \rightarrow \{FeL\}^{3+} \tag{17}$$

$$\{FeL\}^{3+} + h\sqrt{} \rightarrow \{FeL\}^{3+} \bullet \rightarrow Fe^{2+} + L\bullet \tag{18}$$

A study [32] reported that by addition of ethylenediamine-N,N′-disuccinic acid (EDDS), photo-Fenton process was more effective at neutral pH compared to the process at acidic condition. Other study as referred by Clarizia, *et al.* [33] also examined the effect of the adding humic acid to an aqueous solution containing benzene compound in the pH range of 5.0–7.0. The result exhibited that the rate for the oxidation of benzene were as high as those measured at pH 3.0 in absence of humic acid. However, so far, the use of chelating agents in the photo-Fenton for degradation of LAS in wastewater has not been explored. Therefore, there is a great challenge to realize experimentally the use of chelating compounds in the photo-Fenton for laundry wastewater treatment through LAS degradation.

In addition, the other drawback appearing in photo-Fenton is the use of UV light, that is more expensive and hazard for people health and ecosystem [14]. This limits in the large scale application of the photo-Fenton process [14, 21, 34]. Finding solutions of such weakness is obviously essential. An example solution of the weakness is by exploring synthetic or real solar light. The synthetic solar light is represented by wolfram or tungsten lamp [14] emitting visible light, that is low price and environmentally benign.

The results of the nitro-phenols degradation under solar light photo-Fenton, as well as under UV photo-Fenton [33] exhibit that the use of solar light can result in the degradation as high as resulted by UV photo-Fenton process. It is implied that the amount of OH radicals produced by decomposition of H_2O_2 induced by visible light is equal to that of by UV light. In fact, the power of UV light (λ < 350 nm) is higher than the visible one (λ > 350 nm), that should give more OH radicals, as seen in Eq. 8. This fact suggests that OH radicals provided by Fenton's reagent is much more prominent compared to that of by light. With the promising results, the possibility of employing solar energy in photo-Fenton processes helps improving their economic and environmental sustainability.

5. Closing marks

Laundry wastewater containing high linear alkyl benzene sulfonate (LAS) surfactant is disposed into the environment with large volume, that can create serious environmentally and health problems. Removal of LAS in water and laundry wastewater can be successfully conducted through photodegradation mechanism by photocatalysis over TiO_2 and by photo-Fenton process. In order to reach maximal degradation, the process has to be performed by employing the optimal TiO2 mass, or Fe^{2+}/H_2O_2 mole ratio, irradiation time, and pH at a certain LAS concentration. Under the optimal condition, the LAS photodegradation effectively yields smaller and harmless molecules. Moreover, modifications of both methods allow them to be more effective and wider used methods for laundry wastewater treatment. In addition to the simplicity and effectiveness, the processes also suggest the low cost treatment method. It is obvious hence that the photodegradation methods have large potential to be applied in the field and large scale.

Author details

Endang Tri Wahyuni
Chemistry Department, Faculty of Mathematics and Natural Sciences, Universitas Gadjah Mada, Yogyakarta, Indonesia

*Address all correspondence to: endang_triw@ugm.ac.id

References

[1] Patil VV, Gogate PR, Bhat AP, Ghosh PK. Treatment of laundry wastewater containing residual surfactants using combined approaches based on ozone, catalyst and cavitation, Sep. Purif. Technol. 2020;239: 116594. 1. https://doi.org/10.1016/j.seppur.2020.116594

[2] Sugiharto E, Suratman A, Natsir TA, Wahyuni ET. Distribution of detergent waste in the environment and the removal by using photocatalytic degradation and coagulation methods. Am. Chem. Sci. J. 2014; 4(6): 715-725.

[3] Kyzas GZ, Peleka EN, Deliyann EA. Nanocrystalline akaganeite as adsorbent for surfactant removal from aqueous solutions. Materials. 2013; 6: 184-197. doi:10.3390/ma6010184

[4] Makarchuk OV, Dontsova TA. Removal of anionic surfactants from wastewater by magnetic mineral sorbents, J. Wat. Sec., 2016; 2: 1-9. DOI: http://dx.doi.org/10.15544/jws.2016.003

[5] Kaleta J, Elektorowicz M. The removal of anionic surfactants from water in coagulation process. Environ. Tech. 2013; 34(5-8):999-1005 DOI:10.1080/09593330.2012.733415.

[6] Aboulhassan MA, Souabi S, Yaacoubi A, Baudu M, Removal of surfactant from industrial wastewaters by coagulation flocculation process, Int. J. Environ. Sci. Tech. 2006; 3 (4): 327-332.

[7] Korzenowskia C, Martins MBO, Bernardes AM, Ferreira JZ , Duarte ECNF, De Pinhoa MN. Removal of anionic surfactants by nanofiltration : Desalin Water Treat. 2012; 44: 269-275. doi: 10/5004/dwt.2012.3111, .

[8] Kowalska I, Klimonda A. Application of nanofiltration membranes for removal of surfactants from water solutions. E3S Web of Conferences. 2017; 17: 00044. DOI: 10.1051/e3sconf/20171700044

[9] Braga JK, Motteran F, Macedo TZ, Sakamoto IK, Delforno TP, Okada DY, Silva EL , Varesche MBA. Biodegradation of linear alkylbenzene sulfonate in commercial laundry wastewater by an anaerobic fluidized bed reactor. J. Environ. Sci. Heal A : Toxic/Hazardous Substances and Environmental Engineering. 2015; 50 (9): 946-957.

[10] Oliveir LL, Costa RB, Duarte ICS, Silva EL, Varesche MBA, Anaerobic degradation of linear alkylbenzene sulfonate in fluidized bed reactor, Braz. J. Chem. Eng. 2010; 27 (04): 539-543.

[11] Jariyanorasade A, Junyapoon S. Factors affecting the degradation of linear alkylbenzene sulfonate by TiO2 assisted photocatalysis and its kinetics Environ. Asia. 2018; 11(1): 45-60. DOI 10.14456/ea.2018.4.

[12] Ahmari H, Heris SZ, Khayyat MH. Photo catalytic degradation of linear alkyl benzene sulfonic acid. Res. Chem. Intermed. 2016; 42:6587-6606 DOI 10.1007/s11164-016-2483-1

[13] Ghanbarian M, Nabizadeh R, Mahvi AH, Nasseri S, Naddaf K. Photocatalytic degradation of linear alkyl benzene sulfonate from aqueous solution by TiO_2 nanoparticles, Iran. J. Environ. Health. Sci. Eng. 2011; 8(4): 309-316.

[14] Wahyuni ET, Istiningsih I, Suratman A. Use of visible light for photo degradation of linear alkylbenzene sulfonate in laundry wastewater over Ag- doped TiO2., J. Environ. Sci. Technol. 2020; 13: 124-130.

[15] Mehrvar M, Venhuis HS. Photocatalytic treatment of linear

alkylbenzene sulfonate (LAS) in water. J. Environ. Sci. Heal A. 2005; 40(5): 1003-1012, DOI: 10.1081/ESE-200056129

[16] Wahyuni ET, Roto R, Sabrina M, Anggraini V, Leswana NF, and Vionita C. Photodegradation of detergent anionic surfactant in wastewater using UV/TiO_2/H_2O_2 and UV/Fe^{2+}/H_2O_2 processes. Am. J. Appl. Chem.2016; 4: 174-180.

[17] Hassan MAA, Yusof R, Muhamad SHA. Fenton degradation of linear alkylbenzene sulphonates (LAS), JCNaR. 2015; 2:22-30.

[18] Malakootian M, Jaafarzadeh N, Dehdarirad A. Efficiency investigation of photo-Fenton process in removal of sodium dodecyl sulphate from aqueous solutions , Desalin. Water Treat. 2016; 57(51): 24444-24449. https://doi.org/10.1080/19443994.2016.1140082

[19] Kıran I, Bektaş N, Yatmaz HC, Tekbaş M. Photocatalytic Fenton oxidation of sodium dodecyl sulfate solution using iron-modified zeolite catalyst. Desalin Water Treat. 2013. 51 (28-30): 5768-5775 , https://doi.org/10.1080/19443994.2012.759517.

[20] Mousavi SAR, Mahvi H, Nasseri S, Ghafar S. Effect of Fenton Process (H2O2 / Fe^{2+}) on removal of linear alkylbenzene sulfonate using central composite. Iran. J. Environ. Health. Sci. Eng. 2011; 8 (2): 129-138.

[21] Miranzadeh MB, Zarjam R, Dehghani R, Haghighi M, Badi HZ, Marzaleh MA, Tehrani AM. Comparison of Fenton and photo-Fenton processes for removal of linear alkyl benzene sulfonate (LAS) from aqueous solutions. Pol. J. Environ. Stud. 2016; 25 (4), 1639-1648

[22] Wahyuni ET, Yulikayani PY, Aprilita NA. Enhancement of visible-light photocatalytic activity of Cu-doped TiO_2 for photodegradation of amoxicillin in water. J. Mater. Environ. Sci. 2020; 11 (4): 670-683

[23] Miyake M, Yamashita Y. Chapter 24 - Molecular structure and phase behavior of surfactants. Cosmetic Science and Technology: Theoretical Principles and Applications. 2017; 389-414. https://doi.org/10.1016/B978-0-12-802005-0.00024-0

[24] Konstantinou IK, Albanis TA. TiO2-assisted photocatalytic degradation of azo dyes in aqueous solution: kinetic and mechanistic investigations: A review. Appl. Catal.B: Environmental. 2004; 49: 1-14

[25] Akpan UG, Hameed BH. Parameters affecting the photocatalytic degradation of dyes using TiO_2-based photocatalysts: A review. J. Hazard. Mater. 2009; 170:520-529

[26] Hänel A, Moreń P, Zaleska A, Hupka J. Photocatalytic activity of TiO_2 immobilized on glass beads for phenol removal. Physicochem. Probl. Miner. Process. 2010; 45: 49-56

[27] Dimitrakopoulou D, Rethemiotaki I, Frontistis Z, Xekoukoulotakis NP, Venieri D, Mantzavinos D. Degradation, mineralization and antibiotic inactivation of amoxicillin by UV-A/TiO_2 photocatalysis. J. Environ Manage. 2012; : 168-174

[28] Akpan UG, Hameed GH. Parameters affecting the photocatalytic degradation of dyes using TiO_2-based photocatalysts: A review. J. Hazard. Mater. 2009; 170 : 520-529.

[29] Kusi H, Koprivanac N, Bo ˇ zi ˇ AB, Selanec I. Photo-assisted Fenton type processes for the degradation of phenol: A kinetic study. J. Hazard. Mater. 2006; B136 : 632-644.

[30] Pouran SR, Aziz ARA, Daud WMAW. Review on the main advances in photo-Fenton oxidation

system for recalcitrant wastewaters. J Ind Eng Chem. 2015; 21 : 53-69

[31] Torrades F, García-Montaño J. Using central composite experimental design to optimize the degradation of real dye wastewater by Fenton and photo-Fenton reactions. Dyes Pigm. 2014; 100 : 184-189. http://dx.doi.org/10.1016/j.dyepig.2013.09.004

[32] Huang W. Homogeneous and heterogeneous Fenton and photo-Fenton processes : impact of iron complexing agent ethylenediamine-N,N'- disuccinic acid (EDDS), Universit'e Blaise Pascal - Clermont-Ferrand II, 2012.

[33] Clarizia L, Russ D, Di Somma I, Marotta R, Andreozzi R, Homogeneous photo-Fenton processes at near neutral pH: A review. Appl. Catal. Environmental. 2017; 209 : 358-371.

[34] Kavitha V, Palanivelu K. Degradation of nitrofenol by Fenton and photo- Fenton processes. J Photochem Photobiol A Chem A; Chemistry. 2005; 170: 83-95

Chapter 14

Application of Water Quality Index for the Assessment of Water from Different Sources in Nigeria

Ruth Olubukola Ajoke Adelagun, Emmanuel Edet Etim and Oko Emmanuel Godwin

Abstract

Water quality index (WQI) provides a single number that expresses the overall water quality, at a certain location and time, based on several water quality parameters. The objective of WQI is to turn complex water quality data into information that is understandable and usable by the public. A number of indices have been developed to summarize water quality data in an easily expressible and easily understood format. The WQI is basically a mathematical means of calculating a single value from multiple test results. This chapter discusses, in detail, the application of a water quality index for the assessment of water quality to different several water sources in Nigeria.

Keywords: Water Quality Index, Water Quality Indicators, Surface Water, Underground Water, Environmental Health

1. Introduction

Clean, safe and adequate freshwater is of utmost importance to human existence and the survival of all living components in the ecosystem. Water quality issues are complex and diverse, deserving urgent global attention and action [1]. The decline in water quality has become a global issue of concern because of its inherent ability to cause major alterations to the hydrological cycle. The past decade has seen remarkable impact of man on the environment due to unprecedented increase in population and rapid rate of urbanization as well as the intensification and expansion in agricultural practices. This has led to progressive and continual degradation of resources especially surface water. Polluted water is an important vehicle for the spread of diseases. In developing countries about 1.8 million people, mostly children, die every year as a result of water-borne diseases [2]. According to Bullard, [3] inferred that impaired surface water quality always result in an unhealthy socio-economic environment.

The characteristics of water are defined by its composition and are commonly referred to as water quality. Water quality is generally defined as "the chemical, physical and biological characteristic of water usually in respect to its suitability for a designated use" [4]. The assessment of water quality, usually carried out by determining its physico-chemical and biological properties or parameters against a set of standards, is used to determine whether the water is suitable for consumption or safe for the environment. Water quality assessment can be defined "as the evaluation of physical, chemical and biological state of the water in relation with the natural state, anthropogenic effects and future uses" [5]. The water quality parameters are then used as a reference to a set of standards based on the intended usage of the water broadly classified into industrial/domestic use, human consumption (portability) and restoration (in the environment/ecosystem, generally for health of human/aquatic life). Water quality standards are used to protect different designated uses of water. The standards of each one of these designated uses are very different from each others. For example, the water used for drinking requires a higher standard compared to the standard used for agricultural and industrial use (water for domestic purposes should therefore be free from toxic substances and organisms in order to prevent waterborne diseases).

2. Importance of water quality assessment and monitoring

The assessment of water quality is very pertinent to both public health and aquatic life. Water quality has a significant impact on water supply and oftentimes determines supply options [6, 7]. The understanding and monitoring of sources and quality of water used for water supply is of societal, economic and conservational importance since per capita water demand is increasing while accessibility to freshwater availability is continuing to decline. Local water quality can be used to identify the sources and fates of toxic contaminants and pollutants either from ecology, geology, and anthropogenic activities (industrial processes, runoff from agricultural farms etc) in the area [5]. Identifying the source (s) of contamination and developing appropriate management strategies are essential to minimizing potential public health risks [8]. Moreover, data obtained via assessment and monitoring water quality provides empirical evidence to assist health and environmental decision making. In water management practices, water quality values serve as useful and sensitive indicators of changes in the physical, chemical or biological composition of the overall water status [9].

2.1 Water quality indicators

To determine the quality of a water body, the chemical, biological and physical conditions of a water body must be measured. Chemical measurements, biological surveys, and visual observations (physical) provide a "big picture" of what's happening in a water body. The following is a list of indicators (physical, chemical and biological) that are often measured to assess the quality of water.

a. Physical indicators

Some physical indicators of the quality of a water sample from any source include,

- Temperature - Electrical Conductivity - Taste - Total Suspended Solids (TSS)

- Turbidity - Odor - Color - Total Dissolved Solids (TDS)

b. Chemical indicators

Some chemical indicators of the quality of a water sample from any source include,

- pH – Biochemical Oxygen Demand (BOD) – Chemical Oxygen Demand (COD)

- Dissolved Oxygen (DO) – Total Hardness – Phosphates – Pesticides – Nitrates

- Surfactants – Heavy metals

c. Biological indicators

Some biological indicators of the quality of a water sample includes,

- Bacteria (fecal coliform, *Escherichia coli*, Cryptosporidium, *Giardia lamblia*), – Viruses - Fungi protozoa - Parasitic worms - *Pimephales promelas* (fathead minnow) -*Americamysis bahia* (Mysid shrimp) - Benth ic macroinvertebrates (Ephemeroptera or mayfly, Plecoptera or stonefly and Trichoptera or caddisfly – Sea Urchin – Mollusca (Bivalve mollusks – *Americamysis bahia* (Mysid shrimp)

2.2 Water quality standards

Water quality standards imply statements and numeric values that describe water quality and fall within the following three components:

i. Designated uses of the water body as related to water supply, aquatic life, agriculture, or recreation.

ii. Water quality criteria and general statements that describe good water quality and specific numerical concentrations for various parameters.

iii. Anti-degradation policy designed to maintain and protect the existing water uses for each water body.

The standard used for particular water is a function of the expected use of the water. **Table 1** presents some of the established standards of some water quality parameters. What this means is that the established standard used for drinking water is used only in determining the Drinking Water Quality Index while the Aquatic Water Quality Index standards are used to protect aquatic life. Basically, the index can be calculated for three different uses:

a. Drinking Water Quality Index which includes drinking, recreation, irrigation, and livestock watering use.

b. Aquatic Water Quality Index which includes aquatic life protection and use.

c. Overall Water Quality Index which includes the protection of human health, aquatic ecosystems and wildlife.

2.3 Water quality index

The general norm for reporting water quality parameters by comparing the different analyzed parameters with their respective permissible limits and standards set by regulating bodies at local, regional, national or international levels has

Parameters	WHO	CCME
pH (mg/l)	6.5–8.5	8.5
DO (mg/l)	—	5
Temperature (°C)	25	15
Turbidity (NTU)	5	5
TDS (mg/l)	500	500
Ammonia (mg/l)	0.2	1.37
Nitrate (mg/l)	50	48.2
Lead (mg/l)	0.01	0.01
Iron (mg/l)	0.3	0.3
Chromium (mg/l)	0.05	0.05

Sources: [10, 11].

Table 1.
Sets of some established standards.

been deduced to be ineffective in environmental monitoring program by both managers and the general public [12]. Carlos and Alejandra [13] argued that providing statements that summarize the water quality data in a simple expressible format that describes the general health or status of a water body is more preferable to environmental managers and the general public rather than been asked to give a rather biased interpretation to complex and technical environmental data. The Water Quality Index (WQI) was first developed by Horton [14] and presents a mathematical method of calculating a single value to represent water quality from multiple water quality parameters. The index represents the level of quality of a water body such as lake, river or stream by using some of the regularly used water parameters (BOD, temperature, turbidity, conductivity etc.) [15]. The WQI is based on the measurement of different water quality parameters thus providing a mechanism for presenting a cumulatively derived numerical expression for defining water quality [16]. The water quality index reduces water quality data to common scale and combines them into a single number in accordance with a chosen method or model of computation. WQI reflects the composite influence of different water quality parameters and is calculated from the point of view of the suitability of both surface and groundwater for intended usage.

The method follows three steps namely:

i. Selection of parameters

ii. Determination of quality function for each parameter and

iii. Aggregation through mathematical equation.

In order to rank the overall water quality, the Canadian Council of Ministers of Environment CCME [17] established the use of an index that mathematically combines all water quality measures and provides a general and readily understood description of the quality of water. Over the years, many countries have accepted the CCME scheme representing the water quality index for water quality monitoring and assessment of surface and underground water in terms of their chemical, biological and nutrient constituents and overall esthetic condition. The Canadian Council of Ministers of the Environment Water Quality Index (CCME WQI) is preferred as a tool for the work due to its simplicity and ability to combine complex water quality data without compromising its technical integrity [16]. The CCME Water Quality Index is considered the most effective method of measuring water quality to determine its suitability for an intended use [18].

In the United States, the US National Foundation uses a weighted linear system of the WQI as a guideline for defining water quality [19]. Many other countries have used the same concepts to define their water quality status including Malaysia [20], Spain [21], Bangladesh [22], and China [23]. The water quality index reduces the bulk number of water parameters used in an assessment and provides a single value. This value is a simplified and logical form that expresses the average quality of water at a specific time based on the analytical values of physico-chemical parameters. This procedure facilitates a simpler and easier interpretation of the data rather than assessing each parameter and allows easy public access and understanding of the water quality data [12, 24, 25].

2.4 Merits of the water quality index

Several advantages and benefits accrue from the use of the water quality index [22] include:

1. Reduction in the number of parameters required to compare water quality for a definite use

2. Provision of a single number that represents overall water quality at a certain location and time

3. Identification of space and time dynamics in the quality of water.

4. Provision of assurance on the safety of a water body to users such as habitat for aquatic life, irrigation water for agriculture and livestock, recreation and esthetics, and drinking water supplies.

5. It is very effective for water quality monitoring.

6. Provides means of comparisons between different rivers and sampling sites.

7. The indices is one of the most simplified methods of communicating water quality classification to the general public or those in authority.

8. It simplifies a complex dataset into easily understandable and usable information.

9. The single-value output of the index, derived from several parameters, provides important information about water quality that is easily interpretable by the general and non-technical populace.

10. The index is a useful tool for communicating water quality information to the large public and to legislative decision makers.

2.5 Limitations of the water quality index

Despite the benefits attributed to the WQI, it is however besieged with some challenges [26, 27], some of which are stated below,

1. WQI is not an absolute measure of degree of pollution or the actual water quality.

2. Lack of precision and accuracy in classification technique of importance of evaluation of parameters.

3. Inefficiency in dealing with uncertainty and subjectivity in a complex environmental issue such as the incompatibility of observations, uncertainty, imprecision in criteria.

4. Lack of a uniform method for measuring water pollution involving biological parameter.

5. Inadequate to transfer complex environmental data into information.

3. Water quality determinant

The selection of significant water quality parameters is vital and key to having good representation of all indicators of water quality [28, 29]. Water quality parameters commonly used by various researchers include dissolved oxygen, total phosphates, temperature, pH, turbidity, chemical oxygen demand, fecal coliform, total solids, biochemical oxygen demand and nitrates [30–32]. The weight associated with each parameter is based on its respective standards and the magnitude of the assigned weight indicates the parameter's significance and impact on the index. Below is the weighting factors assigned to some of the water quality parameters (**Table 2**).

Water quality parameters	Weight factors
Dissolved Oxygen	0.22
Biological Oxygen Demand	0.19
Chemical Oxygen Demand	0.16
Ammoniacal nitrogen	0.15
Suspended solid	0.16
pH	0.12

Table 2.
Water quality parameters and weight factors.

4. Water quality index calculation

Though, a lot of water quality parameters are used for water assessment, some of the parameters seem to have a common similarity as they have their basis of comparing water quality parameters with their respective regulatory standards with interpretation of the results as good or bad [33].The parameters involved in the weighted arithmetic water index method water quality uses:

i. Degree of purity which is obtained from the most commonly measured water quality variables: temperature, biochemical oxygen demand, fecal coliform, pH, dissolved oxygen, total phosphates, turbidity, nitrates and total solids.

ii. Water quality rating scale, (qi)

iii. Relative weight and (wi)

iv. Overall WQI (Qi)

The WQI is calculated by averaging the individual index values of some or all of the parameters within five water quality parameter categories that depicts the pollution level or status of the water:

i. Water clarity: turbidity (NTU) and/or Secchi disk depth (meters or feet);

ii. Dissolved oxygen: Dissolved oxygen concentration (mg/l);

iii. Oxygen demand: biochemical oxygen demand (mg/l), chemical oxygen demand (mg/l) and/or total organic carbon (mg/l);

iv. Nutrients: total nitrogen (mg/l), and/or total phosphorus (mg/l); and

v. Bacteria: total coliform (per mg/l) and/or fecal coliform (per mg/l)

The numerical value of the quality rating (q_i) is obtained from the water quality data then multiplied by a weighting factor that is relative to the significance of the test to water quality. The formula below is used to obtain q_i:

$$q_i = \frac{c_i}{s_i} \; x \; 100 \tag{1}$$

where,
q_i, = quality rating scale.
c_i, = concentration of i parameter.
s_i = WHO standard value of i parameter.
Relative weight (w_i) is calculated by

$$w = \frac{1}{s_i} \tag{2}$$

The standard value of the i parameter is inversely proportional to the relative weight. The relative weight (w_i) is calculated by

$$w_i = \frac{w_i}{\sum_1^n w_i} \tag{3}$$

Finally, overall WQI was calculated according to the following expression:

$$WQI = \frac{\sum_i^n Q_i w_i}{\sum W_i} \tag{4}$$

The sub-index Si and WQI are computed using the relationship in Eqs. (3) and (4), respectively

$$SIi = w_i \; x \; q_i \tag{5}$$

$$WQI = \sum SIi \tag{6}$$

where SI_i is the sub-index of the *ith* parameter and q_i is the rating based on the concentration of the *ith* parameter.

Ranking of WQI Values.

The Global Environmental Monitoring Systems [34] adopted the Water Quality Index (WQI) developed by the Canadian Council of Ministers of Environment (CCME) and based its development on the combination of three factors into one index. The detailed formulation of the WQI, as documented by CCME [17] and Amir *et al.*, [35] comprises three factors which include:

Scope, **F1** - the number of variables whose objectives are not met and calculated as

$$F_1 = \frac{Number\ of\ failed\ Variables}{Total\ Number\ of\ Variables}\ \text{x}\ 100 \tag{7}$$

Frequency, **F2**, – the frequency with which the objectives are not met.

$$F_2 = \frac{Number\ of\ failed\ Tests}{Total\ Number\ of\ Tests}\ \text{x}\ 100 \tag{8}$$

Amplitude, **F3**, – the amount by which the objectives are not met.
F3 is calculated in three steps:

a. The number of times by which an individual concentration is greater than (or less than, when the objective is a minimum) the objective is termed an "excursion" and is estimated as follows;

b.

$$Excursions_i = \frac{Failed\ test\ values_i}{Objestive_i} - 1 \tag{9}$$

For cases in which the test value must not exceed the objective:

$$Excursions_i = \frac{Objestive_i}{Excursions_i Objestive_i} - 1 \tag{10}$$

c. The collective amount by which individual tests is out of compliance is calculated by summing the excursions of individual tests from their objectives and dividing by the total number of tests (both those meeting objectives and those not meeting objectives). This variable, referred to as the normalized sum of excursions (*nse*), is calculated as:

$$nse = \sum_{i=1}^{n} \frac{excursions_i}{Number\ of\ tests} \tag{11}$$

d. F3 was thereafter calculated by an asymptotic function that scales the normalized sum of the excursions from objectives (*nse*) to yield a range between 0 and 100 as given in Equation

$$F_3 = \frac{nse}{0.01nse + 0.01} \tag{12}$$

The CCME WQI is determined using equation below:

$$\mathrm{WQI} = 100 - \frac{\sqrt{F_{1+}^2 F_{2+}^2 F_3^2}}{1.732} \tag{13}$$

The calculation produces a score value that ranges between 0 and100. The higher the score the better the quality of water. The CCME WQI values ranges between 0 which depicts a worst water quality and 100, the best water quality [36]. The interpretation is that a water body with WQI scores that range between 71 and 100 are very suitable for the expected use, meet the required expectations for water quality and are of lowest concern, scores that ranges between 51 and 70 indicate marginal concern while a water body with WQI values with scores below 50 do not meet expectation and are of highest concern.

The CCME places the WQI values into five categories with the following interpretations [22]:

- *Excellent*: (CCME WQI Value 95–100) – Water quality is protected with a virtual absence of threat or impairment; conditions very close to natural or pristine levels.

- *Good:* (CCME WQI Value 80–94) – Water quality is protected with only a minor degree of threat or impairment; conditions rarely depart from natural or desirable levels.

- *Fair*: (CCME WQI Value 65–79) – Water quality is usually protected but occasionally threatened or impaired; conditions sometimes depart from natural or desirable levels.

- *Marginal*: (CCME WQI Value 45–64) – Water quality is frequently threatened or impaired; conditions often depart from natural or desirable levels.

- *Poor:* (CCME WQI Value 0–44) – Water quality is almost always threatened or impaired; conditions usually depart from natural or desirable levels.

A number of indices have been developed to summarize water quality data in an easily expressible and easily understood format. The scores are then ranked into one of the five categories described below (**Table 3**) [34, 37]:

WQI value	Ratings of water quality
91–100	Excellent
71–90	Good
51–70	Medium
26–50	Bad
0–25	Very Bad

Table 3.
Ratings of water quality indices.

5. Sources of water in Nigeria

Globally, the provision and supply of adequate water to the populace is one of the core responsibilities and duties of the government. This is because water is among the first requirements in the hierarchy of citizens' needs and a failure to guarantee water supplies to those that need it most can lead to serious political setbacks [38]. A good knowledge of the source(s) of water is necessary to improve on the provision and supply of water to the populace. Nigeria is divided into six geological zones, namely, North-east, North-west, North-central, South–south, South-east, and South-west (**Figure 1**). The country has six hydrological basins covering the swampy forest in the south, the dense rainforest in the east, hilly shrub lands in the middle belt, savannah grasslands in the north, and semi-arid areas in the far north [38]. There are two major river systems in the country: the River Niger and River Benue both meet at Lokoja. River Niger enters the country from the northwest and River Benue enters from the northeast [39].

The most available sources of water for most urban–rural communities in developing countries, including Nigeria, are surface waters (rivers, streams, ponds and lakes) and groundwater (in form of boreholes and hand-dug wells). Surface waters in Nigeria are usually contaminated with domestic, agricultural, and industrial wastes and cause many water-related diseases and ill health to living organisms [40, 41] while Nigerian groundwater quality is generally good but these waters are often laden with high contents of heavy metals (e.g., Fe, Mn, Cd, As Hg), nitrates, fluorides or cyanides and can be contaminated with a wide variety of pathogenic organisms (often above recommended WHO levels).

A larger part of the Nigerian populace are self -dependent in meeting their daily water provision from natural sources: rivers, streams, ponds, rain and hand-dug wells or modern supply sources which include public sector supplies or private and commercial borehole businesses [42]. Access to adequate water supply in Nigeria is hampered by geographical, socio-economic and institutional factors. Reports from the WHO/UNICEF [43–45] indicate that 72% of urban dwellers have access to

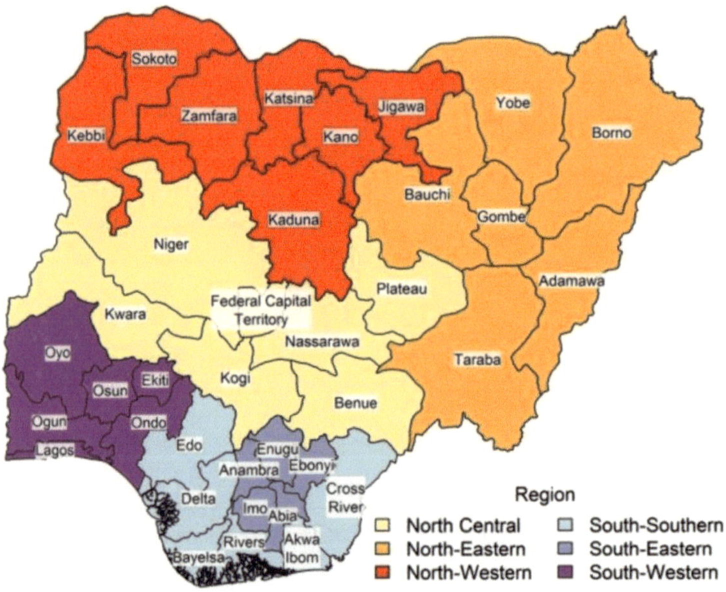

Figure 1.
Map of Nigeria showing the six Hydrological Basin.

Indicators	NE	NW	NC	SE	SW	SS	National	Rural	Urban
Safe water source (%)	30.7	50.64	48.9	40.8	73.5	45.9	51.4	40.4	73.4
Water treatment before drinking (%)	4.6	7.5	14.1	11.4	20.4	5.8	11.3	14.5	9.7

NE: North East, NW: North West, NC: North Central, SE: South East, SW: South West, SS: South–South.
Source: Extracted from Amakom [48].

Table 4.
Evaluation of regional access to water supply in Nigeria.

improved water sources while 43% of the rural populations do not have ready access. Regionally, access to improved drinking water sources in the north central, north eastern and north western zones is 52.2%, 27.3%, 42.5%, respectively. Access to improved drinking water sources is 72.7% and 54.1% in the south western and south eastern zones of Nigeria, respectively. Thus, high disparity exists between the urban and rural populations regarding their access to good water. A similar disparity between the northern and southern regions of Nigeria is clearly shown as depicted in **Table 4**. It can be inferred from the table that the SW and NW zones and the urban areas have demonstrably higher access to a safe water supply. The problem of water pollution arising from petroleum oil exploration in the south tends to limit the availability of freshwater resources from the natural sources [46, 47]. In most parts of the Niger- Delta (SS) region of Nigeria, the major challenge for survival is the availability of good quality (potable) water free of environmental pollution and degradation. According to Raimi [46], the sources and percentage frequencies of water in the oil producing communities in the Central Senatorial District of the Bayelsa State is: rain (61%), rivers (13%), pipe-borne (33%) borehole (91%) and hand-dug well (3%). This distribution of water sources is similar in most cities in the other parts and zones in the country. Thus, it implies that the most frequently used water sources in Nigeria are borehole, rain water and pipe-borne water except in the rural areas where the major sources are hand dug wells and rivers.

6. Water quality indices of surface and underground water sources in Nigeria

This section provides the water quality indices of surface and underground water sources from different part of Nigeria. Several authors have applied water quality index (WQI) to evaluate the quality of water from different water sources especially surface and underground water across the different zones in the country [18, 47, 49–51]. Herein, the country is divided into four regions (south, east, west and north) and the WQI of water from different sources including rivers, boreholes, hand-dung wells etc. of the country is discussed.

Several researchers have assessed and reported the WQI of water bodies in the Southern states of the country [47, 52–55]. Most of the rivers investigated in Bayelsa: Korama, Otamiri, Oramiukwu, Ase, and Orashi Rivers showed poor water quality, and water environment clearly unsuitable for drinking [46]. The Otamiri and Orimiukwu Rivers have very bad water quality based on the WQI while the Ase River was observed to have bad water quality with a high degree of deterioration at the downstream [52]. The Orashi River displayed a marginal level of pollution as about 50% of parameters failed to meet the required standards [54]. The Brass River in Bayelsa State was considered to be far from excellent [53].

Aigberua and Tarawou [56] investigated the WQI of some surface waters (rivers) in the Rivers State along the Taylor Creek area of the state. Their calculated water quality indices (WQIs) scores fall within the range which indicates water quality status tending from "poor water quality" to "unsuitable for drinking". Taylor Creek shows a slightly acidic water environment that contains high levels of nitrate loading, pH, total dissolved solids and *E. coli* in addition to an objectionable level of color and unsightly appearance. The WQI assessment reflects water of poor quality and generally unsuitable for public consumption. The presence of multiple dumpsites mostly from leachates along the stretch of the river may be responsible for the poor degradation in water quality. It was therefore recommended that the water is not fit for human consumption or and recreational purposes. Overall, the WQI assessment of Taylor Creek revealed that water is unsuitable for drinking and may pose serious health risks.

The water quality index (WQI) approach was used to assess the suitability of water from three local government areas in River state by Chinwendu [57]. The result of these assessments indicated that borehole water was unsafe for human and animal consumption. These waters had an acidic pH while the dissolved oxygen, temperature and calcium values were not within the WHO and NSDWQ permissible drinking water standards. The water quality index of the borehole waters in the region exceeded permissible water quality standards in all sampling locations due to groundwater contamination resulting in water that was unsafe for human and animal consumption.

The suitability of water from different sources (stream, borehole and pipe - born water) were assessed using the WQI in the Niger Delta region of Nigeria by Etim et al. [58]. The concentrations of the respective parameters are below the WHO/ICMR standards. The quality of water based on the index number representing overall suitability of the water indicated that the water samples analyzed from pipe born and borehole water were safe for human consumption and domestic purposes while the samples analyzed from stream water are not safe for human consumption.

The quality of some river bodies around the Warri metropolis was evaluated by Godwin and Oborakpororo [59] based on their various physico-chemical parameters. The results obtained from the study showed that all the surface water samples were found to be unfit for human consumption with very high turbidity and suspended solids. The presence of fecal coliform in the various water bodies was much higher than the stated standard of regulatory agencies. The physicochemical parameters of groundwater in 12 cluster boreholes in Enugu North district/region, southeast Nigeria categorized all the water samples within the range of good to excellent [60].

The quality of 12 different water sources and 2 treated water used by peri-urban town in the West region of Nigeria were evaluated to assess their suitability for drinking and domestic us [61]. Water quality parameters included pH, temperature, acidity, total alkalinity, chloride content and coliform. The results indicated that all the physicochemical parameters of the water samples complied with regulatory standards. Similarly, most sites complied with heavy metals criteria. At these sites, fecal coliform and *E. coli* tested positive for all the samples except one tap water sample. The majority of the water samples (86%) were rated as excellent based on the physicochemical parameters. However, the inclusion of microbiological data in the WQI revealed that only 7% of the samples analyzed can be regarded as excellent water. Akoteyan and his team [62] studied the water quality characteristics of Owo river for municipal water supply in Lagos-Nigeria. The study showed that the physical parameters assessed (electrical conductivity, pH, total hardness, anions and cations) were within the maximum permissible limit of WHO standard for drinking water quality. The calculated WQI showed that the water is suitable for human use. Olagbemide [63] applied the WQI for the assessment of Eleyele Lake, Ibadan, to check the quality of the lake water with respect to different physico-chemical parameters using standard methods. Water samples were collected from different river sites (i.e., before the lake, on the lake and after the lake). The results of the Water Quality Index showed that the water quality at these sites was poor. This suggests that the lake is polluted and not totally safe for human consumption without proper treatment. Very high values were obtained for color, turbidity, total solid, total suspended solid, BOD, COD, alkalinity, phosphate, chloride, magnesium, nitrate, total organic carbon, total organic matter were observed and all were above the permissible WHO values.

Murtala and Ahaneku [64], studied some physicochemical parameters (pH, dissolved oxygen, temperature, turbidity, total dissolved solid, nitrate, ammonia, iron, lead and chromium) from the river Asa in Illorin (Kwara state) and presented the complex water quality data of the river as the WQI that can easily be understood by the technical and non-technical personnel. The result of the Water Quality Index showed that three of the four stations investigated should be ranked as poor and the remaining station as marginal. The implication is that the river failed the Drinking Water Quality Index and is not suited as a potable source of drinking water. The seasonal variation of some physicochemical properties of River Asa in Kwara state was also assessed and the river water quality status was evaluated using CCME Water Quality Index. The result of the study revealed the river is not suitable source for drinking water.

Ogbozige and co-workers [65] assessed 12 water quality parameters (turbidity, TDS, pH, Cl^-, EC, DO, BOD_5, COD, total nitrogen, total phosphorus, Fe and Mn) for the River Kaduna, Nigeria on a monthly basis for a period of one year at 15 sampling locations using standard methods. The data were used to develop Water Quality Index (WQI) across the 15 sampling locations. The WQI revealed that the water quality of four (4) sampling locations was poor and the general water quality of the remaining 11 sampling locations was marginal. The water quality assessment of water consumed in Kaduna State revealed that among the 15 rivers, 4 of the rivers (Kutimbi, Kigo, Breweries and Rigasa) recorded poor WQI while the river upstream of Narayi community was marginal. Results indicated that the quality of the rivers at Narayi and Rigasa communities was bad. The water qualities of the remaining 8 rivers were of better quality including River Romi. Based on the results, the WQI of River Kaduna on the Canadian scale is mostly marginal. Yisa [37] evaluated the quality of selected hand-dug wells in Maikunkele area in Niger state using WQI technique. These results indicated that the quality of the samples was marginal while one location was extremely bad. The results also revealed a high contamination of coliform in the samples and nitrate concentration above standard of WHO, EPA, APHA and the Nigeria drinking water standards.

The Water Quality Index and heavy metal contents of underground water sources in Doma Local Government Area, Nasarawa State, Nigeria was investigated to ascertain the suitability of the water for domestic purpose using physicochemical parameters: temperature, turbidity, TDS, TSS, pH, EC, total hardness, alkalinity, chloride, nitrate and sulphates in the water samples [66]. The physicochemical parameters determined for borehole and hand dug well water samples (i.e., temperature, turbidity, total dissolved solids, total suspended solids, electrical conductivity, total hardness, alkalinity, chloride, nitrate, and sulphate) were all within the standards recommended by regulatory bodies NSDWQ and WHO. The mean pH for the hand dug well water was within the recommended standard values; however, the pH value for the borehole was outside the range recommended standards (the water was slightly acidic). The WQI evaluated for both borehole and hand dug well water samples showed the ground water sources presented good water quality.

The results of the mean metal concentrations in borehole and hand dug well water samples shows that the concentrations of Cd, Cu, Pb and Zn are within the permissible limit recommended by regulatory bodies while those of Cr and Fe are higher than standard values. Oko and his crew [67] collected water samples from boreholes and hand dug wells located in two wards in Wukari town in Taraba state and assessed some physico-chemical parameters using analytical methods. The calculated WQI showed that the water samples from the borehole was of better quality for drinking than the hand dug well.

7. Conclusion

In Nigeria, the most frequent water sources are surface waters (rivers, streams, ponds and lakes) and groundwater (borehole and hand-dug wells). The physico-chemical assessments of water samples showed that while some of the parameters are within permissible limits, many exceeded the stipulated standards. Application of the water quality index (WQI) to determine the suitability of the water for an intended use indicated that most water sources in the western part of the country are good and suitable for human consumption except for incidences of high levels of fecal contamination in some rivers. The WQI for most locations in the northern part of the country is either bad or poor and not suitable for human consumption. In the eastern and southern part of the country, the WQI index indicated marginal quality that was not suitable for human consumption without treatment. This marginal quality could be as a result of the high levels of nitrate and acidic pH of most of the waterbodies in the area. In all, it is recommended that prior treatment of the water is very important before consumption so as to avoid water-borne related diseases and illnesses.

Author details

Ruth Olubukola Ajoke Adelagun, Emmanuel Edet Etim*
and Oko Emmanuel Godwin
Department of Chemical Sciences, Federal University Wukari, Nigeria

*Address all correspondence to: emmaetim@gmail.com

References

[1] Breabăn I.G, Gheţeu, D, Paiu, M. Determination of Water Quality Index of Jijia and Miletin Ponds, Bulletin UASVM Agriculture. 2012. 69 (2)/2012

[2] World Health Organization. Water, sanitation and hygiene links to health. 2004. Available at: www.who.int/water_sanitation-health/publications/facts2004/en/index.html

[3] Bullard W. E, Effects of Land use on Water Resources in the Ecology of Man: An Ecosystem Approach, Smith, R. L. (ed), New York, Harper and Row Publisher, 1972.

[4] Johnson D.L, Ambrose, S.H, Bassett, T.J, Bowen, M.L, Crummey, D.E, Isaacson, J.S, Johnson, D.N., Lamb, P., Saul, M. and Winter-Nelson, A.E. (1997). Meanings of Environmental Terms. Journal of Environmental Quality, 26, 581-589.

[5] Chapman D, (ed.). Water Quality Assessments: A Guide to the Use of Biota, Sediments and Water. *Environmental Monitoring*. Second Edition. UNESCO, WHO, and UNEP. E&FN Spon, London UK. 1996.

[6] Ouyang Y, Evaluation of river water quality monitoring stations by principal component analysis. Water Res. 2005; 39: 2621–2635.

[7] Lodder W.J, Van den Berg, H.H., Rutjes, S.A, de Roda Husman, A.M. Presence of enteric viruses in source waters for drinking water production in The Netherlands. Appl. Environ. Microbiol. 2010;76: 5965–5971.

[8] Carroll S.P, Dawes L, Hargreaves, M, Goonetilleke, A. Water Quality Profile of an Urbanising Catchment—Ningi Creek Catchment; Technical Report; School of Urban Development, Queensland University of Technology: Caboolture Shire Council, QLD, Australia, 2006; 1–93.

[9] Cambers G, Ghina, F. Water Quality, an Introduction to Sandwatch. An Educational Tool for Sustainable Development. United Nations Educational, Scientific and Cultural Organization (UNESCO): Paris, France. 2005; p. 25-31.

[10] Ashok L. Doug, H. Tribeni S, Environmental Monitoring and Assessment, 2006;113: 411–429.

[11] WHO, Guidelines for Drinking water Quality. International Standard for Drinking Water Guidelines for Water Quality, 4th ed, Geneva, Switzerland, 2011.

[12] Shweta T, Bhavtosh S, Prashant S, Rajendra D. Water Quality Assessment in Terms of Water Quality Index, American Journal of Water Resources. 2013;1 (3) 34-38

[13] Carlos M, Alejandra V.V, Journal of Urban and Environmental Engineering. 2011;1(3)18 25.

[14] Horton R. K, An index number for rating water quality. Journal of Water Pollution Control Federation, 1965; 37 (3): 300-306.

[15] Kankal N.C, Indurkar, M.M., Gudadhe S.K., Wate, S.R, Water Quality Index of Surface Water Bodies of Gujarat, India, Asian J. Exp. Sci., 2012; 26(1)39-48.

[16] Miller W. W, Joung, H. M, Mahannah, C. N, Garrett J. R. Identification of water quality differences in Nevada through index application J. Environ. Quality. 1986;15: 265-272

[17] CCME. Canadian Council of Ministers of Environment: Water

Quality Index User's Manual. Canadian Water Quality Guidelines for the Protection of Aquatic Life, 2001; 1-5.

[18] Ochuko U, Thaddeus O, Oghenero OA, John EE, A comparative assessment of water quality index (WQI) and suitability of river Ase for domestic water supply in urban and rural communities in Southern Nigeria. Int J Human Soc Sci. 2014; 4 (1):234–45.

[19] Ashok L, Doug H, Tribeni S, Environmental Monitoring and Assessment, 2000;113: 411–429.

[20] DOE WQS Phase 2 Study: Development of water Criteria and standards for Malaysia, Department of Environment, Ministry of Science, Technology and the Environment, Kuala Lumpur.1990

[21] Agencia Catatalanadel Agua (Catalonia, Spain). www.mediambient.gencat.net/aca/ca/inici.jsp 2005.

[22] Akter T, Jhohura F.T, Akter F, Chowdhury, T.R. Mistry, S.K. Dey, D. Barua, M.K. Islam M.A. and Rahman, M.. Water Quality Index for measuring drinking water quality in rural Bangladesh: a cross sectional study Journal of Health, Population and Nutrition. 2016;35: 4-12

[23] Vadde K. K, Wang J, Cao L,Yuan T, Alan J. M. Raju S. Assessment of Water Quality and Identification of Pollution Risk Locations in Tiaoxi River (Taihu Watershed), China . Water. 2018;10, 183; doi:10.3390/w10020183 wD

[24] Hallock D, A Water Quality Index for Ecology's Stream Monitoring Program, 2002 http://www.ecy.wa.gov/biblio/0203052.html

[25] Yogendra K, Puttaiah, E.T. Determination of Water Quality Index and suitability of an urban waterbody in Shimoga Town, Karnataka, Proceedings of Taal. The 12th World Lake Conference: 2008 342-346

[26] Chai LL, River Quality Classification of Sungai Padas Using water Quality indices. FSAS. 1999. p. 319.

[27] Yilmaz Icaga. Fuzzy evaluation of water quality classification. Ecological Indicators. 2007;7(3):710–8.

[28] Roşu C, Piştea I, Călugăr M., Martonoş I, Ozunu A, Assessment of ground water quality status by using water quality index (WQI) method in Tureni village, Cluj county, Aerul si apa componente ale mediului, pg 2013; 111-118, Cluj-Napoca, Romania

[29] Tim H, Msc thesis University of Victoria 2012

[30] Oişte A. M, Breabăn I. G. Water quality index for Rediu, Cacaina and Ciric river in urban area of Iasi city, Present Environment and Sustainable Development, 2012;6(2).

[31] Piştea I, Roşu C, Martonoş I, Ozunu A, Romanian surface water quality: Tarnava Mare river between Medias and Copsa Mica case study, Environmental Engineering and Management Journal, 2013;12 (2) 283-289

[32] Iticescu C, Georgescu L.P, Topa C. M, Assessing the Danube water quality index in the city of Galati, Romania, Carpathian Journal of Earth and Environmental Sciences, 2013 Vol 8, No 4.

[33] United Nations Environment Programme Global Environment Monitoring System/Water Programme, UNEP/GEMS. C/O National Water, 2007. http://www.Gemswater.Org.[24] Hase

[34] Global Environment Monitoring System/Water Programme, UNEP/GEMS. C/O National Water, 2007. http://www.Gemswater.Org.

[35] Amir A. K, Annette T, Renee P, Haseen K, Richard W, Water Quality Resources Journal Canada, 2005,40(4) 448–456.

[36] Saffran K, Cash K, Hallard K, Canadian water quality guidelines for the protection of aquatic life. CCME Water Quality Index 1.0 User's Manual, Canada. 2001

[37] Yisa J, Jimoh T, Analytical studies on water quality index of river Landzu. American Journal of Applied Sciences. 2010; 7(4):453-458.

[38] Akpabio E. M, Water meanings, sanitation practices and hygiene behaviours in the cultural mirror: a perspective from Nigeria. Journal of Water, Sanitation and Hygiene for Development. 2012;02 (3): 168-181.

[39] FGN. National Water Supply and Sanitation Policy. Department of Water Supply and Quality Control, Federal Ministry of Water Resources. Federal Republic of Nigeria. 2000.

[40] Ojekunle I. A, Transport and Urban Environmental Quality in Nigeria in Contemporary to AD 2000, (Edited by Osuntokun, A.; Aworawo and F. Masajuwa), Frankad Publishers, Lagos. 2000

[41] Ayeni A. O, Soneye A.S.O, Balogun I. I, The Arab World Geographer, 2009;12(1-2) 95–104.

[42] Akpabio, Emmanuel M. Water Supply and Sanitation Services Sector in Nigeria: The Policy Trend and Practice Constraints. 2012.

[43] WHO/UNICEF. Joint monitoring programme for water supply and sanitation. Meeting the MDG drinking water and sanitation target: mid-term assessment of progress. WHO; Geneva: UNICEF, New York. 2010

[44] National Bureau of Statistics (NBS). The multiple indicator cluster survey 2007. National Bureau of Statitics, Abuja, Nigeria. 2007

[45] Onabolu, B, Jimoh O. D, Igboro S. B, Sridhar M. K. C, . Onyilo G, Gege A. Ilya R, Source to point of use drinking water changes and knowledge, attitude and practices in Katsina State, northern Nigeria. Physics and Chemistry of the Earth. 2011;36: 1189-1196.

[46] Raimi OM, Vivien OT, Adedoyin OO. The sources of water supply, sanitation facilities and hygiene practices in oil producing communities in central senatorial district of Bayelsa state, Nigeria. MOJ Public Health. 2018; 7(6):304–312. DOI: 10.15406/mojph.2018.07.00265

[47] Raimi MO, Pigha Tarilayun K. Water-Related Problems and Health Conditions in the Oil Producing Communities in Central Senatorial District of Bayelsa State. Imperial Journal of Interdisciplinary Research. 2017;3(6):780–809.

[48] Amakom U. Sanitation sector status and gap analysis: Nigeria, a report. 2009

[49] Akoteyon, I. S, Omotayo A. O, Soladoye O, Olaoye H. O. Determination of Water Quality Index and Suitability of Urban River for Municipal Water Supply in Lagos-Nigeria European Journal of Scientific Research. 2011;54 (2) : 263-271

[50] Samuel O, Olasoji N. O, Oyewole B. A, Joshua N, Edokpayi. Water Quality Assessment of Surface and Groundwater Sources Using a Water Quality Index Method: A Case Study of a Peri-Urban Town in Southwest, Nigeria Environments. 2019;6:23; doi: 10.3390/environments6020023

[51] Ahaneku I. Edwin and Animashaun I. Murtala. Determination of water quality index of river Asa, Ilorin, Nigeria Adv. Appl. Sci. 2013;4 (6):277-284

[52] Oshurhe O, Origho T, Ohwohere-Asuma, O, Ewhuwhe-Ezo J. A Comparative Assessment of Water Quality Index (WQI) and Suitability of River Ase for Domestic Water Supply in Urban and Rural Communities in Southern Nigeria. International Journal of Humanities and Social Science, 2014; 4(1): 234-245.

[53] Leizou K. E, Nduka J. O, Verla A. W. Evaluation of Water Quality Index of the Brass River, Bayelsa State, South-South, Nigeria. International Journal of Research Granthaalayah, 2017;5(8): 277-287.

[54] Ezeilo F. E, Oba K. M. Evaluation of Water Quality Index of Orashi River, Rivers State, Nigeria. International Journal of Environmental Issues, 2016; 12(1-2): 60-75.

[55] Ovrawah L, Hymore FK. Quality of Water from Hand-Dug Wells in the Warri Environs of Niger Delta Region. African Journal of Environmental Studies. 2001; 2:16–17.

[56] Aigberua A, Tarawou T. Water Quality Index (WQI) Assessment along Inland Fresh Waters of Taylor Creek in Bayelsa State, Nigeria Journal of Environmental Treatment Techniques 2019;7(3): 260-269

[57] Chinwendu Emeka, Bright Nweke, Jecinta Osere and Chimaobi K. Ihunwo. Water Quality Index for the Assessment of Selected Borehole Water Quality in Rivers State European Journal of Environment and Earth Sciences 2020; 1(6):1-4

[58] Etim E.E, Odoh R, Itodo,A.U, S, Umoh S.D, Lawal U.Water Quality Index for the Assessment of Water Quality from Different Sources in the Niger Delta Region of Nigeria Frontiers in Science 2013;3(3): 89-95 DOI: 10.5923/j.fs.20130303.02

[59] Asibor G, Oborakpororo O, Surface Water Quality Assessment of Warri Metropolis Using Water Quality Index . International Letters of Natural Sciences. 2019;74: 18-25

[60] Nwachukwu R, Ekere V. E, Agbazue B. U. Ngang Janefrances N. Ihedioha . Hydrochemistry and Water Quality Index of groundwater resources in Enugu north district, Enugu, Nigeria Environmental Monitoring and Asse ssment. 2019;191. DOI:10.1007/s10661-019-7271-0

[61] Edokpayi J.N, Rogawski, E.T, Kahler D.M, Hill C.L, Reynolds C, Nyathi E, Smith J.A, Odiyo J.O, Samie A, Bessong P. Challenges to Sustainable Safe Drinking Water: A Case Study of Water Quality and Use across Seasons in Rural Communities in Limpopo Province, South Africa. 2018; 10: 159.

[62] Akoteyon I. S. Evaluation of groundwater quality using water quality indices in parts of Lagos – Nigeria. Journal of Environmental Geography, 2013;6(1-2): 29-36.

[63] Olagbemide PT. Application of Water Quality Index for the Assessment of Eleyele Lake, Ibadan, Nigeria Archives of Applied Science Research, 2017;9 (3): 16-23

[64] Ahaneku I. E, Murtala A. I. Determination of water quality index of river Asa, Ilorin, Nigeria Adv. Appl. Sci. 2013; 4(6):277-284

[65] Ogbozige FJ, Adie DB, Igboro SB, Giwa A. Evaluation of the Water Quality of River Kaduna, Nigeria Using Water Quality Index J. Appl. Sci. Environ. Manage. 2017; 21 (6) 1119-1126

[66] Opaluwa, O.D, Mohammed Y, Mamman S, Ogah A.T, Ali D, Assessment of Water Quality Index and Heavy Metal Contents of Underground Water Sources in Doma Local Government Area, Nasarawa State,

Nigeria Asian Journal of Applied Chemistry Research. 2020;6(3): 27-40

[67] Oko O.J, Aremu M.O, Odoh R, Yebpella G, Shenge G.A. Assessment of Water Quality Index of Borehole and Well Water in Wukari Town, Taraba State, Nigeria. Journal of Environmental and Earth Sciences. 2014; 4(3):336–344.

Chapter 15

The Effect of Wastewater Treatment Methods on the Retainment of Plastic Microparticles

Rana Zeeshan Habib, Ruwaya al Kindi and Thies Thiemann

Abstract

Microplastics as plastic pieces of ≤5 mm in size, are found in most ecosystems, both terrestrial and aquatic. Many of the microplastics find their way into the environment through the wastewater. For this reason, a knowledge of the microplastic retainment performance of wastewater treatment plants of various design is important. In this regard, several wastewater treatment processes have been studied, including new methods that are still at the development stage. This manuscript reviews the literature on such wastewater treatment methods and their ability to retain microplastics.

Keywords: microplastics, wastewater treatment plants, wastewater treatment method, plastic retainment

1. Introduction

A number of reviews have appeared on the topic of wastewater treatment and microplastic retainment [1–17]. Over the last 10 years, the understanding of microplastics and their impact on the environment has developed as have the analytical techniques to identify and quantify microplastics. In this respect, the focus has shifted to even smaller plastic particles dubbed "nano plastics". The identified sources of micro(nano)plastics have increased to include secondary plastics created by such mundane processes as opening a package [18] or making tea using plastic tea bags [19]. On the other hand, the importance of the individual sources of primary microplastics has shifted, with the ban of plastic microbeads in cosmetics coming into effect in many regions [20–25], changing the attention more to microtyres [26, 27], synthetic fibers [28, 29] and to secondary micro- and nano plastics [30]. There will be a shift of the sources of secondary microplastics as the ban in certain regions of plastic bags [31–33] and single use plastics [34, 35] comes into effect, as both are potential materials for microplastics due to subsequent degradative fragmentation processes. Plastics already existing in the environment degrade very slowly [36]. Furthermore, the examination of food articles and drinking water [37–40] for micro- and nano plastics has increased, as micro- and nano plastics have been found in foods and drink as diverse as table salt [41], soft drinks [42], beer [43, 44], and meat [45].

Microplastics (MPs) are defined as plastic particles of ≤5 mm in size [46–48]. For smaller particles, of size ≤1 μm, the term nanoparticles (NPs) is often used [46, 49]. Some authors define NPs as particles of up to 100 nm in size [50]. Plastic particles include polymeric films and synthetic fibers. Plastic microparticles come from different sources. They can be degraded and fragmented materials from tires (tires and road-wear) [51], clothing [52, 53], plastic bags [30] and packaging [18], where larger pieces of plastic are exposed to wear or weathering [54]. These are secondary MP. Primary MP are materials that are produced industrially at this small size. These include solid micropellets in cosmetic formulations, such as in facial cleaners and body scrubs [20, 55], microspherules in toothpastes (2–5 μm in size) [56], microparticles in washing powder/detergents [57, 58] and scrubbers used for air-blasting surfaces to remove paints and rust [59, 60] in paints and coatings themselves [58], and in drilling fluids in oil and gas exploration [1]. Drug delivery systems have used plastic micro−/nanoparticles, also – these are often biodegradable materials [61]. The amounts of materials used as primary MP and secondary MP stemming from the degradation of meso- and macroplastics on-land have been estimated in different studies commissioned by different European countries [62–65] and by the European Community [66]. Often, sediments of water bodies [67], especially oceans [68, 69], and terrestrial soil are some of the places where MPs may end up when released into the environment. There are a number of ways that MPs can enter the world's oceans that include direct run-offs into the oceans or into rivers that lead to oceans. Additionally, atmospheric transfer of MPs [70], which has been largely neglected until relatively recently, has been found to contribute to the accumulation of MPs in rivers, lakes [71] and oceans [72]. Terrestrial acquisition of MPs in soils can also happen in a number of ways that again includes atmospheric transport, but can also occur through fertilizer and even irrigation water [73]. Plastic mulching also contributes [74]. In both the dispersal of MPs to the aquatic and the terrestrial environment wastewater treatment plants (WWTPs) play a major role. On the one hand, WWTPs play a major part in retaining MPs from the sewage water, on the other hand, MPs can enter the soil through the application of sewage sludge [75, 76]. In the treatment of wastewater, WWTPs themselves can become point sources of MPs [77–80], releasing MPs into the receiving water. Thus, many examples have been found where the concentration of MPs downriver of a WWTP was higher than upriver. As the volume of wastewater is bound to increase over the years with an increase of population, new methods of wastewater treatment are being developed that help retain MPs better. This comes against the background of studies that assess the retaining capabilities of different treatment methods in existing WWTPs. Both are topic of the current review.

2. Studies of retaining microplastics in existing wastewater treatment plants

2.1 Standard functional units of WWTPs

WWTPs are of different design and of different sizes (**Figure 1**). In general, most WWTPs start off with passing the wastewater through bar screens (screening) to remove large solids and an oil and grit removal tank as pre-treatment, before the water is left to settle in a clarifier, where solids are removed in form of sludge sunk to the bottom and in form of scum on the water surface as primary wastewater treatment. This primary treatment is followed by activated sludge treatment as secondary wastewater treatment. The treatment uses flocs of microorganisms that decrease the BOD (biological oxygen demand) of the water due to a decrease of

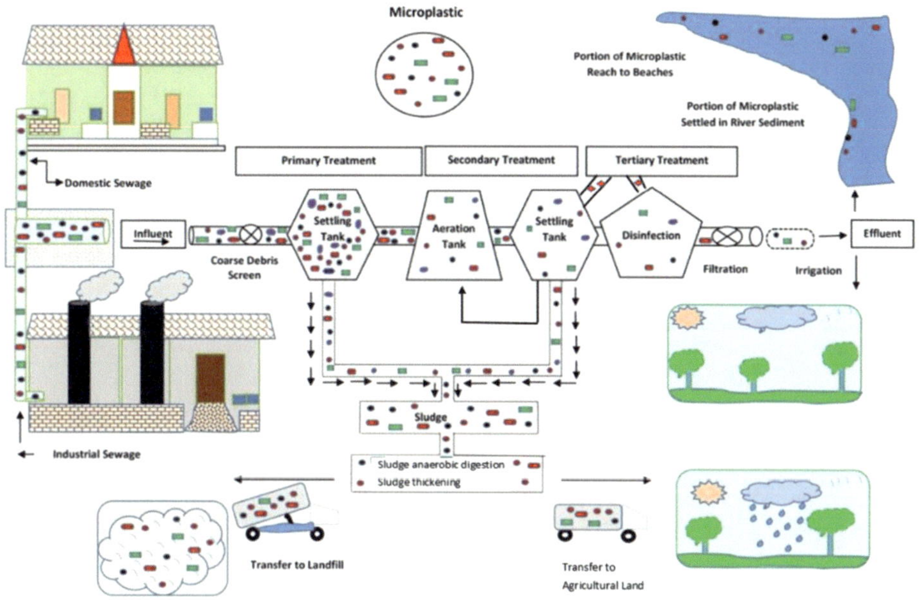

Figure 1.
Schematic flow of microplastic (MP) particles through a wastewater treatment plant (WWTP).

organic components through conversion to CO2 along with a decrease in nitrogen content by conversion of ammonia bound nitrogen to elemental nitrogen through a nitrification process and a subsequent denitrification process, involving both heterotrophic as well as autotrophic microorganisms and both aerobic and anoxic reaction zones. The activated sludge is separated from the treated water in a subsequent clarifier where some of the activated sludge is recycled. Activated sludge treatment can be conducted in a number of different ways. It is also possible to operate this process in a sequential batch reactor (SBR), where bioreactor and clarifier are run in a timed sequence within one vessel. Tertiary treatment methods can be added. They can include rapid sand filtration, membrane filtration, reverse osmosis, advanced oxidation processes and further biological methods [81]. Finally the treated water is disinfected, either by UV irradiation, addition of ozone (O3), of chlorine (Cl2) to produce hypochlorous acid (HOCl) or of chlorine dioxide (ClO2) [82], before it is discharged, mostly into a natural receiving water body such as a lake, river or the ocean. Especially in countries with extreme water scarcity or in large metropolises the water can also be recycled directly for consumption/use [83].

2.2 Early MP retainment studies in WWTPs

More than 70 WWTPs have been studied to date as to their MP retaining capability [1]. Some of these studies were devoted to the assessment of microplastic accumulation in the sludge of the WWTPs [8]. While early studies were carried out in North America [80, 84–90], Europe [78, 90–106], and Australia [79, 107], a larger number of recent studies emanate from East Asia [108–123]. It must be noted that some WWTPs were handling sewerage and storm run-off separately, others were not. In some of the cases where sewerage and storm run-off were handled together, specific note was made of black fragments in the influent that derived from tire abrasion in form of microtires [124]. **Table 1** gives an overview of many of the published studies on MP retention efficiency of WWTPs around the world [77–80, 84–88, 90, 92–94, 97–100, 103–144].

Ref.	MP conc. in influents	MP conc. in effluent	WWTP type	Overall retention/ efficiency	Country
Conley et al., 2019 [77]	147, 126, 146 MP/L	3.7, 17.6 and 17.2 MP/L	Primary and secondary	97.6, 85.2, 85.5%	USA
Talvitie et al., 2015 [78]	610 MP/L	13.5 MP/L (incl. all textile fibers)	Primary, secondary and tertiary	97.6%	Finland Viikinmäki
Browne et al., 2011 [79]	n.a	90 MP/L	Primary, secondary and tertiary	n.a.	Australia
Carr et al., 2016 [80]	1 MP/L	1 MP/L	Primary, secondary and tertiary	95–99%	USA
Carr et al., 2016 [80]	1.10×10^9 MP/day (681 million L./day)	0.88 MP/m^3	Secondary and Tertiary	99.9%	USA
Mason et al., 2016 [84]	n.a	0.05 MP/L	17 WWTPs, Tertiary	n.a.	USA
Michielssen et al., 2016 [85]	367 MP/L	0.5 MP/L	Tertiary (AnMBR)	99.4%	USA (Northfield)
Michielssen et al., 2016 [85]	133.0 MP/L	5.9 MP/L	Primary and secondary	93.8%	USA (Detroit)
Michielssen et al., 2016 [85]	367 MP/L	2.6 MP/L	Primary, secondary and tertiary	97.2%	USA (Northfield)
Estahbanadi and Fahrenfeld 2016 [86]	n.a	0.028 to 0.44 MP/L	Primary and secondary	n.a.	USA
Sutton et al., 2016 [87]	n.a	0.086 MP/L	Primary and secondary	n.a.	USA
Dyachenko et al., 2017 [88]	n.a	0.02 MP/L	Primary, secondary and tertiary	n.a.	USA
Gies et al., 2018 [90]	31.1 MP/L	0.5 MP/L	Primary and secondary	98.3%	Canada
Lares et al., 2018 [92]	57.6 MP/L	1.05 MP/L	Primary and secondary	98.3%	Finland (Mikkeli)
Mintenig et al., 2014 and 2017 [93, 94]	n.a	0.1 to 10.1 MP/L	12 WWTPs: mostly secondary and tertiary	97%	Germany (Oldenburg)
Talvitie and Heinonen 2014 [97]	627 MP/L in addition to 3160 black particles/L	23 MP/L In addition to 125 black particles/L	Secondary	96.1%	Russia (St. Petersburg)

DOI: http://dx.doi.org/10.5772/intechopen.97083

Ref.	MP conc. in influents	MP conc. in effluent	WWTP type	Overall retention/ efficiency	Country
Dris et al., 2015 [98]	n.a	14 to 50 MP/L	Secondary	83–95%	France
Magnusson and Norén 2014 [99]	1.5 X 10^4 MP/m^3 3.2 X 10^6 MP/h	1.77 X 10^3 MP/h	Secondary	>99%	Lysekil (Sweden)
Murphy et al., 2016 [100]	15.70 MP/L	0.25 MP/L	Secondary	98.4%	UK
Leslie et al., 2017 [103]	73 MP/L	9 to 91 MP/L	7 WWTPs	72%	Netherlands
Kalčiková et al., 2017 [104]	n.a	0.021 MP/L	Primary (Mechanical and Biological)	87%	Slovenia
Simon et al., 2018 [105]	7216 MP/L	54 MP/L	—	98.3%	Denmark
Wisniowska et al., 2018 [106]	$19.4 \cdot 10^3$ to $552.2 \cdot 10^3$ MP/1 m^3	0.028 to 0.96 MP/L	n.a.	95–99%	Poland
Ziajahromi et al., 2017 [107]	n.a	0.28 MP/L	Primary, secondary and tertiary	92–99%	Australia
Yang et al., 2019 [108]	12.03MP/L	0.59 MP/L	Primary and secondary	95%	China (Beijing)
Long et al., 2019 [109]	1.57–13.69 MP/L	0.20–1.73 MP/L	Primary and secondary	97.8%	China
Xu et al., 2019 [110]	196.00 MP/L	9.04 MP/L	Primary and secondary	97.2%	China (Changzhu)
Lv et al., 2019 [111]	0.28 mp/L	0.13 and 0.05 MP/L	n.a.	MBR 99.5%,	China
Liu et al., 2019 [112]	80 MP/L	28.4 MP/L	Primary and secondary	64.4%	China
Ren et al., 2020 [113]	16.0 MP/L	2.9 MP/L	Primary, secondary and tertiary	81.9%	China (Zhengzhou)
Wei et al., 2020 [114]	430–2154 MP/m^3	430–2154 MP/m^3	RD-WWTFs	84%	China (Hangzhou)
Tang et al., 2021 [115]	23.3 MP/L and 80.5 MP/L	23.3 to 7.9 MP/L	Primary and secondary	66.1 and 62.7%,	China (Wuhan City)
Nguyen et al., 2021 [116]	n.a	n.a	Primary, Secondary and tertiary	80%	Korea (Seoul)

Ref.	MP conc. in influents	MP conc. in effluent	WWTP type	Overall retention/ efficiency	Country
Yuan et al., 2020 [117]	n.a	10.30 MP/L, 6.10 MP/L	Primary, Secondary and tertiary	97.67% and 98.46%	China (Nanjing)
Park et al., 2020 [118]	10 to 470 MP/L	10 to 470 MP/L	Primary, Secondary and tertiary	98.7–99.99%	Korea
Zhou et al., 2020 [119]	54,100 MFs/L	537.5 MFs/L (MF)	Primary, Secondary and tertiary	85%	China (Keqiao industrial park)
Mak et al., 2020 [120]	n.a	10,816 MP/m^3	Primary, Secondary	86.4%	Hong Kong (Victoria Harbor)
Zou et al., 2020 [121]	n.a	1.719 ± 1.035 MP/L	n.a	n.a	China (Guangzhou)
Hidayaturrahman et al., 2020 [122]	13813 MP/L	132 MP/L	Primary, Secondary and tertiary	> 98%	South Korea (Daegu)
Talvitie et al., 2017 [125]	6.9	0.005 MP/L	4 tertiary WWTPs	99.9%	Finland
Magni et al., 2019 [126]	2.5 MP/L	0.4 MP/L	Primary, secondary and tertiary	84%	Italy
Bayo et al., 2020 [127]	15.70 MP/L	13.04 MP/L	Primary	90.3%	Spain (Cartagena)
Gűndoğdu et al., 2018 [128]	4,825,697/day	7.02 MP/L	Secondary	73%	Turkey (Seyhan)
Gűndoğdu et al., 2018 [128]	2,040,639/day	4.11 MP/L	Secondary	79%	Turkey (Yűreğir)
Bayo et al., 2019 [129]	15.70 MP L^{-1}	0.25 MP/L	Primary	90.3%	Spain (Cartagena)
Blair et al., 2019 [130]	3 and 10 MP L^{-1}	<1 and 3 MP/L	Tertiary	96%	UK
Wolff et al., 2019 [131]	n.a	59 and 30 MP/L	Primary and secondary	n.a	Germany
Ziajahromi et al., 2021 [132]	n.a	22.1×10^6 to 133×10^6 per day	n.a	99.8–98.2%	Australia
Petroody et al., 2020 [133]	12667 MP/m^3	12667 ± 668, 3514 ± 543 and 423 ± 44.9 MP/m^3	n.a	96.7%	Iran (Sari)
Edo et al., 2020 [134]	n.a	12.8 ± 6.3 MP/L	Primary and secondary	>90%	Spain (Madrid)

Ref.	MP conc. in influents	MP conc. in effluent	WWTP type	Overall retention/ efficiency	Country
Ben-David et al., 2021 [135]	28.28 MP/L	1.97 MP/L	Primary, Secondary and tertiary	97%	Israel (Karmiel)
Tagg et al., 2020 [136]	n.a	1.5 MP/L	Primary, Secondary and tertiary	76.9%	UK (East Midlands)
Akarsu et al., 2020 [137]	1.1 and 3.6 MP/L	0.9 MP/L	Primary, Secondary and tertiary	55–97%	Turkey Mersin Bay
Naji et al., 2021 [138]	74 (±11.01, SD) and 67 (±18.35, SD) MP 35/L	70.66 MP/L	Primary and secondary	n.a	Iran (Bandar Abbas City)
Rajala et al., 2020 [139]	n.a .	0.[illegible] mg/L, 6.7 mg/L (used)	Secondary	99.4%	Finland
Alvim et al., 2020 [140]	n.a	11.1 MP/L	Primary, Secondary	n.a	Spain (Valencia)
Pittura et al., 2021 [141]	(12,170,000 MP/h) 3.6 MP/L	1,730,000 MP/h	Primary, Secondary	94%	Italy
Raju et al., 2020 [142]	11.80 ± 1.10 MP/L	2.76 ± 0.11 MP/L	Secondary	76.61%	Australia (New south wales Hunter Region)
Ferreira et al., 2020 [143]	n.a	0.24 ± 0.07 MP/m^3 (Laucala Bay) and 0.09 ± 0.02 MP/m^3 (Suva Harbour)	79 WWTPs'	N.a	Fiji (Suva)
Schmidt et al., 2020 [144]	n.a	$4 * 10^0$ and $4.5 * 10^5$ MP/m^3	Secondary	n.a	Germany

Table 1.
Published studies on the microplastic (MP) retention efficiency of different WWTPs.

In the following, we describe the outcome of some of the earlier studies (2012–2016) in differently-sized WWTPs in various parts of the world. In 2014, Magnusson and Norén [99] studied the MP retention in a smaller WWTP in Lysekil, Sweden (Långeviksverket, serving 14.000 inhabitants, flow rate of 5160 m^3/day). During the study time, the WWTP received per m^3 15.1 ± 0.89·10^3 MP (10.7 ± 0.39·X 10^3 plastic fibers; 2.67 ± 0.77 X 10^3 plastic fragments; 1.78 ± 0.80·X 10^3 plastic flakes). Of these, only 8.25 ± 0.85 MP m^{-3} (4.00 ± 0.58 plastic fibers, 3.75 ± 1.25 plastic fragments, 0.50 ± 0.50 plastic flakes) could be found in the effluent, which was released into the sea. This amounted to 99.96% retention for plastic fibers. While MPs in the effluent were still appreciably higher than in the receiving water, Magnusson and Norén [99] found a steady decrease in fiber concentrations with increasing distance from the discharge point, from 1.82 ± 0.45 fibers/m^3 at 20 m from the discharge point to 1.14 ± 0.38 fibers/m^3 at 200 m from the discharge point. A larger sized WWTP was studied in 2012 by J. Talvite et al. [78] at Viikinmäki, Finland (serving 800.000 inhabitants in the Helsinki metropolitan area, flow rate 270.000 m^3/day). The influent carried 180 textile fibres L^{-1} and 430 synthetic particles L^{-1}. After the primary sedimentation, the wastewater contained an average of 14.2 (±0.7) fibres and 290.7 (±28.2) synthetic particles L^{-1}. After the secondary sedimentation, 13.8 (±1.6) fibres and 68.6 (±6.3) synthetic particles were still present. The remainder of the fibers and particles had settled in the sludge. Thus, most of the fibers were eliminated in the primary sedimentation process; most of the other synthetic particles were removed in the second sedimentation. As a tertiary stage, the WWTP also included a biological filtration, which removed further particles from the treated water. Removal of fibers in the second and third treatment stages was insignificant. The final effluent carried 4.9 (±1.4) fibres and 8.6 (±2.5) synthetic particles L^{-1}. This means that 3.73 x 10^9 fibers were released daily, with the effluent [78] into the Gulf of Finland, Baltic Sea. In 2014, Talvitie and Heinonen [97] published a study on the Central WWTP of Vodokanal in St. Petersburg, Russia (serving about 4 million people, flow rate of 959.000 m^3/day), carried out in collaboration with HSY, Vodokanal of St. Petersburg and Water Research and Control Center. The influent was found to carry 467 fibers L^{-1}, 160 synthetic particles L^{-1}, in addition to 3160 less identified black particles L^{-1}, most likely of synthetic nature. After pre- and primary treatment, these values decreased to 33 fibers L^{-1}; 21 synthetic particles L^{-1}, and 302 black particles L^{-1}. After, secondary treatment, these values decreased further to 16 fibers L^{-1}; 7 synthetic particles L^{-1}, and 125 black particles L^{-1}. Nevertheless, still 153.4 X 10^9 fibers were released daily with the effluent, some of which lastly will reach the Baltic Sea. Murphy et al. studied the MP retaining capability of a secondary WWTP in UK. Here, the influent contained on average 15.70 (±5.23) MP·L^{-1}. This was reduced to 0.25 (±0.04) MP·L^{-1} in the final effluent, which is a decrease of 98.4% [100]. The team reported that about 45% of microplastics were removed in the grease and grit tank, while primary sedimentation in the first clarifier accounted for 34% removal [97]. 20% of the microplastics were removed in the secondary stage [100]. In 2016, M.R. Michielssen et al., investigated the retainment efficiency towards small anthropogenic litter (SAL) of a Detroit WWTP (Great Lakes Water Authority, serving 2.36 million people, flow rate 2.5 million m^3/day). SAL includes both plastic based and cellulose derived materials. Pretreatment at the Detroit WWTP removed 58.6% SAL. Primary treatment retained an additional 25.5%, secondary treatment (with activated sludge) an additional 9.7% SAL for a total of 93.8% SAL removal overall [85]. Nevertheless, the effluent was found to release about 8.94 billion fibers a day [85]. M.R. Michielssen et al. also looked at the much smaller Northfield WWTP (Michigan, serving 100.000 people, flow rate 1700 m^3 day^{-1}), which features sand filtration as a tertiary treatment [85]. Here, pretreatment was found to retain

35.1% SAL. Primary and secondary treatment held back a further 53.3% and 1.4% SAL, respectively. Sand filtration as a tertiary treatment method further reduced SAL by 7.4% of the total. Overall, the Northfield WWTP reduced the SAL load by 97.2%, with 8.9 million fibers released daily with the effluent [85]. In addition, M.R. Michielssen et al. studied the effect of an anaerobic membrane (AnMBR) test reactor, situated at the Northfield WWTP, as a stage directly after the pretreatment of the influent. Here, 99.4% of the inflowing MP were retained – this made for 64.5% of the total MP in the influent. Thus, pretreatment and AnMBR retained 99.6% of the MP in total [85]. Finally, in 2014–2015, Carr et al. [80] studied 7 tertiary WWTPs and one secondary WWTP in Los Angeles County, Southern California. The studied WWTPs had a gravity filtration as the tertiary stage. Bench studies of the group with water spiked with microplastics showed that all MPs could be retained by such filtration processes. This, however, did not hold up in the" real-life" scenario of the WWTPs. Nevertheless, in this study, Carr et al. showed that gravity filtration as a tertiary stage in a WWTP can give up to 99.9% MP retention, calculated over all stages [80].

2.3 Retention of MPs in preliminary and primary treatment stages of WWTPs

P.U. Iyare et al. [5] showed that the bulk of the removal of MPs, at an average of 72%, comes during pre- and primary treatment. Dris et al. reported 69% MP retention in the pre- and primary stages of a WWTP in Paris [98]. Gies et al. saw that the MP concentration decreased from 31.1 ± 6.7 MP L^{-1} in the influent to 2.6 ± 1.4 MP L^{-1} (91.7% MP retention efficiency) in the primary effluent of a major secondary WWTP near Vancouver, Canada [90]. Michielssen et al. reported that screening and primary sedimentation removed 84–88% SAL ([85], see above) in studied US WWTPs. From WWTPs in Russia [97], Finland [78] and Canada [87] it was found that pre- and primary wastewater treatment removed 92–93% of fibres. In 2015, Ziajahromi et al. performed one of few studies on a WWTP with solely a pre- and primary treatment stage, receiving wastewater from over 1 million inhabitants in the Sydney area [107]. The pre- and primary stages were standard screening (mesh size of 5 mm), grit removal and sedimentation, with the effluent discharged in the deep ocean. Here, 1.5 MP L^{-1} were detected in the effluent. With a through-put of 300.000 m^3 day^{-1}, 460 million MP day^{-1} were being discharged from the WWTP into the ocean [107].

2.4 Retention of MPs in the secondary treatment stage of WWTPs

Different studies have looked at the MP retention in the secondary treatment stages of WWTPs, where the MP is then collected in the accumulating sludge. In most cases, the secondary treatment in a WWTP involves an activated sludge process, where different bacteria lower both the organic content as well as the nitrogen content of the water. Due to the different requirements of the bacteria, some of which are autotrophs and some of which are heterotrophs, some operating under aerobic conditions, some under anoxic conditions, the process involves different stages, where the water passes through aerated zones to anoxic zones and back. Different set-ups for such processes have been developed, involving different reaction chambers or a single batch reactor, where the different stages of the process run sequentially in time. After the process, the water needs to be separated from the sludge. This may be through passing the mixture to a settling tank or through a membrane in form of microfiltration or ultrafiltration, where the activated sludge is passed back to the bioreactor. This combination then is called a membrane bioreactor (MBR). MBRs have also be seen as a separate entity as a tertiary treatment method [125].

H. Lee and Y. Kim [124] looked at the efficiency of three different types of activated sludge processes, the A2O (anaerobic-anoxic-aerobic), the sequence batch reactor (SBR) and the Media process. As of 2013, these were the main processes used at public WWTPs in South Koreas with a capacity of over 500m^3/day [A_2O 23.7%, SBR 34.8%, Media 22.8%] with membrane bioreactor (MBR), long term aeration, and special microbial processes accounting for the remainder of the processes [124]. The WWTP running the A2O process, with a throughput of 35.000 m^3 day^{-1} (serving 67.700 inhabitants) reduced 29.85 MP L^{-1} found in the influent (taken before the pre-treatment) to 0.435 MP L^{-1} in the effluent after disinfection (98.5 overall retention efficiency), with 14.9 MP g^{-1} found in the ensuing sludge. The WWTP running the SBR process, with a throughput of 110.000 m^3 day^{-1} (serving 235.700 inhabitants) reduced 16.45 MP L^{-1} in the original influent to 0.14 MP L^{-1} in the final effluent (99.1% overall retention efficiency), with 9.65 MP g^{-1} noted in the sludge. Finally, the WWTP running the Media process, with a throughput of 130.000 m^3 day^{-1} (serving 245.000 inhabitants), reduced 13.86 MP L^{-1} in the original influent to 0.29 MP L^{-1} in the final effluent (98% overall retention efficiency), with 13.2 MP g^{-1} found in the sludge [124]. Based on the MP found in the sludge, the retention efficiencies of the secondary treatment alone were 49.3%, 44.7%, and 49.0% for the A2O process, the SBR process, and the Media process, respectively [121]. It is not clear, if the primary settling tank is included in these numbers as the fate of the sludge from the settling tank has not been discussed. If included, the numbers would compare well with the numbers given by Murphy et al., who reported a 53.8% MP removal in the primary settler and the secondary stage. Still, in the South Korean WWTPs 3 billion, 4 billion, and 11 billion MPs were discharged annually with the final effluent in the A2O process, in the SBR process, and in the Media process, respectively.

2.5 Retention of MPs in the tertiary treatment stage of a WWTP

Tertiary treatment methods have been studied extensively in regard to MP retention. In this regard, sand and gravel filtration is a common tertiary treatment, and quite a few of the early papers looking at the retaining capability of tertiary WWTPs also investigated the performance of such filters. In 2015, New York State had authorized a study of its WWTPs and found that some WWTPs using filtration processes in their tertiary stage still released MPs. The study focused on plastic microbeads, and here it can be said that certain WWTPs using membrane microfiltration, continuous backwash up flow dual sand (CBUDS) microfiltration or rapid sand filtration indeed did not show any plastic microbeads in the effluent at the time. Data on the retention of synthetic fibers was not released, however [120]. Two other studies came from New York State at that time, both citing release of microplastics downstream from WWTPs [84, 145]. One of the studies looked at a WWTP in Western New York State (Lake Erie) (12.000 people served, flow rate 13.000 m^3/day) using granular filtration (sand/anthracite coal) as the tertiary stage, with 0.009 MP/L found in the effluent, leading to a release of 101.000 MP/day, 68% of which were fibers [84]. The effluents of three tertiary WWTPs in the San Francisco Bay area with sand filtration or sand/anthracite coal filtration were found to have higher loadings with 0.064, 0.092, and 0.127 MP/L [84], leading the largest of the WWTPs to release more than 9.6 million MP/day [84]. Also, a later study from a WWTP in Northern Italy [126] showed that sand filtration as the tertiary stage with an overall MP retention rate of 84% can still lead to significant releases in the order of 160 million MPs day^{-1}. For a WWTP in the Murcia region, Spain (serving 29.800 people; flow rate: 12.000 m^3/day) J. Bayo et al. gave a 75.5% MP retention rate for the gravity rapid sand filtration, with 3 sand filters installed in parallel [127]. Here, it was noted that RSF could retain plastic microparticulates (95.5%) better than synthetic fibers

(53.8%) [126]. In contrast, in cases. Rapid sand filtration has also been shown to lead to very significant reduction of MP concentration in the effluent, e.g., from 0.7 (±0.1) to 0.02 MP (±0.007) MP L^{-1} (97% MP retention) at the Kakolanmäki WWTP (Turku Region Waste Water Treatment Plant) in Turku, Southern Finland [122]. On the downside, there has been a report of fragmentation of MP material during sand filtration [146].

Another filtering technique uses disc filters (DF) as a final polishing step, removing particles from biologically treated wastewater. Disc filters are made of a stack of round filter meshes in a closed tank, where the filter mesh is a woven material, made of polypropylene, polyester, or polyamide with a pore size of 10–40 μm. The sludge cake formed from the retained particles is periodically removed by high-pressure back-flushing. M. Simon et al. looked [147] at the efficiency of DF in a WWTP in Grindsted, Denmark (flow rate 10.040 m^3/day). The effluent from the secondary clarifier was noted to carry 20 mg L^{-1} suspended solid. This was reduced by DF to 3–8 mg L^{-1}. When passed through DF of a pore size of 18 μm, the MP content could be reduced from 29 MP L^{-1} to 3 MP L^{-1} (89.7% removal efficiency). Talvitie et al. looked at the filtration of the secondary effluent through a pilot-scale disc filter (Hydrotech HSF 1702-1F) consisting of two discs each composed of 24 filter panels at the Viikinmäki WWTP, located in Helsinki, Finland. Here, DF-10 (10 μm pore size) decreased the MP concentration from 0.5 (± 0.2) to 0.3 (± 0.1) (40% removal efficiency) and DF-20 (20 μm pore size) from 2.0 (± 1.3) to 0.03 (± 0.01) (98% removal efficiency). The results were noted to fluctuate from trial to trial [125].

There are a number of membrane filtration techniques. However, MP removal through micro- and ultrafiltration (UF) has been studied less frequently. Often, UF is used in combination with coagulation and can be used as a secondary or tertiary treatment method. Polymeric or ceramic membranes with a pore size between 1 and 100 nm are used, laid out to retain large organic molecules such as proteins as well as bacteria, protozoa, and viruses. UF is not specifically designed to retain micro- or nano-plastics. UF membranes can be fouled easily. To that effect, a coagulation step as pretreatment with iron-based coagulants has been advocated, especially in combination with an addition of polyacrylamide (PAM), which has been reported to increase the removal efficiency of small-sized polyethene particles ($d < 0.5$ mm) significantly from 13 to 91% [148, 149]. UF can also be used as a pretreatment for a reverse osmosis (RO) separation (see below) to protect the RO membrane. Nevertheless, fouling of membranes due to meso-particles, where MP have the same size, continues to be a problem [150].

An alternative membrane separation technique is that using dynamic membranes (DMs). DMs operate with a layer formed on a supporting membrane by particles in the influent. So, these particles in the influent create a filtration layer that can be supported by a larger pore-sized mesh or by low-cost porous materials. DMs have been run successfully with particles that are of a similar size to microplastics [151].

Reverse osmosis (RO) is the process filtering water from a region of high solute concentration through a semipermeable membrane to a region of low-solute concentration by applying a pressure larger than the osmotic pressure. RO units are used in desalination plants but are also used in drinking water treatment plants and in some WWTPs. Ziajahromi et al. [107] have looked at a WWTP in the Sydney area operating with a reverse osmosis (RO) unit (13.000 m^3 day^{-1}) as a tertiary treatment. Here, the MP concentration decreased from 2.2 MP L^{-1} in the primary effluent to 0.21 MP L^{-1}, after the reverse osmosis (RO) process. This still leads to a discharge of 10 million MP day^{-1} into the tributary of a major urban river in Australia. It is thought that the occurrence of larger sized pores on the membrane, the membrane material and other membrane imperfections may contribute to the passage of the MP through the membrane [107].

Finally, dissolved air flotation (DAF) as a flocculation process can be used as a tertiary treatment method. It was found to remove 95% of MP remaining from the secondary treatment [125]. In this case, dissolved air flotation (DAF) was studied as a full-scale tertiary treatment at Paroinen WWTP (Hameenlinna Region Water Supply and Sewerage Ltd) located in city of Hameenlinna, Southern Finland. In DAF, water is saturated with air at high pressure and then pumped to a flotation tank at 1 atm, forming dispersed water. The formed air bubbles (typically 20–70 μm in size) in the dispersed water adhere to the suspended solids causing them to float to the surface, from which they are removed by skimming. The process necessitates only a small retention time of the treated water. At the Paroinen WWTP, before the flotation, flocculation chemical polyaluminum chloride was added to the wastewater with a dosage of 40 mg L^{-1} to enhance flocculation [125]. Y. Wang et al. studied DAF with three common types of MP in freshwater and found the hydrophilic-hydrophobic interaction not to be ideal for an efficient separation of MPs without additives, citing a removal of 32–38% of MPs, only. The efficiency could be increased by 13.6–33.7%, however, with two additives that modified the surface of the air bubbles [152].

3. Conclusion

Microplastic is a serious pollutant in our aquatic and terrestrial ecosystems. WWTPs play a major role in limiting the dispersal of MP in the environment. Nevertheless, as waste streams flow through WWTPs, these in turn become point sources of MPs, where MPs are released in the millions into rivers, lakes and lastly into the sea. Studies on different WWTPs around the world have given a good indication of the retaining efficiency of different wastewater treatment stages and methods. Usually, a large part of MP is retained in the preliminary and primary treatment stages. However, the amount of MP released in the final effluent is often a function of the tertiary treatment method used. In this regard, a further development of membranes and techniques used in combination with membranes for the filtration of MP seems of interest.

Author details

Rana Zeeshan Habib[1], Ruwaya al Kindi[1] and Thies Thiemann[2*]

1 Department of Biology, College of Science, UAEU, Al Ain, UAE

2 Department of Chemistry, College of Science, UAEU, Al Ain, UAE

*Address all correspondence to: thies@uaeu.ac.ae

References

[1] R. Habib, T. Thiemann, R. al Kendi. Microplastics and wastewater treatment plants. J. Water Res. Protect. 2020;**12**:1-35.

[2] S. Uddin, S. W. Fowler, M. Behbehani. An assessment of microplastic inputs into the aquatic environment from wastewater streams, Mar. Pollut. Bull. 2020;**160**:111538.

[3] A. Cristaldi, ,M. Fiore, P. Zuccarello, G. O. Conti, A. Grasso, I. Nicolosi, C. Copat, M. Ferrante. Efficiency of wastewater treatment plants (WWTPs) for microplastic removal: a systematic review. Int. J. Environ. Res. Public Health. 2020;**17**:8014.

[4] *D. Sol*, A. Laca, A. Laca, M. Díaz. Approaching the environmental problem of microplastics: Importance of WWTP treatments. Sci. Total Environ. 202**0**;**740**;140016.

[5] P.U. Iyare, S.K. Ouki, T. Bond. Microplastics removal in wastewater treatment plants: a critical review. Environ. Sci.-Water Res. & Techn. 2020;**6**:2664-2675.

[6] X.T. Bui, T.D.H. Vo, P.T. Nguyen, V.T. Nguyen, T.S. Dao, P.D. Nguyen. Microplastics pollution in wastewater: Characteristics, occurrence and removal technologies. Environ. Technol. Innov. 2020;**19**:101013.

[7] S. Freeman, A. M. Booth, I. Sabbah, R. Tiller, J. Direking, K. Klun, A. Rotter, E. Ben-David, J. Javidpour, D.L. Angel. Between source and sea: The role of wastewater treatment in reducing marine microplastics. J. Environ. Managem. 2020;**266**:110642.

[8] Z.Q. Zhang, Y.G. Chen. Effects of microplastics on wastewater and sewage sludge treatment and their removal: A review. Chem. Eng. J. 2020;**382**:122955.

[9] P.L. Ngo, B.K. Pramanik, K. Shah, R. Roychand. Pathway, classification and removal efficiency of microplastics in wastewater treatment plants. Environ. Pollut. 2020;**255**:113326.

[10] X. Zhang, J. Chen, J. Li. The removal of microplastics in the wastewater treatment process and their potential impact on anaerobic digestion due to pollutants. Chemosphere 2020;251:126360.

[11] M. Wu, W. Tang, S. Wu, H. Liu, C. Yang. Fate and effects of microplastics in wastewater treatment process. Sci. Total Environ. 2021;757:143902.

[12] Y. Hu, M. Y. Gong, J.Y. Wang, *A. Bassi*. Current research trends on microplastic pollution from wastewater systems: a critical review. Rev. Environ. Sci. Technol. 2019;**18**:207-230.

[13] G. Gatidou, O.S. Arvaniti, A.S. Stasinakis. Review on the occurrence and fate of microplastics in Sewage Treatment Plants. J. Hazard. Mat. 2019;**367**:504-512.

[14] J. Sun, X.H. Dai, Q.L. Wang, M.C.M. van Loosdrecht, B.J. Ni. Microplastics in wastewater treatment plants: Detection, occurrence and removal. Water Res. 2019;**152**:21-37.

[15] S. Raju, M. Carbery, A. Kuttykattil, K. Senathirajah, S.R. Subashchandrabose, G. Evans, P. Thavamani, Transport and fate of microplastics in wastewater treatment plants: implications to environmental health. Rev. Environ. Sci. Biotechnol. 2018;**17**:637-653.

[16] M. Enfrin, L. Dumée, J. Lee. Nano/microplastics in water and wastewater treatment processes – orgin, impact, and potential solutions. Water Res. 2019;161:621-638.

[17] R. Z. Habib, R. al Kindi, T. Thiemann. The effect of wastewater treatment plants on retainment of plastic microparticles to enhance water quality– a review. J. Environ. Protect., submitted.

[18] Z. Sobhani, Y. Lei, Y.H. Tang, L.W. Wu, X. Zhang, R. Naidu, M. Mgharaj, Microplastics generated when opening plastic packaging. Sci. Rep. 2020;10:4841.

[19] L. M. Hernandez, E.G. Xu, H.C.E. LLarsson, R. Tahara, V.B. Maisura, N. Tufenkji, Plastic teabags release billions of microparticles and nanoparticles into tea. Environ. Sci. Technol. 2019, 53, 12300-12310

[20] R. Z. Habib, M. Abdoon, R. Al Meqbaali, F. Ghebremedhin, M. Elkashlan, W. F. Kittaneh, N. Cherupurakal, A.-H. I. Mourad, T. Thiemann, R. Al Kindi. Analysis of Microbeads in Cosmetic Products in the United Arab Emirates. Environ. Pollut., 2020;**258**:113831.

[21] J.P. McDevitt, C.S. Criddle, M. Morse, R.C. Hale, C.B. Bott, C.M. Rochman. Addressing the issue of microplastics in the wake of the microbead-free Waters Act - a new standard can facilitate improved policy. Environ. Sci. Technol. 2017; **51**:6611-6617.

[22] United States. Microbead-Free Waters Act of 2015. Pub.L. 114-114 (Dec. 28th, 2015).

[23] P. Dauvergne. The power of environmental norms: marine plastics: pollution and the politics of microbeads. Environ. Pol. 2018;**27**:579-597.

[24] C. Guerranti, T. Martellini, G. Perra, C. Scopetani, A. Cincinelli, Microplastics in cosmetics: Environmental issues and needs for global bans. Environ. Toxicol. Pharmacol. 2019;68: 75-79.

[25] L. Anagnostia, A. Varvaresou, P. Pavlou, E. Protopapa, V. Carayanni. Worldwide actions against plastic pollution from microbeads and microplastics in cosmetics focusing on European policies. Has the issue been handled effectively? Mar. Pollut. Bull. 2021;**162**:111883.

[26] B. Baensch-Baltruschat, B. Kocher, F. Stock, G. Reifferscheid, Tyre and road wear particles (TRWP) - A review of generation, properties, emissions, human health risk, ecotoxicity, and fate in the environment. Sci. Total Environ., 2020;**733**:137823.

[27] P.J. Kole, A.J. Löhr, F.G.A.J. Van Belleghem, A.M.J. Ragas Wear and tear of tyres: a stealthy source of microplastics in the environment. Int. J. Environ. Res. Public Health, 2017;**14**:1265-1296.

[28] F.S. Cesa, A. Turra, J. Baroque-Ramos. Synthetic fibers as microplastics in the marine environment: A review from textile perspective with a focus on domestic washings. Sci. Total Environ. 2017;**598**:1116-1129.

[29] R. Rathinamoorthy, S.R. Balasaraswath. A review of the current status of microfiber pollution research in textiles. Int. J. Cloth. Sci. Technol. 2020. DOI: 10.1108/IJCST-04-2020-0051.

[30] T. O'Brine, R.O. Thompson, T. Brine. Degradation of plastic carrier bags in the marine environment. Mar. Pollut. Bull. 2010;60:2279-2283.

[31] K. Borkowski. Plastics waste litter in oceans as a driving force for regulations plastics. Polimery. 2019;**64**: 759-763.

[32] T.D. Nielsen, K. Holmberg, J. Stripple.Need a bag? A review of public policies on plastic carrier bags – Where, how and to what effect? Waste Managem. 2019;**87**:428-440.

[33] M. Kasidoni, K. Moustakas, D. Malamis. The existing situation and challenges regarding the use of plastic carrier bags in Europe. Waste Managem. & Res. 2015;**33**;419-428.

[34] Y. Chen, A.K. Awasthi, F. Wei, Q. Tan, J. Li. Single-use plastics: Production, usage, disposal, and adverse impacts. Sci. Total Environ. 2021;**752**:141772.

[35] K.E.K. Vimal, K. Mathiyazhagan, V. Agarwal, S. Luthra, K. Sivakumar. Analysis of barriers that impede the elimination of single-use plastic in developing economy context. J. Clean. Prod. 2020;**272**:122629.

[36] A. Chamas, H. Moon, J. Zheng, Y. Qiu, T. Tabassum, J.H. Jang, M. Abu-Omar, S.L. Scott, S. Suh. Degradation rates of plastics in the environment. ACS Sustainable Chem. Eng. 2020;**8**:3494-3511.

[37] D. Elkhatib, V. Oyanedel-Carver. A critical review of extraction and identification methods of microplastics in wastewater and drinking water. Environ. Sci. Technol. 2020;54:7037-7049.

[38] K. Novotna, L. Cermakova, L. Pivokonska, T. Cajthaml, M. Pivokonsky. Microplastics in drinking water treatment – current knowledge and research needs. Sci. Total Environ. 2019;667:730-740.

[39] M. Shen, B. Song, Y. Zhu, G. Zheng, Y. Zhang, Y. Yang, X. Wen, M. Chen, H. Yi. Removal of microplastics via drinking water treatment: current knowledge and future directions. Chemosphere 2020;251:126612.

[40] D. Eerkes-Medrano, H.A. Leslie, B. Quinn. Microplastics in drinking water: a review and assessment of emerging concern. Curr. Opin. Environ. Sci. Health, 2018;7:69-75.

[41] M. Renzi, A. Blašković. Litter & microplastics features in table salts from marine origin: Italian versus Croatian brands. Mar. Pollut. Bull. 2018;**135**:62-68.

[42] V.C. Shruti, F.Pérez-Guevara, I. Elizalde-Martínez, G. Kutralam-Muniasamy. First study of its kind on the microplastic contamination of soft drinks, cold tea and energy drinks - Future research and environmental considerations. Sci. Total Environ. 2020;**726**:138580.

[43] D.W. Lachenmeier, J. Kocareva, D. Noack, T. Kuballa. Microplastic identification in German beer - an artefact of laboratory contamination? Deutsche Lebensmittel-Rundschau: Zeitschrift für Lebensmittelkunde und Lebensmittelrecht 2015;**111**:437-440.

[44] G. Liebezeit, E. Liebezeit. Synthetic particles as contaminants in German beers. Food Addit. Contam. Part A Chem. Anal. Control Expo Risk Assess. 2014;**31**:1574-1578.

[45] R. Z Habib, R. al Kindi, T. Thiemann, unpublished results.

[46] A.L. Andrady. Microplastics in the marine environment. Mar. Pollut. Bull. 2011;**62**: 1596-1605.

[47] D. Eerkes-Medrano, R.C. Thompson, D.C. Aldridge. Microplastics in freshwater systems: a review of emerging threats, identification of knowledge gaps and prioritization of research needs. Water Res.2015;**75**: 63-82.

[48] A.A. Koelmans, E. Besseling, W.J. Shim. Nanoplastics in the aquatic environment. Critical Review Marine Anthropogenic Litter. Springer International Publishing, 2015:325-340.

[49] A. Cózar, F. Echevarría, J.I. González-Gordillo, X. Irigoien, B. Úbeda, S. Hernández-Leon, A.T.

Palma, S. Navarro, J. Garcia-de-Lomas, A. Ruiz, M.L. Fernández-de-Puelles, C.M. Duarte. Plastic debris in the open ocean. Proc. Natl. Acad. Sci. 2014;**111**:10239-10244.

[50] A. Ter Halle, L. Jeannau, M. Martignac, E. Jardé, B. Pedrono, L. Brach, J. Gigault. Nanoplastic in the North atlantic subtropical gyre. Environ. Sci Technol. 2017;**51**:13689-13697.

[51] M. Bondelind, A. Nguyen, E. Sokolova, K. Björklund. Transport of Traffic-Related Microplastic Particles in Receiving Water. Mannina G. Springer International Publishing, 2009:317-321. https://doi.org/10.1007/978-3-319-99867-1_53.

[52] K.J. Kapp, R.Z. Miller. Electric clothes dryers: An underestimated source of microfiber pollution. PLOS One 2020;15: e0239165.

[53] I.E. Napper, R. C. Thompson. Release of synthetic microplastic plastic fibres from domestic washing machines: Effects of fabric type and washing conditions. Mar. Pollut. Bull. 2016;**112**:39-45.

[54] S. Kubowicz, A.M. Booth. Biodegradability of plastics: challenges and misconceptions. Environ. Sci. Technol. 2017;51:12058-12060.

[55] Q. Sun, S.Y. Ren, H.G. Ni. Incidence of microplastics in personal care products: An appreciable part of plastic pollution. Sci. Total Environ. 2020;**742**:140218.

[56] G.S. Ustabasi, A. Baysal. Occurrence and risk assessment of microplastics from various toothpastes. Environ. Monitor. Assessm. 2019;191:438.

[57] European Chemicals Agency (ECA) Annex to the annex XV restriction report - microplastics, 2019 https://echa.europa.eu/documents/10162/db081bde-ea3e-ab53-3135-8aaffe66d0cb.

[58] A. Scudo, B. Liebermann, C. Corden, D. Tyrer, J. Kreissig, O. Warwick. Intentionally added microplastics in products – final report. Amec Foster Wheeler Environment & Infrastructure UK Limited, 2017. https://edi3-fs1/shared/data/projects/39168 ppchem microplastics restriction/c000 client/reports/final report/39168_intentionally added microplastics - final report_20171020_clean.docx

[59] M. Cole, P. Lindeque, C. Halsband, T.S. Galloway. Microplastics as contaminants in the marine environment: a review. Mar. Pollut. Bull. 2011;**62**:2588-2597.

[60] S. Sharma, S. Chatterjee. Microplastic pollution, a threat to marine ecosystem and human health: a short review. Environ. Sci. Pollut. Res. 2017;**24**:21530-21547.

[61] Y. Wang, P. Li, T. Truong-Dinh Tran, J. Zhang, L. Kong. Manufacturing techniques and surface engineering of polymer based nanoparticles for targeted drug delivery to cancer. Nanomaterials. 2016;**6**:26.

[62] P. Sundt, P-E. Schulze, F. Syversen, F. Sources of microplastic pollution to the marine environment. Report no M-321/2015, Mepex for the Norwegian Environment Agency (Miljødirektoratet, 86) 2014/2015.

[63] K. Magnusson, K. Eliasson, A. Fråne, K. Haikonen, J. Hultén, M. Olshammar, J. Stadmark, A. Voisin, A. Swedish sources and pathways for microplastics to the marine environment. Report C183, Swedish Environmental Research Institute, Stockholm, 2016. (revised 2017).

[64] C. Lassen, S.F. Hansen, K. Magnusson, F. Norén, N.I.B. Hartmann, P.R. Jensen, T.G. Nielsen, A. Brinch.

Microplastics – occurrence, effects and sources of releases to the environment in Denmark, Environmental project No. 1973. Danish Ministry of the Environment – Environmental Protection Agency (Denmark) 204, 2015.

[65] R. Essel, L. Engel, M. Carus, R.H. Ahrens. Sources of microplastics relevant to marine protection in Germany. Texte 64/2015 Project No. 31969 Report No. (UBA-FB) 002146/E, Federal Environment Agency, Germany, 2015.

[66] European Commission (DG Environment) Intentionally added microplastics in products: final report (European Commission, Brussels) Doc Ref. 39168 Final Report 17271i3. 2017.

[67] G. Peng, B. Zhu, D. Yang, L. Su, H. Shi, D. Li. Microplastics in sediments of the Changjiang estuary. China. Environ. Pollut. 2017;**225**:283-290.

[68] E. Cunningham, S.M. Ehlers, J.T.A. Dick, J.D. Sigwart, K. Linse, J.J. Dick, K. Kiriakoulakis. High abundances of microplastic pollution in deep-sea sediments: evidence from Antarctica and the Southern Ocean. Environ. Sci. Technol. 2020;**54**:13661-13671.

[69] J. Barrett, Z. Chase, J. Zhang, M.M.B. Holl, K. Willis, A. Williams, B.D. Hardesty, C. Wilcox. Microplastic pollution in deep-sea sediments from the Great Australian Bight. Front. Mar. Sci. 2020. https://doi.org/10.3389/fmars.2020.576170

[70] Y. Zhang, S. Kang, S. Allen, D. Aleen, T. Gao, M. Sillanpää. Atmospheric microplastics: A review on current status and perspectives. Earth-Sci. Rev. 2020:**203**;103118.

[71] S. Allen, D. Allen, V.R. Phoenix, G. Le. Roux, P. Durańtez Giminéz, A. Simmoneau, S. Binet, D. Galop, Atmospheric transport and deposition of microplastics in a remote mountain catchment. Nature Geoscience. 2019;**12**:339-344.

[72] N. Evangeliou, H. Grythe, Z. Klimont, C. Heyes, S. Echkhardt, S. Lopez-Aparicio, A. Stohl. Atmospheric transport is a major pathway of microplastics to remote regions. Nature Commun. 2020:**11**;3381.

[73] J-J. Guo, X.P. Huang, L. Xiang, Y.Z. Wang, Y.W. Li, H. Li, Q.Y. Cai, C.H. Mo, M.H. Wong. Source, migration and toxicology of microplastics in soil. Environ. Int. 2020;**137**:105263.

[74] Y. Huang, Q. Liu, W. Jia, C. Yan, J. Wang. Agricultural plastic mulching as a source of microplastics in the terrestrial environment. Environ. Pollut. 2020;**260**:114096.

[75] P. van den Berg, E. Huerta-Lwanga, F. Corradini, V. Geissen. Sewage sludge application as a vehicle for microplastics in eastern Spanish agricultural soils. Environ. Pollut. 2020;**261**:114198.

[76] Q. Li, J. Wu, X. Zhao, X. Gu, R. Ji. Separation and identification of microplastics from soil and sewage sludge. Environ. Pollut. 2019;**54**:113076.

[77] K. Conley, A. Clum, J. Deepe, H. Lane, B. Beckingham. Wastewater treatment plants as a source of microplastics to an urban estuary: removal efficiencies and loading per capita over one year. Water Res. X 2019;**3**:100030.

[78] J. Talvitie, M. Heionen, J.P. Pääkkönen. E. Vahtera, A. Mikola, O. Setälä, R. Vahala. Do wastewater treatment plants act as a potential point source of microplastics? Preliminary study in the coastal Gulf of Finland, Baltic Sea. Water Sci. Technol. 2015;**72**:1495-1504.

[79] M. A. Browne, P. Crump, S. J. Niven, E. Teuten, A. Tonkin, T. Galloway,

R. Thompson. Accumulation of microplastic on shorelines worldwide: sources and sinks. Environ. Sci. Technol. 2011;**45**: 9175-9179

[80] S.A. Carr, J. Liu, A. G. Tesoro. Transport and fate of microplastic particles in wastewater treatment plants. Water Res., 2016;**91**:174-182.

[81] S. Majumder, Poornesh, M.B. Reethupoonar, R. Mustafa. A review on working, treatment, and performance evaluation of sewage treatment plant. Int. Eng. Res. Appl. 2019;**9**:41-49.

[82] M.M. Amin, H. Hashemi, A.M. Bovini, Y.T. Hung. A review on wastewater disinfection. Int J. Env. Health Eng. 2013;2:22.

[83] A.N. Angelakis, T. Asano, A. Bahri, B.E. Jiminez, G. Tchobanoglous. Water reuse: from ancient to modern times and the future. Front. Environ. Sci. 2018;11: https://doi.org/10.3389/fenvs.2018.00026.

[84] S.A. Mason, D. Garneau, R. Sutton, Y. Chu, K. Ehmann, J. Barnes, P. Fink, D. Papazissimos, D.L. Rogers. Microplastic pollution is widely detected in US municipal wastewater treatment plants effluent. Environ. Pollut. 2016;**218**: 1045-1054.

[85] M.R. Michielssen, E.R. Michielssen, J. Ni, M.B. Duhaime. Fate of microplastics and other anthropogenic litter (SAL) in wastewater treatment plants depends on the unit processes involved. Environ. Sci. Water Res. Technol., 2016;**2**:1064-1073.

[86] S. Estahbani, N.L. Fahrenfeld. Influence of wastewater treatment plant discharges on microplastic concentrations in surface water. Chemosphere. 2016;**162**:277-284.

[87] R. Sutton, S. A. Mason, S. K. Stanek, E. Willis-Norton, I.F. Wren, C. Box. Microplastic contamination in the San Francisco bay, California, USA. Mar. Pollut. Bull. 2016;**109**:230-235.

[88] A. Dyachenko, J. Mitchell, N. Arsem. Extraction and identification of microplastic particles from secondary wastewater treatment plant (WWTP) effluent. Anal. Meth., 2017;**9**:1412-1418.

[89] K.A. Zubris, B.K. Richards. Synthetic fibers as an indicator of land application of sludge. Environ. Pollut. 2005;138:201-211.

[90] E.A. Gies, J.L. LeNoble, M. Noël, A. Etemadifar, F. Bishay, E.R. Hall, P.S. Ross. Retention of microplastics in a major secondary wastewater treatment plant in Vancouver, Canada. Mar Pollut Bull 2018;**133**:553-561.

[91] A. M. Mahon, B. O'Connell, M. G. Healy, I. O'Connor, R. Officer, R. Nash, L. Morrison. Microplastics in sewage sludge: effects of treatment. Environ. Sci. Technol. 2017;51:810 – 818.

[92] M. Lares M.C. Ncibi M. Sillanpää, M. Sillanpää. Occurrence, identification and removal of microplastic particles and fibers in conventional activated sludge process and advanced MBR technology. *Water Res.*, 2018;**133**:236-246.

[93] S.M. Mintenig, I. Int-Veen, M. Löder, G. Gerdts. Mikroplastik in ausgewählten Kläranlagen des Oldenburgisch- Ostfriesischen Wasserverbandes (OOWV) in Niedersachsen. Probenanalyse mittels Mikro-FTIR Spektroskopie, 2014. [MikroPlastik in selected sewage treatment plants of the Oldenburgisch-Ostfriesischen Water Board in Niedersachsen. Sample analysis using micro-FTIR spectroscopy] Alfred-Wegener-Institut, Helmholtz-Zentrum für Polar- und Meeresforschung (AWI), Biologische Anstalt Helgoland (in German).

[94] S. M. Mintenig, I. Int-Veen, M.G.J. Löder, S. Primpke, G. Gerdts. Identification of microplastic in effluents of waste water treatment plants using focal plane array-based micro-Fourier-transform infrared imaging. Water Res., 2017 (Suppl. C);**108**:365-372.

[95] J. Talvitie, A. Mikola, O. Setälä, M. Heinonen, A. Koistinen. How well is microlitter purified from wastewater?–a detailed study on the stepwise removal of microlitter in a tertiary level wastewater treatment plant. Water Res., 2017;**109**:164-172.

[96] J. Talvitie. Wastewater treatment plants as pathways of microliter to the aquatic environment, Doctoral thesis, Aalto University, Department of Built Environment. Helsinki. 2018;106. ISBN 978-952-60-7980-6.

[97] J. Talvitie, M. Heinonen. Preliminary study on synthetic microfibers and particles at a municipal waste water treatment plant, HELCOM BASE Project – Implementation of the Baltic Sea Action Plan in Russia, 2014.

[98] R. Dris, J. Gasperi, V. Rocher, M. Saad, N. Renault, B. Tassin. Microplastic contamination in an urban area: a case study in Greater Paris. Environ. Chem. 2015;**12**:592-599.

[99] K. Magnusson, F. Norén, Screening of microplastic particles in and down-stream a wastewater treatment plant. Report C55, Swedish Environmental Research Institute, Stockholm, 2014.

[100] F. Murphy, C. Ewins, F. Carbonnier, B. Quinn. Wastewater Treatment Works (WwTW) as a source of microplastics in the aquatic environment. Environ. Sci. Technol., 2016;**50**:5800-5808.

[101] A. Lusher, R. Hurley, C. Vogelsang, L. Nizzetto, M. Olsen. Mapping microplastics in sludge, Technical Report L.NR. 7215-2017 (NIVA), 2018.

[102] S. Sujathan, A.-K. Kniggendorf, A. Kumar, B. Roth, K. H. Rosenwinkel, R. Nogueira. Heat and bleach: a cost-efficient method for extracting microplastics from return activated sludge, Arch. Environ. Contam. Toxicol. 2017;73:641-648.

[103] H.A. Leslie, S.H. Brandsma, M.J.M. van Velzen, A.D. Vethaak. Microplastics en route: field measurements in the Dutch river delta and Amsterdam canals, wastewater treatment plants, North Sea sediments and biota. Environ. Int. 2017;**101**:133-142.

[104] G. Kalčíková, B. Alič,, T. Skalar, M. Bundschuh, A. Žgajnar Gotvajn. Wastewater treatment plant effluents as source of cosmetic microbeads to freshwater. Chemosphere, 2017;**188**:25-31.

[105] M. Simon, N. van Alst, J. Vollertsen. Quantifiaction of microplastic mass and removal rates at wastewater treatment plants applying Focal Plane Array (FPA) – based Fourier Transform Infrared (FT-IT) imaging. Water Res. 2018;**142**:1-9.

[106] E. Wisniowska, K. Moraczewska-Majkut, W. Nocon. Efficiency of microplastics removal in selected wastewater treatment plants - preliminary studies. Desalination Water Treatm. 2018;**134**:316-323.

[107] S. Ziajahromi, P. A. Neale, L. Rintoul, F.D.L. Leusch. Wastewater treatment plants as a pathway for microplastics: development of a new approach to sample wastewater-based microplastics. *Water Res.*, 2017 (Suppl C);**112**:93-99.

[108] L. Yang, K. Li, S. Cui, Y. Kang, L. An, K. Lei. Removal of microplastics in municipal sewage from China's largest

water reclamation plant. Water Res. 2019;**155**:175-181.

[109] Z. Long, Z. Pan, W. Wang, J. Ren, X. Yu, L. Lin, H. Lin, H. Chen, X. Jin. Microplastic abundance, characteristics, and removal in wastewater treatment plants in a coastal city of China. Water Res. 2019;**155**:255-265.

[110] X. Xu, Y. Jian, Y. Xue, Q. Hou, L. Wang. Microplastics in the wastewater treatment plants (WWTPs): Occurrence and removal. Chemosphere 2019;**235**:1089-1096.

[111] X. Lv, Q. Dong, Z. Zuo, Y. Liu, X. Huang, W.M. Wu. Microplastics in a municipal wastewater treatment plant: Fate, dynamic distribution, removal efficiencies, and control strategies. J. Clean. Prod. 2019;**225**:579-586.

[112] X. Liu, W. Yuan, M. Di, Z. Li, J. Wang. Transfer and fate of microplastics during the conventional activated sludge process in one wastewater treatment plant of China. Chemical Engineering Journal, 2019;**362**:176-182.

[113] P.J. Ren, M. Dou, C. Wang, G.Q. Li, R.P. Jia. Abundance and removal characteristics of micro¬plastics at a wastewater treatment plant in Zhengzhou. Environmental and Science Pollution Research, 2020;**27**:36295-36305.

[114] S. Wei, H. Luo, J. Zou, J. Chen, X. Pan, D.P.L. Rousseau, J. Li. Characteristics and removal of microplastics in rural domestic wastewater treatment facilities of China. Sci. Total Environ. 2020;**739**:139935.

[115] N. Tang, X. Liu, W. Xing. Microplastics in wastewater treatment plants of Wuhan, Central China: Abundance, removal, and potential source in household wastewater. Sci. Total Environ. 2020;**745**:141026.

[116] N.B. Nguyen, M.-K. Kim, Q.T. Le, D.N. Ngo, K.-D. Zoh, S.-W. Joo. Spectroscopic analysis of microplastic contaminants in an urban wastewater treatment plant from Seoul, South Korea. Chemosphere, 2021;**263**:127812.

[117] F. Yuan, H. Zhao, H. Sun, J. Zhao, Y. Sun. Abundance, morphology, and removal efficiency of microplastics in two wastewater treatment plants in Nanjing, China. Environ. Sci. Pollut. Res. 2020; https://doi-org.uaeu.idm.oclc.org/10.1007/s11356-020-11411-w.

[118] H.J. Park, M.-J. Oh, P.-G. Kim, G. Kim, D.H. Jeong, B.K. Ju, W.-S. Lee, H.M. Chung, H.J. Kang, J.H. Kwon. National reconnaissance survey of microplastics in municipal wastewater treatment plants in Korea. Environ. Sci. Techn. 2020;**54**:1503-1512.

[119] H. Zhou, L. Zhou, K. Ma. Microfiber from textile dyeing and printing wastewater of a typical industrial park in China: Occurrence, removal and release. Sci. Total Environ. 2020;**739**:140329.

[120] C.W.D. Mak, Y.Y. Tsang, M.L.M. Leung, J.K.H. Fang, K.M. Chan. Microplastics from effluents of sewage treatment works and stormwater discharging into the Victoria Harbor, Hong Kong. Mar.Pollut. Bull. 2020;**157**:111181.

[121] Y. Zou, C. Ye, Y. Pan. Abundance and characteristics of microplastics in municipal wastewater treatment plant effluent: a case study of Guangzhou, China. Environ. Sci. and Pollut. Res. 2020; https://doi-org.uaeu.idm.oclc.org/10.1007/s11356-020-11431-6.

[122] H. Hidayaturrahman, T.G. Lee. A study on characteristics of microplastic in wastewater in South Korea: identification, quantification, and fate of microplastics in treatment process. Mar. Pollut. Bull. 2019;**146**:696-702.

[123] NYS AOG, Discharging Microbeads to our Waters: An Examination of Wastewater Treatment Plants in New York, Paper Environmental Science: Water Research & Technology Open Access Article. Published on 14 October 2016. Downloaded on 2/13/2021 3:50:05 AM.

[124] H. Lee, Y. Kim. Treatment characteristics of microplastics at biological sewage treatment facilities in Korea. Mar. Pollut. Bull., 2018;**137**:1-8.

[125] J. Talvitie, A. Mikola, A. Koistinen, O. Setälä. Solutions to microplastic pollution – removal of microplastics from wastewater effluents with advanced wastewater treatment technologies. Water Res. 2017;**123**:401-407.

[126] S. Magni, A. Binelli, L. Pittura, C.G. Avio, C. della Torre, C.C. Parenti, S. Gorbi, F. Regoli. The fate of microplastics in an Italian wastewater treatment plant. Sci. Total Environ. 2019;652:602-610.

[127] J. Bayo, J. López-Castellanos, S. Olmos. Membrane bioreactor and rapid sand filtration for the removal of microplastics in an urban wastewater treatment plant. Mar. Pollut. Bull. 2020;**156**:111211.

[128] S. Gündoğdu, C. Çevik, E. Güzel, S. Kilercioğlu. Microplastics in municipal wastewater treatment plants in Turkey: a comparison of the influent and secondary effluent concentrations. Environ. Monit. Assess., 2018;**190**:626-632.

[129] J. Bayo, S. Olmos, J. López-Castellanos. Microplastics in an urban wastewater treatment plant: The influence of physicochemical parameters and environmental factors. Chemosphere 2019;**238**:124593.

[130] R.M. Blair, S. Waldron, C. Gauchotte-Lindsay. Average daily flow of microplastics through a tertiary wastewater treatment plant over a ten-month period. Water Res. 2019;**163**:114909.

[131] S. Wolff, J. Kerpen, J. Prediger, L. Barkmann, L. Müller. Determination of the microplastics emission in the effluent of a municipal wastewater treatment plant using Raman microspectroscopy. Water Res. X 2019;**2**:100014.

[132] S. Ziajahromi, P.A. Neale, I.T. Silveira, A. Chua, F.D.L.Leusch. An audit of microplastic abundance throughout three Australian wastewater treatment plants. Chemosphere 2021;**263**:128294.

[133] S.S.A. Petroody, S.H. Hashemi, C.A.M.van Gestel. Factors affecting microplastic retention and emission by a wastewater treatment plant on the southern coast of Caspian Sea. Chemosphere. 2020;261:128179.

[134] C. Edo, M. Gonzalez-Pleiter, F. Leganes, F. Fernandez-Pinas, R. Rosal. Fate of microplastics in wastewater treatment plants and their environmental dispersion with effluent and sludge. Environ. Pollut. 2020;**259**:113837.

[135] E. Ben-David, M. Habibi, E. Haddad, D,L. Angel, A.M. Booth, I. Sabbah. Microplastic distributions in a domestic wastewater treatment plant: Removal efficiency, seasonal variation and influence of sampling technique. Sci. Total Environ. 2021;**752**:141880.

[136] A.S. Tagg, M. Sapp, J.P. Harrison, C.J. Sinclair, E. Bradley, Y. Ju-Nam, J.J. Ojeda. Microplastic monitoring at different stages in a wastewater treatment plant using reflectance micro-FTIR imaging. Front. Environ. Sci. 2020. https://doi.org/10.3389/fenvs.2020.00145.

[137] C. Akarsu, H. Kumbur, K. Gödağ, A.E. Kideys, A. Sanchez-Vidal,

Microplastics composition and load from three wastewater treatment plants discharging into Mersin Bay, north eastern Mediterranean Sea. Mar. Pollut. Bull. 2020;**150**:110776.

[138] A. Naji, S. Azadkhah, H, Farahani, S. Uddin, F.R. Khan. Microplastics in wastewater outlets of Bandar Abbas city (Iran): A potential point source of microplastics into the Persian Gulf. Chemosphere 2021;**262**:128039.

[139] K. Rajala, O. Grönfors, M. Hesampour, A. Mikola. Removal of microplastics from secondary wastewater treatment plant effluent by coagulation/flocculation with iron, aluminium and polyamine-based chemicals. Water Res. 2020;**183**:116045.

[140] C.B. Alvim, M.A. Bles-Piá, J.A. Mendoza-Roca. Separation and identification of microplastics from primary and secondary effluents and activated sludge from wastewater treatment plants. Chem. Eng. J. 2020;**402**:126293.

[141] L. Pittura, A. Foglia, C. Akyol, G. Cipolletta, M. Benedetti, F. Regoli, A.L. Eusebi, S. Sabbatini, L.Y.Tseng, E. Katsou, S. Gorbi, F. Fatone. Microplastics in real wastewater treatment schemes: Comparative assessment and relevant inhibition effects on anaerobic processes. Chemosphere, 2021;**262**:128415.

[142] S. Raju, M. Carbery, A. Kuttykattil, K. Senthirajah, A. Lundmark, Z. Rogers, Suresh SCB, G. Evans, T. Palanisami. Improved methodology to determine the fate and transport of microplastics in a secondary wastewater treatment plant. Water Res. 2020;**173**:115549.

[143] M. Ferreira, J. Thompson, *A. Paris*, D. Rohindra, C. Rico. Presence of microplastics in water, sediments and fish species in an urban coastal environment of Fiji, a Pacific small island developing state. Mar. Pollut. Bull. 2020;153:110991.

[144] C. Schmidt, R. Kumar, S. Yang, O. Büttner. Microplastic particle emission from wastewater treatment plant effluents into river networks in Germany: Loads, spatial patterns of concentrations and potential toxicity. Sci. Total Environ. 2020;737:139544.

[145] C. Martin, O. Eizhvertina, Quantitative Analysis of Microplastics in WWTP Effluent in the Niagara Region, Niagara College Canada, Niagara-on-the-Lake, Canada.Technical Report published for Niagara College Environmental Technician Field and Lab (co-op): Final Term Project, 2014.

[146] J. C. Prata. Microplastics in wastewater: state of knowledge on sources, fate and solutions, Marine Poll. Bull. 2018;**129**:262-265.

[147] M. Simon, A. Vianello, J. Vollertsen. Removal of > 10 μm microplastic particles from treated wastewater by a disc filter. Water 2019;**11**:1935.

[148] B. Ma, W. Xue, Y. Ding, C. Hu, H. Liu, J. Qu. Removal characteristics of microplastics by Fe-based coagulants during drinking water treatment. J. Environ. Sci. 2019;**78**:267-275.

[149] B. Ma, W. Xue, C. Hu, H. Liu, J. Qu, L. Li. Characteristics of microplastic removal via coagulation and ultrafiltration during drinking water treatment. Chem. Eng. J. 2019;**359**:159-167.

[150] Q. Ding, H. Yamamura, N. Murata, N. Aoki, H. Yonekawa, A. Hafuka, Y. Watanabe. Characteristics of meso-particles formed in coagulation process causing irreversible membrane fouling in the coagulation-microfiltration water treatment. Water Res. 2016;**101**:127-136.

[151] L. Li, G. Xu, H. Yu, J. Xing. Dynamic membrane for micro-particle removal in wastewater treatment: performance and influencing factors. Sci. Total Environ. 2018;**627**:332-340.

[152] Y. Wang, Y. Li, L. Tian, L. Ju, Y. Liu. The removal efficiency and mechanism of microplastic enhancement by positive modification dissolved air flotation. Water Environ. Res. 2020. https://doi.org/10.1002/wer.1352

Chapter 16

Microwave Digestion of Hazardous Waste Sludge in Geothermal Hot Waters by Char/Fly Ash Granule Composts-Hazardous Sludges and Industrial Waste Water Treatment

Yıldırım İsmail Tosun

Abstract

Most of the previous were regarding characteristics of sludge from urban/municipal activities concerning environmental issues on industrial sludge discharges causing fatal disasters in the lakes and water streams. The washing treatment of mud was searched. This research study concentrated over oxidative heavy metal dissolution and sterilization washing of muddy sludge of chemical, steel and copper refinery plants. The hazardous Hg and Pd contents using washing dissolution provided recovery of metals and treated sludge as the feedstock for digestion process. The research used hazardous sludge which is the by-product of the heat treated steel manufacturing process of CN baths and sludge from pulp washing industries. However, there is a sterilization washing by microwave radiation was reported on various sludge metal contamination characteristics in wastewater treatment stage. The results of a limited number of bench-scale sludge washing experiments conducted in the tube reactor study confirmed high radiation trends for washing dissolution with H2O2 in soil samples obtained from different locations in the north lake area of discharge of at the Plant Site. In general the contaminants in waste pond soils partitioned preferentially to the fine fraction of the soil (<150 μm however, the sand fraction (−0,5 mm + 150 μm) still contained significant contamination. These tests also showed that the heavy metal contaminants were highly dissolved at 45–76% in the wash water, which will reduce washing toxicity and improve metal recovers.

Keywords: fly ash, microwave radiation, salt slurries, metal sorption, energy toxic risk assessment, stochastic cost estimation, treatmen sorbent simulation, hybrid sorbent, waste sludge, salt slurries, Microwave activation waste water treatment, heavy metal, fly ash composts, shale

1. Introduction

All of the individual domestic wastewater streams contribute different amounts to the total nutrient and element potential beneficial to plant and farming. The contaminating seepages to streams should be control by local governors including the discharged areas for wastewater. However, industrial wastewater is commonly defined as wastewater from dairy factories, cheese factories, nut mills, pulp, paper, petrochemical flow, as well as industrial wastes such as various chemicals, salts and tanning acids, mining leachates. These sources vary widely in composition and often require special teritoriary processes to comply with discharge regulations. The simulation of hydrological flow discharge and fresh water wells and seepage contacts with irrigations is critical for agricultural farming and urbanization. The geotermal hot waters near Tigris River and even sulphide ore seepages, waste leachates of gold mining may deteriorate fresh water sources and agricultural land in the local area. Theuse of geothermal waters for precipitation of contaminated effluents with neutralizationh will protect the environment and agricultural fields in the South Eastern Anotalian region. The chemical analysis of the geothermal waters and given in **Table 1** [1]. The rivers, stream and flow waters in the South Eastern Anotalian region with contaminated area are illustrated in **Figure 1**.

Hazardous digestion of sludge tends to occur highly common as industrial waste streams or seepage, dissolved matters resulting from tanning or mining material oxidation of toxic species, react with geothermal bicarbonate and producing a strong hydroxyl precipitates, also neutralize the alkalinity of waste water streams as given in Eqs. (1)–(3). The basic alkali matter of geothermal waters may neutralize the acidic waste streams in production at higher pH levels over 5 with digested heavy metals in sulphide minerals. The neutralization by alkali matters govern the toxic seepage control by precipitation reactions in geothermal hot saline waters containing ammonia and bicarbonate as given below;

Derince Çayı	**BATMAN**
Batman Çayı	BATMAN
Dicle Nehri	BATMAN
Yanarsu Çayı (Garzan Çayı)	BATMAN
Dicle Nehri	BATMAN
Bitlis Çayı	SİİRT
Sutopu Deresi	BİTLİS
Ulu Çay (Botan Çayı)	SİİRT
Ulu Çay (Botan Çayı)	SİİRT
Dicle Nehri	SİİRT
Dicle Nehri	SİRNAK
Yukarısaksan Deresi	SİRNAK
Çığlı Suyu(Zap Suyu)	VAN
Çığlı Suyu(Zap Suyu)	HAKKARİ
Şemdinli Çayı	HAKKARİ

Table 1.
The streams and ground water area in the Şırnak, Batman and south eastern region [1, 2].

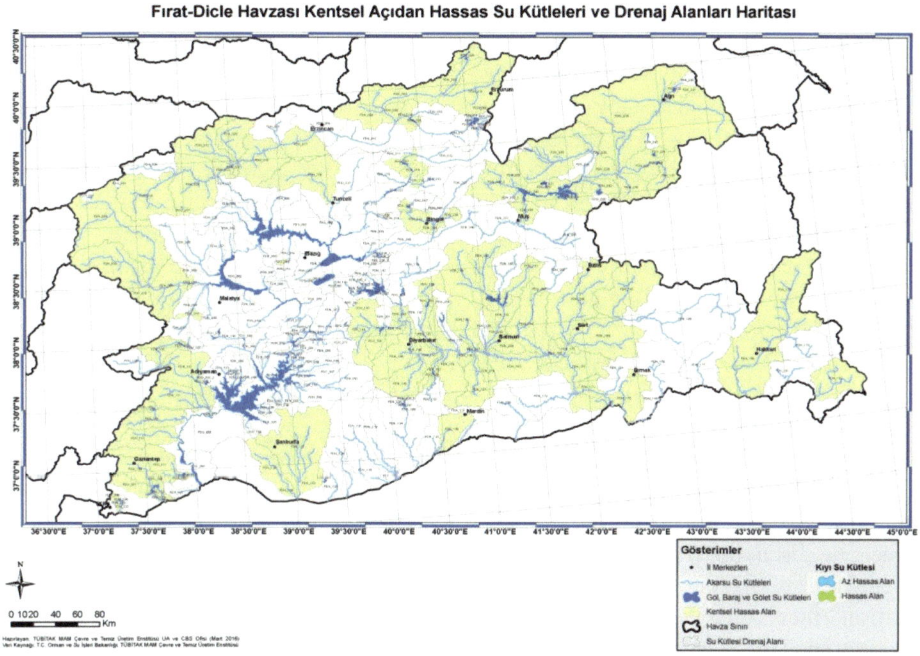

Figure 1.
Satellite view of precise and geothermal water areas in South eastern Anatolia, Euprate and Tigris Zone [2].

Bicarbonate and sulfite hot waters fall pH and hydrolysis as below.

$$MeSO_4^{=} + 3HCO_3^{=} + H_2O + H^+ \rightarrow Me^{2+}\ H_2S + 3CO_2 + 2H_2O \tag{1}$$

The sulfide matteris highly reactive to heavy metals as given Eqs. (2) and (4):

$$Fe^{2+} + H_2S \rightarrow FeS + 2H^+, \tag{2}$$

$$Zn^{2+} + H_2S \rightarrow ZnS + 2H^+, \tag{3}$$

which form very stable sulfide metals. The further reaction is an oxidation of S^{2-} to S^-, as given in Eq. (6):

$$FeS + S \rightarrow FeS_2 \tag{4}$$

generated in the late muds close to the settled mud - water interface. ZnS and PbS, in the soil are much stable and gives oxidation S in the −2 state to effluent. However, the sulfate part of reaction (2–3) may cause redox effect an oxidation. Then

$$H_2S + 4H_2O \rightarrow S_2O_3^2/SO_{42\_} + 10H^+ + 8e^-. \tag{5}$$

The oxidation electropotantial matter of waste waters provides toxic heavy metal dissociation to stream seepages near the mining leaching area. The geothermal waters provide the hydroxide precipitates in solid matter as mud. Even jarosite precipitates by sulfihite rich geothermal hot water streams area showing an output view in the land as redish brownish colors. However, the land may become dangareous with higher heavy metal contamination for fish farming and stream fishing. Batman province copper and lead sulphide deposits and hot streams of high sulphate come out high nitrate potential contamination of fresh waters soueces as given in **Figures 1** and **2** [2].

Some of the drinking water needs of Siirt Center, Kurtalan, Tillo (Aydınlar) Districts and Kayabağlar, Gökçebağ and Atabağı Districts are provided by natural spring water called Hesko in the countryside of Şirvan District and caisson wells on Botan Stream. In addition, there are underground water drillings for agricultural purposes, water drillings opened to meet the utility water needs of individual industrial facilities, and many water drillings with or without registration (without groundwater usage permit) in rural (villages) to meet the utility and drinking water needs. There are two healing geothermal springs in the province, namely the Sağlar (Billoris) Hot Spring at 15 km on the Siirt-Eruh Road and the Fiber Spa at the banks of the Reşan Stream in the Kışlacık Village.

Wastewater arises in Şırnak as any rainwater runoff, as well as coal mining and geotermal hot saline discharge, domestic or commercial wastewater acidic seeepage or any combination of these carried by sludge to fresh water resources. The type of wastewater generated is changed by both the population and the combination of geotermal seepage surrounding, domestic industrial activities. Hazardous sludge affect the discharge patterns as well as the fresh water chemical condition. The wastewater management need efficient waste treatment system. The proper identification and characterization of the contents entering a wastewater treatment plant is essential. This is based on the physical, chemical and biological properties of the flow; on the receiving environment where the treated wastewater will be discharged. The direct and subsequent impacts are important, as well as the environmental and discharge standards already established. Four main types of wastewater can be listed as domestic, industrial, agricultural and urban. Urban wastewater is defined as a combination of domestic and industrial wastewater, as well as environmental wastewater infiltration and rainwater, while agricultural

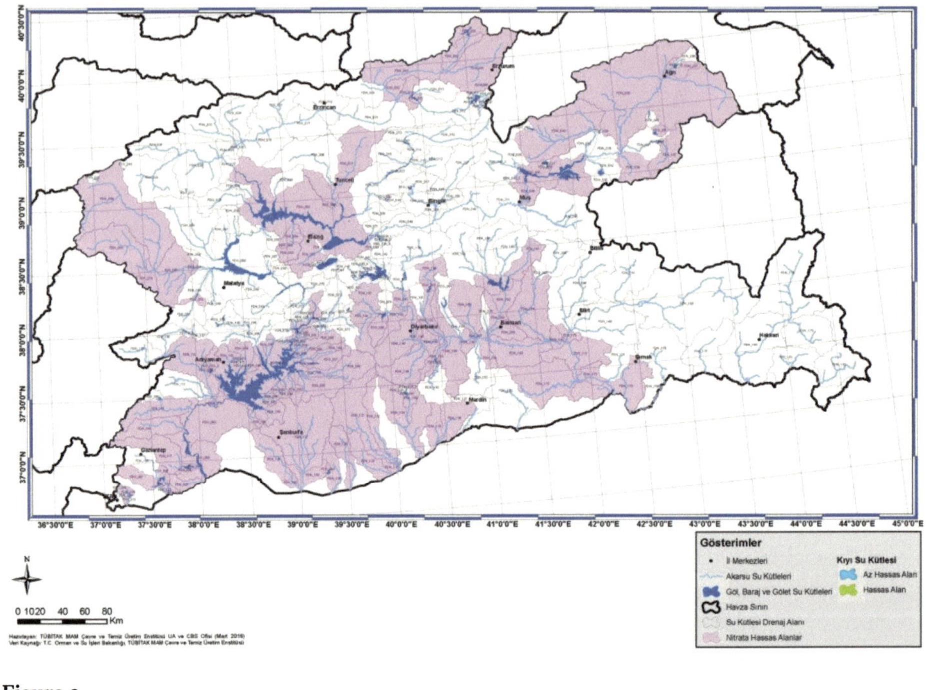

Figure 2.
Euphrate - Tigris Basin nitrate sensitive ground water areas and ground water flow areas.

wastewater consists of wastewater obtained from processes from surrounding farms, agricultural activities and sometimes polluted groundwater. **Tables 2** and **3** showed the parameter values permitted in the region in which high agricultural wheat fields, cattle and fish farming are affected by ground water change and contamination rgarding to legislative precautions [3, 4].

Generally, the main contamination occurs mainly on domestic and industrial sewage, where plants are affected and the source of contamination. However, the agricultural irrigation and farming chemicals is becoming increasingly important due to the high amount of use pesticides and fertilizers in agricultural fields especially dry stony lands exhousted by high amonunt remnant chemical toxic mattersin the land. The composition of industrial wastewater varies according to the type of environmental industry, with the pollutant and pollutant composition related to the general classification into inorganic and organic industrial wastewater [5–10]. Wastewater was directly discharged to the sea water surfaces for a long time as a result of the dislocations of many stream discharge structures and the land deterioration of the discharge transmission pipes, which is completely crossed from the Şırnak.

Raw wastewater in Şırnak due to coal mining quarry creates a risk if these systems fail and the suspended sludge wastewater becomes more difficult to treat. There is no technical and technological relationship between the collection of wastewater through the sewerage network, its transmission to decantation, its treatment and discharge. Therefore, compliance with the criteria of projecting in neutralization and decantation added to the end of the sewage is very weak and non-existent. Thus, current and future technologies will eventually have to deal with mining leachate area and tailing pond areas such as the following control mechanisms [11–22]:

- water decantation treatment and neutralization recycling;
- precipitation heavy and/or precious metals, anions, adsorption residual organic chemicals, complexes or chelates;
- oxidation cyanide and arsenic without destruction;
- collection oil spills, separation by solvent extraction liquors
- precipitation neutraklizing of acidic mine waters which is certain amounts of base metals such as copper, zinc, lead in addition to ferric iron and sulfate;

Discharge Parameter, unit	Composite sample, 2 hours	Composite Sample, 24 hours
Chemical oxygen requirement (COD)(mg/L)	250	100
Oil and grease(mg/L)	20	10
Ammonium nitrogen (NH4-N)(mg/L)	150	100
Chrome (Cr + 6)(mg/L)	0.5	0.5
Lead (Pb)(mg/L)	2	1
Total cyanide (CN^-)(mg/L)	0.5	0.1
pH	6–9	6–9

Table 2.
*The discharge legislation (**RG-13/2/2008-26786**) values for waste water treatment plant's discharge of spare parts, machine manufacturing, electric machines and equipment [3, 4].*

Parameter	Sewerage systems in wastewater infrastructure results with full treatment	Sewage systems in wastewater infrastructure facilities resulting by deep sea discharge
Temperature (°C)	40	40
pH	6.5–10.0	6.0–10.0
Suspended solid (mg/L)	500	350
Oil and grease (mg/L)	250	50
Tar and petroleum based oils (mg/L)	50	10
Chemical oxygen demand (COD) (mg/L)	4000	600
Biochemical Oxygen Demand (BOD5) (mg/L)	—	400
Sulphate (SO4 =) (mg/L)	1700	1700
Total sulfur (S) (mg/L)	2	2
Phenol (mg/L)	20	10
Free chlorine (mg/L)	5	5
Total nitrogen (N) (mg/L)	—	(a) 40
Total phosphorus (P) (mg/L)	—	(a) 10
Arsenic (As) (mg/L)	3	10
Total cyanide (Total CN^-) (mg/L)	10	10
Total lead (Pb) (mg/L)	3	3
Total cadmium (Cd) (mg/L)	2	2
Total chromium (Cr) (mg/L)	5	5
Total mercury (Hg) (mg/L)	0.2	0.2
Total copper (Cu) (mg/L)	2	2
Total nickel (Ni) (mg/L)	5	5
Total zinc (Zn) (mg/L)	10	10 Total tin (Sn) (mg/L)
Total silver (Ag) (mg/L)	5	5
Cl^- (Chloride) (mg/L)	10000	—
Surfactants (MBAS) reacting with methylene blue (mg/L)	Not involved	Not involved

(a) These parameters will not be taken into account in wastewater evaluation. b) For strong organic wastewater containing more than 2% inert COD and a total COD value of more than 5000 mg/L, the BOD5 value is taken as basis instead of COD.

Table 3.
*The discharge parameter values at environmental legislation (**RG-13/2/2008-26786**) for wastewater regarding Turkish standards intended for discharging wastewater to wastewater infrastructure [4].*

- absorption control of residual effluent matters in flotation such as frothers, flotation collectors and modifiers (activators or depressing agents, pH regulators);
- precipitaion pasting of radioactive waste waters,
- decantation control in aqueous effluents and soils [7, 8].

DOI: http://dx.doi.org/10.5772/intechopen.97656

Groundwater can be contaminated by the failure of on-site wastewater systems that can contaminate nearby water sources and wells. One way community work to protect the health of the public is to establish a very good concurrent management program for on-site wastewater treatment systems. The idea behind these programs is to assist homeowners by monitoring and managing centralized systems to make sure they are always working correctly and that the health of the community is never at risk.

In this research, prospected tests that the formation of Fe^{3+} by geothermal ozon reactivity on the oxidation Fe^{3+} surface in turn forms CN, AsO^{-3}, OH^{-} complexes [8, 11]:

The As (III) oxidation reaction then proceeds as Eq. (6)

$$Fe^{2+} + H_2O_2 \rightarrow Fe^{III}OH^{2+} + OH^{-}Fe^{2+} + H_2O_2 \rightarrow Fe^{3+}OH^{2+} + OH - (2OH\bullet + H_3AsO_3 \rightarrow H_2AsO_4 - +H_2O + H+) \quad (6)$$

Toxic intermediates may be sorped by char and shale clay may be generated as neutralization and precipitation heavy metals as organic complexs from this geotermal digestion technique. Also, the barrier-integrity verification, effective emplacement of barriers and modeling were found to be quiet difficult [13–15].

$$2Ca0.3(Fe2(III)Al1.4\ Mg0.6)Si8O20(OH)4nH2O + 2Na + +S2O42 - +2H2O \leftrightarrow 2NaCa0.3(Fe(III)Fe(II)Al1.4\ Mg0.6)Si8O20(OH)4nH2O + SO32 - +4H+ \quad (7)$$

$$Clay - Fe(III) + 4SO2 - \bullet \leftrightarrow Clay - Fe(II) + 2S2O42 - +H2O \rightarrow 2SO32 - +S2O32 - +2H+ \quad (8)$$

$$Fe2 + +CrO42 - +4H2O \rightarrow (FeX2Cr1 - x)(OH)3 + 5OH- \quad (9)$$

The dissolution kinetics of soil mud particle for Cr heavy metal is followed by equation

$$\frac{dc_{CrO4}}{dt} = k_i e^{-tic} dc \quad (10)$$

Where CN, AsO_3 contamination mg/l, k the rate of dissolution of ciyanide and chromate, i is the reaction style, t is time,

The digested soil mud and accumulated metal in effluent of lake streams as regarding contamination is followed by equation, where n is kinetic order type

$$\frac{dc_{AsO3}}{dt} = k_i c^{tin} dc \quad (11)$$

The saturation amount affect digested level of toxic CN concentration. The accumulated effluent activity of oxygen and oxidation reactions of hot water streams as precipitates metals and even iron cyanide chealates in contaminated streams. The sulfite and sulphate hot waters react as followed by equation, where SO_4^{-2} sulphate and sulfite concentration in effluent, $.f_i$ is concentration rate of sulphate in total effluent

$$\frac{dc_{Pb}}{dt} = k_i c^{tin} .dc.f_i \left(SO_4^{-2}\right)^{tin} \quad (12)$$

The dissolution concentration of accumulated metal in aliquate of limestone rocks dissolution by hot water streams in subground lakes with high CO2 gas dissolved streams as regarding Pb heavy metal contamination is followed by equation, where HCO_3^{-2} bicarbonate concentration in effluent

$$\frac{dc_{Pb}}{dt} = k_i c^{tin}.dc.f_i\left(HCO_3^{-2}\right)^{tin} \tag{13}$$

The digested soil mud and heavy metal in effluentof high fertilizer digestion by wrong amount of fertilizer use in farming discharges to fresh water streams as decayed mud with lack of COD and Pb heavy metal contamination is followed by equation, where HNO_3^{-2} nitrate concentration in effluent

$$\frac{dc_{Pb}}{dt} = k_i c^{tin}.dc.f_i\left(HNO_3^{-2}\right)^{tin} \tag{14}$$

Fish farming in the lakes and streams in the region require lower concentrations for breeding below 1 mg/l Pb Cu and Cd and Zn in contact to basaltic rocks and sulphide matter near copper ore deposites. The water could monthly oxidize slightly surfaces of sulfides resulting in seepages contained highly around %1–2 Pb and 200mgCu at high attitude deposits in Siirt and Şırnak. Even gold mining heap

Effluent, mg/l	**Şırnak Coal Mine Pool**	**Şırnak, Hezil Stream**	**Güçlükonak Hot Stream**	**Batman Hot Stream**	**Şırnak Kasrik Laguun**	**Ilısu Dam Laguun1**	**Ilısu Dam Laguun2**
Hg	8,11	4,71	12,3	14,11	4,71	4,71	4,71
Pb	10,58	14,53	23,2	12,58	11,53	5,7	5,2
Fe	40,33	70,62	59	93,3	56,2	60,62	67,62
K + Na	7,52	8,46	8,7	8,52	8,6	≥70	≥50
Cd	24,72	19,56	14,1	14,72	19,56	16	15
Mn	33,3	24,1	24,2	43,3	24,1	≤25	≤25
Cu	27,2	30,2	15,7	7,2	10,2	≤15	≤15
As	1,10	2,44	2,8	2,10	2,44	≤5	≤5
SO_4	0,57	0,37	1,9	0,67	0,55	≤15	≤15
Soil, ppm							
Hg	34,11	48,71	52,3	54,11	40,71		
Pb	10,58	24,53	23,2	20,58	11,53		
Fe	4,33	7,62	5,9	9,33	5,62		
K + Na	74,52	81,46	81,7	84,52	88,6	≥70	≥50
Cd	24,72	9,56	10,1	4,72	19,56		
Mn	2,72	3,02	1,5	0,72	1,02	≤5	≤5
Cu	3,33	2,41	2,4	4,33	2,41	≤5	≤5
As	1,10	2,44	2,8	2,10	2,44		
SO_4	0,57	0,37	1,9	0,67	0,55		

Table 4.
Şırnak and Batman province reveals the potential geothermal hot waters and contaminated soil.

Biosorbent	Co (mg L–1)	te(exp) (min)	qe(exp) (mg g–1)
Rice husk	5–300	150	17.87
Wheat shell	250	120	10.84
Pine cone powder	120	15	19.02
Tea leaves	20	30	62.80
Sawdust	10	30	1.50
Seed powder	10	30	4.82

Table 5.
Equilibrium time for Cu adsorption capacities studied with different biosorbents.

leachates also leaks to ground water streams by flooding [15, 16]. The contamination of some accumulated heavy metal contents of hot streams and soils in the region are given in **Table 4** [22, 23].

The geotermal saline water digestion approach is based on washing the entire sludge and fluid that extracts the contaminants from all size fractions. The pilot microwave digestion and geotermal hot water assisted sludge decantation techniques is used floculants, polyelectrolytes, chelants, inorganic acids, or surfactants depending on site.

The industrial waste water cleaning areas, non-burning, slippery to create a safe working environment absorbant fly ash is used. In the waste water treatment, the amount of fly ash used for hazardous waste water treatment is more than 180,000 tons/year. The sorbent matter variation may eliminate the contaminated effluent levels at equilibrium ambient concentrations given in **Table 5** [9].

Heavy metal containing mud or with lack of COD and scarcity of water forces to control fresh water resorces by using decantation and neutralization with adsorbants such as clay in the chemical industry and in tanning discharges [6–9]. The clay, fly ash and sepiolite is good absorbant [10–14]. The activated bentonite or montmorillonite is good sorbent for fresh fruit drinks and brewery, water tretmeants in Europe [15–19]. The use of clay is exceeded 2 million tons/year in the world waste water treatment [20–21]. The fly ash is stable in neutralization on the layered clay cages at hot stream temperatures [22–28]. This research work was carried out on geothermal hot waters and microwave radiated digestion of waste waters in Şırnak for heavy metal sorption and reduction. In this study, fly ash, char, shale and marly shale of Şırnak and absorbant properties, was improved by saline water digestion. Fly ash and char absorbance by cavity passing through certain bed properties was performed and the absorbance was measured at the mechanical strength change has been studied.

2. Material and methods

2.1 Fly ash - char compost with Geotermal water digestion

The geotermal water digestion technique digests the hazardous species and reduces the amount of metal content in sludge [29],

In this study, the effect of water quality) was subjected to the concentration and further alkali activation tests with mixed type The geothermal water quality on concentration and alkali activation were declared based on the pH, CEC (Cation exchange capacity), sludge viscosity (**Table 6**).

Sorbent Type	SiO_2	Al_2O_2	Fe_2O_2	MgO	K_2O	CaO	MnO	LOI*
Zeolite (%)	47.85	24.30	0.32	1.7	0.77	2.7	0.10	10.27
Şırnak Asphaltite Char Shale	27.50	7.70	10.83	2.17	1.97	10.5	1.74	5.47
Bentonite	50.45	17.80	6.83	12.17	4.97	3.57	0.20	7.37
Marly Shale	17.85	11.60	0.83	5.17	3.97	20.57	0.40	5.27
Fly ash	27.80	13.60	17.83	4.17	2.97	10.70	1.50	16.27

**LOI: Loss on Ignition at 1000°*

Table 6.
Sorbent types for washing treatment.

2.2 Microwave digestion and heavy metal sorption

Microwave assisted organic reactions hydratation, synthesis of complexing chelates by alkali matters in clays and geotermal waters. The fly ash content:

$2MgFe\ Si_8O_{20}\ (OH)_4$. The chlorite, aqueous magnesium, aluminum silicate.

Sepiolite is $6Mg_9\ Si_{12}O_{30}(OH)4\ 6H2O$ group is hydrolized Mg silicate. The pore water is scarcely bound to crystal sliding layer of magnesia and hydroxyl base. Fly ash and clay minerals under microwave radiation is heated easily [10]. Microwave digestion could activate digestion use as heating as studied in this study over slurries illustrated in **Figure 3**.

2.3 Sorbent clay/oak wood char - waste sludge

Clay minerals are used as activated, compositions are closely dehydrated. The particle size, particle shape, surface chemistry, ion exchange capacity, color, etching, viscosity, plasticity, sorption ion, adsorption surface are the main parameters in waste water treatment [30–39]. The properties of clay minerals significantly impact on the use of. Absorption can be carried out in the presence of water or other liquid the pores of the mass (solid material) [40–44]. Absorbents material in waste water and other chelates is a sponge sorbent as material containing pores adsorp the contaminating heavy metals and cyanide durings neutralization affect [45–49] (**Figure 4**).

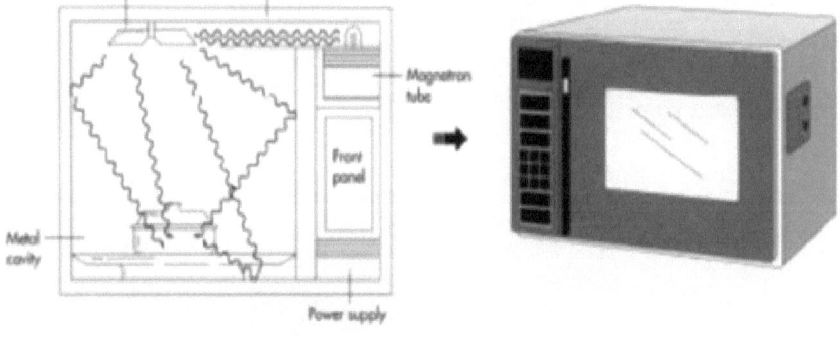

Figure 3.
Microwave radiation heating for waste water digestion.

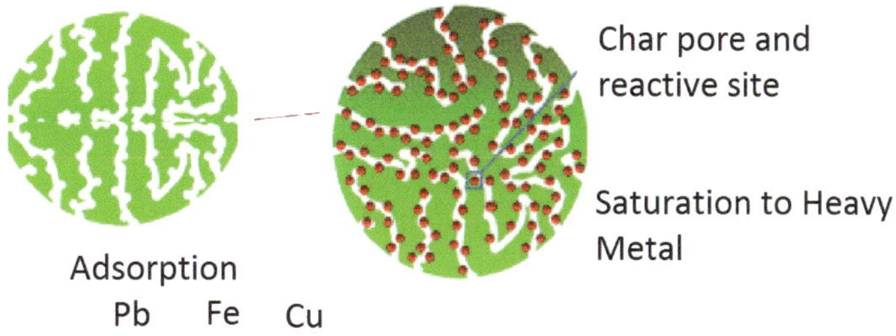

Figure 4.
The micro pictures Şırnak marly shale char shale as sorbent.

Sorbent	Bulk Density, kg/cm3	Porosity,%
Shale chlorite	800–980	45
Şırnak Coal Char	530	66
Oak wood char	220	78
Silopi fly ash	700	33

Table 7.
Shale and Marly shale and fly ash properties.

	Specific gravity, g/cm^3	Specific surface area, m^2/g	Micro porosity, %
Zeolite (%)	2,56	34	44
Şırnak Asphaltite Char Shale	2,03	45	32
Bentonite	2,62	32	66
Marly Shale	2,71	11	12
Fly ash	2.23	67	47

Table 8.
The chemical properties of different sorbents and Şırnak shale and Marly shale and fly ash used in neutralization and adsorption treatment.

Component	Güçlü Konak 1 (mg/L)	Çermik (mg/L)	Batman 3 (mg/L)	Beytüşsebap (mg/L)	St (mg/L)	Ya (mg/L)
Ca2+	55	222	17	953	3760	221,40
Fe2+	0.1	n.d.**	n.d.	n.d.	n.d.	834
K+	6.1	6.4	5.0	97.2	430	1370
Mg2+	10	33	6.1	276.8	1270	2860
Mn2+	0.02	n.d.	n.d.	n.d.	n.d.	n.d.
Na +	15	109	365	16,785	90,100	94,900
F^-	0.3	0.5	0.5	n.d.	n.d.	n.d.
Cl^-	19	121	310	27,541	14,300	18,000
NO3Ã	1.3	114	0.9	n.d.	n.d.	n.d.
SO42Ã	13	250	14	1132	3600	283
HCO3Ã	210	502	512	185.4	40	67

Table 9.
The chemical composition of geothermal hot waters in Şırnak and Batman [1, 2].

2.4 Fly ash/Asphaltite shale char composite

This compost is especially sorbent used waste water treatment. The fly ash compost granules use as hazardous industrial waste waters and the sorbent types and desirable properties of activated and cleaned sorbents in the experiments are given in **Table** 7 and chemistry are given in **Table 8** for geothermal waters used in neutralization (**Table 9**).

Material to be used as powder is packed distributed in tube bed permeable granule grain size, basic as absorption capacity by the cat should be accepted. High absorption capacity having fly ash should avoid digestion waste waters passed through cavities. The grain size distribution of sorbent granules is important and between 1 and 6 mm. The sorbent beds are rounded surfaces not to create dust mass transport during mechanical stick and fouling cavity.

3. Results and discussion

3.1 Neutralization decantation and precipitation

High salt content in the hot brines may deteriorate electrolyte potential at low levels. The dissolved silisium contents may even bound bicarbonate and calcium rective component. Low level salt contents of Şırnak and Batman province will be highly advantageous for precipitation of heavy metals and metal cyanides (**Figure** 5).

3.2 Microwave digestion with geothermal water- washing

The required ozone dosage depends on a number of factors, most notably the type of wastewater treated. The Şırnak solid wastes can threaten surface water and groundwater resources. Strategies are needed to identify and solve local pollution

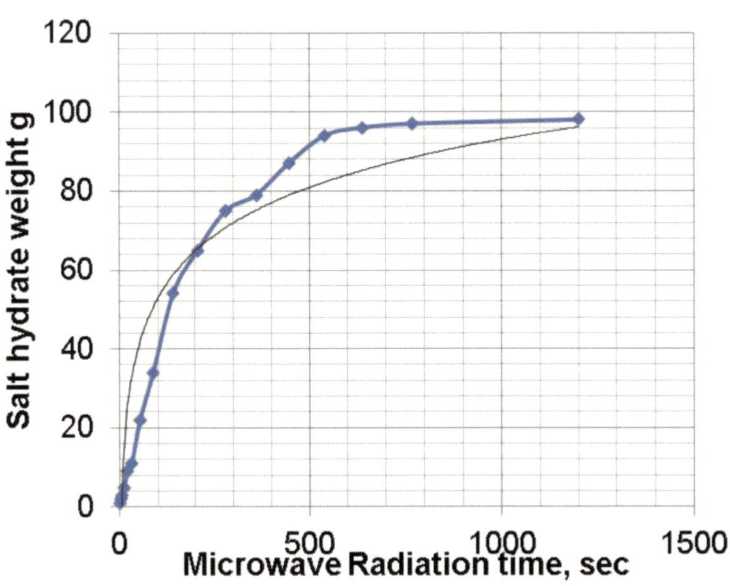

Figure 5.
The composite sorbent use, Fly ash distribution in salt hydration under microwave radiation.

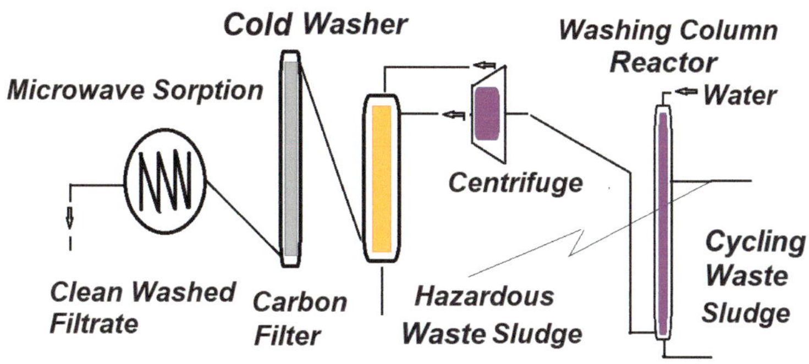

Figure 6.
Schematic view of an washing with microwave recycled by microwave sorption technique.

problems, and residents, businesses and industry need to be educated about the health hazards associated with untreated wastewater. Safe drinking water and proper treatment have been recognized as indispensable factors for sustaining life. Increasing poverty is exacerbated by factors such as increasing population growth and rapid urbanization, as well as hydrological variability and climate change. These socio-economic and environmental factors come about by increasingly valuing water and wastewater infrastructure. Reliable wastewater treatment systems serve together with the degree and quality of the wastewater, which determines the effects of these treatment plants on the environmental water resources released, as a good indicator of the level of development in municipal and public health. In the last few years, the amount of municipal wastewater generated has increased significantly due to the steady increase in the population with increasing dependence on decreasing water resources (**Figure 6**).

3.3 Char carbon compost washing technology - sorbent applications

The slurries is stand for 30 minutes after being conditoned. At the end of the period, the settled mud was removed by decantation method and dried.

The effluent by adding salts at concentrations ranging from 30,125 ppm to 1000 ppm, until the precipitates were obtained in sufficient quantities with pure water in the waste water.

The step wise washing cycle is practised in the geotermal waste water slurry: there is no water–effluent washing column unit connected to the waste sludge, and the washing sorbent bed contained one single microwave radiation column is used to wash in the three decantation washing norms: roughing, scraping and cleaning. The variation of the third cycle washing is also slowed recycling by microwave act.

The simple production presented as adapted and optimized depending on the target application. The main applications are briefly described in the following sections. Although this review only focuses on state-of-the art commercially available pellet plants, it should be noted that some prospective advanced applications for heat melting of binder are currently being studied, mainly in the form of prototypes or proof-of-concepts. These innovative applications include:

- Compost systems, in which the extrusion mold system takes advantage of temperature gradients in wet gradient.

- Compression press systems, where the high load press is used to drive the forming sludge in plant.

- Continous conversion systems, utilizing the high temperature binding gradients 75°C and amounts (of at least 200°C) in slurries to drive a recycle.
- Hot production, where the scraping power of the load system is used to drive the compressive form of hot system.

3.4 Langmuir absorption model by hot water streams

Langmuir and Freundlich concentrations for As(V) adsorption onto sorbent at different temperatures Non-linear pseudo-first-order [11–12]:

$$qt = q_{e,1}\left(1 - e^{-k_1 t}\right) \quad (15)$$

Non-linear pseudo-second-order:

$$qt = \frac{qe^2 kt}{1 + qekt} \quad (16)$$

Linear pseudo-second-order:

$$\frac{1}{qt} = \frac{1}{kqe} + \frac{1}{qe^2} \quad (17)$$

For an overview of these more innovative and prospective applications, the general common method can be given in **Tables 10** and **11**.

it is found that the amounts of neutralization ions is hydrated and formed in the mud at higher microwave temperatures with even the H + ions. Mg, and Fe atoms in the octahedral clay layers and the Al atoms in the octahedral centers manage stable porous sorbent compost, as well as the Al atoms in the tetrahedral layer, as well as Al_2O_3, MgO and Fe_2O_3.

The first order sorption concentration at three stage cycling counted by the eqution regarding the posphate sulfite, and nitrate concentrations in the waste waters as studied stepwise concentration weight increase determined by weight regressed exponantial rate change as given below:

Run,	C, mg/l	k_1	a	b
1	28	0,3	0,15	1,2
2	20	0,24	0,22	1,7
3	12	0,21	0,27	2,4

Table 10.
The activated bentonite compost with char shale of Şırnak materials.

Run,	C, mg/l	k_1	a	b
1	28	0,3	0,15	1,2
2	20	0,24	0,22	1,7
3	12	0,21	0,27	2,4

Table 11.
The activated fly ash compost with char shale of Şırnak materials.

$$\ln c_{Pb}{}^{NO3,PO3,S2O3} = 1 + \frac{k_{NO3}t}{1!} + \frac{k_{PO3}t^2}{2!} + \frac{k_{S2O3}t^3}{3!}, 3ppm < c < 300ppm \quad (18)$$

Cation exchange abilityof clay was so effective parameter in metal sorption manner. The alkali pH provided efficient washing criteria in the column sorption.

It is illustrated in the **Figure 7**, the lower pH decreases inversely limiting sorption and reduce the amount of neutralizing salt added to oak wood char sorbent. When $FeCl_3$ is used, iron ion fouling is observed in the sorption with coal char with low coal porosity and metal iron content.

Wood char is known to have a considerable dependence on the layer charge and edge charge pH. Therefore, a decrease in the cation exchange capacity should be expected in parallel with the decrease in pH. Neutralization supresses the oxidation of sludge and precipitation of chelates may improve the adsorption to wood char with high mesopore structure as seen in **Figure 8**.

As seen **Figure 9**, The bicarbonate hot waters affect the waste seepage in the copper mining leachate zone. The high level of iron and lead show the contamination at char load change, high level fly ash suspensions at 10% weight rate with 22% volume rate obtained microwave washing stepwise cycled- depending on the neutralization salt concentration added at 10 g-100 mg.

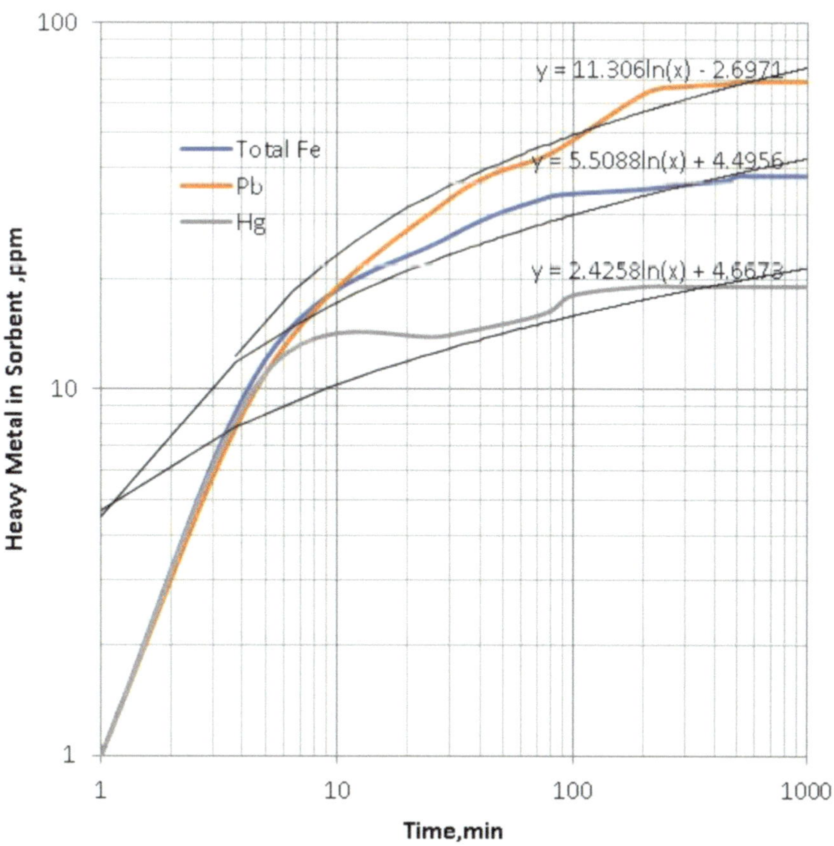

Figure 7.
The change in metal sorption depending on the metal concentration incorporated in the coal char/fly ash suspensions.

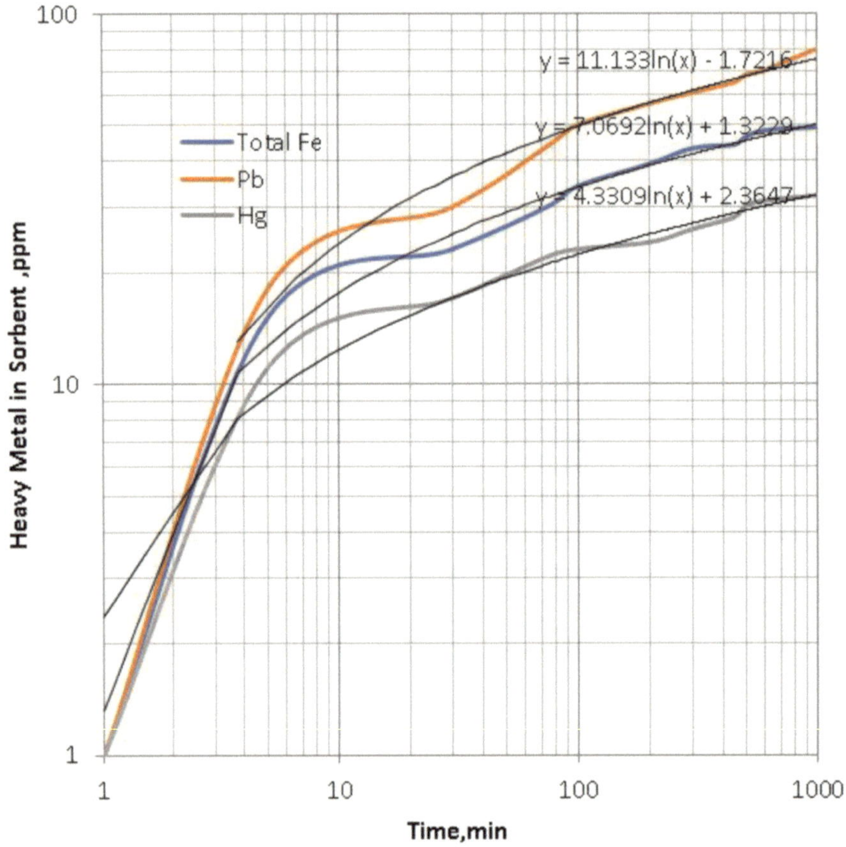

Figure 8.
The change in metal sorption depending on the metal concentration incorporated in the wood char/fly ash suspensions.

The high level of geotermal spa waters and heap leachate lead and cyanide metal complexs could be adsorbed by coal char and wood char susupensions by microwave act decantation unit, cycled washed and enhanced with use of hot waters. The outcome effluent of clarification treatment and injection to subground as shown in **Figure 10**, rehabiliate the irigation area with the contamination at cyanide or metal load at high level even by means of fly ash suspensions at low weight rate, 10%.

4. Conclusions

This waste geotermal source combined with the discharge of inefficiently treated wastewater discharge into surface water sources. The method protect farming and irrigation sources imposed a direct threat not only to existing macro and microflora and fauna, but also to the provision of good quality water required for all socio-economic functions. For this reason, continuous monitoring of the operating status of geotermal wastewater treatment plants is increasing the importance of the environment and fresh water source become a key factor in determining the amount and quality of wastewater clarified and decanted by the relevant municipalities.

It is understood that public institutions and organizations have almost completely lost their functions in the field of gold heap leaching, mining waste production and the private sector has difficulties in finding financing and making

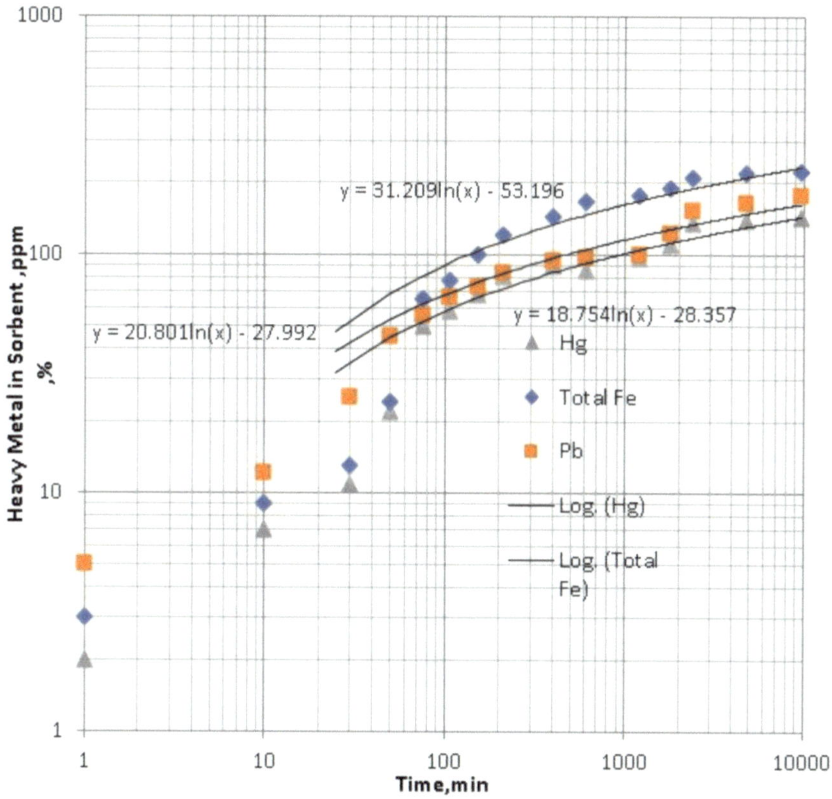

Figure 9.
The geotermal hot water of Güçlükonak in Şırnak changed neutralization and metal sorption depending on the metal concentration incorporated in the wood char/fly ash suspensions.

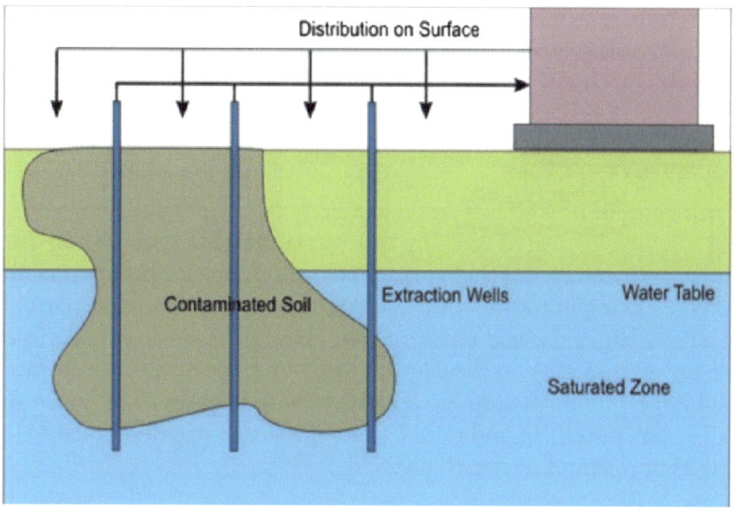

Figure 10.
The proposed feed flow into ground water, stability change in contaminated metal level, the sorption manner depending on the active site incorporated in the wood char/fly ash suspensions.

investments, necessary work should be done to create an alternative energy policy based on the country's domestic and renewable energy options, which are more environmentally sensitive. Waste materials of coal mining may promote the neutralization and fertile carbon healing of soil.

Since water in our country has scarce and hard environmentally undesirable chemical content, R&D studies that positively affect treatment efficiency in treatment model determination and cost for aim to minimize environmental problems should be focused on, these studies should be planned and encouraged by the government. License splitting practices, which have negative consequences in terms of production and efficiency, should be abandoned immediately in the evaluation of reservoirs in ground controlled basins.

Water management, ground water control, waste pool design, slime control and hydro power plant projects significantly need that; waste water management and ground waters should be analyzed in environmental, social and economic terms as a whole, and it should be decided whether the projects will be implemented considering the social benefit.

In resoirvoir basins, hydro power electricity generation activities, priority should be given to the people of the region in employment and the needs that arise during these activities should be met from the region as much as possible and focus on regional development.

In the use of fresh water, the current and advanced technologies in the world should be followed and the use of these technologies should be encouraged by legal regulations.

The waste water Law, the Environmental Impact Assessment Legislation, the legislation that has a direct or indirect connection with the electricity generation from hydro power plants, should be rearranged by a structure that includes professional associations, universities and non-governmental organizations with a steadfast understanding that takes into account the needs of the society.

The energy that society needs; It should be offered to the public with equal opportunities regardless of population density, water scarcity and region, and it should be ensured that energy, which is a human right, is cheap, reliable and accessible.

Heap leaching applications for gold and copper productions in the area is planned the feasibility reports regarding environmental contamination shows some degree of contamination and certain collection pools and seepage area will be highly contaminated by atmospherical dry contions. That hard winter conditions oxidized the contaminating oxidation products inthe ground water streams and dry weather conditions with scarcity of water may affect the cattle breeding and fish farming by evaporation of clean fresh water sources in the summer term.

In the pH measurements made, the pH value of 7,3 in washing hazardous waste water finally at the last washing column decreased to 5, depending on the concentration of salt content of sorbents in the water.

Stepped microwave heated washing test measurements showed that washed waste waters obtained after cycled at third steps in durated 2 hours time in slow decantation flow using hot water with alkali bicarbonate and sulfite matter flow with 1 mm sorbent packages showed reductions in Pb, Hg and Fe at 37% performance.

In clean water aliquate had the 22 ppm Pb, 5 ppm Hg and 67 Fe chelate and precipitated Pb reduction at step model with nitrate is observed. The clean washed water is 0,73 ppm/min.l, Hg and total Fe reduction rate are decreased to 0,43 ppm/min.l and 0,23 ppm/min.l, respectively.

Abbreviations

Greek symbols

a	affinity parameter of the Langmuir isotherm (L mg^{-1})
b	stoichiometric constant defined by

B	reactant solid defined
Bi_m	Biot number for mass transfer
C_i	concentration of manganese in the bulk external phase of stage i (mg L^{-1})
C_0	feed concentration of manganese in the column (mg L^{-1})
D_{ef}	effective diffusion coefficient (m^2 s^{-1})
F	objective function
h	fixed bed height (m)
k_e	mass transfer coefficient in the bulk external phase (m s^{-1})
k_r	reaction rate constant for heterogeneous systems (m s^{-1})
N	number of stages
Q	volumetric flowrate (m^3 s^{-1})
q_i	concentration of immobilized manganese within the adsorbent particle at stage i (mg g^{-1})
q_m	theoretical maximum adsorption capacity of the Langmuir isotherm (mg g^{-1})
r	radial distance from the center of the particle, $0 < r < R_p$(m)
R	radius of column (m)
R_p	radius of adsorbent particle (m)
R^2	determination coefficient (−)
$r_{c,i}$	unreacted core radius at stage i (m)
t	time (s)
V_i	volume of stage i (L)
α	backmixing coefficient (−)
φ	column hold-up (−)
ρ	density of adsorbent particle (g m^{-3})
τ	mean residence time of fluid in the column (s)

Author details

Yıldırım İsmail Tosun
Mining Engineering Department, Engineering Faculty, Şırnak University, Şirnak, Turkey

*Address all correspondence to: yildirimismailtosun@gmail.com

References

[1] Anonymous, 2016, Development Report, TC6 Region DSİ

[2] Anonymous, 2019, DSİ, Water Reports, 10th Regional Directorate, 2018

[3] Anonymous, 2016, JMO, Hassas su kütleleri ile bu kütleleri etkileyen alanlarin belirlenmesi ve su kalitesinin iyileştirilmesi hakkinda yönetmelik, Resmi Gazete 29227, 23 Aralık 2016

[4] Anonymous, 2016, JMO, Su kirliliği kontrolü yönetmeliği, Resmi Gazete Tarihi: 31.12.2004 Resmi Gazete Sayısı: 25687

[5] M. Abatal, V. C. Quiroz, M. T. Olguín, A. R. Vázquez-Olmos, J. Vargas, F. Anguebes-Franseschi, G. Giácoman-Vallejos, 2019, Sorption of Pb(II) from Aqueous Solutions by Acid-Modified Clinoptilolite-Rich Tuffs with Different Si/Al Ratios, Appl. Sci. 2019, 9(12), 2415; https://doi.org/10.3390/app9122415

[6] X. Xu, L. Lin, C. Papelis and Pei Xu, 2019, Sorption of Arsenic from Desalination Concentrate onto Drinking Water Treatment Solids: Operating Conditions and Kinetics, Water 2018, 10 (2), 96; https://doi.org/10.3390/w10020096

[7] Yıldırım İ.Tosun, 2016, Microwave Roasting of Pyrite and Pyrite Ash for Sponge Iron Production, SWEMP 2016, 16th International Symposium on Environmental Issues and Waste Management in Energy and Mineral Production, October 5–7, Istanbul, Turkey

[8] Bajpai, S. K.; Jain, A. Removal of copper (II) from aqueous solution using spent tea leaves (STL) as a potential sorbent. Water SA 2010, 36, 221–228

[9] Chen, H.; Dai, G.; Zhao, J.; Zhong, A.; Wu, J.; Yan, H. Removal of copper (II) ions by a biosorbent—Cinnamomum camphora leaves powder. J. Hazard. Mater. 2010, 177, 228–236, DOI: 10.1016/j.jhazmat.2009.12.022

[10] S. Ahamed, A. Hussam, A.K.M. Munir,2009, Groundwater Arsenic Removal Technologies Based on Sorbents: Field Applications and Sustainability, Handbook of Water Purity and Quality, Academic Press, Amsterdam (2009) 379–417

[11] J.S. Ahn, C.M. Chon, H.S. Moon, K. W. Kim, 2003, Arsenic removal using steel manufacturing by-products as permeable reactive materials in mine tailing containment systems, Water Research, 37 (2003), pp. 2478–2488

[12] *P.,Atkins. 2006, PHYSICAL. CHEMISTRY. Eighth Edition. Peter Atkins*. Professor of *Chemistry*,. University of Oxford. *Eighth Edition. 2006* by *Peter Atkins* and *Julio* de *Paula* ,. 23 *The kinetics of chemical reactions*. 850

[13] Shaobo Liu, Binyan Huang, Liyuan Chai, Yunguo Liu, Guangming Zeng, Xin Wang, Wei Zeng, Meirong Shang, Jiaqin Deng and Zan Zhou, 2017, Enhancement of As(V) adsorption from aqueous solution by a magnetic chitosan/biochar composite, RSC Adv., 2017, 7, 10891–10900, DOI:10.1039/C6RA27341F

[14] Whitten K.W., Galley K.D. and Davis R.E. *General Chemistry* (4th edition, Saunders 1992), p.638–9

[15] I, Oboh, E. Aluyor,T. O. K. Audu, 2013, Second-order kinetic model for the adsorption of divalent metal ions on Sida acuta leaves, International Journal of Physical Sciences 8:1722–1728

[16] P. Senthil Kumar*, C. Vincent, K. Kirthika, and K. Sathish Kumar, 2010,

Kinetics and equilibrium studies of Pb2+ in removal from aqueous solutions by use of nano-silversol-coated activated carbon,Braz. J. Chem. Eng. vol.27, no.2, São Paulo, https://doi.org/10.1590/S0104-66322010000200012

[17] H Medhi, P R Chowdhury, P D. Baruah, and K G. B, 2020, Kinetics of Aqueous Cu(II) Biosorption onto Thevetia peruviana Leaf Powder, ACS Omega 2020, 5, 23, 13489–13502, https://doi.org/10.1021/acsomega.9b04032

[18] Barrera, H.; Ureña-Núñez, F.; Bilyeu, B.; Barrera-Díaz, C. Removal of chromium and toxic ions present in mine drainage by Ectodermis of Opuntia. J. Hazard. Mater. 2006, 136, 846–853, DOI: 10.1016/j.jhazmat.2006.01.021 [Crossref], [PubMed], [CAS], Google Scholar

[19] Ahamed, J. A.; Begum, A. S. Adsorption of copper from aqueous solution using low-cost adsorbent. Arch. Appl. Sci. Res. 2012, 4, 1532–1539Google Scholar

[20] Medhi, H.; Bhattacharyya, K. G. Kinetics of Cu(II) Adsorption on Organo-Montmorillonite. J. Surf. Sci. Technol. 2015, 31, 150–155[CAS], Google Scholar

[21] Bhattacharyya, K. G.; Gupta, S. S. Adsorption of a few heavy metals on natural and modified kaolinite and montmorillonite: A review. Adv. Colloid Interface. Sci. 2008, 140, 114–131, DOI: 10.1016/j.cis.2007.12.008 [Crossref], [PubMed], [CAS], Google Scholar

[22] Gupta, S. S.; Bhattacharyya, K. G. Adsorption of Ni(II) on clays. J. Colloid. Interface. Sci. 2006, 295, 21–32, DOI: 10.1016/j.jcis.2005.07.073 [Crossref], [PubMed], [CAS], Google Scholar

[23] Bhattacharyya, K. G.; Gupta, S. S. Adsorption of Fe(III), Co(II) and Ni(II) on ZrO-kaolinite and ZrO-montmorillonite surfaces in aqueous medium. Colloids Surf. A Physicochem. Eng. Asp. 2008, 317, 71–79, DOI: 10.1016/j.colsurfa.2007.09.037 [Crossref], Google Scholar

[24] J.P. Allen, I.G. Torres, 1991, Physical separation techniques for contaminated sediment, N.N. Li (Ed.), Recent Developments in Separation Science, CRC Press, West Palm Beach, FL (1991)

[25] S.J. Allen, L.J. Whitten, M. Murray, O. Duggan,1997, The adsorption of pollutants by peat, lignite and activated chars, Journal of Chemical Technology & Biotechnology, 68 (1997), pp. 442–452

[26] E. Álvarez-Ayuso, H.W. Nugteren, 2005, Purification of chromium(VI) finishing wastewaters using calcined and uncalcined Mg-Al-CO3-hydrotalcite, Water Research, 39 (2005), pp. 2535–2525

[27] Groundwater Discharge - The Water Cycle, https://water.usgs.gov/edu/wusw.html

[28] Groundwater Storage - The Water Cycle https://water.usgs.gov/edu/watercyclegwstorage.html

[29] Assouline S, Tavares J, Tessier D. 1997. Effect of compaction on soil physical and hydraulic properties: Experimental results and modeling.Soil Science Society of America Journal61: 390–398.

[30] Gee GW et al. 1994. Variations in water balance and recharge potential at three western desert sites. Soil Sci Soc Am J 58: 63–72.

[31] Brunauer, S., Emmett, P.H., Teller, E. 1932. Adsorption of gases in multimolecular layers. Journal of the American Chemical Society, 60, 309–319.

[32] Christidis, G.E., Scott, P.W., Dunham, A.C. 1997. Acid activation and

bleaching capacity of bentonites from the islands of Milos and Chios, Aegean, Greece. Applied Clay Science, 12, 329–347.

[33] Gregg, S.J., Sing, K.S.W. 1982. Adsorption, surface area and porosity, Academic Press, London, 52 pp.

[34] Lopez-Gonzalez, J.D., Deitz, V.R. 1952. Surface changes in an original and activated bentonite. Journal of Research of the National Bureau of Standards, 48, 325–333.

[35] Marshall, C.E. 1935. Layer lattices and base-exchange clays. Zeıtschrift Fur Kristallographie, 91, 433–449.

[36] Murray, H.H. 1999. Applied clay mineralogy today and tomorrow. Clay Minerals, 34, 39–49.

[37] Murray, H.H. 2000. Traditional and new applications for kaolin, smectite and palygorskite: a general overview, Applied Clay Science, 17, 207–221.

[38] Nguetnkam, J.P., Kamga, R., Villieras, F., Ekodeck, G.E., Razafitianamaharavo, A., Yvon, J. 2005. Assessment of the surface areas of silica and clay in acid-leached clay materials using concepts of adsorption on heterogeneous surfaces. Journal of Colloid and Interface Science, 289, 104–115.

[39] Novak, I., Cicel, B. 1978. Dissolution of smectites in hydrochloric acid; II, Dissolution rate as a function of crystallochemical composition. Clays and Clay Minerals, 26, 341–344.

[40] Önal, M., Sarıkaya, Y., Alemdaroğlu, T., Bozdoğan, İ. 2002. The effect of acid activation on some physicochemical properties of a bentonite. Turkish Journal of Chemistry, 26, 409–416

[41] Srasra, E., Bergaya, F., van Damme, H., Arguib, N.K. 1989. Surface properties of an activated bentonite. Decolorization of rape-seed oil. Applied Clay Science, 4, 411–421.

[42] Srivastava, R.V. 2003. Controlling of SO2 Emissions, A Review of Technologies, Nova Science Publishers, Inc., New York, 1 pp.

[43] Venaruzzo, J.L., Volzone C., Rueda M.L., Ortiga J. 2002. Modified bentonitic clay minerals as adsorbents of CO, CO2 and SO2 gases. Microporous and Mesoporous Materials, 56, 73–80.

[44] Volzone, C., Ortiga, J. 2009. Adsorption of gaseous SO2 and structural changes of montmorillonite. Applied Clay Science, 44, 251–254

[45] Alemdaroğlu, T., Akkuş, G., Önal, M., Sarıkaya, Y. (2003). Investigation of the Surface Acidity of a Bentonite Modified by Acid Activation and Thermal Treatment. Turk. J. Chem., 27, 675–681.

[46] Benesi, H.A. (1956). Acidity of Catalyst Surfaces I. Acid Strength from Colors of Adsorbed Indicators. J. Phys. Chem., 78, 5490–5494.

[47] Benesi, H.A. (1957). Acidity of Catayst Surfaces II. Amine Titration Using Hammett Indicators. J. Phys. Chem., 61, 970–973.

[48] Caglar, B., Afsin, B., Tabak, A. (2007). Benzamide Species Retained by DMSO Composites at a Kaolinite Surface. J. Therm. Anal. Cal., 87, 429–432.

[49] Rodriguez, M.A.V., Barrios, M.S., Gonzalez, J.D.L., Munoz, M.A.B. (1994). Acid Activation of a Ferrous Saponite (Griffithite): Physico-Chemical Characterization and Surface Area of the Products Obtained. Clays Clay Miner., 42, 724–730.

Chapter 17

Treatment of Tannery Effluent of Unit Bovine Hides' Unhairing Liming by the Precipitation

Anass Omor, Karima Elkarrach, Redouane Ouafi, Zakia Rais, Fatima-Zahra ElMadani and Mustafa Taleb

Abstract

The tannery effluents are characterized by high toxic pollutants such as sulfides; used in the tanning of animal's skin. The mean objective of this work is the evaluation of the pollution degree of various operating units, and the treatment of tannery effluent generated from unhairing-liming unit. According to physicochemical characterization, this effluent was largely basic and very loaded in sulfides, which have harmful effects on human health and the environment as well. Otherwise, the microbiological characterization showed an absence of pathogenic bacteria and a low concentration of mesophilic aerobic flora, because of this effluent toxicity. Thus, the treatment of this effluent is indispensable before its reject into the environment. In fact, chemical precipitation is a promising approach for the treatment of this effluent. In this regard, ferric chloride was used as chemical agent to reduce and removal sulphide ions from this effluent. As result, this treatment gave an excellent abatement rate of sulphide, which reached more than 90% using a pH of 8.5 and a ferric chloride concentration of 1.4 mol/L.

Keywords: Tannery, sulfides, characterization, chemical precipitation, ferric chloride

1. Introduction

The leather industry plays an important role in the global economy, particularly in African countries [1, 2], particularly in Morocco [3]. Besides, the tanning industry is an important activity, which involves the processing of leather animal skin by removing fat and hair through different operations namely unhairing-liming, rinsing, deliming-bating and tanning … etc. This tanning process led skins unalterable and rigid [4]. Two methods of tanning are used, chrome tanning and vegetable tanning. At a global level, between 70% and 80% of leather is produced by chrome tanning [5, 6]. Tannery industries use a lot of chemicals and produce huge volumes of wastewater and solid waste [7]. Consequently, tanning industries have been known as a pollution source in the whole world, including Morocco. In fact, they always reject into the environment a large amount of wastewaters, which is loaded with toxic pollutants such as sulfides.

Sulfides can be reduced to hydrogen sulfide (H2S). This toxic gas can poison all living beings, especially humans. Indeed, prolonged sulfide inhalation may cause degeneration of the olfactory nerve and cause death just after few breaths. Plus, the inhalation of this gas even in small amounts can lead to a loss of consciousness [8–13].

Thus, the discharge of these tannery effluents, without prior treatment, harms human health and the environment too. For that, the treatment of these effluents has been very necessary.

Several previous research works have proven some treatment processes for these effluents such as activated carbon adsorption [14], ions exchange [15–18], chemical precipitation using ferric chloride [19, 20], coagulation [21], electrocoagulation [7, 22], sequencing batch reactor [23], bio-augmentation [24] ... etc. However, all these studies are only focused on chromium removal from these effluents, and they ignored the elimination of sulfides even they also very toxic.

In this regard, the objective of this chapter was the evaluation of the quality of tannery effluents rejected from different tanning operations. Then, the treatment of wastewater loaded in sulfide and generated from unhairing-liming operation.

2. Chemical consumption and generated pollution of modern tanning industry

Leather is obtained by treating the skin to keep it in good condition. The tanning process consumes a large quantity of chemical products and water according to tanning

Operation	Products used	Quantity (Kg)	Quantity of water consumed (m^3)	Quantity of water discharged (m^3)
Preparation	Savon	3	2,5	2
Unhairing-liming and rinsing	Sodium sulfide	42,5	20	19,5
	Sodium carbonate	6		
	Lime	20		
Deliming-bating	Sulfate of ammonia	5		
	Sodium metabisulphite	50		
	Sulfuric acid	56		
	Salt	140		
Chrome tanning	Chromium	100	2	1,5
	Sodium bicarbonate	12,5		
	Tanning oil	6		
Retanning and finishing	Soap	0,5	4,5	4
	Formic acid	0,5		
	Sodium bicarbonate	34		
	Tannins	64		
	Oil	73		
	dyes	8		
	Pigments, Resins, Waxes	17	0,02	0,01
	Matting agents, Touching agents	11		
	Lacquers	26		
Total	**680 Kg Chemical products**		29	27

Table 1.
Quantity of chemicals and water consumed during the treatment of 1000 kilograms bovine hides in tannery.

type (Traditional or modern). For a modern tannery, the treatment of 1000 kg of bovine hides consumes around 680 kg of chemicals (**Table 1**) and 29 m^3 of water. These amounts were used during several operations such as unhairing-liming, rinsing, deliming-bating and chrome tanning; in which 46% of these amounts of chemicals and water were used during correspond unhairing-liming unit (**Table 1**). Otherwise, this tannery rejects around 27 m^3 of wastewaters, which is a huge amount. These wastewaters are loaded with excess chemical products, which are not absorbed by skins, and organic matter eliminated from skins during unhairing-liming operation.

3. Wastewaters of modern tannery

A modern tannery was selected to study the quality of tannery wastewater. This industries located in industrial area of Doukkarat in Fez city, Morocco. This latter trait bovine hides only. As mentioned above, the tanning process involves several operations namely unhairing-liming (R1), rinsing (R2), deliming-bating (R3) and chrome tanning (R4) (**Figure 1**). So, the samples of wastewater were collected from the fuller of these units. The samples were monthly collected starting from February 2015. The sampling and conservation conditions were performed according to the ISO 5667-2 standard [25]. The physical parameters: temperature, pH and conductivity were directly measured after sampling. All samples were stored in a refrigerator at a temperature of 4°C according to AFNOR standards set by Rodier [25].

4. Characterization of industrial tannery wastewaters

Tannery wastewater contains significant content of chemical substances, including toxic compounds. Thus, several parameters were carried out to characterize these tannery wastewaters. Among these parameters, there are physicochemical analyses such as pH, temperature, electrical conductivity, turbidity, suspended solids (SS), chemical oxygen demand (COD), sulphate ions (SO_4^{2-}), nitrite (NO_2^-), nitrate (NO_3^-), ammonium (NH_4^+), orthophosphate (PO_4^{3-}) and sulfide ions (S^{2-}).

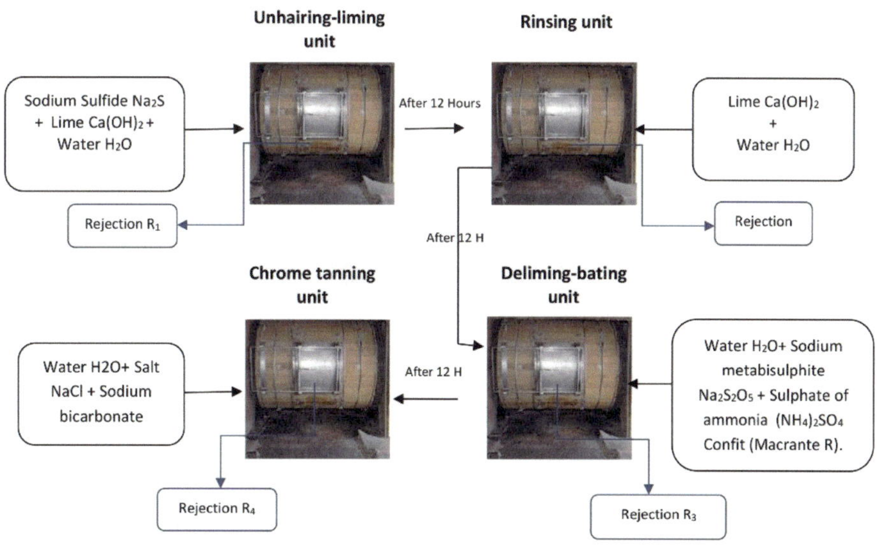

Figure 1.
Tanning process and sampling method of wastewater rejected by a modern tannery.

The pH was measured using a pH meter HANNA with a type electrode Senti X 22 according to NF T90.008 [25]. Conductivity and turbidity were measured by ORION type conductivity. Suspended solids (SS) were determined by centrifuging a wastewater volume according to standard NF T90.105 [25]. Sulfide ions were measured by the indirect method according to standard NF T 60–203 [25]. Ammonium, orthophosphate, nitrate, nitrite, sulfate, BOD_5 and COD were carried out by the spectrophotometric method using a DR/2005HACH at a fixed wavelength and according to AFNOR standards issued by Rodier J. et al. [25]. The bacteriological parameters were also evaluated, especially total coliform (TC), fecal coliform (FC), fecal sterptocoques (SF), total aerobic mesophilic (FMAT) and staphylococci. The enumeration of these bacteria was performed using desoxycolate lactose, slanetz, lauriabertani and Chapman respectively [25].

In general, the average temperature of R1, R2, R3 and R4 was between 24 and 27°C. In fact, these effluents take usually the environmental temperature, because of the absence of heating or cooling operations. The **Figure 2** shows the results of pH, conductivity and suspended solids for the fourth rejects (R1, R2, R3 and R4). So, R1 was largely basic, R2 had also a basic pH, R3 was slightly basic, whereas R4 was acidic (**Figure 2a**). The use of sodium carbonate and bicarbonate, during the first operations, could explain this basic pH of R1, R2 and R3. Concerning R4, the use of acids, for the solubilization ofchromium salt, could explain its acidic pH. The values obtained are comparable to those found in previous work on wastewater from tanneries that have a weakly basic pH [4, 26–28].

As for the conductivity, its average value was around 10 and 30 ms/cm, and then it was largely exceeding the Moroccan standards [29]. The largest values are recorded for unhairing-liming reject (R1), deliming-bating reject (R3), and

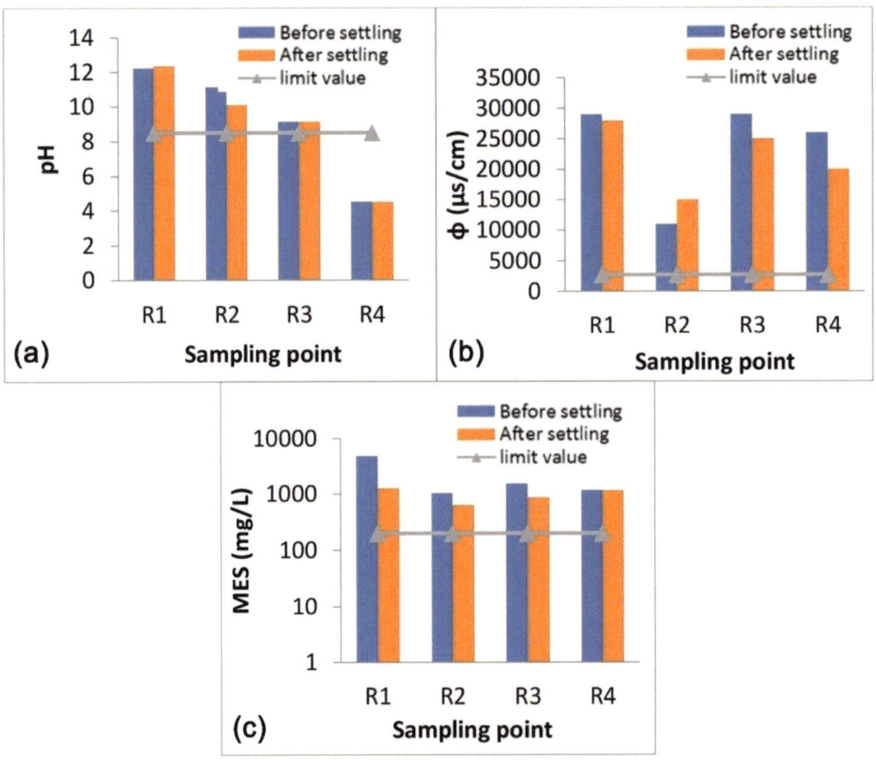

Figure 2.
Results of physical parameters within the fourth rejects of a modern tannery before and after settling: (a) pH; (b) conductivity; (c) suspending matter.

chromium tanning reject (R4) (**Figure 2b**). These high values of conductivity show significant use of salt during these tanning operations (**Table 1**). For suspended solids (SS), they had a high amount, which achieved more than 5000 mg/L for all effluents (R1, R2, R3 and R4) (**Figure 2c**). However, R1 had the highest amount of SS; this could be justified by the huge organic matter (Proteins, hair, fat ....) eliminated from animal skin during this step.

Concerning nitrogen compounds, ammonium, nitrate and nitrite have been followed and their average concentration has shown in **Figure 3**. R1 had again a high concentration of nitrate, nitrite and ammonium compared to R2, R3 and R4 (**Figure 3a–c**). This could be explained by the use of ammonium during unhairing-liming unit (**Table 1**). Indeed, nitrate amount was higher than nitrite and ammonium amount, this could be explained by the reduction of ammonium to nitrate passing by nitrite. This reduction could be through chemical reactions or bacteria such as *Nitrobacter, Nitrosomonas* ... etc. [29]. On the other hand, these fourth rejects were conformed to Moroccan standards of discharge in term of nitrogen compounds [29]. These results of nitrate, nitrite and ammonium ions are consistent with those of some authors [30, 31].

Figure 4 presents the result of phosphate ions. As shown, their amount was largely lower than the Moroccan standards reject for all tannery effluents (R1, R2, R3 and R4). Furthermore, R1 had the highest phosphate amount due to the use of some phosphate chemicals in this unit.

Concerning sulfate ions, **Figure 5a** shows that R3 and R4 were very loaded in sulfate than R1 and R2. This could be due to the use of sulfate chemicals in these units. Nevertheless, the tannery effluents were exceeded the Moroccan standards except for the reject R2 (**Figure 5a**). These high loads were due to the use of many

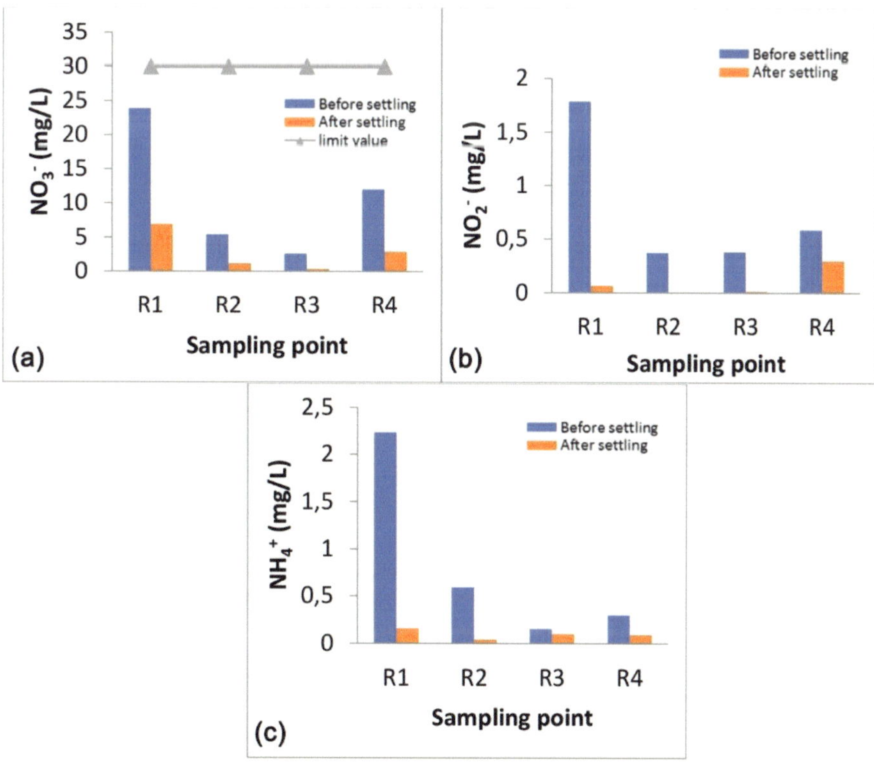

Figure 3.
Nitrogen compounds of tannery effluents before and after settling: (a) nitrate; (b) nitrite; (c) ammonium.

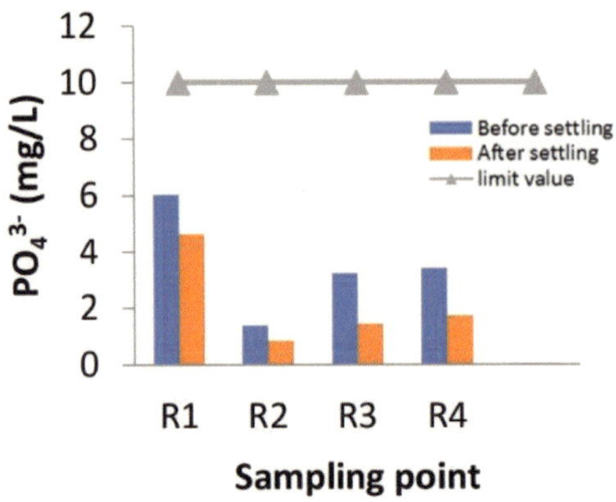

Figure 4.
Composition of the effluent studied before and after settling: Orthophosphate ions.

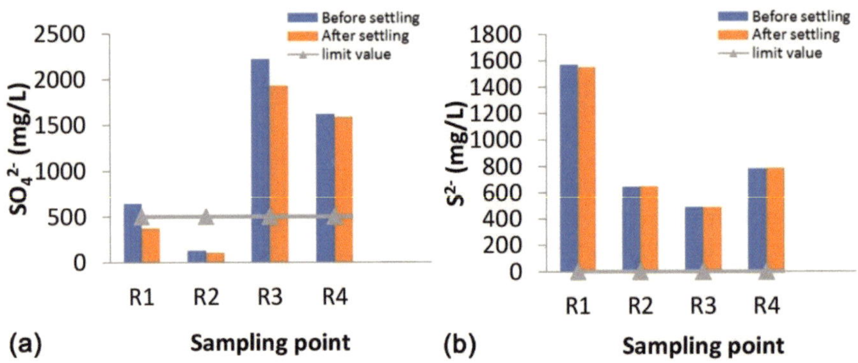

Figure 5.
Composition of the effluent studied before and after settling: (a) Sulfate ions; (b) Sulfide ions.

sulfate products during deliming and tanning units as chromium sulfate [32]. Concerning sulfide, the R1 had the highest load which approximates 1600 mg/L (**Figure 5b**). This huge amount could be justified by the use of the sulfide and sulfuric acid during the unhairing-liming step to eliminate hair of the animal skin. Indeed, the effluent of R1 is largely alkaline; which proves the presence of hydrosulfide HS- ions according to the Pourbaix diagram [33]. The results obtained are consistent with those found by [30] for the final rejection and those found by [34].

On the other hand, the organic load of the effluent is evaluated by measuring the chemical oxygen demand (COD) (**Figure 6a**) and biological oxygen demand BOD5 (**Figure 6b**). The **Figure 5** shows that R1 is very loaded in COD and BOD_5 than others (R2, R3 and R4) exceeding Moroccan standards reject [29]; this could be justified by the high amount of chemicals used in this first unit and the organic matter eliminated as well. The same figure reveals that all effluents are non-biodegradable because of the report COD/BOD_5, which was higher than 4 (**Figure 6c**). The concentrations found in COD are comparable to results obtained by several authors [30, 31].

Even if the characterization of these four rejects after settling shows a slight reduction of all physicochemical parameters, their amount did not meet Moroccan standards.

The microbiological analyses showed a low concentration of the total aerobic mesophilic flora (FMAT), which the average value was 300, 400 and 700 CFU/mL

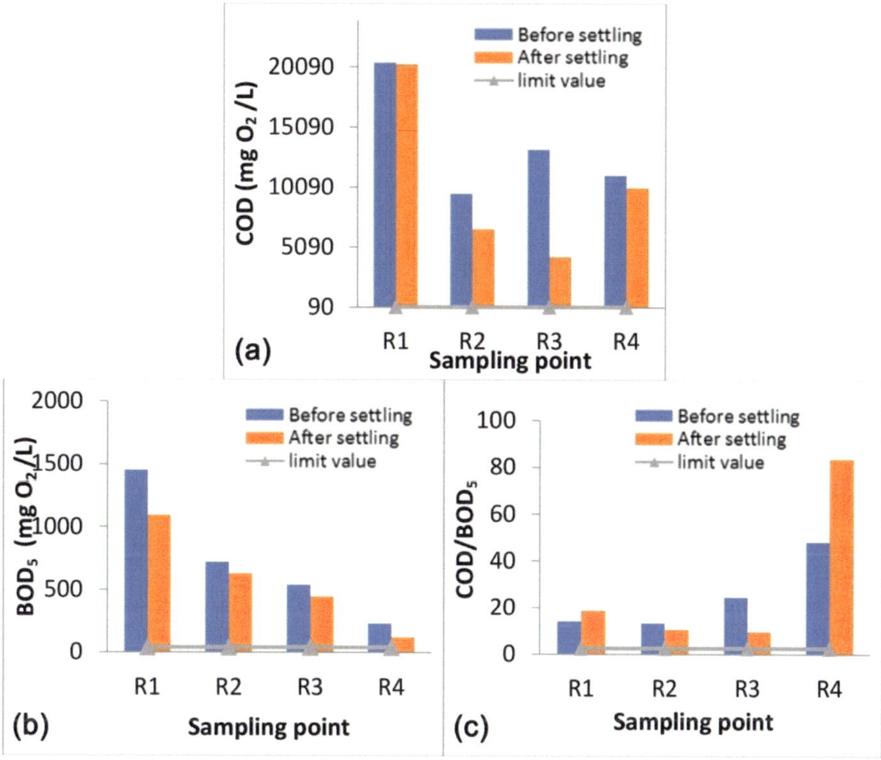

Figure 6.
Composition of the effluent studied before and after settling: (a) chemical oxygen demand; (b) biological oxygen demand; (c) ratio COD/BOD$_5$.

respectively for R1,R2 and R4 (**Table 2**). For pathogenic and fecal germs (Staphylococci, fecal streptococci, and the fecal coliform), **Table 2** reveals an absence of all of them. This could be explained by the high concentration of salts that may inhibit bacterial growth [35], and also by the toxicity of chromium which is present with a very high concentration [36]. However, we can conclude that the obtained germs (MTAF) may be halophilic and chromium bacteria.

In conclusion, tannery effluents are very complex, toxic and loaded in organic and inorganic matter, especially the first reject R1 of the unhairing-liming unit. Furthermore, R1 had a huge amount of sulfide, which could easily reduce to hydrogen sulfide under anaerobic conditions. As mentioned above, this toxic gas harms all living organisms, including human health. Therefore, the treatment of R1 has been very essential to remove sulfide ions.

Effluents	FC	TC	SF	FMAT	Staphylococci
R_1 (CFU/mL)	0	0	0	300	0
R_2 (CFU/mL)	0	0	0	400	0
R_3 (CFU/mL)	0	0	0	700	0
R_4 (CFU/mL)	0	0	0	0	0

CFU, colony forming units; FC, fecal coliform; TC, Total coliform; SF, Fecal sterptocoques; FMAT, total aerobic mesophilic.

Table 2.
Microbiological characterization of the wastewater of different operating tanning units.

5. Treatment of the unhairing-liming unit wastewater by precipitation

Several processes have been studied for the treatment of tannery wastewater, using simple and advanced methods. These processes include physicochemical treatments such as electrochemical methods [37, 38], filtration [28, 39], ion exchange [40], membrane filtration [41, 42], precipitation [43, 44], coagulation [5, 45], solvent extraction [46, 47], reverse osmosis [48, 49], adsorption [45, 50] and aerobic or anaerobic biological systems [51–53].

However, the high operating costs, the large amount of used chemicals, and the production of sludge are the main disadvantages of traditional chemical processing [2, 54]. On the other hand, advanced treatment techniques, such as reverse osmosis, ion exchange and membrane filtration are very expensive and generate another waste [54–56].

In fact, a dechromatization station was performed to remove the chromium from R4 of tannery industries in Doukkarat area in Fez city, Morocco. Nevertheless, there is not any plant to reduce sulfide toxicity in this Moroccan city. Thus, the elimination of sulfide ions is mandatory, but the removing process should be non-expensive and efficient. As mentioned above, R1 is non-biodegradable, and then, the physicochemical treatment is the best adequate treatment.

The chemical precipitation process is relatively simple and inexpensive. There are many precipitant agents such as ferric chloride, aluminum sulfate ... etc. [21]. The principle of this treatment is based on the production of the insoluble complex from pollutants and chemical agent. Furthermore, the conventional chemical precipitation processes include hydroxide precipitation [57] and sulfide precipitation [58, 59]. According to the literature, ferric chlorides can react with sulfide ions to produce a complex compound. For this reason, chemical precipitation using ferric chloride may be a great process to remove sulfide ions from R1.

Khatoon et al. [60], showed that COD and chromium could be treated by coagulation with an elimination rate of 38 to 46% for the suspended matter and 30 to 37% for the Total COD. The chromium elimination rate is 74 to 99% for an initial concentration of 12 mg/L using a coagulant dose of 800 mg/L with an optimal pH of around 7.5. This study showed also that ferric chloride gave better results than aluminum sulfate.

Other studies [61, 62] consist of the elimination of sulfur compounds from unhairing-liming effluent after a preliminary settling for one hour, following by filtration in a sintered glass. This glass had a porosity of 10 microns and a diameter of 70 mm.

For the treatment of tannery effluent [61, 63], particularly unhairing-liming effluent, a volume of a $FeCl_3$ solution was gradually added to 200 ml of this effluent until the formation of an insoluble complex. Afterward, these two phases (Liquid/solid) are separated mechanically and the liquid phase was only analyzed.

This treatment is based on the reduction of sulfide ions by ferric chloride $FeCl_3$ in a slightly basic medium according to these reactions, which were established by those authors [20, 64–67].

$$2Fe^{3+} + HS^- \rightarrow 2Fe^{2+} + S^0 + H^+ \qquad (1)$$

$$Fe^{2+} + HS^- \rightarrow FeS + H^+ \qquad (2)$$

$$2Fe^{3+} + 3HS^- \rightarrow 2FeS + S^0 + 3H^+ \qquad (3)$$

$$FeS + S^0 \rightarrow FeS_2 \qquad (4)$$

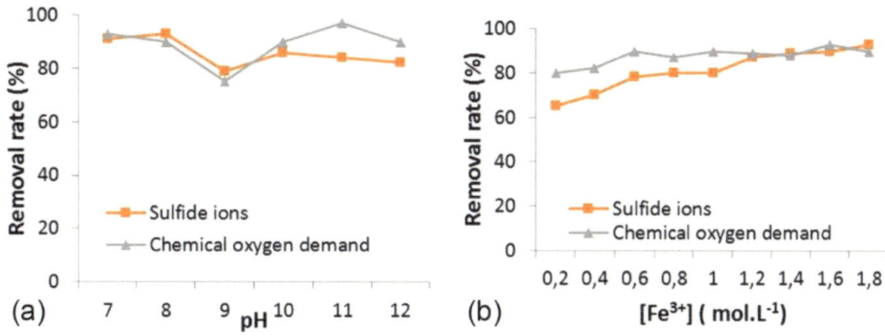

Figure 7.
(a) Effect of pH on the removal of sulfide ions and COD of effluent unhairing-liming; (b) elimination rate of sulfide ions in terms of the concentration of ferric ions at the pH of the medium (operating conditions: pH = 8.5, T = 24°C, $[S^{2-}]_0$ = 1570.94 mg/L).

According to the first Eq. (1), ferric ions react with sulfide ions to produce elemental sulfur. Afterward, the product ferrous ions will also react with sulfide ions to produce FeS precipitate. Otherwise, the reaction between ferric ions and sulfide ions may produce FeS and elemental sulfur according to the third Eq. (3). Finally, the FeS is converted to pyrite (FeS2) according to the fourth reaction (4). The precipitation depends strongly on the medium pH and the concentration of ions ferric [37–40].

Figure 7a reveals that the best removal rate of sulfide ions and the chemical oxygen demand was at the pH 8.5. Meanwhile, **Figure 7b** shows that the abatement rate of sulfide ions and COD increases when ferric ions (Fe^{3+}) concentrations increase too, and then this removal stabilizes at a value around 85% and 90% respectively for sulfide ions and COD, starting a ferric ions concentration of 1.4 mol/L. This could be explained by the high presence of hydrogen sulfide ions (HS^-)at pH of 8.5 according to the Pourbaix diagram [33].

The adjustment of pH effluent was performed by the addition of sulfuric acid (H_2SO_4) at a concentration of 1 N to obtain pH values between 7 and 12. For optimization of ferric chloride concentration, different concentrations were carried out ranging from 0.2 to 1.8 mol/L and using a pH effluent of 8.5 (**Figure 7b**).

The results show that the COD and sulfide ions had the same evolution of elimination depending on the pH and concentration of ferric ions. COD removal reached 90% at pH 7, 8 and 11 for ferric ion concentrations of 1, 1.2 and 0.8 mol/L respectively. As to sulfide ions, their removals achieved 90% and 84% at pH 7 and 8 using ferric ion concentrations of 1.6 and 1.8 mol/L respectively.

6. Conclusion

The main objective of this chapter was the characterization of different effluents of a modern tannery, giving a Moroccan modern tannery of Fez city as an example, and the treatment of unhairing-liming effluent, which was very loaded in sulfide.

The physicochemical characterization, of the fourth rejects of this modern tannery, showed a huge organic and inorganic pollution of these effluents, particularly unhairing-liming effluent that is largely alkaline and characterized by a huge organic and mineral pollution such as sulfides. However, the biological characterization revealed that these four effluents were empty from fecal and pathogenic germs due to their high inorganic toxicity. Otherwise, chemical precipitation using ferric chloride could remove a big amount of COD and sulfide ions during the treatment of unhairing-liming effluent. The abatement rate of sulfide ions reached 90% using a pH effluent of 8.5 and a ferric ions concentration of 1.4 mol/L.

In conclusion, the treatment of unhairing-liming wastewater could contribute to the protection of wildlife from the toxicity of sulfide ions through the reduction of the emission of greenhouse gases (Hydrogen Sulfide). Furthermore, chemical precipitation may be the best treatment for this type of effluent due to the high sulfide removal.

Acknowledgements

The authors gratefully acknowledge the Tannery Adam Group, the Biotechnology Laboratory of the Faculty of Sciences Dhar El Mahraz and process engineering laboratory of the Superior School of Technology, University Sidi Mohamed Ben Abdallah of Fez city, Morocco; for their cooperation and technical assistance during this study.

Author details

Anass Omor[1*], Karima Elkarrach[2], Redouane Ouafi[1], Zakia Rais[1], Fatima-Zahra ElMadani[1] and Mustafa Taleb[1]

1 Engineering Organometallic Materials, Molecular and Environmental Laboratory, University Sidi Mohamed Ben Abdallah, Fez, Morocco

2 Biotechnology Laboratory, University Sidi Mohamed Ben Abdallah, Fez, Morocco

*Address all correspondence to: anassomor@gmail.com

References

[1] Mwinyihija M, Quiesenberry W. Review of the challenges towards value addition of the leather sector in Africa. 2013; 2(11):518-528.

[2] Cecilia WM. Evaluation of Banana Peels , Pumice and Charcoal Potential to Adsorb Chromium Ions from Tannery Wastewater Cecilia Wangechi Muriuki A Thesis submitted in partial fulfillment for the Degree of Masters in Environmental Engineering and Management in the Jomo. 2015.

[3] FEDIC. Le secteur du cuir au MAROC. 2015.

[4] Sawadogo R, Guiguemde I, Diendere F, Diarra J, Bary A. Caractérisation physico-chimique des eaux résiduaires de tannerie : cas de l'usine TAN ALIZ à Ouagadougou/Burkina Faso. Int J Biol Chem Sci. 2012;6(6): 7087-7095.

[5] Aboulhassan MA, El Ouarghi H, Ait Benichou S, Ait Boughrous A, Khalil F. Influence of experimental parameters in the treatment of tannery wastewater by electrocoagulation. Sep Sci Technol [Internet]. 2018;53(17):2717-26. Available from: https://doi.org/10.1080/01496395.2018.1470642

[6] Bajza Z, Vrcek IV. Water Quality Analysis of Mixtures Obtained from Tannery Waste Effluents. Ecotoxicol Environ Saf [Internet]. 2001;50(1):15-18. Available from: http://www.sciencedirect.com/science/article/pii/S0147651301920858

[7] Babu RR, Bhadrinarayana NS, Begum KMMS, N.Anantharaman. Treatment of tannery wastewater by electrocoagulation. J Environ Sci. 2007; 19(12):1409-1415.

[8] Hossain MA. Effects of Occupational Exposure on Allergic Diseases and Relationship with Serum IgE Levels in the Tannery Workers in Bangladesh. Bioresearch Commun [Internet]. 2016;02(01):158-163. Available from: http://www.bioresearchcommunications.com/pdf/v2i1-jan-2016-158-163.pdf

[9] Islam LN, Rahman F, Hossain A. Serum Immunoglobulin Levels and Complement Function of Tannery Workers in Bangladesh. J Heal Pollut. 2019;9(21).

[10] Vaskova H, Kolomaznik K. A preliminary study of Raman spectroscopy potential for chromium detection. Proc 2018 19th Int Carpathian Control Conf ICCC 2018. 2018;5-8.

[11] Green A. Health Communication, Policy Adherence, and Marketing on Indoor Tanning Facility Websites in Ontario, Canada [Internet]. 2018. Available from: https://atrium2.lib.uoguelph.ca/xmlui/handle/10214/14163

[12] Xu J, Zhao M, Pei L, Zhang R, Liu X, Wei L, et al. Oxidative stress and DNA damage in a long-term hexavalent chromium-exposed population in North China. A cross-sectional study. BMJ Open. 2018;8(6):1-10.

[13] Sarwar F, Malik RN, Chow CW, Alam K. Occupational exposure and consequent health impairments due to potential incidental nanoparticles in leather tanneries: An evidential appraisal of south Asian developing countries. Environ Int. 2018;117(April):164-174.

[14] Ouafi R, Rais Z, Taleb M, Benabbou M, Asri M. Sawdust in the treatment of heavy metals-contaminated wastewater. In: Stefan E OK, editor. Sawdust: Properties, Potential Uses and Hazards. Nova Science Publishers, Incorporated; 2017. p. 147-81.

[15] Fahim NF, Barsoum BN, Eid AE, Khalil MS. Removal of chromium(III) from tannery wastewater using activated

carbon from sugar industrial waste. J Hazard Mater. 2006;136(2):303-309.

[16] Sahu SK, Meshram P, Pandey BD, Kumar V, Mankhand TR. Removal of chromium(III) by cation exchange resin, Indion 790 for tannery waste treatment. Hydrometallurgy [Internet]. 2009;99(3-4):170-174. Available from: http://dx.doi.org/10.1016/j.hydromet.2009.08.002

[17] Louarrat M, NtiecheRahman A, Bacaoui A, Yaacoubi A. Removal of Chromium Cr(Vi) of Tanning Effluent with Activated Carbon from Tannery Solid Wastes. Am J Phys Chem. 2018; 6(6):2327-2449.

[18] Mella B, Benvenuti J, Oliveira RF, Gutterres M. Preparation and characterization of activated carbon produced from tannery solid waste applied for tannery wastewater treatment. Environ Sci Pollut Res. 2019;

[19] Gutierrez O, Park D, Sharma KR, Yuan Z. Iron salts dosage for sulfide control in sewers induces chemical phosphorus removal during wastewater treatment. Water Res [Internet]. 2010;44(11):3467-3475. Available from: http://dx.doi.org/10.1016/j.watres.2010.03.023

[20] Nielsen AH, Lens P, Vollertsen J, Hvitved-Jacobsen T. Sulfide-iron interactions in domestic wastewater from a gravity sewer. Water Res. 2005; 39(12):2747-2755.

[21] Song Z, Williams CJ. Treatment of tannery wastewater by chemical coagulation. Desalination. 2004;164: 249-259.

[22] Ayhan S, Özacar M. Treatment of tannery liming drum wastewater by electrocoagulation. 2009;167:940-946.

[23] Elkarrach K, Merzouki M, Laidi O, Biyada S, Omor A, Benlemlih M. Sequencing batch reactor: Inexpensive and efficient treatment for tannery effluents of fez city in Morocco. Desalin Water Treat. 2020;202:71-77.

[24] Elkarrach K, Merzouki M, Biyada S, Benlemlih M. Bioaugmentation process for the treatment of tannery effluents in Fez, Morocco: An eco-friendly treatment using novel chromate bacteria. J Water Process Eng. 2020;38(June):101589.

[25] Rodier J, Bernard L, Nicole M. Analyse de l'eau . 9ème edition [Internet]. 2009. p. 1579. Available from: https://numerique.dunod.com/70567/L-analyse-de-l-eau--9e-ed-.ebook

[26] Cooman K, Gajardo M, Nieto J, Bornhardt C, Vidal G. Tannery wastewater characterization and toxicity effects on Daphnia spp. Environ Toxicol. 2003;18(1):45-51.

[27] Hamsatou MMD. Caractéristiques physico-chimiques, bactériologiques et impact sur les eaux de surface et les eaux souterraines. UNIVERSITE DE BAMAKO Faculté de Médecine de Pharmacie et d'Odonto-Stomatologie; 2005.

[28] Chowdhury M, Mostafa MG, Biswas TK, Saha AK. Treatment of leather industrial effluents by filtration and coagulation processes. Water Resour Ind [Internet]. 2013;3:11-22. Available from: http://dx.doi.org/10.1016/j.wri.2013.05.002

[29] Minister of the Interior M of ME. Valeurs Limites de Rejet à respecter par les déversements [Internet]. 2014. Available from: http://www.water.gov.ma/wp-content/uploads/2016/01/4.3.3.Valeurs-Limites-de-Rejet.pdf

[30] Sundarapandiyan S, Chandrasekar R, Ramanaiah B, Krishnan S, Saravanan P. Electrochemical oxidation and reuse of tannery saline wastewater. J Hazard Mater [Internet]. 2010;180(1-3):197-203. Available from: http://dx.doi.org/10.1016/j.jhazmat.2010.04.013

[31] Mendoza-Roca JA, Galiana-Aleixandre M V., Lora-García J,

Bes-Piá A. Purification of tannery effluents by ultrafiltration in view of permeate reuse. Sep Purif Technol. 2010;70(3):296-301.

[32] Bosnic M, J. Buljan, R.P. Daniels. Pollutants in tannery effluents [Internet]. 2003. Available from: http://leatherpanel.org/sites/default/files/publications-attachments/polutants.pdf

[33] Choudhary L, Macdonald DD, Alfantazi A. Role of Thiosulfate in the Corrosion of Steels: A Review. CORROSION. 2015 Sep;71(9):1147-1168.

[34] Vidal G, Nieto J, Cooman K, Gajardo M, Bornhardt C. Unhairing effluents treated by an activated sludge system. J Hazard Mater. 2004;112(1-2): 143-149.

[35] Rene ER, Kim SJ, Park HS. Effect of COD/N ratio and salinity on the performance of sequencing batch reactors. 2008;99:839-846.

[36] Jobby R, Jha P, Yadav AK, Desai N. Chemosphere Biosorption and biotransformation of hexavalent chromium [Cr (VI)]: A comprehensive review. Chemosphere [Internet]. 2018;207:255-66. Available from: https://doi.org/10.1016/j.chemosphere.2018.05.050

[37] Sharma S, Simsek H. Treatment of canola-oil refinery effluent using electrochemical methods: A comparison between combined electrocoagulation + electrooxidation and electrochemical peroxidation methods. Chemosphere [Internet]. 2019;630-9. Available from: https://doi.org/10.1016/j.chemosphere.2019.01.066

[38] Ghasem A, Mahya M, Davood N. Combined Electrocoagulation/Electrooxidation Process for the COD Removal and Recovery of Tannery Industry Wastewater. Environ Prog Sustain Energy. 2017;00(00):1-8.

[39] Abdel-Shafy HI, El-Khateeb MA, Mansour MSM. Treatment of leather industrial wastewater via combined advanced oxidation and membrane filtration. Water Sci Technol. 2016; 74(3):586-594.

[40] Korak JA, Huggins RG, Arias-Paić MS. Nanofiltration to Improve Process Efficiency of Hexavalent Chromium Treatment Using Ion Exchange. J Am Water Works Assoc. 2018;110(6):E13–E26.

[41] Mouiya M, Abourriche A, Bouazizi A, Benhammou A, El Hafiane Y, Abouliatim Y, et al. Flat ceramic microfiltration membrane based on natural clay and Moroccan phosphate for desalination and industrial wastewater treatment. Desalination. 2018; 427(November 2017):42-50.

[42] Zouboulis A, Peleka E, Ntolia A. Treatment of Tannery Wastewater with Vibratory Shear-Enhanced Processing Membrane Filtration. Separations [Internet]. 2019;6(2):20. Available from: https://www.mdpi.com/2297-8739/6/2/20

[43] Ramírez-Estrada A, Mena-Cervantes VY, Fuentes-García J, Vazquez-Arenas J, Palma-Goyes R, Flores-Vela AI, et al. Cr(III) removal from synthetic and real tanning effluents using an electro-precipitation method. J Environ Chem Eng [Internet]. 2018;6(1):1219-1225. Available from: http://dx.doi.org/10.1016/j.jece.2018.01.038

[44] Wang D, Ye Y, Liu H, Ma H, Zhang W. Effect of alkaline precipitation on Cr species of Cr(III)-bearing complexes typically used in the tannery industry. Chemosphere [Internet]. 2018;193(Iii):42-9. Available from: https://doi.org/10.1016/j.chemosphere.2017.11.006

[45] Mella B, Barcellos BS de C, da Silva Costa DE, Gutterres M. Treatment of

Leather Dyeing Wastewater with Associated Process of Coagulation-Flocculation/Adsorption/Ozonation. Ozone Sci Eng [Internet]. 2018;40(2): 133-40. Available from: https://doi.org/10.1080/01919512.2017.1346464

[46] Bharagava RN, Saxena G, Mulla SI, Patel DK. Characterization and Identification of Recalcitrant Organic Pollutants (ROPs) in Tannery Wastewater and Its Phytotoxicity Evaluation for Environmental Safety. Arch Environ Contam Toxicol [Internet]. 2018; 75(2):259-72. Available from: https://doi.org/10.1007/s00244-017-0490-x

[47] Karmakar S, Bhowal A, Das P. Waste Water Recycling and Management [Internet]. Vol. 336, Waste Water Recycling and Management. Springer Singapore; 2019. 15-26 p. Available from:http://dx.doi.org/10.1007/978-981-13-2619-6_2

[48] Gupta SK, Gupta S. Closed loop value chain to achieve sustainable solution for tannery effluent. J Clean Prod. 2019;213:845-846.

[49] Roopa D, Divya R, Nathiya S. Management of RO reject water from the tannery industry by solar tunnel dryer. Int J Adv Res Dev. 2019;4(2):15-20.

[50] Puchana-Rosero MJ, Lima EC, Mella B, Da Costa D, Poll E, Gutterres M. A coagulation-flocculation process combined with adsorption using activated carbon obtained from sludge for dye removal from tannery wastewater. J Chil Chem Soc. 2018;63(1):3867-3874.

[51] Munz G, Gori R, Cammilli L, Lubello C. Characterization of tannery wastewater and biomass in a membrane bioreactor using respirometric analysis. Bioresour Technol J. 2008;99(01): 8612-8618.

[52] Jemec A, Zupanc GD. Bioresource Technology Anaerobic digestion of tannery waste : Semi-continuous and anaerobic sequencing batch reactor processes ˇ ic. Bioresour Technol. 2010;101:26-33.

[53] Yusif BB, Bichi KA, Oyekunle OA, Girei AI, Garba PY, Garba FH. A Review of Tannery Effluent Treatment. Int J Appl Sci Math Theory. 2016;2(3):29-43.

[54] Abdulla HM, Kamal EM, Mohamed AH, El-bassuony AD. CHROMIUM REMOVAL FROM TANNERY WASTEWATER USING CHEMICAL AND BIOLOGICAL TECHNIQUES AIMING ZERO. In: PROCEEDING OF FIFTH SCIENTIFIC ENVIRONMENTAL CONFERENCE. 2010. p. 171-83.

[55] Nabila B. Epuration et reconcentration de l'acide chromique contenant des impuretés métalliques par un procédé associant l'électrodialyse à l'échange d'ions. 2009.

[56] Abdillahi MM. Assemblage et séparation de polyélectrolytes pour le traitement d ' eaux contaminées par des cations métalliques. 2016.

[57] Fu F, Wang Q. Removal of heavy metal ions from wastewaters: A review. J Environ Manage. 2011 Mar;92(3): 407-418.

[58] Contestabile M, Panero S, Scrosati B. A laboratory-scale litium battery recycling process. J Power Sources. 1999;83:75-78.

[59] Kurniawan TA, Chan GYS, Lo WH, Babel S. Physico-chemical treatment techniques for wastewater laden with heavy metals. Chem Eng J. 2006; 118(1-2):83-98.

[60] Khatoon J. TREATNENT OF TANNERY WASTEWATER USING ACTIVATED SLUDGE PROCESS. 2012.

[61] Omor A, Rais Z, El Rhazi K, Merzouki M, El Karrach K, Elallaoui N, et al. Optimization of the method

wastewater treatment of unit bovine hides's unhairing liming. J Mater Environ Sci. 2017;8(4).

[62] Omor A, El K, Elallaoui N, Rais Z, Chetouani A. Characterization and treatment of effluents loaded with sulphides from two tanneries : Modern and Artisanal. 2019;1:61-72.

[63] Omor A, El Rhazi K, Elallaoui N, Taleb M, Taleb A, Rais Z, et al. Characterization and treatment of effluents loaded with sulphides from two tanneries: Modern and Artisanal. Moroccan J Chem. 2019;7(1).

[64] Zhang L, Yuan Z. Inhibition of sulfate-reducing and methanogenic activities of anaerobic sewer biofilms by ferric iron dosing. 2009;43:4123-4132.

[65] Yang S, Bae J. A feasibility of coagulation as post-treatment of the anaerobic fluidized bed reactor (AFBR) treating domestic wastewater. J krean Soc Water Wastewater. 2014;28(6):623-634.

[66] Querol X, Chinchon S, Lopez-Soler A. Iron sulfide precipitation sequence in Albian coals from the Maestrazgo Basin, southeastern Iberian Range, northeastern Spain. Int J Coal Geol. 1989;11(2):171-189.

[67] Jin R, Yang G, Zhang Q, Ma C, Yu J, Xing B. The effect of sulfide inhibition on the ANAMMOX process. Water Res [Internet]. 2012;47(3):1459-1469. Available from: http://dx.doi.org/10.1016/j.watres.2012.12.018

Chapter 18

Performance of Chitosan as Natural Coagulant in Oil Palm Mill Effluent Treatment

Man Djun Lee and Pui San Lee

Abstract

This chapter presents the study on pollutant removal in palm oil mill effluent using chitosan as natural coagulant. Up until today, palm oil mill effluent (POME) considered one of the significant sources of environmental pollution. The characteristics of POME include contaminating the source of drinking water, which also harmful to the aquatic ecosystem by creating a highly acidic environment or causing eutrophication. With increasing public awareness of environmental pollution, it creates the need to address this issue. Chitosan is non-polluting food-based anionic and biodegradable biopolymer that are environmentally friendly useful in wastewater treatment. The critical parameter to determine the effectiveness of pollutants removal is chemical oxygen demand, color, and total suspended solids. This chapter also presents and discusses some of the significant findings to provide proper understandings and implications in this topic.

Keywords: wastewater treatment, oil palm industry, chemical oxygen demand, total suspended solids, color removal

1. Introduction

Palm oil industry is a significant industry sector and plays a significant role in Malaysia's economy as one of the largest palm oil producers in the entire world [1]. The palm oil industry in Malaysia contributes about 39% of the world palm oil production and also 44% of palm oil world export [2]. Due to this importance, a large area of land has been converted into oil palm plantation estate. At the same time more and more palm oil mill has been built to process the increasing amount of oil palm fresh fruit bunch (FFB) into crude palm oil [1]. The growth of the industry at the same time indicates the increase of wastewater or palm oil mill effluent (POME) produced and released into the watercourse, which will bring harm to the environment.

The process of extracting crude palm oil from the fresh fruit bunch consumes much water and therefore produces a large volume of wastewater. In Malaysia, a record of 0.67 cubic meters of POME generated in order to process one ton of FFB [1]. Approximately 5–7.5 tons of water is required to produce one ton of crude palm oil. Eventually, more than 50% of these water would become POME which is shown in **Figure 1** [2].

It is approximately 48–72 million tons, and 49–71 million tons of POME was generated in the year 2013 and 2014, respectively. In the year 2014, it estimated 19.66 million tons of crude palm oil produced with roughly 44 million cubic meters

Figure 1.
Palm oil mill effluent.

of POME generated [3]. In POME generated by processing 1 ton of FFB, it contains about 29-30 kg of 30°C, 3-days Biochemical Oxygen Demand (BOD_3) [1]. From the data of POME produced in the year 2014, if the raw POME discharged into the environment without any further treatment, the BOD discharged is equal to the waste generated by 75 million people which is the 2.5 times of current Malaysia's population [3]. POME is also said to be 100 times polluting than domestic sewage [1]. According to the Department of Environment (DOE) practice, there are two ways of discharging treated POME, which are into water course or land. For the discharge into the watercourse, there are seven contaminants contained in the POME regulated. The regulated parameters are BOD_3, suspended solids (SS), oil and grease (O&G), ammoniacal nitrogen (AN), total nitrogen (TN), pH and temperature. For the discharge onto the land, the only parameter is BOD_3 which set at 5000 mg/L. **Table 1** shows the characteristic of raw and treated POME obtained from the discharge point of the local palm oil mill in Malaysia and DOE discharge limit [2].

The most popular method to treat the POME in Malaysia is the ponding system due to low equipment cost and the system is easy to operate. In Malaysia, there are more than 85% of palm oil mills that are currently adopting this method to reduce the BOD of POME into an acceptable limit which is less than 100 mg/L in West Malaysia and 50 mg/L in East Malaysia. In the ponding system, the POME undergoes biological treatments which include anaerobic digestion process followed by aerobic ponding with the hydraulic retention time of 40 days or above. However, the ponding system also causes some drawbacks which are long hydraulic retention time (HRT), vast land needed and the release of greenhouse gases (methane). There are also many palm oil mills which are unable to achieve the discharge limit only by using the ponding system [3].

If untreated POME discharges into the watercourse, it will undergo biodegradation process and consume dissolved oxygen in the water which eventually

Parameters	Raw POME	DOE Discharge Limit
Temperature (°C)	85	45
pH	4.2	5.0–9.0
Oil & Grease (mg/L)	6000	50
BOD (mg/L)	25,000	100
COD (mg/L)	51,000	—
TS (mg/L)	40,000	1500
TSS (mg/L)	18,000	400
TVS (mg/L)	34,000	—
TN (mg/L)	750	200
Color (ADMI)	Above 500	200

Table 1.
Characteristic of raw POME and DOE discharge limit [4].

will kill the marine animals, especially fish in the river. The untreated POME, which is acidic, will cause the watercourse to turn acidic and affect the aquatic life. Moreover, the oil content in untreated POME tends to form a thick layer on the water surface that will prevent the absorption of oxygen. The dark brown color and unpleasant smell of POME will turn the stream into brownish and unacceptable for public consumption [2]. Apart from that, the high concentration of suspended solids will remain at the bottom of the river and undergo biodegradation, which will produce sludge oxygen demand (SOD) and deplete the dissolved oxygen [4]. In order to protect the environment, DOE Malaysia establishes a standard where the final discharge of treated POME that came out from the mill must be less than 100 mg/L of COD. Hence, for POME to have the minimum or no impact on the environment when discharging and to comply with the discharge limits, the palm oil mill must have an effective POME wastewater treatment system. The cost of maintenance and operation of the POME wastewater treatment system, availability of land and location of mill greatly influencing the choice and selection of POME wastewater treatment systems in Malaysia. In return, it will stress the industry players, especially small and medium scale palm oil mill financially. Therefore, the central idea of this study is to provide an inexpensive and uncomplicated method for small and medium scale palm oil industries to process POME before discharging to the watercourse. This study provides insights into utilizing chitosan and polyglutamic acid in the POME treatment process to remove pollutants that contribute to high COD, color, and TSS of POME.

2. Palm oil mill effluent

2.1 Source of POME in palm oil mill

The most common way in extracting palm oil from fresh fruit bunches (FFB) is the wet palm oil milling process. Several stages of wet palm oil milling process required a tremendous amount of water and steam for washing and sterilizing. As a result, this generates a considerable quantity of wastewater or better known as POME from palm oil mill. **Figure 2** shows a simplified process flow diagram to produce palm oil.

In a palm oil mill operation using a conventional manufacturing process, there are three primary processing operations responsible for producing the

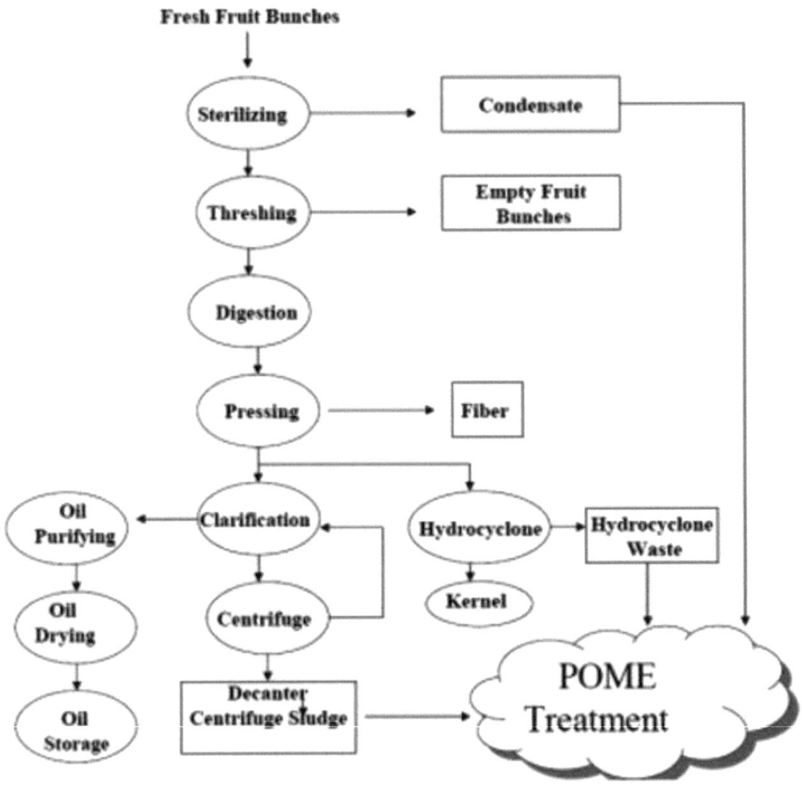

Figure 2.
Typical palm oil process flowchart [5].

POME. These three primary processes are sterilization of FFB, clarification of the extracted crude palm oil and hydro-cyclone separation of the cracked mixture of kernel and shell. Sterilization process customarily carried out in horizontal cylindrical autoclaves known as sterilizers where the FFBs are cooked with steam at the pressure about 3 atm for 1 to 1.5 hours. The sterilization process aims to inactivate the natural enzymes in the fruits (lipases) and inhibit the splitting of fat into free fatty acid (FFA) and cause oil loss. Besides, the steam sterilization process loosens the fruits from the bunch and soften the mesocarp to ease the oil extraction. This station contributes approximately 36% of total POME [1]. The clarification process is to separate the oil produced from the press station, which is mixed with water and solid from the bunch fiber. The oil usually is separated from the mixture in the clarifier tank by using gravity, de-sander and also decanter. This station contributes the majority part of the POME, which is 60% [6]. The nuts from the nut silo will be cracked by nutcracker in the ripple mill. These cracked kernel and shell mixture are separated in air columns and by a water bath in hydrocyclone. This station only produces around 4% of POME. The POME generated from sterilizer condensate, clarification of oil and hydro-cyclone is in the ratio of 9:15:1 (36%:60%:4%). **Table 2** shows the characteristics of different source of wastewater in palm oil mill that combined to produce POME [1].

2.2 Characteristics of POME

The POME from different mills would have different characteristics due to different oil extraction technique, FFB quality, climate, condition of palm oil processing and mill requirement on POME discharge limit [6]. POME is a mixture of water (up to 95%), oil and fine suspended solids [7]. The suspended solid (TSS) is the vegetative matter such as cell walls, organelles, short fibers, water-soluble carbohydrates (glucose, reducing sugar and pectin), nitrogenous compound (protein and amino acid), free organic acid, lipids and also combined small organic and mineral constituents [8]. POME is considered as non-toxic waste as the palm oil mills usually do not use any harmful chemical in the entire milling process [1]. The dark color of POME is usually caused by the decomposition of lignocellulosic materials; which produces lignin, tannin, humic acids, carotene and other organic matter that are recalcitrant to conventional treatment [9]. These suspended solids will eventually contribute to the high BOD of POME [1].

In term of organic content, based on the biochemical oxygen demand (BOD), raw POME has an average BOD of 25,000 mg/L. Raw POME is highly acidic. Biodegradability of effluent can be determined from the BOD/COD ratio. COD stands for chemical oxygen demand. BOD/COD ratio indicates the fraction of chemically oxidized organics which is eligible for biological degradation. In East Malaysia, the POME discharged when the BOD is less than 50 mg/L as required by Department of Environment (DOE). The pollution load of POME generated in a palm oil mill in a day can be calculated based on the following Eq. (1) and (2):

$$\textbf{Pollution Load,}\frac{\textbf{kg}}{\textbf{d}} = \frac{\text{POME Flow Rate,}\frac{\text{m}^3}{\text{d}}\text{x Concentration, mg / L}}{1000} \tag{1}$$

$$\begin{aligned}\textbf{POME Flow Rate,}\frac{\textbf{m}^3}{\textbf{d}} &= \text{Process capacity}\frac{\text{ton FFB}}{\text{h}} \times \text{Process Efficiency, \%} \\ &\times \text{Operating Hours,}\frac{\text{h}}{\text{d}} \times \text{POME generated,}\frac{\text{m}^3}{\text{ton}}\end{aligned} \tag{2}$$

2.3 Biochemical oxygen demand (BOD)

Biochemical oxygen demand (BOD) is the measure of the amount of oxygen that bacteria will consume during the decomposition of organic matter content under aerobic conditions. BOD test should be carried out according to APHA Standard Method 5510B [10]. BOD is determined by incubating a sealed sample of water for five days and measuring the loss of oxygen from the beginning to the

Parameter	Sterilizer condensate	Clarification wastewater	Hydro-cyclone wastewater
pH	5.0	4.5	—
Oil and grease; mg/L	4000	7000	300
Biochemical oxygen demand (BOD) 3 days, 30°C; mg/L	23,000	29,000	5000
Chemical oxygen demand (COD); mg/L	47,000	64,000	15,000
Suspended solids; mg/L	5000	23,000	7000
Dissolved solids; mg/L	34,000	22,000	100
Total nitrogen; mg/L	600	1200	100
Ammoniacal-nitrogen; mg/L	20	40	—

Table 2.
Characteristic of different sources of wastewater [1].

end of the test. The samples are usually diluted before the incubation because the bacteria could deplete all of the oxygen in the bottle before the test is complete [11]. It is essential to determine the sample size and dilution ratio before the BOD test, as this will ensure valid BOD results. The pH value of the samples should be in the range of 6.0–8.0, as alkalinity or acidity of samples can prevent bacteria from growing during the BOD test. pH can adjust by adding sodium hydroxide (NaOH) and sulfuric acid (H_2SO_4) [10]. When the test carries out in this way, the BOD usually abbreviated as BOD_5. BOD is a severe problem in natural waters because the dissolved oxygen (DO) of the water can be stressed by BOD oxidation [12].

2.4 Chemical oxygen demand (COD)

Chemical oxygen demand is a measure of the amount of oxygen required to oxidize all organic material into carbon dioxide and water. COD values usually are higher than BOD values, but COD measurements can be obtained in a few hours while BOD measurements will take around five days [11]. Samples heated for 2 hours with sulfuric acid and strong oxidizing agent potassium dichromate. The reduction reaction is shown in Eq. (3).

$$6Cl^{-}+Cr_2O_7^{2-}+14H^{+}\rightarrow 3Cl_2+2Cr^{3+}+7H_2O \quad (3)$$

The amount of Cr^{3+} produced is measured at wavelengths and reflected in mg/L of COD.

2.5 Total suspended solids (TSS)

Total suspended solids are a measure of suspended matter contained in the wastewater. Suspended solids contain BOD and can impair water quality by adding turbidity and reducing esthetics. Discharges of SS also caused deposits that developed at the bottom of waterways. In the laboratory, standard filtration and drying method used to measure SS, where the increase of weight of a container/filter is measured, for a known volume of wastewater filtered [12]. The TSS before and after the experiment measured according to Standard Methods Section 2540 D, and total solids dried from 103–105°C. The treated and the untreated POME samples were evaporated in a weighed dish and dried to a constant weight in an oven from 103–105°C. The increase in weight over the empty dish represents the total solids. TSS calculation is shown in Eq. (4).

$$TSS\left(\frac{mg}{L}\right)=\frac{(\text{Weight of dried residue}+\text{dish}-\text{weight of dish})\,\text{mg x }1000}{\text{sample volume, ml}} \quad (4)$$

2.6 Conventional method in POME treatment

The natural chemical properties of the POME make it easily treated by a biological approach. Currently, there are three biological processes employed in the palm oil industry which are anaerobic, facultative anaerobic, and aerobic treatments. The anaerobic treatment is the major part which is removing pollutant (BOD). It can remove BOD up to 95% [13]. There are four main stages which are hydrolytic, acidogenic, acetogenesis and methanogenic. The hydrolysis process begins with bacteria of insoluble organic polymers (carbohydrate) and complex organic compound (protein and lipid) to make them available for other bacteria. Hydrolytic microorganisms will secrete extracellular enzymes for hydrolysis. This process will convert organics into simpler molecule such as amino acids, glycerol, triglycerides, sugar and fatty acids. Meanwhile, in acidogenesis process, the hydrolyzed or soluble products from the first stage are further broken down by acidogen into simpler organic compound such as volatile fatty acid (VFA), ammonia, carbon dioxide, hydrogen and hydrogen sulfide. In the acetogenesis process, the simple molecule from the previous stage is further digested by acetogens to produce carbon dioxide, hydrogen and acetic acid. For the methanogenesis process, the acetic acid, hydrogen and VFA are converted to methane, carbon dioxide and water by methanogens.

The ponding system is a combination of a series anaerobic, facultative, and algae (aerobic) ponds, as shown in **Figures 3** and **4**. Ponding system primarily anaerobic and facultative ponds require less energy as it does not need mechanical mixing, operation control or monitoring. The major drawback of the ponding system is a large area of land is needed to accommodate a series of ponds to achieve the discharge limit [13]. In constructing the ponds, depth is the primary consideration for different types of ponds. However, the optimum depth for the anaerobic pond is 5-7 m, the facultative anaerobic pond is 1–1.5 m and aerobic pond is 0.5-1 m. The sufficient hydraulic retention time (HRT) of anaerobic, facultative anaerobic and aerobic ponds are 45, 20 and 14 days, respectively [13]. The problems arise from the ponding system is the formation of scum. Scum form when the bubbles rise to the surface together with the fine suspended solids. It is caused by the presence of oil and grease in the POME. Another drawback of the ponding system is the solid

Figure 3.
Series of ponds for POME treatment.

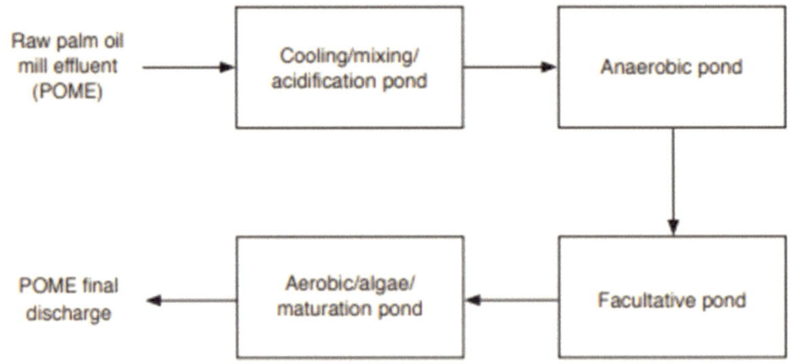

Figure 4.
Typical configuration for ponding treatment system for POME [14].

sludge accumulates at the bottom of the ponds. It will affect the effectiveness of the pond as it decreases the volumetric capacity and hydraulics retention time (HRT) [13]. Therefore, de-sludging is required when the sludge is more than one-third of the pond. About 85% of the palm oil mills that POME in Malaysia adopted ponding system because it is inexpensive, low capital, simple and easy to handle [14]. The palm oil industry is widely favored to the ponding system as only clay lining of ponds is needed and can be constructed easily by excavating hence low marginal cost [4].

The combination of open digester and ponding system is another type of conventional POME treatment system that combines an open digester tank with a series of ponds. **Figure 5** shows a typical open digester tank. Digester tank may build with various volumetric capacities ranging from 600 until 3600 m^3. In this treatment method, digester has the same function as the anaerobic pond. It carries out the anaerobic digestion. The output of the POME from the digester will then enter facultative anaerobic ponds and then algae (aerobic) ponds. The digester can decrease the BOD in a shorter time than the pond. The HRT for digester is only 20–25 days which is a lot shorter than anaerobic ponding system. Although it is proven that the digester is more effective than anaerobic ponds, it brings some drawback to the operator. The disadvantages of digester include scum formation on the top, sludge accumulation at the bottom and the corrosion of the steel structure of the digesters due to prolonged exposure to hydrogen sulfide. There are incidents which reported that the digester burst and collapsed [13].

Figure 5.
Typical open digester tank.

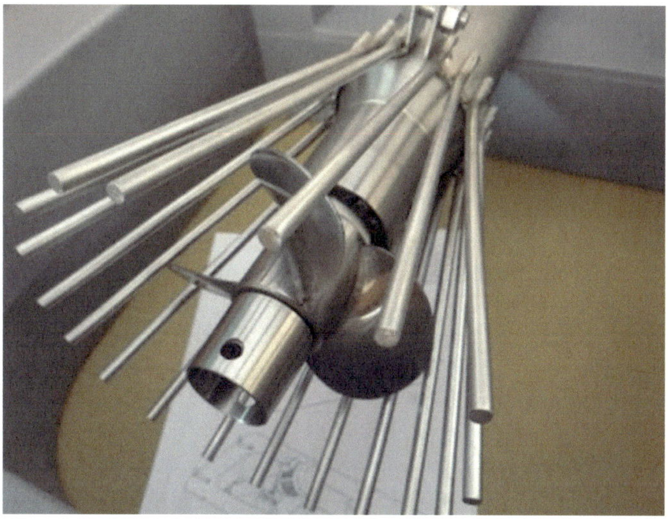

Figure 6.
Surface aerator for POME ponds.

Extended aeration is to complement the previous conventional treatment system, which shown in **Figure 6**. In this treatment method, mechanical surface aerators are introduced at the aerobic ponds to supply more oxygen to the ponds. It can reduce the BOD in POME effectively by aerobic processes. Usually, the surface aerators are installed at the end of the ponds before discharging the POME. This treatment method is useful only used when the land area is a constraint and does not permit extensive wastewater treatment [13].

3. POME polishing technologies

In recent years, many studies conducted to investigate alternative POME treatment technologies, especially in secondary and tertiary treatment. The technologies that are widely investigated are adsorption, coagulation or flocculation, membrane filtration and advanced oxidation processes. Most of these investigations are in laboratory scale, but they show potential to overcome the drawback of conventional ponding system [2]. **Figure 7** shows an overview of recent POME polishing technologies.

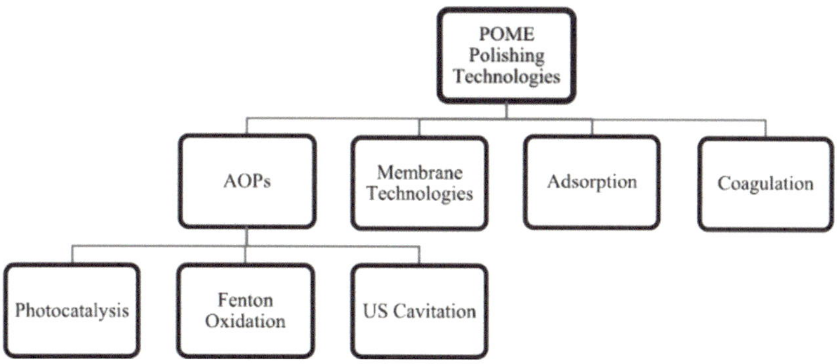

Figure 7.
Overview of recent POME polishing technologies [2].

3.1 Advanced oxidation process (AOP)

AOPs are the processes which degrade the organic pollutant by the powerful and reactive hydroxyl radical (OH•). Hydroxyl radical (OH•) generated would have an oxidation potential of 2.8 eV OH• can generate through either one or a combined of chemical oxidation by using H_2O_2, ozone, ultrasound and radiation assisted source (ultraviolet) [15]. During the treatment of wastewater, OH• will attack the organic pollutants and convert them to CO_2, H_2O and inorganic salt [16]. AOPs can effectively degrade the pollutants and have its advantages of non-selectively, mineralization of pollutants and ease of operation compared to the conventional methods. The most popular AOPs are Fenton oxidation, photocatalysis, ultrasound cavitation and ozonation. Fenton oxidation uses the reaction between Fe^{2+} and H_2O_2 to produce OH• [17]. Photocatalysis applies metal oxide (such as TiO_2) in the presence of irradiation (UV and Vis) to produce OH• [18]. Ozonation uses the ozone, which is a powerful oxidant with high thermodynamic oxidation potential [19]. Ultrasound (US) cavitation uses ultrasound to oxidize the pollutants. AOPs are more effective when combined two or more AOPs due to more OH• is generated, lower catalyst consumption and shorter process time [20]. AOPs have successfully adopted as tertiary treatment of wastewater such as olive oil mill wastewater (OOWW), agrochemicals, pulp and paper, textile wastewater and pharmaceutical [2].

3.2 Membrane technologies

Besides advance oxidation processes (AOPs), membrane technology is also one of the popular polishing methods of POME. The advantages of membrane technologies are high removal rate, modularity, and ability to integrate with other water treatment method. However, the main drawback of membrane technologies is that the membrane fouling will cause significant reduction in permeate flow due to the surface and pore-blocking of the membrane. The high initial capital and maintenance costs have also limited the application of the membrane. The most commonly used membrane in membrane technologies are microfiltration (MF), ultrafiltration (UF), nano-filtration (NF) and reverse osmosis (RO) [2]. There is an argument about membrane technologies in removing COD from POME compared to other technology. Higher pressure might have provided higher treatment efficiency but at the same time also contributes to the increasing rate of membrane fouling. The effectiveness of membrane technologies in POME treatment depends on the properties of the membrane. Nano-filtration performs better than ultrafiltration, but it has a higher level of fouling compared to ultrafiltration. Membrane technologies can be combined with other technology such as coagulation and flocculation to increase their treatment effectiveness [2]. **Table 3** shows some of the application of membrane technologies used in POME polishing.

3.3 Adsorption technologies

For adsorption technologies, it is a physicochemical separation process involving inter-phase transfer between an adsorbent and a solution. The pollutant in the solution (adsorbate) absorb onto the surface of the adsorbent. Adsorption can be a reversible process which offers the possibility of adsorbent regeneration through desorption [28]. Adsorption mechanism mainly influenced by the physical forces (physisorption) or chemical interactions (chemisorption) between the adsorbent and adsorbate. The adsorption is also influenced by characteristics of the adsorbent such as specific surface area, porosity and surface charge. Chemical structure of

Process	Pressure (bar)	pH	Time (hr)	COD removal (%)	BOD removal (%)	Solids removal (%)	Color Removal (%)	Ref.
Membrane NF + UF	1–5	—	4	—	—	—	97.9	[21]
Membrane UF	5	9.05	4	88	—	80	—	[22]
Membrane UF	2	8	1.5	90	90	—	—	[23]
Membrane UF	—	6.6–7	—	95	—	79	—	[24]
Membrane UF	0.2–0.8	—	4.5	57	—	97.7	—	[25]
Membrane UF	1–7	1–14	—	60	60	—	97	[26]
Membrane UF + RO	UF: 0–7 RO: 0–60	5.5–6 0.5	—	98.8	99.4	—	—	[27]

Table 3.
The application of membrane technologies for POME polishing.

the adsorbate and environmental condition such as temperature, pH, solubility and ionic strength will influence the adsorption performance [29]. **Table 4** shows some of the application of adsorption process in POME polishing.

3.4 Coagulation and flocculation technologies

Coagulation process commonly used to remove the organic matter and suspended solids (SS) from the wastewater. During the coagulation process, the chemical is added into the wastewater to enhance the flocculation and sedimentation. It will help in removing dissolved solids and suspended solids from the wastewater. Aluminum and iron-based compounds coagulant are often used in water treatment, as they are simple, easy to handle, cheap, and have excellent removal efficiency [37].

Process/Adsorbent dosage	pH	Time (min)	COD removal (%)	Color removal (%)	TSS removal (%)	Ref.
BP/Natural clay: 5 g/L	7	90	95	95	95	[30]
BP/GAC: 150 g/L	4	87.9	—	90	—	[31]
BP/Fly ash: 90 g/L	4	60	89	97	96	[2]
FB/Resin: 0.3 mBH	9.28	—	88	98	—	[2]
BP/AC: 10 g/L	8.5	30	98	—	99	[32]
BP/Banana peel: 300 g/L	2	1800	100	95.96	100	[33]
BP/AC: 200 g/L	—	120	98.99	79.3	98.45	[34]
FB/Resin: 0.3 mBH	3	—	72	64	—	[35]
BP/Zeolite: 10 g/L	3	50	—	—	—	[36]

Table 4.
Application of adsorption for POME polishing.

Coagulant type/ dosage	pH	Mixing rate (rpm)	Time (min)	COD removal (%)	TSS removal (%)	Oil and Grease removal (%)	Ref.
Calcium lactate: 50 mg/L	—	258	23	58	58	—	[43]
Alum: 2124 mg/L	6	—	20	59	—	—	[44]
Chitosan: 0.5 g/L	4	100	15	—	95	95	[45]
Mango Pit: 50 g/L Fly ash: 90 g/L	4	200	60	89	96	—	[2]
PAC: 0.6 g/L + AC: 10 g/L	8.5	50	30	98	99	—	[32]
PAC: 2 g/L	—	150	5	93	—	—	[46]

Table 5.
Application of coagulation-flocculation in POME treatment.

Nevertheless, the residual aluminum and iron concentrations may inhibit the biological treatment process in wastewater due to the reduction of microorganism respiration rate and low organic matter elimination [38]. The drawbacks of these chemicals are high cost, non-biodegradable and possible adverse effect of the chemical. Recently, interests have shift to natural and biodegradable coagulants such as PGA, cotton, chitosan, natural seed gum, *Jatropha curcas* seeds, and *Moringa oleifera* [39–42]. It is because chemical coagulants are non-biodegradable, costly and not environmental-friendly. The coagulation technologies can also combine with other polishing technologies such as adsorption, membrane filtration and AOPs to achieve better pollutant removal. **Table 5** shows the application of coagulation-flocculation for POME polishing.

4. Chitosan

Chitosan is a biopolymer coagulant which is non-toxic, biodegradable, renewable and environmental friendly [47]. Chitosan is a type of marine polymer which has widely used in practical fields such as wastewater management, pharmacology, biochemistry and biomedical. Chitosan is a cellulose-like polyelectrolyte biopolymer which derived from de-acetylation of chitin, as shown in **Figure 8**. Chitin

Figure 8.
Derivation of chitosan from chitin [47].

can easily found in marine nature, and it is occurring in the insects, yeast, fungi and exoskeletons of crustaceans [45]. Chitosan contains a high amount of amino functions that provide novel binding properties for heavy metals in wastewater [48]. Chitosan can coagulate effectively at pH less than 4.5 as strong acidic condition exaggerates POME to form unstable flocs [49]. The mechanisms involved in the coagulation can divide into two main categories which are charge neutralization or electrostatic interaction and sweep coagulation/co-precipitation [50]. The chitosan coagulation process is charge neutralization while synthetic coagulant such as ferric chloride ($FeCl_3$) is sweep coagulation as shown in **Figure 9** [51]. The flocs formed by charge neutralization are smaller than the flocs formed through sweep coagulation [52]. The smaller sized flocs could bring fouling risk to the membrane if membrane technologies are used.

4.1 Performance of chitosan in POME treatment

The optimum condition for coagulation treatment with chitosan as a coagulant is about 100 ppm (mg/L) of POME at pH 4.5. The removal percentage for the COD, the color and the TSS is 15.39%, 85.79 and 97%, respectively. The results are shown in **Figure 10**. The graph in **Figure 10** showed that any further increase in dosage does not increase the color and the TSS removal significantly. However, a further increase in dosage causes the COD to increase. The negative result of the COD

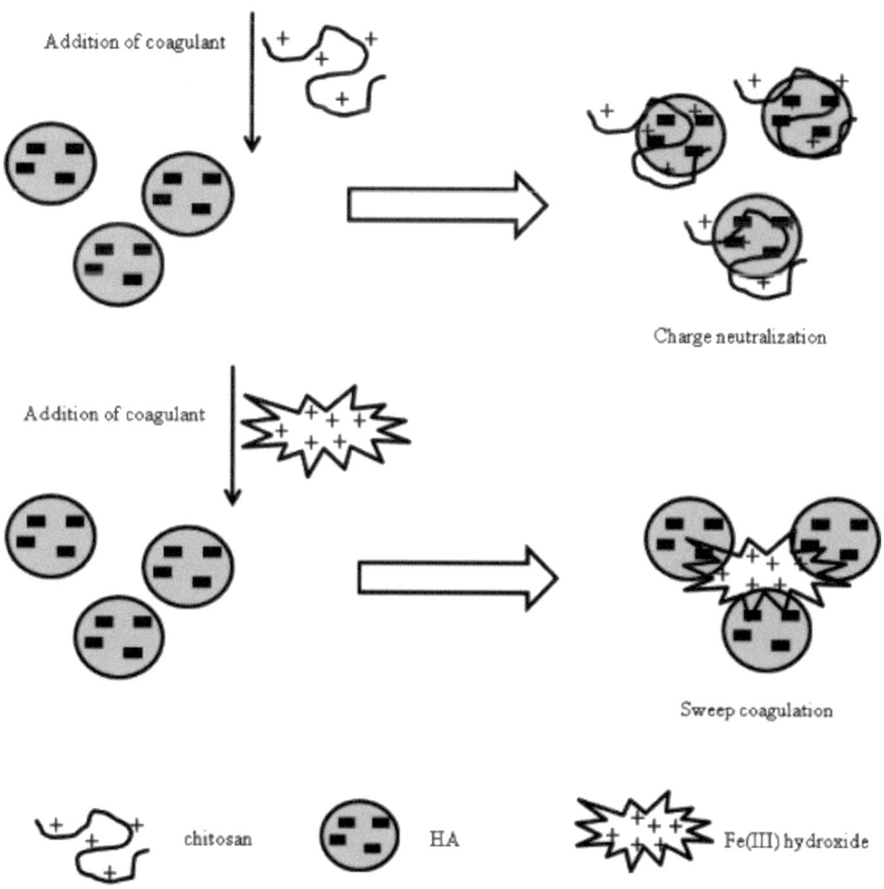

Figure 9.
Chitosan charge neutralization and $FeCl_3$ sweep coagulation.

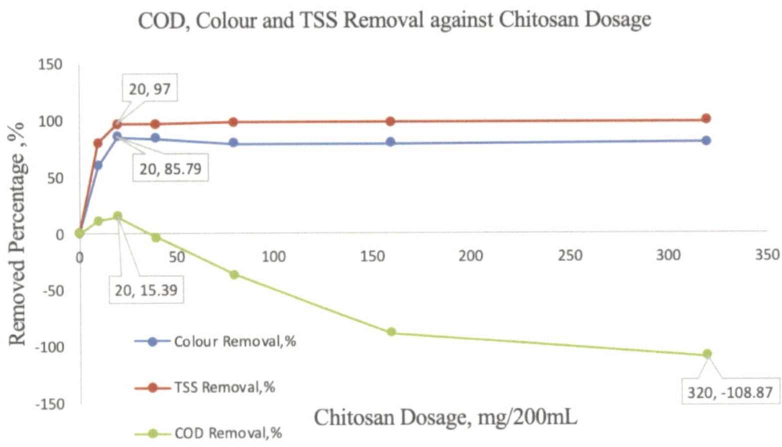

Figure 10.
Effect of chitosan dosage in pollutant removal (COD, color and TSS).

removal observed with the addition of chitosan, which is a natural biopolymer coagulant (an impurity) that causes the COD to increase when the dosage exceeds its saturation point. In essence, the low COD removal was due to natural properties of POME as chitosan is not effective in removing dissolved solid [53]. Typical raw POME has a total solid of 40,000 mg/L, while 34,000 mg/L of it is dissolved solid [2]. Furthermore, TSS removal is very effective at low chitosan dosage [45]. However, chitosan is effective in removing suspended solids that contributes to the COD but not total dissolved solids (TDS).

4.2 Performance of chitosan paired with other method in POME treatment

This study was done by combining ultrasound (US) cavitation, chitosan and ferric chloride ($FeCl_3$) in different ways to determine the best combination and order of treatment. Every treatment method is conducted by following the result of optimum condition obtained from previous studies. The result is shown in **Figure 11**.

From the graph in **Figure 11**, the COD removal for the combination of ultrasound (US) cavitation, followed by ferric chloride coagulation treatment, is the highest, at

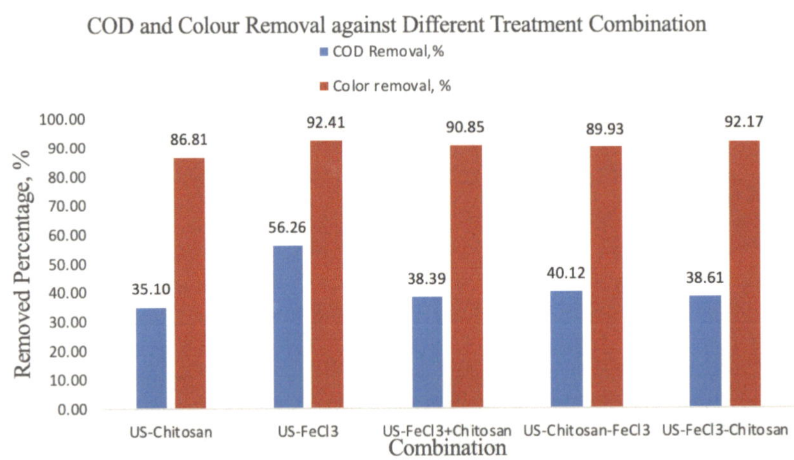

Figure 11.
COD and color removal percentage with the different treatment combination.

56.26%. Besides, the color removal for this combination is the highest among other combinations, at 92.41%.

Furthermore, among all these combinations where chitosan is present, the COD removal percentage is less substantially, ranging from 35.1% to 40.12%, which can be observe at the combination of US- $FeCl_3$ (ultrasound and ferric chloride) that shows the COD removal at 56.26%. However, when chitosan added after it, the COD removal percentage drops to 38.61%, due to chitosan being a natural coagulant that is biodegradable [47]. Chitosan is not very useful in coagulating the organic pollutant (COD), which dissolved in the POME [53]. Therefore, chitosan will become the pollutant, contributing to COD and causing the COD removal percentage to decrease. Even though ferric chloride performs better when paired with other polishing methods, the dosage usage of chitosan in POME treatment is lesser and hence more superior in terms of cost-effectiveness and environmentally friendly method for palm oil mills in dealing with wastewaters.

5. Concluding remarks

This chapter presents the treatment performance of palm oil mill effluent by utilizing chitosan as natural coagulant. Chitosan is natural, food-based and environmentally friendly biopolymer which has enormous potential to be used to treat POME before discharged to watercourse. Some of the methods for combinations, as suggested in the study. The pollutant removal performance measure in terms of COD, color, and TSS removal percentage. The main contribution of this chapter is to provide a low cost and simple method to small and medium oil palm processing industry in processing their wastewater before discharge to the environment. For chitosan, the main advantage is that low dosage would contribute to high removal of suspended solids in POME. However, the disadvantage is that if paired with other methods such as ultrasound cavitation and ferric chloride, it would not have significant improvement in terms of pollutant removal percentage. On the other hand, ferric chloride could work with other methods to improve pollutant removal significantly. Nevertheless, utilizing chitosan would not contribute to significant increment in the overall treatment cost, which would encourage palm oil mill to adapt this method in treating their wastewater.

Author details

Man Djun Lee* and Pui San Lee
School of Engineering and Technology, University College of Technology Sarawak, Sibu, Sarawak, Malaysia

*Address all correspondence to: man.djun@ucts.edu.my

References

[1] Liew WL, Kassim MA, Muda K, Loh SK, Affam AC. Conventional methods and emerging wastewater polishing technologies for palm oil mill effluent treatment: A review. J Environ Manage. 2014;149:222-35.

[2] Bello MM, Abdul Raman AA. Trend and current practices of palm oil mill effluent polishing: Application of advanced oxidation processes and their future perspectives. J Environ Manage. 2017;198:170-82.

[3] Parthasarathy S, Mohammed RR, Fong CM, Gomes RL, Manickam S. A novel hybrid approach of activated carbon and ultrasound cavitation for the intensification of palm oil mill effluent (POME) polishing. J Clean Prod. 2016;112:1218-26.

[4] Hassan MA, Yacob S, Shirai Y, Hung Y-T. Treatment of palm oil wastewaters. Sci Total Environ. 2006; 366(1):187-96.

[5] Lam MK, Lee KT. Renewable and sustainable bioenergies production from palm oil mill effluent (POME): Win-win strategies toward better environmental protection. Biotechnol Adv. 2011;29(1):124-41.

[6] Wu TY, Mohammad AW, Jahim JM, Anuar N. Pollution control technologies for the treatment of palm oil mill effluent (POME) through end-of-pipe processes. J Environ Manage. 2010;91(7):1467-90.

[7] Manickam S, Zainal Abidin NB, Parthasarathy S, Alzorqi I, Ng EH, Tiong TJ, et al. Role of H2O2in the fluctuating patterns of COD (chemical oxygen demand) during the treatment of palm oil mill effluent (POME) using pilot scale triple frequency ultrasound cavitation reactor. Ultrason Sonochem. 2014;21(4):1519-26.

[8] Edewor JO. A comparison of treatment methods for palm oil mill effluent (POME) wastes. 2000;212-8.

[9] Tan YH, Goh PS, Lai GS, Lau WJ, Ismail AF. Treatment of aerobic treated palm oil mill effluent (AT-POME) by using TiO 2 photocatalytic process. 2014;2:61-3.

[10] American Public Health Association. 5210 Biochemical Oxygen Demand (BOD). 2001.

[11] Brown and Caldwell. Description of commonly considered water quality constituents. Watershed Prot Plan Dev Guideb. 2001;1-11.

[12] Doran MD, Dee PE, Techknowledge MD. Wastewater and wastewater treatment very basics. 2008;2008(October):1-7.

[13] Hassan MA, Yacob S, Shirai Y, Hung Y. Treatment of palm oil wastewaters. Waste Treat Food Process Ind. 2005;101-17.

[14] Zainal NH, Jalani NF, Mamat R, Astimar AA. A review on the development of palm oil mill effluent (POME) final discharge polishing treatments. J Oil Palm Res. 2017;29(4): 528-40.

[15] Soon AN, Hameed BH. Heterogeneous catalytic treatment of synthetic dyes in aqueous media using Fenton and photo-assisted Fenton process. DES. 2011;269(1-3):1-16.

[16] Neyens E, Baeyens J. A review of classic Fenton's peroxidation as an advanced oxidation technique. J Hazard Mater. 2003;98(1-3):33-50.

[17] Gar M, Tawfik A, Ookawara S. Comparison of solar TiO_2 photocatalysis and solar photo-Fenton for treatment of pesticides industry wastewater :

Operational conditions, kinetics, and costs. J Water Process Eng. 2015;8:55-63.

[18] Dong S, Zhang X, He F, Dong S, Wang B. Visible-light photocatalytic degradation of methyl orange over spherical activated carbon-supported and Er 3 + : YAlO 3 -doped TiO 2 in a fluidized bed. 2014;(December 2013).

[19] Aparicio MA, Eiroa M, Kennes C, Veiga MC. Combined post-ozonation and biological treatment of recalcitrant wastewater from a resin-producing factory. 2007;143:285-90.

[20] Buthiyappan A, Raman A, Aziz A, Mohd W, Wan A. Recent advances and prospects of catalytic advanced oxidation process in treating textile effluents. 2016;32(1):1-47.

[21] Amat NAA, Tan YH, Lau WJ, Lai GS, Ong CS, Mokhtar NM, et al. Tackling colour issue of anaerobically-treated palm oil mill effluent using membrane technology. J Water Process Eng. 2015;8:221-6.

[22] Said M, Ahmad A, Wahab A, Tusirin M, Nor M. Blocking mechanism of PES membrane during ultrafiltration of POME. J Ind Eng Chem. 2014;1-7.

[23] Azmi NS, Yunos KFM. Wastewater treatment of palm oil mill effluent (POME) by ultrafiltration membrane separation technique coupled with adsorption treatment as pre-treatment. Agric Agric Sci Procedia. 2014;2:257-64.

[24] Azmi NA, Yunos KFM, Zakaria R. Application ofsandwich membrane for the treatment of palm oil mill effluent (POME) for water reuse. In: Procedia Engineering. 2012. p. 1980-1.

[25] Wu TY, Mohammad AW, Jahim J, Anuar N. Palm oil mill effluent (POME) treatment and bioresources recovery using ultrafiltration membrane : Effect of pressure on membrane fouling. 2007;35:309-17.

[26] Ahmad AL, Ismail S, Bhatia S. Membrane treatment for palm oil mill effluent: Effect of transmembrane pressure and crossflow velocity. Desalination. 2005;179(1-3 SPEC. ISS.): 245-55.

[27] Latifahmad A, Ismail S, Bhatia S. Water recycling from palm oil mill effluent (POME) using membrane technology. 2003;157(May):87-95.

[28] Fu F, Wang Q. Removal of heavy metal ions from wastewaters : A review. J Environ Manage. 2011;92(3):407-18.

[29] Kyriakopoulos G, Doulia D, Hourdakis A. Effect of ionic strength and pH on the adsorption of selected herbicides on Amberlite. Int J Environ. (October 2014):37-41.

[30] Said M, Hasan HA, Tusirin M, Nor M. Removal of COD, TSS and colour from palm oil mill effluent (POME) using montmorillonite. 2015;3994(October):0-8.

[31] Alkhatib MF, Mamun AA, Akbar I. Application of response surface methodology (RSM) for optimization of color removal from POME by granular activated carbon. 2014;

[32] Ridzuan M, Ali M, Shirai Y, Samsu A, Amiruddin A, Ali M, et al. Treatment of effluents from palm oil mill process to achieve river water quality for reuse as recycled water in a zero emission system. J Clean Prod. 2013;10-3.

[33] Fong M. Treatment and decolorization of biologically treated Palm Oil Mill Ef fl uent (POME) using banana peel as novel biosorbent. 2014;132:237-49.

[34] Mohammed RR, Ketabachi MR, Mckay G. Combined magnetic field and adsorption process for treatment of biologically treated palm oil mill effluent (pome). Chem Eng J. 2014;

[35] Belo MM, Nourouzi MM, Abdullah LC, Choong TSY, Koay YS, Keshani S. Ac ce pt ed us t. J Hazard Mater. 2013;

[36] Shavandi MA, Haddadian Z, Ismail MHS, Abdullah N, Abidin ZZ. Removal of residual oils from palm oil mill effluent by adsorption on natural zeolite. 2012;1:4017-27.

[37] Keeley J, Smith AD, Judd SJ, Jarvis P. Reuse of recovered coagulants in water treatment : An investigation on the effect coagulant purity has on treatment performance. Seperation Purif Technol. 2014;131:69-78.

[38] Lees EJ, Noble B, Hewitt R, Parsons SA. RESPIRATION RATE AND SLUDGE CHARACTERISTICS OF AN ACTIVATED MICROBIAL BIOMASS. 2001;79(September).

[39] Nourani M, Baghdadi M, Javan M, Bidhendi GN. Production of a biodegradable flocculant from cotton and evaluation of its performance in coagulation-flocculation of kaolin clay suspension : Optimization through response surface methodology (RSM). Biochem Pharmacol. 2016;

[40] Ang WL, Mohammad AW, Benamor A, Hilal N. Chitosan as natural coagulant in hybrid coagulation-nanoflitration membrane process for water treatment. Biochem Pharmacol. 2016;

[41] Pui K, Shak Y, Wu TY. Coagulation – flocculation treatment of high-strength agro-industrial wastewater using natural *Cassia obtusifolia* seed gum : Treatment efficiencies and flocs characterization. Chem Eng J. 2014;256:293-305.

[42] Bhatia S, Othman Z, Ahmad AL. Coagulation – flocculation process for POME treatment using *Moringa oleifera* seeds extract : Optimization studies. 2007;133:205-12.

[43] Zahrim AY, Nasimah A, Hilal N. Pollutants analysis during conventional palm oil mill effluent (POME) ponding system and decolourisation of anaerobically treated POME via calcium lactate-polyacrylamide. J Water Process Eng. 2014;4:159-65.

[44] Taylor P, Malakahmad A, Chuan SY. Desalination and water treatment application of response surface methodology to optimize coagulation – flocculation treatment of anaerobically digested palm oil mill effluent using alum. 2013;(December 2014):37-41.

[45] Ahmad AL, Sumathi S, Hameed BH. Coagulation of residue oil and suspended solid in palm oil mill effluent by chitosan, alum and PAC. Chem Eng J. 2006;118(1-2):99-105.

[46] Poh PE, Ong WYJ, Lau E V, Chong MN. Journal of Environmental Chemical Engineering Investigation on micro-bubble flotation and coagulation for the treatment of anaerobically treated palm oil mill effluent (POME). Elsevier. 2014;2(2):1174-81.

[47] Crini G, Badot PM. Application of chitosan, a natural aminopolysaccharide, for dye removal from aqueous solutions by adsorption processes using batch studies: A review of recent literature. Prog Polym Sci. 2008;33(4):399-447.

[48] Ravi Kumar MN. A review of chitin and chitosan applications. React Funct Polym. 2000;46(1):1-27.

[49] Ahmad AL, Sumathi S, Hameed BH. Adsorption of residue oil from palm oil mill effluent using powder and flake chitosan: Equilibrium and kinetic studies. Water Res. 2005;39(12):2483-94.

[50] Ng M, Liu S, Chow CWK, Drikas M, Amal R, Lim M. Understanding effects of water characteristics on natural organic matter treatability by PACl and a novel PACl-chitosan coagulants. J Hazard Mater. 2013;263:718-25.

[51] Ang WL, Mohammad AW, Teow YH, Benamor A, Hilal N. Hybrid chitosan/FeCl3 coagulation–membrane processes: Performance evaluation and membrane fouling study in removing natural organic matter. Sep Purif Technol. 2015;152:23-31.

[52] Jarvis P, Jefferson B, Parsons SA. Floc structural characteristics using conventional coagulation for a high doc, low alkalinity surface water source. Water Res. nnnnn40(14):2727-37.

[53] Musonge P. Influence of effluent type on the performance of Chitosan as a coagulant. 2014;2(1):1-6.

Chapter 19

Tertiary Treatment for Safely Treated Wastewater Reuse

Nebil Belaid

Abstract

The tertiary treatment of resulting water from a conventional biological treatment process is envisaged in the aim to obtain a high quality of water that can be reused for different purposes. This treatment is based on the integration of the membrane-based technologies in the total process of wastewater treatment. The experimental studies are carried out on a small pilot, equipped with different mineral membranes of micro and ultrafiltration. These membranes are used for the different tested processes (MF, MF-UF and cogulation-MF). The results obtained make it possible to attend a complete elimination of the total flora and an additional reduction of the other parameters such as turbidity, suspended matter, COD and BOD. Tests on a large scale are then carried out on a semi-industrial pilot, equipped with the same type of membranes. The optimization of the operating conditions made allow the obtaining under the conditions of transmembrane pressure 0.85 bar, a cross flow velocity of 2.25m/s and with ambient temperature a filtrate flux of about 200 L/hm ^{2}. The coupling of a stage of coagulation in the membrane process allows the reduction of the effect of the membrane fouling and an improvement of 36% of the filtrate flux.

Keywords: treated wastewater, tertiary treatment, microfiltration, ultrafiltration, reuse

1. Introduction

Population growth and economic development are putting pressure on water resources, especially in arid regions. Indeed, most MENA countries will have annual renewable water resources of less than 1000 m^3/capita by the year 2025, according to estimates and projections of country-based populations and annual renewable water resources [1]. Moreover, the majority of MENA countries were classified as having a water deficit in 2010 (less than 500 m^3/capita) [1]. Consequently, there is a need for new non-conventional water resources, such as water desalination, wastewater and rain harvesting, to meet the increasing demand. Wastewater reuse is gaining increasing attention for groundwater recharge and irrigation, since agriculture is the dominant water user in the region [2, 3].

Indeed, wastewater reuse for irrigation offers some attractive environmental and socioeconomic benefits [4–6]. In fact, the irrigation with treated wastewater leads to supply nutrients as fertilizer [7] as well as improvement crop production during the dry season [8, 9]. However, planners are aware of the potential disadvantages of wastewater reuse for irrigation which are, aside from pathogenic contamination of irrigated crops, mainly related to their specific chemical composition being somewhat different from most natural waters used in irrigation [10]. Wastewater generally contains high concentrations of suspended and dissolved solids, both organic and inorganic (e.g. chloride, sodium, boron and selected heavy metals), that are added to wastewater during domestic and industrial usage [11]. Most of the salts added are only partially removed during conventional sewage treatment (secondary and tertiary), so they remain in the irrigation water [12]. Its content of trace elements, pathogens and high nitrogen may present a risk to the receiving environment. Additional treatment, particularly at the microbiological level, therefore appears necessary to ensure both user safety and reduce the impact on the receiving environment [13]. In this regard, membrane processes appear very promising for the complementary treatment of treated wastewater (TWW) [14]. Indeed, there is growing interest in direct filtration of wastewater treatment process for water reclamation to ease global water shortages [15–19]. Nowadays, integrated membrane systems treatment is becoming widely popular due to its feasibility, process reliability, commercial availability, relative insensitivity in case of wastewater treatment and lower operating costs [20]. Especially, direct filtration using ceramic membrane has been considered as an attractive option due to properties of ceramic membrane (e.g. a high durability and a high chemical resistance) [21–23].

In Tunisia, the reuse of treated wastewater (mainly secondary treated wastewater) for agricultural purpose is restricted for forages crops irrigation only. In Sfax (center east of Tunisia) where the average annual potential evaporation of 1200 mm, combined with the low rainfall and high temperatures, irrigation proves to be essential for crop production. Therefore, the treated wastewater has been used for forages irrigation since 1989. This practice had led to soil fertility improvement [24]. By contrast, an increase of soil salinity [25] and metallics elements accumulation has been detected [26]. In order to minimize health and environmental risks and for unrestricted irrigation reuse, the final treated wastewater quality should be improved.

The aim of this work is to study the feasibility of membrane techniques application, in particular microfiltration and ultrafiltration, for the tertiary treatment of the treated wastewater. Our study involves qualitative optimization trials, to define the appropriate treatment process. Three methods were tested, MF alone, MF-UF coupling and coagulation-MF coupling. Other tests to optimize the operating conditions and permeate flow are carried out, to evaluate the selected membrane process.

2. Materials and methods

2.1 Effluent origin

The treated wastewater is collected at the outlet of the Wastewater treatment plant (WWTP) of Mahrès (40 km south of the city of Sfax) which mainly treats domestic wastewater. Samples are collected during the period from January to May. After each companion, part of the sample is stored at −4° C for characterization. The WWTP of Mahrès treats the urban wastewater of the city and that of the Chafar seaside area. It is designed for a capacity of 13,000 equivalent inhabitants, which corresponds to a daily flow of 780 m^3/day. The average daily flow was 800 m^3/day with peak flow rates exceeding 1400 m^3/day, especially in summer. The station treatment process includes pretreatment (grit screening), biological treatment with activated sludge in an oxidation channel. A settling basin and sludge treatment (thickening, de-watering). The treated wastewater is, at the end, rejected in the sea.

2.2 Experimental methods and setups

2.2.1 Conduct of membrane filtration tests

Filtration is a physical process that involves the separation (removal) of a particulate and colloidal matter from a liquid. Indeed, membranes serve as selective barriers that allow the passage of some constituents and retain others. Based on pore size, shape and chemical/physical properties, membranes can separate different particles, organisms and chemical species. In this study, the conduct of the tests is based on determining the efficiency of two membranes processes, microfiltration (MF) and ultrafiltration (UF), for the removal of residual pollutants of secondary treated wastewater. Two aspects have been developed, one qualitative (final quality of the treated water) and the other one is quantitative (density of the permeate flow).

2.2.1.1 Qualitative filtration tests

In filtration process, the quality of treated water, permeate, depends on initial effluent quality and membranes properties. While, the quantity of produced permeate depends mainly of operating conditions.

In the first part of filtration tests the objective is to find the best permeate quality in terms of physic-chemical and biological properties. Thus, different membranes processes were tested (**Figure 1**):

- MF only: one single stage of microfiltration (0.2 μm) is performed to the effluent.
- MF-UF coupling: this process is composed by two stages. The effluent is filtered firstly, by a microfiltration membrane (0.2 μm) and secondly by an ultrafiltration one (15 KDa).
- Coagulation-MF coupling: after coagulation and settling, the effluent is than filtered by a microfiltration membrane (0.2 μm).

A small pilot scale "Kerasep" was used during the experiments (**Figure 2**). The system is equipped by different ceramic membranes modules. These ones have a nominal surface area of 370 cm^2 and 400 mm of length (**Table 1**). This driver is easy to handle and allows operation on small volumes. Indeed, about 3 L of effluent were used for each filtration test. The operating parameters (pressure and frequency of pump rotation) are set at random. After each test, a characterization of the permeate is carried out.

2.2.1.2 Quantitative filtration tests

Once the best process is chosen, the operating conditions must be optimized. In fact, the best conditions give the maximum of permeate quantity with the least energy (low pressure).

The optimizing of operating conditions is carried out on a semi-industrial "Kerasep" pilot (**Figure 3**). The membranes modules of this system have a nominal surface area of 800 cm^2 and 865 mm of length (**Table 1**). In this pilot, all operating parameters are controlled (transmembrane pressure, circulation speed and temperature). Thus, a 50 L of effluent is filtered through membrane module at fixed

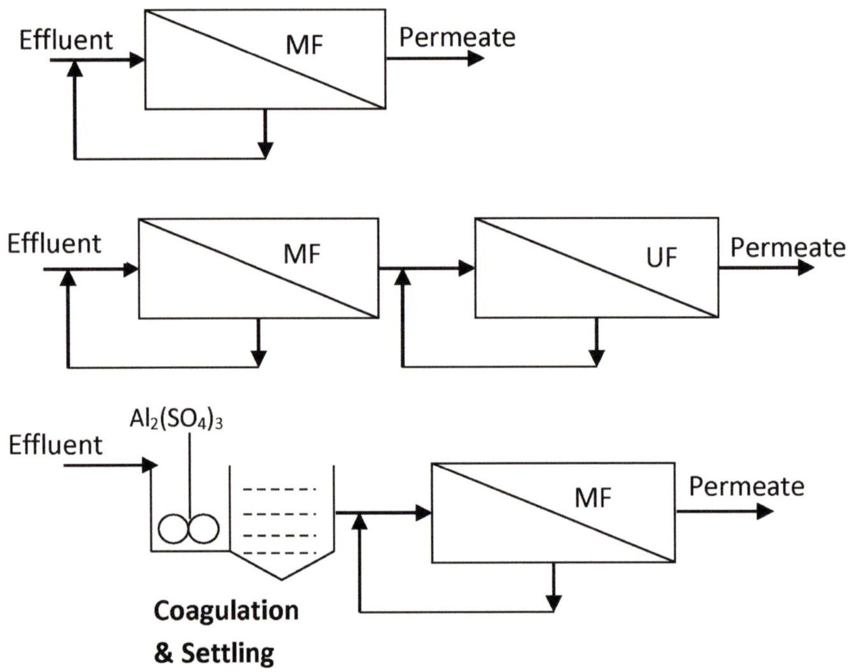

Figure 1.
Experimental design of tested processes.

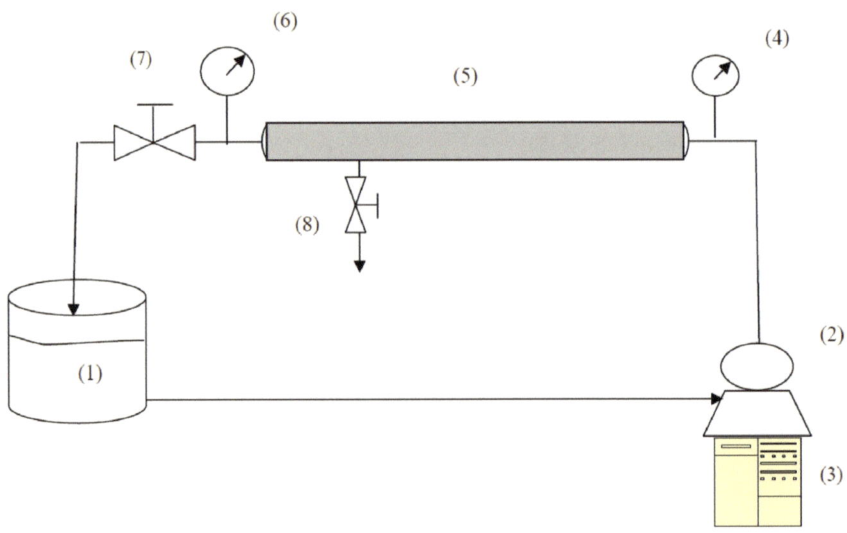

Figure 2.
Experimental small scale pilot. (1) effluent thank; (2) volumetric pump with adjustable frequency; (3) control block; (4) inlet pressure gauge; (5) membrane module; (6) outlet pressure gauge; (7) pressure valve adjustment; (8) permeate.

condition. This operation is repeated many times until reaching the best operating condition. Indeed, this optimization goes through two stages:

- The search for optimal conditions by varying the circulation speed (U) and the transmembrane pressure (TMP). For a given circulation speed (U), the PTM is varied and the permeate flow is measured. The plotting of the permeate flow

Characteristics	MF membranes		UF membrane
Average pore diameters	0.1 μm	0.2 μm	15 KDa
Length (mm)	865	400	400
Diameter (mm)	20	20	20
Surface area (cm2)	800	370	370
Number of channels	7	7	7
Diameter of canals (mm)	4.5	4.5	4.5

Table 1.
Specification of the used membranes.

curves as a function of circulation speeds and transmembrane pressures makes it possible to choose the best operating conditions.

- Monitoring the evolution of the permeate flow and the volume reduction factor (VRF) as a function of time, makes it possible to identify the nature and state of the membrane clogging.

The VRF is calculated as follows: $\frac{Vi - Vp}{Vp.}$

with
Vi: initial volume of the effluent
Vp: permeate volume

2.2.2 Membranes characteristics and cleaning procedure

The tubular membranes used are of the mineral type made of monolithic ceramic. They are consisted of an aluminum oxide support and a titanium oxide filtration layer. These characteristics facilitate effective cleaning with acidic or alkali solutions. The specifications of the different membranes used are shown in the **Table 1**.

The membrane and module cleaning protocol is most often provided by the manufacturer. However, this protocol has been modified to making it adapted to the nature of the effluent treated in this work. **Table 2** summarizes the adapted procedure.

2.3 Characterization of TWW

Treated effluents were sampled at the outlet of the Mahres wastewater treatment plant at different times and conserved at −4°C before characterization. Effluent samples were analyzed for pH and electrical conductivity using a pH meter [27] (AFNOR standard method N° NF T 90–008, see AFNOR, 1997) and a conductimeter (AFNOR N° NF EN 27888) respectively. Chemical oxygen demand (COD), suspended solids (SS), biochemical oxygen demand (BOD) and total phosphorus were measured according to standard methods (AFNOR N° NF T 90–018, NF EN 872, NF T 90–103, NF EN 1189). Cations and anions were measured using ion chromatography and trace metals by using Furnace Atomic Absorption

Spectrometry after aqua regia acid digestion (AFNOR N°NF EN ISO 15587-1). Carbonates and bicarbonates were estimated by titration with HCl of an aliquot of the effluent samples (AFNOR N° NF EN ISO 9963-2). Turbidity was determined at 860 nm by using a spectrophotometer DR/4000 U. The apparent color was determined by transmittance between 400 and 700 nm with the same

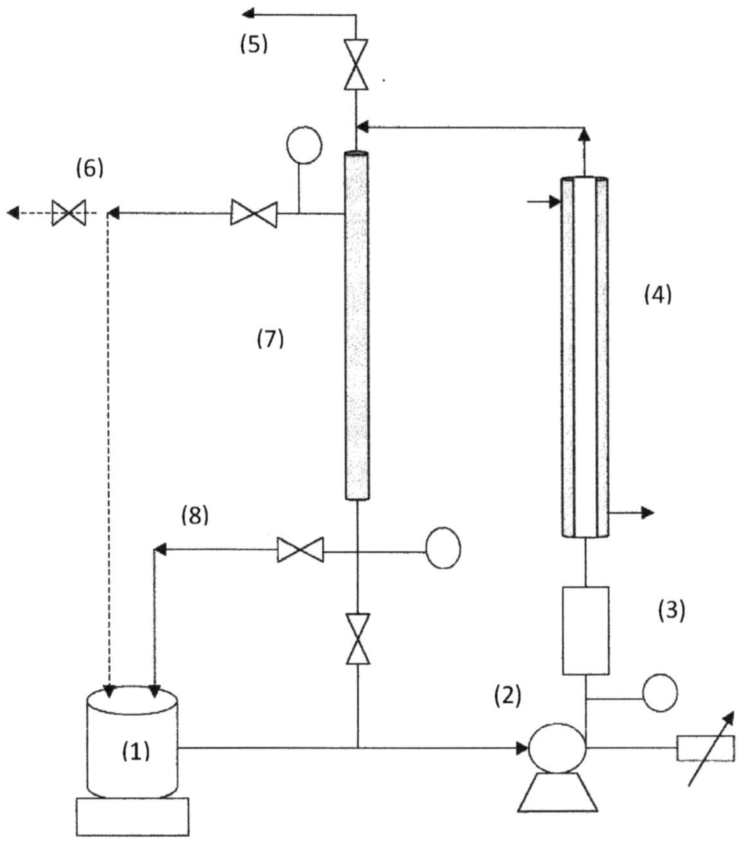

Figure 3.
Experimental semi industrial pilot. (1) effluent thank; (2) volumetric pump with adjustable frequency; (3) flowmeter; (4) exchanger; (5) purge; (6) permeate; (7) membrane module; (8) Retentate.

Rinsing steps	Reagents	T (C°)	Duration (min)	P (bar)
Rinsing/draining	water	—	—	1
Basic wash with recycling	NaOH (10 g/L)	80–85	60	3–4
Rinsing until neutral	water	—	—	2–3
Acid wash with recycling	HNO3 (5 ml/L)	55–60	30	3–4
Rinsing until neutral	water	—	—	2–3
Water flow	water	—	10	2–4

Table 2.
Membranes cleaning procedure.

spectrophotometer. The count of the total flora is carried out on Plate Count Agar (PCA) medium by inoculation on the surface and incubation at 37° C for 24 hours.

3. Results

3.1 Treated wastewater quality

The WWTP of Mahrès mainly receives and treats domestic wastewater. The quality of this water remains generally stable throughout the year except during the summer period which corresponds to an increase in the affluent flow. The results obtained show that the temperature and the pH of the water, leaving the station, increase from January to May (**Table 3**). The treated wastewater always remained alkaline with an average pH of 7.5. The mean electrical conductivity (EC) of the effluents reached 4.27 $mS \cdot cm^{-1}$, which places the TWW in the class of high salinity according to the FAO legislation. The elevated EC values of the studied effluent are mainly explained by the abundance of free ions such as Na^+, Cl^- and SO_4^{2-} which exceed the standards (**Table 3**). Turbidity and SS drop after January. The COD and the BOD are slightly elevated compared to the standards of discharges into nature. Moreover, TWW also contain large amounts of nitrate, phosphate and potassium, which are crucial nutrients for plant growth and soil fertility whatever this water is reused for irrigation. However, excepting Cr concentrations, the heavy metal contents are low. Whereas, the values of the total flora are high (**Table 3**).

Parameters	Effluent	Standards*
Temperature, °C	15–23	< 25 °C
pH	7.25–7.84	6.5 < pH < 8.5
CE, mS/cm	3.6–4.27	
Turbidity, NTU	1–141	
TDS, g/l	1.51–2.13	
Color, ADMI	26–32	70
SS, mg/l	3–121	30
COD, mg/L	115–231	90
BOD, mg/L	30–50	30
NH_4^+, mg/L	16	1
NO_3^-, mg/L	15–24	50
P total, mg/L	1.07–6.7	0.05
Cl-, mg/L	572–693	600
SO_4^{2-}, mg/L	646–844	600
HCO_3^-, mg/L	335–433	
Na^+, mg/L	434–538	500
Mg^{2+}, mg/L	60–66	200
K^+, mg/L	21–59	50
Ca^{2+}, mg/L	183–253	500
Cd, mg/L	0.02	0.005
Cr, mg/L	0.12	0.01
Cu, mg/L	0.04	0.5
Fe, mg/L	1.14	1
Zn, mg/L	6.5	5
Ni, mg/L	< 0,008	0.2
Total flora, ufc/mL	1.610^6–610^6	—

**Tunisian standards for irrigation reuse NT 106.03.*

Table 3.
Treated wastewater quality.

By referring to the discharge standards (**Table 3**), a gradual improvement in the physicochemical parameters of the water during the study period is observed. In fact, the high values of turbidity and SS observed during the January campaign were subsequently greatly reduced. This can be attributed to the appearance of the phenomenon of bulking (expansion of sludge) in the station during the month of January and its disappearance thereafter. The effluent remains difficult to biodegrade since the BOD/COD ratio is usually less than 0.3. The heavy metal contents are low, which leads to the conclusion that there are no industrial discharges in the station. Thus, if the physical and chemical qualities of the treated wastewater are generally close to the standards, the biological quality still remains high. Tertiary treatment could then complete the treatment process and leads to water quality that meets all the requirements.

3.2 Tertiary treatment

The improvement of the final quality of the effluent and in particular the biological quality is studied by applying membrane processes. Two axes are developed. First, improving the final water quality by testing different processes. Then, define the operating conditions which ensure the best flow of permeate.

3.2.1 Qualitative study

3.2.1.1 MF-UF coupling

The process is composed by two-stage (**Figure 1**). In the first, the effluent undergoes microfiltration with recovery of the permeate which is then treated in a second stage by ultrafiltration. Two tests were carried out for this process using each time different microfiltration membranes.

During the first test, microfiltration and ultrafiltration are carried out on a bench-top pilot equipped with membranes of pore size 0.1 μm and 15 KDa respectively. The main results obtained as well as the retention efficiency (RE) are reported in the **Table 4**.

Qualitatively, the first treatment with MF leads to an elimination of more than 90% of the turbidity and the SS. The COD and the BOD are also reduced to values lower than those of the Tunisian standard of discharge in the receiving environment (respectively 90 and 30 mg/L). After the second treatment with UF, most of the parameters analyzed undergo an additional reduction (**Table 4**). Moreover, UF is more suitable for removal COD particles [7]. For this test the analysis of biological parameters was not performed.

The evolution of the permeate flow during the filtration test shows that the flows are lower in MF than those obtained after MF-UF. Indeed, the values obtained are respectively in the order of 25L/h m^2 and 150 L/h m^2 (**Figure 4**). In fact, the decrease of the permeate flux to more than the half during the first 20 minute of the filtration is caused by the clogging phenomenon of the membrane due to the colloidal fraction in the effluent. In addition, the importance of membrane fouling leads to the drop of MF flow to a relatively low value at the end of experiment. However, in the second step of UF; the permeate flux is higher despite the small size of membrane pores. In fact, the majority of particles and colloids have been already eliminated after the first step of MF.

In order to improve the permeate flux of the first microfiltration step, a membrane of greater porosity (0.2 μm) was used during a second test of the MF-UF coupling, while keeping the same characteristics of the UF membrane of the second stage. During this test, complete elimination of turbidity, MES and total

Parameters	Effluent	MF	RE (%)	MF-UF	RE (%)
pH	7.25	7.48	—	7.76	—
Turbidity, NTU	141	7	95	5	96.5
SS, mg/L	121	9	92.5	4	96
Color, ADMI	29	22	24	20	31
COD, mg/L	231	60	74	25	89
BOD, mg/L	40	10	75	10	75
P total, mg/L	5.16	3.99	22.5	3.45	33
Total flora, ufc/mL	610^6	0	100	0	100

RE: retention efficiency.

Table 4.
Treated wastewater quality after MF and MF-UF treatment.

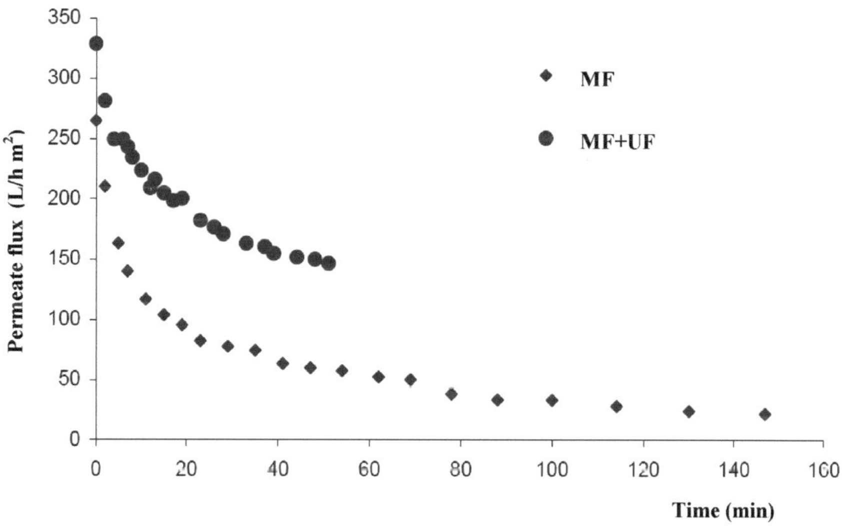

Figure 4.
Permeate flux decline for MF (0.1 μm) step and MF-UF step (small scale pilot).

flora was observed. However, the reduction in COD was lower, not exceeding 40%. The existence of small organic particles, not filtred, may be the cause of the low reduction in COD. However, a significant improvement in the permeate flux of the MF was observed. Thus, stabilized flow rates of around 90 L/h m^2 are obtained. In the second stage of UF, the same performance of the previous test is obtained, ie a stabilized flow rate of around 150 L/h m^2 (**Figure 5**).

Likewise, significant clogging of the first stage MF membrane was observed. In order to limit the consequences of this problem, an additional pre-treatment step appears essential.

3.2.1.2 Coagulation-MF coupling

In this process, microfiltration was preceded by coagulation pretreatment (**Figure 1**). Alumina sulfate is chosen as the coagulant.

In order to optimize the dose of used coagulant, varying amounts of alumina sulfate are added (between 20 and 100 mg/L). Stirring is performed with a Jar Test.

After settling, COD measurements are taken. The best reduction in COD is obtained with a dose of 40 mg/L (**Figure 6**).

Indeed, a dose of 40 mg/L of coagulant was added to the raw effluent. After stirring and settling for 24 hours, the water is microfiltered through a 0.2 μm membrane. This coagulation pretreatment has led to an improved of turbidity and a 35% reduction of COD value (**Table 5**). After MF, a reduction in SS, color and BOD values is observed with retention efficiency of 76%, 31% and 75% respectively.

3.2.2 Comparative study of the different processes

The quality of the different treated water from the various processes tested is compared to the initial quality of the effluents as well as to Tunisian standards for reuse in irrigation (**Table 6**).

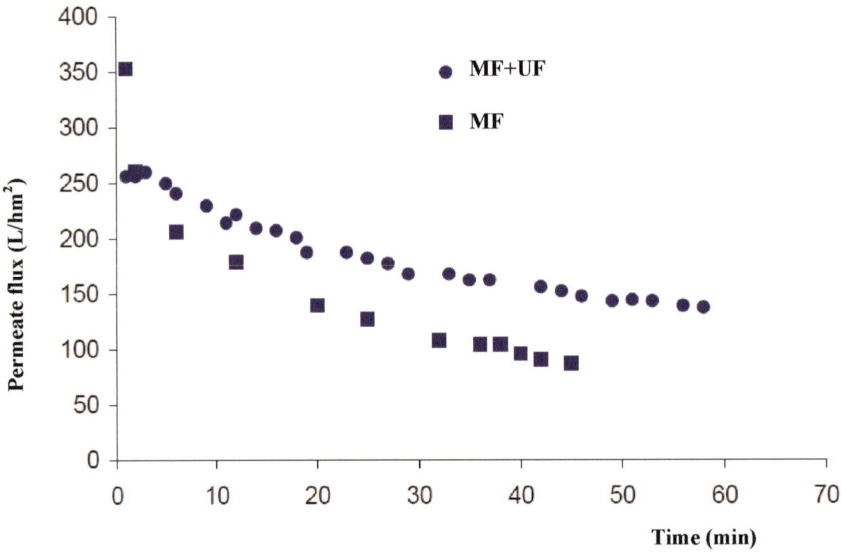

Figure 5.
Permeate flux decline for MF (0.2 μm) step and MF-UF step (small scale pilot).

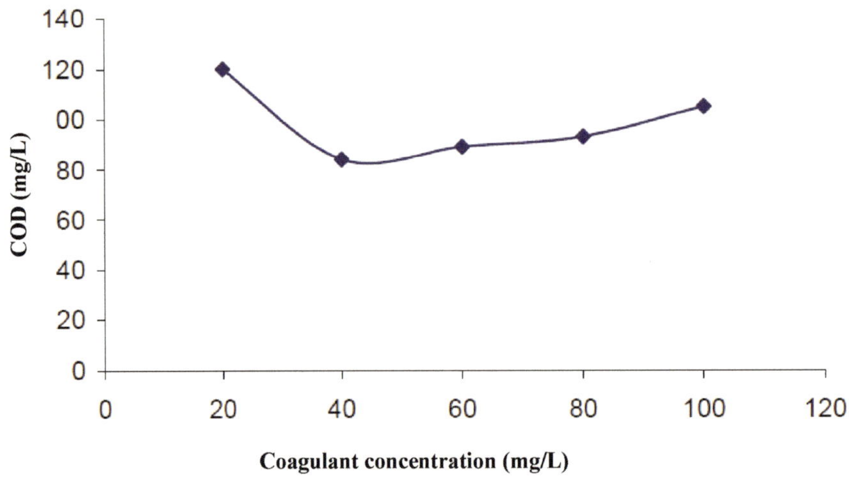

Figure 6.
Optimization of coagulant dose.

Parameters	Effluent	Coagulation	RE %	Coag + MF	RE %
Turbidity, NTU	2	0	100	0	100
MES, mg/L	8,5	8	6	2	76
Color, ADMI	32	31	4	22	31
COD, mg/L	124	80	35	58	53
BOD, mg/L	50	30	25	10	75

RE: retention efficiency.

Table 5.
Treated wastewater quality after coagulation and coagulation-MF treatment.

Parameters	Effluent	MF	MF-UF	Coag + MF	Standards*
pH	7.25–7.84	7.48–8	7.76–8.42	7.65	6.5–8.5
CE, mS/cm	3.6–4.27	3.02–4.05	3.1–4.38	3.68	7
Turbidity, NTU	1–141	0–7	0–5	0	—
TDS, g/L	1.51–2.13	1.51–2.02	1.55–2.18	1.84	
CoLor, ADMI	26–32	21–25	20–21	22	70
MES, mg/L	3–121	3–9	0–4	2	30
COD, mg/L	115–231	60–98	25–90	53	90
BOD mg/L	30–50	10–30	0–20	20	30
Nitrates, mg/L	15–24	5–27	5–33	28	—
P total, mg/L	1.07–6.7	0.7–3.99	0.6–3.45	1.01	—
Cl^-, mg/L	572–693	511–603	514–642	622	2000
SO_4^{2-}, mg/L	646–844	604–726	615–724	764	—
HCO_3^-, mg/L	335–433	335–372	331–360	354	—
Na^+, mg/L	434–538	406–521	432–500	544	—
Mg^{2+}, mg/L	60–66	60–66	45–58	75	—
K^+, mg/L	22–59	17–47	17–27	19	—
Ca^{2+}, mg/L	183–253	197–240	170–221	285	—
Cd, mg/L	0.02	< 0.004	<0.004	—	0.01
Cr, mg/L	0.12	0.08	0.05		0.1
Cu, mg/L	0.04	<0.01	< 0.01	—	0.5
Fe, mg/L	1.14	0.29	0.06	—	5
Zn, mg/L	6.5	0.03	0.02	—	5
Ni, mg/L	< 0.008	—	—	—	0.2
Total Flora, ufc/mL	$1.6\ 10^6 - 6\ 10^6$	0	0	0	—

**Tunisian standards for irrigation reuse NT 106.03.*

Table 6.
Quality of raw effluent and treated by the different processes.

Membrane techniques (MF and UF) do not have a great influence on EC and TDS. Their selectivity is far from stopping mono- and bivalent ions. However, they are effective in removing turbidity and SS. The use of these techniques also leads to an improvement in color, especially after use of UF which provides effluent whitening [28].

In addition, more than 50% of COD and BOD are eliminated during water treatment by the various methods used. The ranges of variation of these values are quite wide and reflect the influence of the quality of the water to be treated on these two parameters. In particular, the residual COD values may reflect the existence of small particles that escape filtration [29]. In general, the final quality of the treated wastewater, whether by the membrane technique or by coagulation-MF coupling, meets Tunisian standards for agricultural irrigation (NT 106–03).

It was also found that the involvement of a membrane technique in the treatment process eliminates the total flora from the treated water. This is because the size of bacteria on the one hand and the clogging of the surface of the membranes used on the other hand combine to greatly reduce the passage of bacteria through the pores of the membranes. However, there can be easy contamination of treated water due to the presence of nutrients such as nitrates and phosphorus [15, 28].

The results obtained also make it possible to observe that there is a reduction in the concentrations of heavy metals in the treated water despite their low concentrations in the initial effluent. These results can be attributed not to the membrane technique used but rather to the retention of organic colloids with which metals are generally associated [14].

All the results obtained show that, despite the variation in the quality of the water collected at the outlet of the Mahrès station, additional treatment by membrane filtration allows the elimination of the total flora and the improvement of other physico-chemicals parameters. In fact, the use of microfiltration alone ensures good quality treated water, complying with standards and can be reused without restriction. On the other hand, the coupling of MF to UF or to coagulation rather has an effect on the quantity of treated water and not on the quality.

3.2.3 Quantitative study

The optimization of the operating parameters is carried out on a semi-industrial pilot equipped with the same type of membrane with a porosity of 0.1 μm and a filtering surface of 800 cm^2. Two optimization trials were performed. The first is to do a single microfiltration step while a coagulation-MF coupling was tested in the second test.

The MF test is carried out with an initial volume of 20 L and under the following operating conditions: a transmembrane pressure TMP of 0.85 bar, a circulation speed U of 2.25 m/s and at room temperature. The initial flux is very high, around 800 L/hm^2 which, after 40 min, stabilizes at a value of 200 L/hm^2. The volume reduction factor (FRV) reaches a value of around 3.5, thus reducing the volume treated to 5.7 L.

For the second MF-coagulation test, 36 L of wastewater was pretreated by adding 40 mg/L of alumina sulfate. The supernatant is then microfiltered under the same operating conditions as the previous test. After 70 minutes, 31 liters of permeate are recovered which corresponds to an FRV of around 8.5. The stabilized permeate flux is approximately 200 L/hm^2 (**Figure 7**).

It appears that the coagulation step led to, on the one hand, reduce the major part of the colloids present in the raw effluent and on the other hand, to achieve very high FRV values. This coupling therefore results in an improvement in permeate flow of around 36% compared to MF alone (**Figure 7**). However, the introduction of this step in an overall sanitation process introduces two drawbacks, one relates to the use of coagulant and the other to the contact time required for this operation.

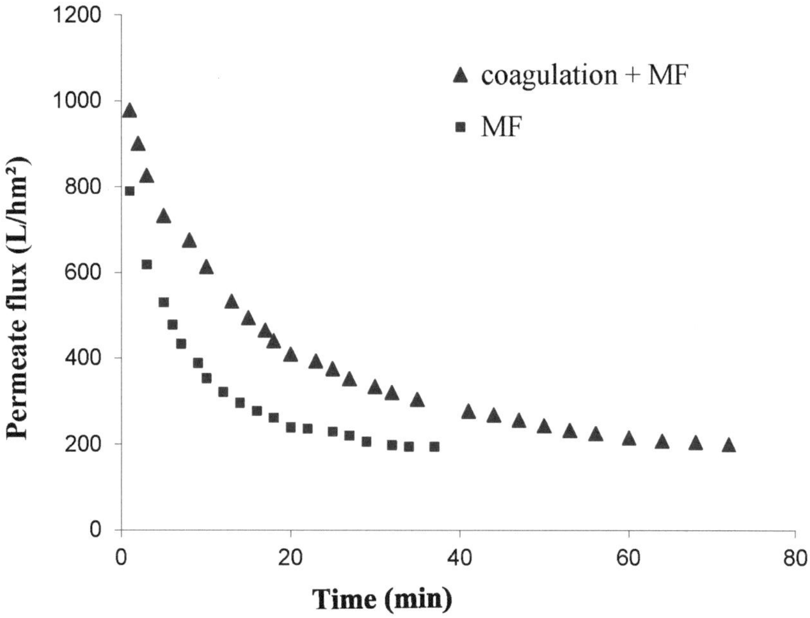

Figure 7.
Permeate flux decline for MF and coagulation-MF coupling (semi-industrial pilot).

It was also observed that during the coagulation-MF coupling, the unclogging of the membrane became easier than before. In fact, the membrane returns to its initial state, after a simple circulation of distilled water. Indeed, it was found that coagulation is the more efficient pretreatment before filtration processes [30–32]

4. Conclusion

The results obtained show that, despite the variation in the quality of the treated wastewater, additional treatment involving a membrane separation technique (MF) made it possible to eliminate the microbiological danger and the improvement of other parameters (biological and physicochemical). Coupling this technique with another process (coagulation or UF) does not lead to a significant improvement in the quality of the treated water. However, such a coupling can have an influence on the amount of water to be treated. Indeed, the pretreatment with coagulation before microfiltration improves the permeate flow and decrease membrane fouling compared to MF alone.

The integration of this microfiltration step on the scale of the wastewater treatment plant makes it possible to increase the available reserves of good quality water and to widen the fields of their uses. In fact, the unrestricted reuse of this treated water for the irrigation of crops of high economic profitability makes it possible to amortize investment costs while guaranteeing the health security of farmers.

Author details

Nebil Belaid*
École Nationale d'Ingénieurs de Sfax, Laboratoire de Radio-Analyses et Environnement, ENIS, BP W, 3038 Sfax, Tunisie

*Address all correspondence to: belaidnebil1@gmail.com

References

[1] Jeuland, M. : Challenges to wastewater reuse in the Middle East and North Africa. Middle East development Journal, 2015; 7-1. doi:10.1080/17938120.2015.1019293

[2] Qadir M., Bahri A., Sato, T., Al-Karadsheh E. : Wastewater production, treatment, and irrigation in Middle East and North Africa. Irrig Drainage Syst, 2010, 24; 37-51. doi:10.1007/s10795-009-9081-y

[3] Salgot M., Folch M.: Wastewater treatment and water reuse. Current Opinion in Environmental Science & Health, 2018, 2; 64-74. doi:10.1016/j.coesh.2018.03.005

[4] Kam G., Ding C.: Wastewater Treatment and Reuse—The Future Source of Water Supply. Encyclopedia of Sustainable Technologies, 2017, 43-52. doi:10.1016/B978-0-12-409548-9.10170-8

[5] Garcia X., Pargament D. : Reusing wastewater to cope with water scarcity: Economic, social and environmental considerations for decision-making. Resources. Conservation and Recycling, 2015, 101; 154-166. doi:10.1016/j.resconrec.2015.05.015

[6] Alkhudhiri A., Darwish NB., Hilal N. : Analytical and Forecasting Study for Wastewater Treatment and Water Resources in Saudi Arabia. J. Water Process Eng. 2019, 32. doi:10.1016/j.jwpe.2019.100915

[7] Oron G., Gillerman L., Bick A., Buriakovsky N., Manor Y., Ben-Yitshak E., Katz L., Hagin J. : A two stage membrane treatment of secondary effluent for unrestricted reuse and sustainable agricultural production. Desalination, 2006, 187; 335-345. doi:10.1016/j.desal.2005.04.092

[8] Pescod M.B. :Wastewater treatment and use in agriculture. Bull. FAO, 1992, 47, Rome, Italy, 125 p.

[9] Yadav RK, Goyal B., Sharma RK., Dubey SK., Minhas PS.: Post-irrigation impact of domestic sewage effluent on composition of soils, crops and groundwater - A case study. Environ. Internat., 2002, 28;481-486. https://doi.org/10.1016/S0160-4120(02)00070-3

[10] De Gisi S., Casella P., Cellamare CM., Ferraris M., Petta L., Notarnicola M. : Wastewater Reuse. Encyclopedia of Sustainable Technologies, 2017, 53-68. https://doi.org/10.1016/B978-0-12-409548-9.10528-7

[11] Levine, A. and Asano T.: Recovering sustainable water from wastewater. Environ. Sci. Technol., 2004, 38; 201A-208A. https://doi.org/10.1021/es040504n

[12] Tarchitzky J., Golobati Y., Keren R., Chen Y.:Wastewater effects on montmorillonite suspensions and hydraulic properties of sandy soils. Soil Sci. Soc. Am. J., 1999 63; 554-560. https://doi.org/10.2136/sssaj1999.03615995006300030018x

[13] Vojtěchovská Šrámková M., Diaz-Sosa V., Wanner J. : Experimental verification of tertiary treatment process in achieving effluent quality required by wastewater reuse standards. Journal of Water Process Engineering, 2018, 22; 41-45. https://doi.org/10.1016/j.jwpe.2018.01.003

[14] Wintgens T., Melin T., Schiller A.,. Khan S , Muston M., Bixio D. , Thoeye C. : The role of membrane processes in municipal wastewater reclamation and reuse. Desalination, 2005, 178; 1-11. https://doi.org/10.1016/j.desal.2004.12.014

[15] Abdessemed, D., Nezzal, G.: Treatment of primary effluent by coagulation-adsorption-ultrafiltration for reuse. Desalination, 2002,152 ; 367-373. https://doi.org/10.1016/S0011-9164(02)01085-8.

[16] Abdessemed, D., Nezzal, G., Ben aim, R.: Treatment of wastewater by ultrafiltration. Desalination, 1999, 126; 1-5. https://doi.org/10.1016/S0011-9164(99)00149-6.

[17] Fujioka, T., Nghiem, L.D.,. Fouling control of a ceramic microfiltration membranefor direct sewer mining by backwashing with ozonated water. Separ. Purif. Technol., 2015,142; 268-273. https://doi.org/10.1016/j.seppur.2014.12.049.

[18] Gong, H., Jin, Z., Wang, X., Wang, K.: Membrane fouling controlled by oagulation/adsorption during direct sewage membrane filtration (DSMF) for organic matter concentration. J. Environ. Sci. (China), 2015, 32 ; 1-7. https://doi.org/10.1016/j.jes.2015.01.002.

[19] Makropoulos, C., Rozos, E., Tsoukalas, I., Plevri, A., Karakatsanis, G., Karagiannidis, L.,Makri, E., Lioumis, C., Noutsopoulos, C., Mamais, D., Rippis, C., Lytras, E. : Sewer-mining: a water reuse option supporting circular economy, public serviceprovision and entrepreneurship. J. Environ. Manag., 2018, 216; 285-298. https://doi.org/10.1016/j.jenvman.2017.07.026.

[20] Hakami M W., Alkhudhiri A. , Al-Batty S., Zacharof MP., Maddy J., Hilal N. : Ceramic Microfiltration Membranes in Wastewater Treatment: Filtration Behavior, Fouling and prevention. Membranes, 2020, 10; 248. https://doi:10.3390/membranes10090248

[21] Shang, R., Verliefde, A.R.D., Hu, J., Zeng, Z., Lu, J., Kemperman, A.J.B., Deng, H.,Nijmeijer, K., Heijman, S.G.J., Rietveld, L.C.: Tight ceramic UF membrane as RO pre-treatment: the role of electrostatic interactions on phosphate rejection. Water Res. 2014, 48; 498-507. https://doi.org/10.1016/j.watres.2013.10.008.

[22] Weber, R., Chmiel, H., Mavrov, V.: Characteristics and application of new ceramic nanofiltration membranes. Desalination, 2003, 157 ; 113-125. https://doi.org/10.1016/S0011-9164(03)00390-4.

[23] Fane AG.: Sustainability and membrane processing of wastewater for reuse. Desalination, 2007;202: 53-58. https://doi.org/10.1016/j.desal.2005.12.038

[24] Belaid N., Neel C., Kallel M., Ayoub T., Ayadi A. and Baudu M.: Long term effects of treated wastewater irrigation on calcisol fertility: A case study of Sfax-Tunisia. Agricultural Sciences, 2012, 3(5); 702-713. http://dx.doi.org/10.4236/as.2012.35085

[25] Belaid N., Neel C., Kallel M., Ayoub T., Ayadi A., Baudu M.: Effects of treated wastewater irrigation on soil salinity and sodicity of Sfax (Tunisia): A case study. Journal of Water Science. 2010, 23(2); 133-145. https://doi.org/10.7202/039905ar

[26] Belaid N., Neel C., Lenain JF., Buzier R., Kallel M., Ayoub T., Ayadi A. Baudu M.: Assessment of metal accumulation in calcareous soil and forage crops subjected to long-term irrigation using treated wastewater: Case of El Hajeb-Sfax, Tunisia. Agriculture, Ecosystems & Environment, 2012, 158(1); 83-93. https://doi.org/10.1016/j.agee.2012.06.002

[27] AFNOR, Evaluation de la qualité des sols, vol. 1. AFNOR Editions, 2004, Paris; p 461

[28] Alonso E., Santos A., Solis GJ., Riesco P. : On the feasibility of urban

wastewater tertiary treatment by membranes: a comparative assement. Desalination, 2001, 141; 39-51. https://doi.org/10.1016/S0011-9164(01)00387-3

[29] Ahn KH.and Song KG. : Treatment of domestic wastewater using microfiltration for reuse of wastewater. Desalination, 1999; 126: 7-14. https://doi.org/10.1016/S0011-9164(99)00150-2

[30] Im D., Nakada N., Kato Y., Aoki M., Tanaka H.: Pretreatment of ceramic membrane microfiltration in wastewater reuse: A comparison between ozonation and coagulation. Journal of Environmental Management, 2019, 251: 109555. https://doi.org/10.1016/j.jenvman.2019.109555

[31] Fan L., Nguyen T., Roddick FA., John L. Harris JL. : Low-pressure membrane filtration of secondary effluent in water reuse: Pre-treatment for fouling reduction. Journal of Membrane Science, 2008; 320: 135-142. https://doi.org/10.1016/j.memsci.2008.03.058

[32] Hatt JW., Germain E., Judd SJ.: Precoagulation-microfiltration for wastewater reuse. Water Research, 2011, 45; 6471-6478. https://doi.org/10.1016/j.watres.2011.09.039

Chapter 20

Immobilization of Powdered Coal Fly Ashes (CFAs) into CFA Beads and Column Studies on Color Removal from Pulp Mill Effluents Using These CFA Beads

Musfiques Salahin and George Yuzhu Fu

Abstract

In this study, immobilization process of the three (3) powder CFAs was studied. The major results on immobilization process were briefly presented. A total number of fifteen (15) column studies from the combination of the five (5) types of CFAs beads and the three (3) PMEs samples were performed. In each column study, a set of aggregate parameters of flow rate, empty bed contract time, operational time, and throughput volume was studied, and the data was fitted to existing modeling of breakthrough curves. The overall operational time was 12–24-hour, color removal efficiencies were 40–90%, and throughput volume of treated PMEs was 10–14 bed volume. For the column study, the correlation coefficient R^2 value for each combination indicated that the Thomas model had a better fit with the observed data than the Adams-Bohart model, and the color adsorption capacities of CFA beads varied in a wide range of 0.31–28.23 mg/g.

Keywords: coal fly ash (CFA), pulp mill effluent (PME), color removal, immobilization, column study

1. Introduction

As one of the dominant industries in Georgia, USA, the pulp and paper industries consume huge amounts of fresh water and a wide variety of chemicals during pulp production processes. A significant amount of water and these chemicals are released as high pollutant load with intense color effluent into surface water bodies [1]. Pulp and paper mill effluents (PPMEs) transport high concentrations of organic/inorganic pollutants and color compounds like lignocellulosic compounds, tannins, hemicelluloses, pectin, resin acids, unsaturated fatty acids, carboxylic acid, and other substances [2]. These untreated effluents are responsible for increasing the levels of chemical oxygen demand (COD), biochemical oxygen demand (BOD), total organic carbon (TOC), adsorbable organic halides (AOXs), toxic contaminants, and heavy metals in the water ecosystem [3]. Therefore, PPMEs must be treated before they are discharged into receiving water bodies.

IntechOpen

Physical treatment can efficiently remove 80% of suspended solids (SS) from pulp mill effluents (PMEs) [4]; Aerobic lagoon and anaerobic treatment studies showed potentially removing 50% color and 60% COD from PMEs [4, 5]. However, these fungal treatment studies are still ongoing and in the early stage of industrial implementation [6]. The coloration of receiving water bodies from PMEs causes negative impacts including an unpleasing esthetic appearance, reduction of dissolved oxygen (DO) level, and reduction of sunlight transmission into bodies of water which may adversely affect aquatic life. Additionally, insufficient sunlight reduces photosynthetic activity, therefore, making it difficult to remove organic pollutants and causing an increase of water/wastewater treatment costs downstream [7, 8]. Hence, tertiary treatment is essential for the further treatment of PMEs.

Adsorption, a proven and widely used treatment process, removes a variety of contaminants from industrial wastewater. Activated carbon, the most common adsorbent, is widely used to remove contaminants from industrial wastewater for a long time [9, 10]. However, the high production costs of activated carbon have motivated researchers to explore alternative low-cost adsorbents like coal fly ash (CFA), rice husk ash (RHA), bagasse fly ash (BFA), etc. industrial for wastewater treatment.

Energy sector is another dominant industry in Georgia, USA. It produced 6.1 million tons of CFA as byproducts in 2011 and 60% of CFA were disposed of in ash ponds. Investigating other beneficial uses of CFA in environmental engineering is necessary. The results from our Batch Studies have shown that CFA can effectively remove color from PME as a low-cost adsorbent [11]. However, these results were achieved in a batch operation using the powdered CFA samples, which are rarely used in a column in practice for a continuous operation due to their low permeability.

Therefore, the objectives of this study were to explore the cost-effective immobilization processes of powdered CFAs with water and addition of binders, to produce the CFAs beads with strength and high adsorption capacity for color removal from PME, and to use these CFA beads in column studies to remove color from the PME under a continuous operation.

2. Material and methods

2.1 Coal fly ash (CFA)

Three (3) raw CFA samples were obtained from three individual coal-based power plants units of Georgia Power Company and labeled as CFAs #1 - #3. The first one was Class "C" subbituminous ash, whereas the last two were Class "F" bituminous ashes. These CFAs were stored in plastic containers at room temperature (20 ± 10°C) in the Water and Environmental Research Lab (WERL) at Georgia Southern University (GSU) for further study. The chemical composition of CFA#2 is summarized in **Table 1**.

2.2 Pulp mill effluent

For this study, three (3) biologically treated secondary PMEs samples were collected from three different pulp mill factories from surrounding areas in Georgia, USA. They were labeled as PMEs #1 - #3, respectively. All these PMEs were collected from secondary clarified effluent outlets. These PMEs were stored in plastic container in a refrigerator at temperature of 4 ± 1°C in the WERL at GSU for further study. The primary properties of these PMEs are summarized in **Table 2**.

Chemical compositions	% w/w
Silicon dioxide (SiO_2)	50.96
Aluminum oxide (Al_2O_3)	21.00
Titanium dioxide (TiO_2)	1.11
Iron oxide (Fe_2O_3)	14.32
Calcium oxide (CaO)	4.39
Magnesium oxide (MgO)	0.9
Potassium oxide (K_2O)	2.49
Sodium oxide (Na_2O)	1.07
Sulfur trioxide (SO_3)	1.94
Loss on ignition	1.82
Total	100%

Table 1.
The chemical composition of CFA#2.

PMEs	Received date	Number of buckets	Color	pH	COD	TOC
			(mg/L (Pt-Co) Color Units)		(mg/L)	(mg/L)
PME 1	16-Mar-15	One (1) 5 – gallon	157	6.92	255	
	9-Jun-16	Two (2) 5 – gallon	182	5.62	222	164.9
PME 2	26-Aug-15	One (1) 5 - gallon	355	8.21	86.6	
	17-Jun-16	Two (2) 5 – gallon	834	6.39	135.5	90.8
PME3	1-Jan-16	Three (3) 1 – gallon	723	7.51	248	
	12-Oct-16	Three (3) 1 – gallon	968	5.89	298.5	110.5

Table 2.
The primary properties of the three (3) PMEs.

2.3 Chemical reagent

For immobilization process, N type hydrated lime was collected from the Material Laboratory at GSU.

2.4 Immobilization process

The immobilization study of powdered CFAs was started by operating the key equipment of pelletizer (DP-14 Agglo-Miser Disc Pelletizer supplied by Mars Mineral) with each of the three CFA samples and water only without adding any binders. Multiple immobilizing parameters, such as RPM and the vertical angle of the pan of the pelletizer, and the ratio of CFA sample to water, as well as optimum curing conditions on humidity and duration, were investigated. The optimal rotational speed of pelletizer and vertical angle of the pan were found 32 RPM and 45 degree, respectively, and they were maintained the same during the whole immobilization process. For each batch, 2500 gm of CFA powder was pelletized by adding hydrated lime or Class "C" type CFA directly while water was added using spraying bottle. After palletization, the fresh pellets had to be cured with thin layer covered by wet cloth at room temperature (20 ± 1°C) in the WERL at GSU for two weeks and exposed to atmosphere for air dry for one week before being used in column studies (12).

2.5 Fixed bed column

For each column study, the fixed-bed column was set-up using a glass column with a 15 mm internal diameter and 750 mm bed height and a peristaltic pump with medium flow range (1 mL/min). The schematic diagram of the fixed-bed column system is illustrated in **Figure 1**. For each column study, the column was packed by 2–4 mm diameter CFA beads, one type at a time. The mass of the CFA beads typically varied in a range of 50–75 g with a varied height of the bed of CFA beads in a range of 505–655 mm, which is dependent on the size distribution of the CFA beads produced from the above immobilization process. A layer of washed sand with a height of 25 mm was placed at the bottom of the CFA beads to protect any loss of beads as well as give mechanical support in the fix-bed column system. The top of the bed of CFA beads was submerged at least 50 mm below the water surface of the PME sample in the column by maintaining a high position of the outlet in treated effluent tubing (inverse "U" shape) shown in **Figure 1**.

All of these column studies were performed at room temperature (20 ± 1°C) at the WERL at GSU and the treated PME samples were collected at different intervals: every 1 h from the 1st hour to 12th hour, every 2 hours from the 13th hour to 24th hour until the column of the CFA beads was exhausted. These samples were analyzed for color. Meanwhile, all the treated PME samples were collected in the treated PME container as an accumulated sample and its color was tested in the end of a column study. The treated PME sample was separated by vacuum-filtering the mixture through a Supor-450 47 mm 0.45 μm membrane paper. Immediately after filtration, the filtrated samples were collected to test color. Duplicate flasks with identical CFA and effluent mixtures were used to represent each sample. The average reading of color and pH measurements between the two duplicate samples were documented. The color of PME samples was measured using HACH DR 5000 Spectrophotometer and HACH Method 8025.

Color removal efficiency of CFA can be calculated using following equation:

$$Removal\ Efficiency\ (\%) = \frac{(C_i - C_t) * 100}{C_i} \quad (1)$$

where E = Removal Efficiency (%), C_i is the initial color concentration (mg/L), C_t is the color concentration at time t (mg/L).

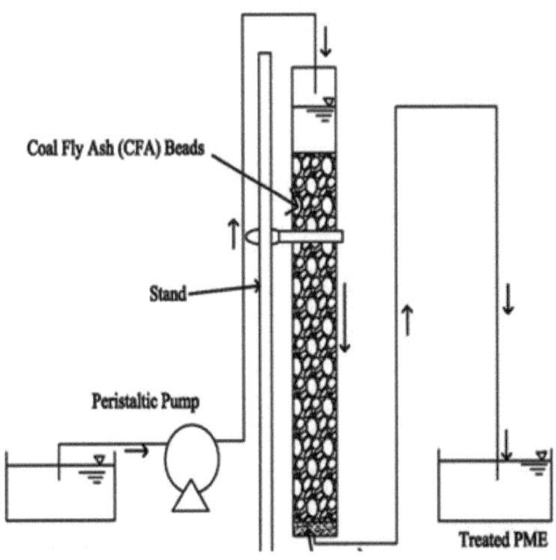

Figure 1.
A schematic diagram of column.

The effluent volume (V_{eff}) can be expressed as:

$$V_{eff} = Q \times t_{total} \tag{2}$$

Where, Q stands for volumetric flow rate (mL/min) and t_{total} stands for the time of exhaustion (min).

The area under the breakthrough curve (A) can be determined by integrating color adsorbed concentration (C_{Ad}) vs. time (t). For a given volumetric flow rate (Q) and initial color concentration (C_0) of PMEs, the total amount of color adsorbed (q_{total}) can be calculated by the following equation:

$$q_{total} = \frac{Q}{1000} \int_{t=0}^{t_{total}} C_{Ad} dt \tag{3}$$

Total amount of color, M_{total} (mg) passed through the fixed-bed column system is determined by following equation:

$$M_{total} = \frac{C_0 \times Q \times t}{1000} \tag{4}$$

Where, C_0 is denoted as initial color concentration (mg/L) of PME, Q is the volumetric flow rate (mL/min) of PMEs.

The fixed-bed column system color removal efficiency with respect to the flow volume can be calculated by following equation:

$$Removal\ Efficiency = \frac{q_{total}}{M_{total}} \times 100 \tag{5}$$

3. Results and discussions

3.1 Immobilization of powder CFAs

In each batch trial during the immobilization process, CFA beads were produced with small spherical shape and maintained CFA grayish original color. It was observed that the RPM and inclined vertical angle of the pan of pellitizer, and water mixing ratio with CFAs were important factors that affected beads size distribution. In the immobilization process, at high RPM of the pan reduced the size of CFA beads while at low RPM beads diameters were increased. The optimal rotational speed of pelletizer and vertical angle of the pan were found to be 32 RPM and 45 degree, respectively, and were maintained the same during the whole immobilization process. It is also found out that CFA1 (Class "C") or hydrated lime can be used as binder to produce CFA beads. The combination of hydrated lime and CFAs required more water while the combination of CFA1 with CFA2or CFA3 required less.

For each batch of immobilization process, cost effective blinders were used to produced relatively uniform and consistent size and shape. The newly produced beads were put into large foil tray with thin layer and covered with a soaked wet cloth at room temperature of $20 \pm 1°C$ for two weeks. The cloth was kept wet during these two weeks and then was removed. The beads were continually exposed to the atmosphere for air dry at room temperature of $20 \pm 1°C$ for one week. In this study only the amount of water used for immobilization was documented while the water used during 2-week "wet" curing period was not documented. It was observed that about 10–20% of water contents could produce the CFA beads with a high strength. In general, it was observed that the strength of the beads increased with in increased weight percentage of hydrated lime, adsorbed more water during curing period, and much more pores formed. It was observed that about 10–20% of water content could produce the beads with a high strength. In this study, it was particularly important for the CFA beads to hold their shape in PMEs. Those CFA beads, which could hold their shape in PMEs at least 24 h, were considered to be beads with a high strength. A summary table for total five (5) types of CFA beads is shown in **Table 3**.

The five (5) types of CFA beads produced from the three (3) CFA samples are shown in **Figure 2**.

3.2 Column studies

3.2.1 Type of beads of CFA1 only with PMEs

In these column studies, the raw PMEs samples were passed through 75 g of CFA1 only beads within the column. The flow rates of Q = 0.78–0.89 mL/min, empty bed contact time (a measure of the time during which water to be treated is in contact with CFA beads in a contact column), EBCT = 125 min with an operation time t = 12–24 h. During this operational period, almost 5–13 bed volumes of PMEs were treated in fixed bed column. **Figure 3** shows the physical properties of PMEs and fixed bed column. At the end of each column study, the apparent color ratio of C/C_0 reached up to 0.546, which corresponded to 45.4% apparent color removal.

3.2.2 Types of CFA beads of CFA2 + Lime and CFA2 + CFA1 with PMEs

In these column studies, the raw PMEs sample were passed through 50 ~ 75 g of CFA2 + Lime (Mass Ratio of CFA2 to Lime = 4:1) and CFA2 + CFA1 beads with a

Type No.	Name of CFA beads	CFA samples	Cost-effective binders	Mass ratio of CFA to binder
1	CFA3 + Lime	CFA3	Hydrated Lime	4:1
2	CFA3 + CFA1	CFA3	CFA1	1:2
3	CFA1	CFA1	Water Only	8.1:1
4	CFA2+ Lime	CFA2	Hydrated Lime	4:1
5	CFA2+ CFA1	CFA2	CFA1	1:2

Table 3.
Summary of the finalized cost-effective binders and a Total of five (5) types of CFA beads produced.

Figure 2.
The five (5) types of CFA beads produced from the three (3) CFA samples (from left to right: 1 CFA3 + 2 CFA1; 1 CFA2 + 2 CFA1; 4 CFA3 + 1 Lime; 4 CFA2 + 1 Lime; and CFA1. Note: the numbers of 1, 2, and 4 are referred to mass ratios).

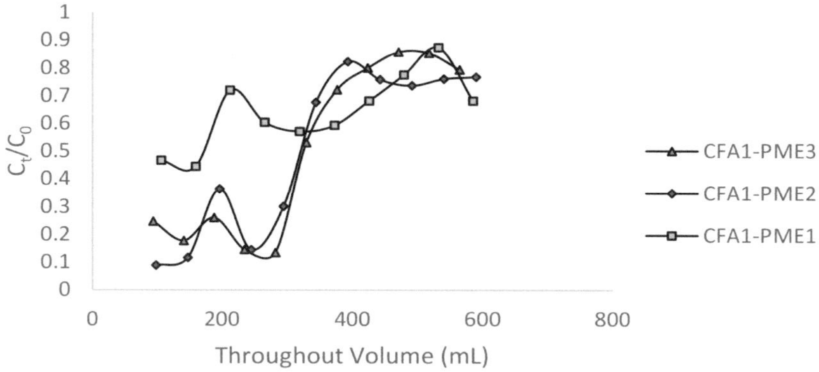

Figure 3.
Breakthrough of color column experiments using three different PME and CFA1 only beads. Experimental setup: Initial color concentration: PME1: 182; PME2: 834 and PME3: 968 (mg/L Pt-Co); bed height: 51 ~ 61 cm; flow rate: 0.78 ~ 0.89 mL/min; at room temperature.

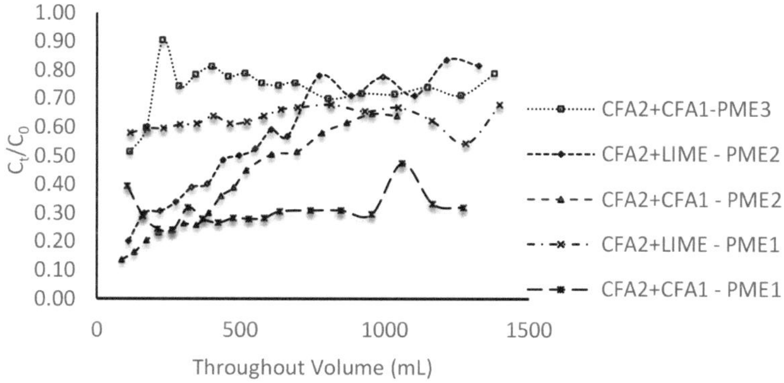

Figure 4.
Breakthrough of color column experiments using three different PMEs, CFA2 and additive materials CFA beads. Experimental setup: initial color concentration: PME1: 182; PME2: 834 and PME3: 968 (mg/L Pt-Co); bed height: 56 ~ 63 cm, flow rate: 0.72 ~ 0.96 mL.

height of 562–660 mm in the column, respectively. The flow rates of Q = 0.72–0.97 mL/min, EBCT = 125 min with an operation time t = 24 h. **Figure 4** shows the physical properties of PMEs and fixed bed column. During this operational period, almost 6–12 bed volumes of PMEs were treated in fixed bed column. At the end of each column study, the apparent color ratio of C/C_0 reached up to 0.27, which corresponded to 73% apparent color removal. The curve of CFA2 + Lime - PME1 shows a different characteristic compare with others. Color might be added due to leaching out of chemicals from the CFA beads and breakdown some chemical reagents present in PME1.

3.2.3 Type of CFA beads of CFA3 + Lime and CFA3 + CFA1 with PMEs

In these column studies, the raw PMEs sample was passed through 50 ~ 75 g of CFA3 + Lime, and CFA3 + CFA1 beads with a height of 582–645 mm in the column, respectively. The flow rates of Q = 0.75–0.96 mL/min, EBCT = 125 min with an operation time t = 24 h. **Figure 5** shows the physical properties of PMEs and fixed bed column. During this operational period, almost 6–14 bed volumes of PMEs were treated in fixed bed column. At the end of each column study, the apparent color ratio of C/C_0 reached up to 0.473, which corresponded to 52.7% apparent color removal.

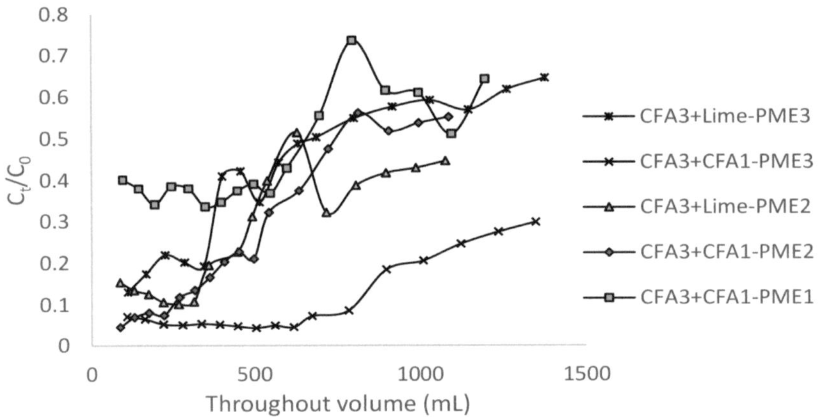

Figure 5.
Breakthrough of color column experiments using three different PME and CFA3 and additive materials CFA beads. Experimental setup: initial color concentration: PME1: 182; PME2: 834 and PME3: 968 (mg/L Pt-Co); bed height: 58 ~ 65 cm; flow rate: 0.75 ~ 0.96.

3.3 Modeling of breakthrough curves from column studies

3.3.1 Adams-Bohart model

The Adams-Bohart model is typically applied to check the dynamic behavior of the column which describes the relationship between $\frac{C_t}{C_0}$ and t in a continuous fixed-bed column system. This model is eminent to predict the initial phase of the breakthrough curve. The linear equation is expressed as:

$$\ln\left(\frac{C_t}{C_0}\right) = K_{AB} \times C_0 \times t - K_{AB} \times N_0 \times \left(\frac{Z}{U_0}\right) \quad (6)$$

Where, C_0 and C_t are denoted as the color concentration (mg/L Pt-Co) of influent and effluent of column system; K_{AB} represents the kinetic constant (L/mg. min); N_0 is the saturation concentration (mg/L); Z is denoted as bed depth of the fixed-bed column and U_0 stands for superficial velocity (cm/min). By plotting $\ln\left(\frac{C_t}{C_0}\right)$ vs. t, N_0 is obtained from the intercept and K_{AB} can be calculated from slope of the graph.

The column studies experimental data obtained from the five different combinations of CFA beads with PME1 were used for linear regression analysis and their corresponding parameters kinetic constant K_{AB} and saturation constant N_0 were obtained along with the correlation coefficient (R^2) given in (Eq. (6)). The model is shown in **Figure 6**. At initial color concentration, C_0 = 182 mg/L, the R^2 value of Adams-Bohart model for different combinations were varied in a range of 0.54–0.80. Under the similar condition, this model corresponding parameters kinetic constant K_{AB} values were varied 2.20–4.40 (10^{-6} L/mg.min), and N_0 = 10.23–0.78 (10^3 mg/L) respectively.

The column studies experimental data obtained from the four different combinations of CFA beads with PME2 were used for linear regression analysis and their corresponding parameters kinetic constant K_{AB} and saturation constant N_0 were obtained along with the correlation coefficient (R^2) given in (Eq. (6)). The model is shown in **Figure** 7. At initial color concentration, C_0 = 834 mg/L, the R^2 value of Adams-Bohart model for different combinations varied in a range of 0.70–0.94.

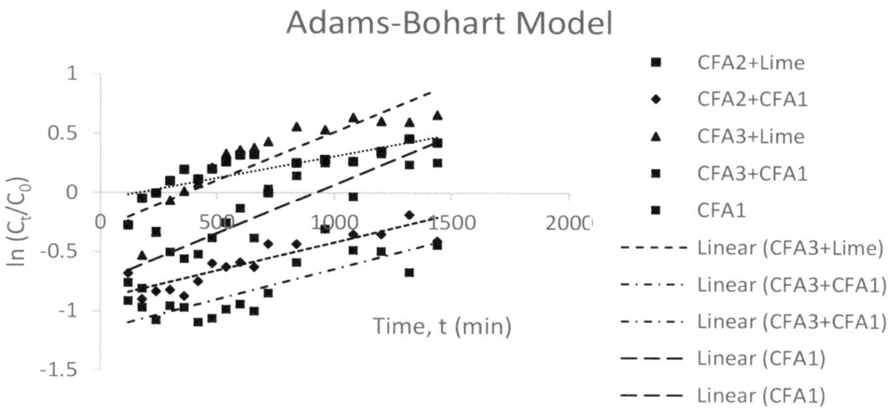

Figure 6.
Application of the Adams-Bohart model to the experimental data from column study for PME1.

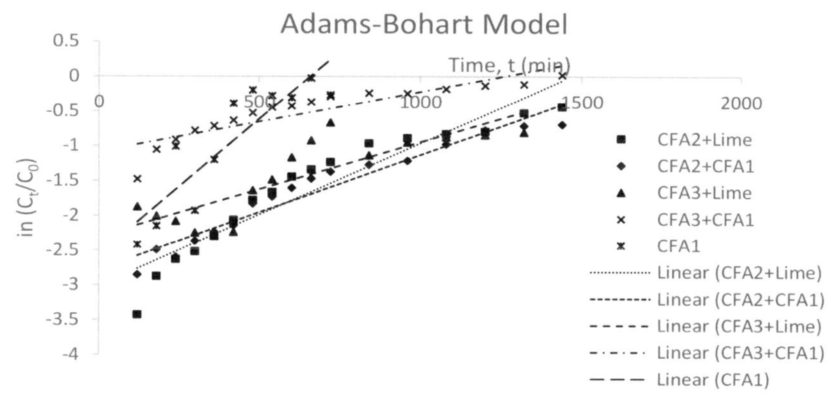

Figure 7.
Application of the Adams-Bohart model to the column study experimental data for PME2.

Under the similar condition, this model's corresponding parameters kinetic constant K_{AB} values was varied 1.68–4.68 (10^{-3} L/mg.min). N_0 = 12.63–4.23 (10^3 mg/L) respectively.

The column studies experimental data obtained from the four different combinations of CFA beads with PME3 were used for linear regression analysis and their corresponding parameters kinetic constant K_{AB} and saturation constant N_0 were obtained along with the correlation coefficient $\left(R^2\right)$ given in (Eq. (6)). The model was shown in **Figure 8**. At initial color concentration, C_0 = 968 mg/L, the R^2 value of Adams-Bohart model for different combinations varied in a range of 0.67–0.85. Under the similar condition, this model's corresponding parameters kinetic constant K_{AB} values was varied 1.14–3.2 (10^{-3} L/mg.min). N_0 = 55.22–16.06 (10^3 mg/L) respectively.

3.3.2 Thomas model

Thomas model is one of the widely used models in fixed-bed continuous column operation. This model based on three assumptions: i) follows Langmuir Isotherm model; ii) obeys the second order reversible reaction kinetics; and iii) there is no axial depression of the adsorbent. The linear equation is expressed as follows

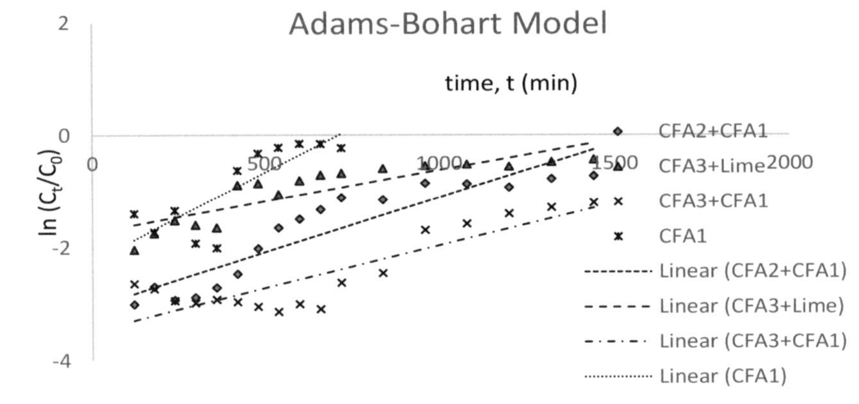

Figure 8.
Application of the Adams-Bohart model to the column study experimental data for PME3.

$$\ln\left(\frac{C_0}{C_t}-1\right)=\frac{K_{Th}\times q_0\times M}{Q}-K_{Th}\times C_0\times t \tag{7}$$

Where, C_0 and C_t are denoted as the color concentration (mg/L Pt-Co) of influent and effluent of column system; K_{Th} stands for Thomas rate constant (L/(min.mg)); q_0 is the maximum color adsorption capacity for CFA beads (mg/g), M is the mass of CFA beads (g), Q is the flow rate (mL/min). By plotting $\ln\left(\frac{C_0}{C_t}-1\right)$ versus t, Thomas rate constant K_{Th} can be obtained from the slope and q_0 can be calculated from the interception of the plot.

The experimental data obtained from the three selected combinations of three types of CFA beads with PME1 column studies were used for linear regression analysis and their corresponding Thomas rate constant K_{Th} and maximum color adsorption capacity q_0 were calculated along with the correlation coefficient (R^2) given in (Eq. (7)). The model is shown in **Figure 9**. At initial color concentration C_0 = 182 mg/L, the R^2 value of Thomas model for different combination varied in a range of 0.54–0.79. The Thomas rate constants were calculated for different combinations and were 6.593, 14.286 and 5.495 ($\times 10^{-6}$ L/min.mg) for the CFA2 + CFA1, CFA3 + CFA1 and CFA1 only beads, respectively. The maximum color adsorption capacity q_0 was obtained from the plot and their values were in a range of 0.31–1.72 mg/g. On the other hand, the color adsorption capacities after 20 hours' operation of pump started were obtained and they varied in a range of 0.72–1.67 mg/g, which

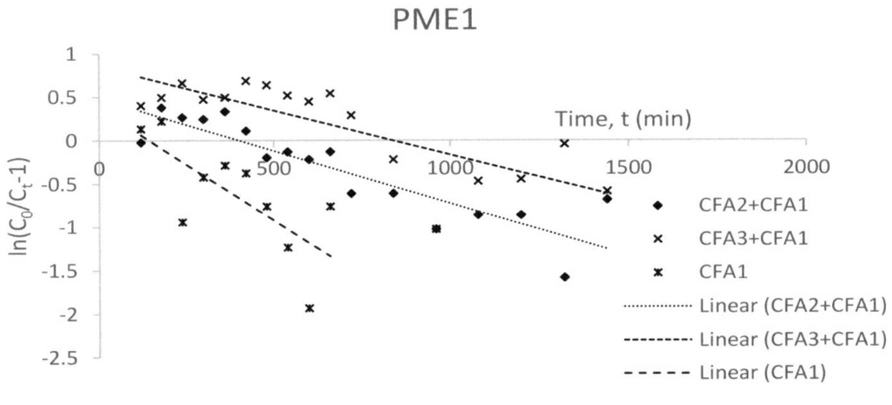

Figure 9.
Application of Thomas model to the column study experimental data for PME1.

were very close to the range of the Thomas model adsorption capacities. However, these values were very low compared with the results from the batch studies. For the CFA beads (Class "F" type CFA + binder), the maximum color removal capacity increased with decreased the volumetric flow rate of column operation.

The experimental data obtained from the five selected combinations of CFA beads with PME2 column studies were used for linear regression analysis and their corresponding Thomas rate constant K_{Th} and maximum color adsorption capacity q_0 were calculated along with the correlation coefficient (R^2) given in (Eq. (7)). The model is shown in **Figure 10**. At initial color concentration C_0 = 834 mg/L, the R^2 value of Thomas model for different combination varied in a range of 0.71 to 0.97. Thomas rate constants were calculated for different combinations and were 3.36, 2.64, 9.71, 2.16 and 3.16 (x 10^{-6} L/min.mg) for the CFA3 + Lime, CFA3 + CFA1, CFA1 only, CFA2 + Lime, CFA2 + CFA1 beads, respectively. The maximum color adsorption capacity q_0 was obtained from the plot and their values were in a range of 3.65 to 14.02 mg/g. On the other hand, the color adsorption capacities after 20 hours' operation of pump started were obtained and they varied in a range of 4.04–12.95 mg/g which were very close to the range of the Thomas model adsorption capacities. However, these values were quite high compared with the results from the batch studies. For the CFA beads (Class "F" type CFA + binder), the maximum color removal capacity increased with decreased the volumetric flow rate of column operation.

The experimental data obtained from the four selected combinations of CFA beads with PME3 were used for linear regression analysis and their corresponding Thomas rate constant K_{Th} and maximum color adsorption capacity q_0 were calculated along with the correlation coefficient (R^2) shown in (Eq. (7)). The model is shown in **Figure 11**. At initial color concentration C_0: 968 mg/L, R^2 value of Thomas model for different combination were varied in between 0.76 to 0.85. Thomas rate constant were calculated for different combination 23.1, 1.86, 1.76 and 35.71 ($x10^{-6}$ L/min.mg). The maximum color adsorption capacity q_0 was obtained from the plot and their values were varied in between 0.8 ~ 28.23 mg/g. Where, color adsorption capacity after 20 hours' operation of pump started were obtained and their value varied in a range of 4.4 ~ 12.71 mg/g that had some difference with Thomas model constant values. Again, these values were quite high compared with the batch studies results. For the CFA beads (Class "F" type CFA + binder), the maximum color removal capacity increased with decreased the volumetric flow rate of column operation.

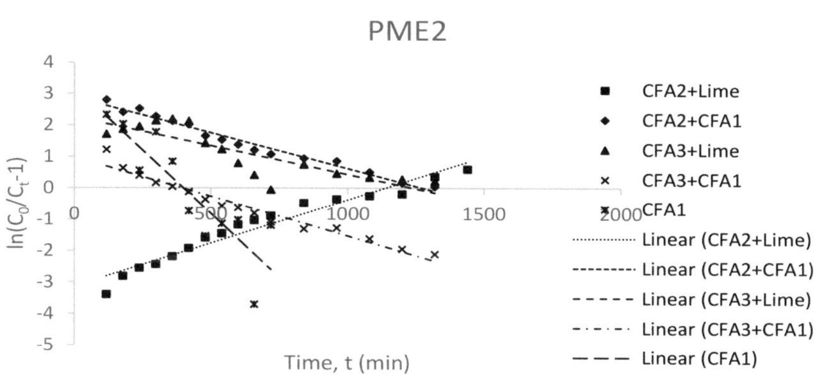

Figure 10.
Application of Thomas model to the column study experimental data for PME2.

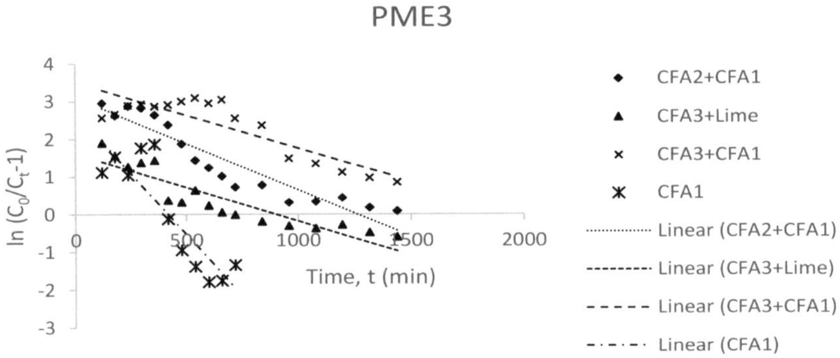

Figure 11.
Application of Thomas model to the column study experimental data for PME3.

4. Conclusions

Based on the significant results obtained from the column studies, the following conclusions can be reached:

- From the immobilization studies, both hydrated lime and the Class "C" of CFA1 were found to be cost-effective binders for Class "F" CFA2 or CFA3 in addition to water, while water only was the cost-effective binder for Class "C" CFA1.

- To produce CFA beads with strength and capability to removal color, the effective mass ratio of Class "F" CFA2/CFA3 sample to hydrated lime was 4:1 while that to Class "C" CFA1 was 1:2.

- From column studies, the CFA beads were able to be packed into columns and used to remove color from the PMEs under continuous operation. They could hold the good shape after they contacted and submerged in PME samples for 12–24 h. The overall color removal efficiencies were in a range of 45–90% while the general throughput volume of treated PME was in a range of 10–14 bed volumes of CFAs beads in the columns.

- From column studies, based on true color removal, for the three PMEs, the CFA3 + CFA1 and CFA2-CFA1 beads were good ones with high adsorption capacities.

- The correlation coefficient of R^2 value for each column study indicated that the Thomas model had a better fit with the observed data than the Adams-Bohart model did both for PME2 and PME3, whereas, the Adams-Bohart model fitted better than the Thomas model for PME1.

Acknowledgements

This study was funded by the Georgia Power 2014 GRAPE Program. The support from the Department of Civil Engineering and Construction at GSU is acknowledged. The representatives from Georgia Power and the three pulp mills in Georgia are thanked to obtain and arrange shipment of CFAs and PMEs samples.

Author details

Musfiques Salahin and George Yuzhu Fu*
Georgia Southern University, Statesboro, USA

*Address all correspondence to: gfu@georgiasouthern.edu

References

[1] Tunay O, Kabdasli I, Arslan-Alaton I, Olmez-Hanci T. Chemical Oxidation Applications for Industrial Wastewaters | IWA Publishing [Internet]. IWA. 2010 [cited 2018 Feb 7]. p. 360. Available from: https://www.iwapublishing.com/books/9781843393078/chemical-oxidation-applications-industrial-wastewaters

[2] Malik A, Grohmann E, editors. Environmental Protection Strategies for Sustainable Development [Internet]. Dordrecht: Springer Netherlands; 2012 [cited 2018 Feb 7]. Available from: http://link.springer.com/10.1007/978-94-007-1591-2

[3] Sajab MS, Chia CH, Zakaria S, Sillanpää M. Removal of Organic Pollutants and Decolorization of Bleaching Effluents from Pulp and Paper Mill by Adsorption using Chemically Treated Oil Palm Empty Fruit Bunch Fibers. BioResources [Internet]. 2014 Jun 13 [cited 2018 Jan 9];9(3):4517–27. Available from: http://ojs.cnr.ncsu.edu/index.php/BioRes/article/view/5373

[4] Thompson G, Swain J, Kay M, Forster CF. The treatment of pulp and paper mill effluent: A review. Bioresource Technology [Internet. 2001 May [cited 2018 Jan 9;77(3):275-286 Available from: http://www.ncbi.nlm.nih.gov/pubmed/11272013

[5] Pokhrel D, Viraraghavan T. Treatment of pulp and paper mill wastewater—a review. Sci Total Environ [Internet]. 2004 Oct 15 [cited 2018 Jan 9];333(1–3):37–58. Available from: http://www.sciencedirect.com/science/article/pii/S0048969704004279?via%3Dihub

[6] Johnson AP, Johnson BI, Gleadow P, Robilliard P. Reducing Kraft mill colour. In: Pulp and Paper Canada. Melbourne: Beca AMEC; 2012. pp. 42-45

[7] Namasivayam C, Radhika R, Suba S. Uptake of dyes by a promising locally available agricultural solid waste: coir pith. Waste Manag [Internet]. 2001 Jul 1 [cited 2018 Jan 4];21(4):381–7. Available from: http://www.sciencedirect.com/science/article/pii/S0956053X00000817?via%3Dihub

[8] Orori BB, Etiégni L, Rajab M, Situma M, Ofosu-Asiedu K. Decolorization of a pulp and paper mill effluent in Webuye Kenya by a combination of electrochemical and coagulation methods DESCRIPTION OF STUDY AREA. [cited 2018 Jan 9]; Available from: https://www.pulpandpapercanada.com/paptac/PDFs/Apr05/wastewater.pdf

[9] Aysu T, Kü Çü @bullet M M. Removal of crystal violet and methylene blue from aqueous solutions by activated carbon prepared from Ferula orientalis. Int J Environ Sci Technol [Internet]. [cited 2018 Jan 3]; Available from: http://www.bioline.org.br/pdf?st15214

[10] Gueu S, Yao B, Adouby K, Ado G. Kinetics and thermodynamics study of lead adsorption on to activated carbons from coconut and seed hull of the palm tree. Int J Environ Sci Technol [Internet]. 2007 Dec 1 [cited 2018 Jan 4];4(1):11–7. Available from: http://link.springer.com/10.1007/BF03325956

[11] Fu GY, Salahin M. Batch studies on color removal from pulp mill effluent (PME) using coal fly ash (CFA) produced from Georgia coal combustion power pants. In: 91st Annual Water Environment Federation Technical Exhibition and Conference, WEFTEC 2018. 2019

Chapter 21

Downflow Hanging Sponge System: A Self-Sustaining Option for Wastewater Treatment

Namita Maharjan, Choolaka Hewawasam, Masashi Hatamoto, Takashi Yamaguchi, Hideki Harada and Nobuo Araki

Abstract

Need of self-sustaining wastewater treatment plants (WWTPs) has become critical to cope up with dynamics of the environmental regulations and rapid advancements in the contemporary technologies. At present there are limited number of self-sustaining WWTPs around the world. The aim of this chapter is to present state -of- art of Downflow Hanging Sponge (DHS) system which was developed as a post treatment unit of Upflow Anaerobic Sludge Blanket (UASB) from sustainability perspective. DHS system is a non-submerged fixed bed trickling filter (TF) that employs a core technology of polyurethane sponges as a media where the microorganisms thrive and major treatment processes take place. This chapter reviews the introduction of DHS system (UASB+DHS) summarizes the quantitative analysis of environmental, economic and social sustainability using indicators. Furthermore, self-sustaining prospects of DHS system are assessed and discussed by comparing with conventional TF (UASB+TF).

Keywords: downflow hanging sponge, trickling filter, self-sustainability, indicators

1. Introduction

Wastewater treatment plants (WWTPs) are integral part of our society. Lately, WWTPs and its management have become an important issue in the world and also listed in many of sustainable development goals [1]. The WWTPs were initially designed to remove the pollutants depending on the flow variability to meet the certain discharge standards. However, emerging concepts and practices in WWTPs field have extended its application to energy recovery, reuse and nutrient recycling. Also, majority of WWTPs around the world are not designed with these multiple functions in mind and depend only on conventional technologies to solve these problems and fails to strike the balance between demand and supply of water. Especially, this is quite evident from the water scarcity clock which shows that there are about 2 billion of world population still living in water scarce areas [2]. Moreover, the changing dynamics of world such as accelerating energy dependent lifestyles, sudden global pandemics, economic fragmentations, climate change patterns and other major environmental concerns have affected the selection of suitable

WWTPs. Since many WWTPs have life cycle of 50–100 years, or even longer, the selection of WWTP will affect the development of that particular area. Several studies have shown that selecting and deciding the WWTPs are mostly based on their technical, economic, environment and social aspects [3]. Though extensive researches are being conducted and major strategies have been formulated, the search for long lasting technology in the wastewater treatment field is still on-going.

2. Downflow hanging sponge system

Aerobic biological treatment process can be traced back to the late nineteenth century. The biological process uses oxygen to break down organic contaminants and nutrients from wastewater. Oxygen is continuously mixed using aeration device (air blower or compressor) into the wastewater. Aerobic microorganisms then feed on the wastewater's organic matter converting it into carbon dioxide and biomass which is later removed. There are several types of aerobic treatment processes based on their designs such as fixed film system, continuous flow suspended growth aerobic system, retrofit aerobic system and composting toilets [4]. For this chapter, fixed film aerobic treatment systems called Downflow Hanging Sponge (DHS) and Trickling filter (TF) have been chosen and discussed. The rationale behind choosing these systems is the similarity in their working principle.

TFs is the second most widely used aerobic biological wastewater treatment system after activated sludge process (ASP) around the world [5]. TF is non-submerged fixed-bed, aerobic biological reactor which was applied for sewage treatment for the first time in England in 1893 [6]. Pre-settled wastewater is continuously trickled or sprayed from the top with the help of a rotating sprinkler. As the water moves through the pores of the filter, organics are aerobically degraded by the biofilm covering the filter material. The trickling filter consists of a cylindrical tank and is filled with different packing material such as stones, rocks, gravels or special pre-formed plastic filter media. Since couple of decades, various improvements have been made in TF and it has found its application as a combination unit with other treatment systems. There are 129 TFs in Latin America being operated as a post treatment unit of Upflow Anaerobic Sludge Blanket (UASB) [7]. UASB is an anaerobic treatment system originally developed for industrial wastewater treatment. With due course of time, it became popular in developing countries for domestic wastewater treatment due to its affordability, simple construction, easy operation and maintenance [6–8]. Recent studies on TF following anaerobic sewage treatment system revealed that 25% of UASB reactor employed TF as post treatment system [8–10]. The combination of anaerobic and aerobic treatment is advantageous, and this system is simpler than those involving ASP, and leads to much lower energy consumption [11]. The combination of UASB and TF exhibits the high treatability and also is economically advantageous over other treatment systems in developing countries. It has been adopted whenever compact systems were required.

Employing the working concept of TF, in 1995 a research team of Professor Hideki Harada came up with the first concept of promising sewage treatment technology referred to as a DHS system in Japan [12]. DHS system is comparatively a new aerobic, post-treatment process where a simple polyurethane sponge act as a medium for all biological removal processes. DHS system consists of the sponge modules arranged along its height unlike the TF which use gravel, plastics, rocks as supporting media. There are six variations of sponges developed and tested through the rigorous improvement in its shape, arrangements and packing method [13]. The first generation was cube type DHS (G1), second was curtain type DHS (G2), third was similar to TF with sponge supported by polypyrene plastic net (G3), fourth was arrayed sponge type (G4), fifth was improved design of G2 and sixth was similar with G3 but with hardened sponges (G6). The other technical details of the configurations are discussed in other study [14].

The working principle of DHS system is similar to TF. The wastewater is supplied to the top of the DHS system with the help of distributor, which trickles down through the sponge module and finally exits the system through a clarifier at the bottom. The influent with its organic matter is trapped and flows down through the sponge modules in the reactor where the biomass within the sponge degrades the organic matter. External aeration is not necessary for the operation since there is natural diffusion of air as it flows through the DHS system.

For almost two decades, DHS system was researched as the post treatment of UASB for domestic wastewater and implementation of its full-scale have justified it [15–18]. Modifications to DHS systems are mainly conducted to eliminate the shortcoming of reactor and improve the nutrient and pathogen removal efficiencies [19, 20]. At early stages, DHS reactor was developed to treat domestic wastewater, however the potential of DHS reactor for treating different types of wastewater such as aquaculture [21], industrial wastewater [22–24] (textile, arsenic, rubber processing etc.) leachates [25] are other trends observed from literatures. The full scale and pilot scale DHS systems are being operating in Japan [10, 16], Thailand [26], India [18, 27, 28], Egypt [20], and Vietnam [22, 23] for various kinds of wastewater. Besides, DHS as a standalone bioreactor for rare metal recovery [29] phosphate recovery [30], gas scrubbing [31] and methane recovery [32] have also been investigated. Since, tremendous amount of researches are being carried out, more emphasis on its self-sustainability would help to validate its application for developing countries for domestic wastewater treatment.

3. Concept of wastewater treatment sustainability

With the on-going stress on selection of WWTPs, sustainability assessment of wastewater has become standard in developed countries and aspiration for the developing countries [33]. The concept of sustainability of wastewater treatment plants is based on the observation that economy, environment and social well-being are interlinked. The term sustainability has various interpretations, however, the World Commission on Environment and Development (WCED, 1987) quoted "Sustainable Development is the development that meets the needs of the present generation without compromising the ability of future generations to meet their own needs" [34]. To assess the sustainability of a system, various dimensions based on the short- or long-term goals have been taken into consideration. From the classic definition of sustainability indicators, it should always incorporate the three main pillars of sustainability i.e. economic, environmental and societal for holistic assessment [35, 36]. In case of developing countries, the studies were more focused on the economic affordability, convenience of end user and stakeholder, health risks, technology sustainability, environmental impacts by products, natural resource optimization and sanitation [37–39]. This pertains to the fact that choosing sustainability indicators should be contextualized to the local requirements for the decision makers to ascertain WWTPs for specific areas.

Since there is no comprehensive definition of self - sustainable WWTP, it could be defined as "a state of treatment system which can sustain itself without or less use of energy or resources from external source without causing harm or less harm to the surrounding environment". This definition is restated with reference to the definition of appropriate technology for water sanitation for developing provided in these studies [40, 41]. The following could be some of the features that indicate the self-sustainability of WWTPs for developing countries;

- Simple design and construction
- Less carbon footprint and economic costs
- Simple operation and maintenance
- Least amount or no chemical use
- Stable and reliable performance meeting all the discharge standards
- Having energy sufficiency potential or energy recovery potential
- Productive reuse of biosolids and treated wastewater
- Promote institutional development (environmental agencies, policy makers and regulation agencies, service providers)

Merely saying DHS is a 'sustainable' system is not possible until and unless sustainability indicators indicate progress towards or away from sustainability [42]. The main goal of this chapter is to present the state-of- - art of DHS system based on sustainability indicators. Additionally, the self - sustainability potential of DHS system was compared and discussed with the similar kind of technology i.e. TF for future application of this technology in developing countries. TF is a well-known technology since ages and comparing DHS system with TF would assist in its proof of concept, scalability and deployment for its validation in the field of the sustainability science. So far, there is only one study which has addressed the sustainability of the full-scale DHS system [43]. However, self-sustainability of DHS system has not been explored yet. From here on wards, DHS system is rephrased as UASB +DHS system as majority of researches on DHS system are presented as post treatment unit of UASB system. Similarly TF is also rephrased as UASB+TF. Apparently, the literatures on performance of UASB+TF systems are scarce so some discussions are presented with only TF data.

This chapter collects and analyzes the pre-requisites of self-sustainability indicators for UASB+DHS system. To address the self-sustainability of UASB+DHS system multiple indicators are considered from literature review for the holistic assessment which is guided predominantly by these studies [44, 45]. The indicators considered for this review are discussed henceforth and are summarized in **Table 1**.

3.1 Treatment performance

Right from the first prototype of UASB+DHS system, its treatment efficiency for organic, nitrogen and pathogen have shown impressive results for domestic wastewater treatment [10, 12–20]. There are plethora of studies reporting the treatment performance of UASB+DHS system. For comparison, the treatment efficiencies for the parameters such as Total suspended solids (TSS), biological oxygen demand (BOD), ammonia (NH_4^+-N) and fecal coliform (FC) are collected and tabulated in **Table 2**. Full scale UASB+DHS system till date have shown significant TSS and BOD efficiencies of 94% and 96% respectively [18, 27]. While some of the selected UASB +TF system indicated a slightly reduced efficiency i.e. (TSS: 88–93% and BOD:89–93%). For most of the developing countries, BOD standards are regarded as the

	Indicators	Sub-indicator	Description	Calculation formula	Values	Suggested units
Environmental Sustainability	**Treatment performance**		Contaminant removal efficiency to mitigate environmental and health risks	$\left(\frac{Cinf.-Ceff.}{Cinf.}\right) \times 100$		Removal % (log removal)
				$C_{inf.}$ = Influent concentration		
				$C_{eff.}$ = Effluent concentration		
	Land area		Land area required for the wastewater treatment facility	Total area occupied /Population equivalence		m^2/p.e.
				CO_2 emissions from COD oxidation	COD removal (kg COD $m^{-3}d^{-1}$) $\times$ 0.08 kg CO_2 / kg COD	
				CO_2 emissions from CH_4 combustion		
		Global warming potential	Emissions of greenhouse gases (CO_2 and N_2O) into the atmosphere	N_2O emissions	CH_4 emission (m^3 CH_4) $\times$ 3.5 (kg CO_2 / m^3 CH_4	(kg CO_2 equivalent $m^{-3}d^{-1}$) Atkins et al. (2005) [46, 47]
					$N_{inflent} \cdot EF_{effluent} \cdot$ 44/ 28 $\times$ 298 (Emission Factor effluent =0.005)	Campos et al. (2016); IPCC [48, 49]
	Carbon Footprint	Energy sufficiency	Energy consumed during emission	CO_2 emissions from energy consumption	Energy consumption ($kWhm^{-3}$ d^{-1}) $\times$ 0.391 (kg $CO_{2\ equivalent}$ / kWh)	(kg $CO_{2\ equivalent}$ $m^{-3}d^{-1}$)
		Reduced Sludge reduction potential	Sludge amount produced, treated water for irrigation, nutrients	$S_s = Q_{ef}(W_{in}-W_{eff})/Q(C_{in}-C_{eff})$		(kg SS kg^{-1} COD removed) [18]
				$Q_{ef(}$= flow rate to settler ($m^{-3}d^{-1}$)		
				Q = flow rate to the reactor ($m^{-3}d^{-1}$)		
				W_{in} = SS influent conc. (kg SS m^{-3})		

	Indicators	Sub-indicator	Description	Calculation formula	Values	Suggested units
				W_{eff} = SS effluent conc. (kg SS m^{-3})		
				C_{in} = COD influent conc. (kg COD m^{-3})		
				C_{eff} = COD effluent conc. (kg COD m^{-3})		
Economic sustainability	Capex		Cost of construction and installation of the WWTP	$TAEC = \left[\frac{r(1+r)^t}{(1+r)^t-1}\right] Capex + Opex$		$/m^3
				t = expenses at time t		$/p.e.-year
				r = discount rate (5%)		
	Opex		Operating costs per volume unit of wastewater treated	$\frac{\sum_{t=1}^{TOpex}}{(1+r)^t}$		$/p.e.
				t = expenses at time t		
				r = discount rate (5%)		
Social sustainability	Public acceptance		Opinion of the local population affected by the plant.			Qualitative
	Esthetics		Measured level of nuisance deriving from e.g. odor, noise, visual impact, insects and other pests.			Qualitative
	System manageability		Ease of construction, complexity of O & M; professional skills required for O & M			Qualitative

Table 1.
Multiple indicators chosen for assessing the sustainability. Adapted from [44, 45].

	Support media	Land area	HRT (h)	Influent (mgL^{-1})				Removal efficiency (%)				References
		m^2/p.e		BOD	TSS	NH_4^+-N	FC(MPN/100 ml)	BOD	TSS	NH_4^+-N	FC (log)	
UASB + TF	Sponge bed	0.3	3	123	79	30	—	89	88	44	3.5	[6]
UASB+DHS	G3 sponge	0.03	1.5	151	228	25	7.71×10^6	96	94	79	3.0	[27]
UASB+DHS	G3.4 sponge	0.05	1.5	161	228	16	1.22×10^9	96	94	83	3.0	[43]
UASB+TF	High rate stones	0.2	2.0	250	150	20	1.8×10^8	93	93	50	2.8	[50]

Table 2.
Treatment performance and land required for UASB+DHS system and UASB+TF system.

basic compliance discharge standard [12] which might have increased the popularity of UASB+DHS system. The data clearly shows that UASB+DHS system has benefits over UASB+TF system attributed by its unique sponges, improving the quality of effluent in terms of organic matter. Similarly, a noticeable ammonia removal efficiency ranging from 79–83% was showcased by UASB+DHS system whereas UASB+TF displayed decreased efficiency below 50%. Studies on molecular microbiology of UASB+DHS have highlighted slow growers such as nitrifiers, denitrifies and even active annamox bacteria in the inner aerobic niches of the sponges facilitating the nitrification and denitrification reaction for nitrogen removal [16, 51]. The other studies also reported that TFs have poor consistency in the removal of nitrogen and phosphorus compared to other conventional treatment systems [6]. Likewise, UASB+DHS system is also efficient for removing pathogenic bacteria from wastewater which was due to high DO condition which prevented growth of bacteria and allowed the higher micro-organisms (protozoa and metazoan) to predate on pathogens such as *E. coli* and total coliforms [52]. Moreover, the other factor for removal of pathogen in the UASB+DHS system reported was adsorption onto biomass [20]. While on the other hand, pathogen removal by UASB +TF system is promising in this case. However, the pathogen removal capacity in TFs have been observed inconsistent and varied from 1.0 log to 3 logs, depending on the operating conditions when compared to ASP [53].

3.2 Land requirements

The increasing land prices and scarcity of available land resources are becoming one of the bottlenecks for WWTP management issues [54]. Land requirement for the WWTPs directly affects its performance and costs. The land requirement per m^2/p.e. for UASB+DHS and UASB+TF systems are shown in **Table 2**. The available literatures show that the land required for the construction of UASB+DHS is almost 10 times less than UASB+TF. Though having the same external design, this difference could be explained by the supporting media. The DHS sponges are comparatively smaller and light weight in comparison to the most frequently used media such as stones, gravel, plastics etc. which implies much higher tank volumes and areas. Nonetheless, having the similar working principle, the packing of the media in DHS system resulted in smaller footprint.

3.3 Carbon footprint

Carbon footprint is relatively a new measure of sustainability in WWTPs to determine its overall impact on climate change and as a result WWTPs performance has recently been evaluated based on carbon minimization [55, 56]. To address sustainability, carbon footprint minimization has become an important environmental indicator [57]. For carbon footprint, assessment, all relevant forms of the energy demand in WWTPs, sludge production and common GHGs emissions are accounted. This review aims to investigate previously unexplored relationships between carbon footprint and sustainability in the context of UASB+DHS system, focusing particularly on the impact of energy minimization measures.

3.3.1 GHGs emission

Global Warming Potential (GWP) is generally used as a metric for weighting the climatic impact of emissions of different greenhouse gases [58]. Among GHGs stated by Kyoto Protocol, the most common GHGs emitted during operation and on-site anthropogenic activities in WWTPs are carbon dioxide (CO_2), methane

(CH_4), nitrous oxide (N_2O) [49]. According to USEPA, WWTPs are the 7th largest contributors of CH_4 and nitrous N_2O emissions in the atmosphere [59]. Particularly, WWTPs produce GHGs during the biological wastewater treatment processes. For calculation, all GHGs emission can be expressed as CO_2 equivalents (CO_2e) with respect to their GWP. CH_4 and N_2O have 28 and 265 times greater GWP compared to CO_2 in a 100-year time horizon [60]. Therefore, more stringent regulatory efforts, mandatory reports and measurements on GHGs emissions from WWTPs are being enforced to control GHGs emissions.

Since, UASB+DHS system is also an anaerobic and aerobic biological treatment process, this information would be vital for the wastewater specialists. For almost all WWTPs, CO_2 production is attributed to two main factors: biological treatment process and electricity consumption. In UASB+DHS system, CO_2 is emitted during the production of the energy required for the plant operation. Emission of N_2O is generated by nitrification and denitrification processes used to remove nitrogenous compounds from wastewater. Similar to the mainstream WWTPs, the organic carbon of wastewater is either incorporated into biomass or oxidized to CO_2. During anaerobic digestion in UASB, it is mainly converted to CO_2 and CH_4. It is assumed that all the CH_4 produced is oxidized to CO_2 during biogas combustion. Estimation of (CO_2e) is attained using units and equations summarized in **Table 1**. For the calculation of GHGs, considering the total treatment process is important. Therefore, GHGs emissions of both the system were estimated based on CO_2 emission from COD oxidation, CH_4 combustion and N_2O emission as presented in **Table 3**. The data for GHGs calculation for UASB+DHS system and UASB+TF were taken from these studies [27, 60, 62]. The value for GWP by UASB+DHS system was 0.59 kg CO_2 equivalent m^{-3} d^{-1} and that for UASB+TF was 0.50 kg CO_2 equivalent m^{-3} d^{-1}. It is to be noted that for the calculation of GWP of UASB+TF, N_2O emissions value was not available as there were no literatures reporting its values. Nevertheless, other studies associated with GHGs emission of TF + ASP system and TF+ Lagoon system showed GWP values of 1.232 kg CO_2 equivalent m^{-3} d^{-1} and 0.898 kg CO_2 equivalent m^{-3} d^{-1} respectively. Therefore, it could be assumed that the UASB+TF system might show fairly higher values compared to UASB+DHS system. The another reason behind assuming the lower GWP values by UASB+DHS system could be justified by its higher solid retention time (SRT) values of almost 92–101 days [17] compared to 2–4 days of TF [65]. Higher values of SRT supports endogenous respiration of biomass which increases the amount of COD oxidized to CO_2 thus decreasing the overall sludge production [17]. This decrease of sludge production reduces the methane production and therefore, a decrease in CO_2 emissions is associated with its combustion. Similarly, higher SRT capacity of DHS system helps to maintain low ammonia and nitrite concentrations in the media which leads to minimum N_2O emissions to the atmosphere. Despite the accuracy of estimated GWP value is not exact, conclusive potential of operating UASB+DHS system at low GHGs emission levels has been assured. Hence, the analysis of GWP revealed the potential of UASB+DHS system to become a sustainable option in the future of wastewater treatment.

	Units	UASB+DHS system	UASB+TF
Global warming Potential	(kg CO_2 equivalent $m^{-3}d^{-1}$)	0.59 [27, 61]	0.50 [62]
Sludge production	(kg SS kg^{-1} COD removed)	0.06 [10]	0.38 [63]
Energy consumption	($kWhm^{-3}$)	0.12 [27]	0.65 [64]

Table 3.
Carbon footprint assessment of UASB+DHS system and UASB+TF.

3.3.2 Sludge production

For most of the WWTPs, one of the biggest challenges is its sludge production, its post treatment and disposal. Being an aerobic system, DHS system has advantage over other biological treatment system for sludge management [61]. Any sludge accumulated in the clarifier of DHS is called as excess sludge. The sludge production in DHS reactor is calculated by taking the sum of SS volumes in the DHS effluent and the settled excess sludge in the clarifier and relative to the COD or BOD removed by the system. For bench scale experiment, the excess sludge produced by UASB+DHS system was 0.02 kg SS/kg COD removed which is basically 2.5% of the total COD removed or 7% of the total SS load removed [17]. Further, excess sludge from UASB usually varied from 0.03 to 0. 2 kg SS/kg COD removed [8]. Therefore, a total sludge from UASB+DHS system was 0.06 kg SS/kg COD removed. While, excess sludge production from UASB+TF system was 0.38 kgSS/kgCOD removed [63]. The sludge production from UASB+TF system was almost 6 times higher than UASB+DHS system. The DHS sponges are designed with the high void ratio and reticulated structure which cater as a favorable site for the attachment, adsorption and growth of active biomass [25, 61]. Further, the profiling data from same researchers stated that the majority of organic removal especially SS occurred at highest part of the reactor, however after attaining stable state, uniform distribution of sludge was observed along its height. In real scale DHS, for every liter of wastewater treated, about 0.04 kg-COD was wasted as excess sludge which is quite negligible as compared to the other treatment systems. The basic mechanism for the sludge removal in DHS is the physical entrapment of the sludge inside and outside of the sponge which lengthens the solid retention time and provide ample time for self-degradation of sludge minimizing the excess sludge production [61, 66].

3.3.3 Energy consumption

Nowadays, for developing countries energy efficiency has become the first priorities in the WWTPs hierarchy [67]. Minimizing net energy consumption for WWTPs has become mandatory [68]. Generally, for aerobic treatment processes, the aeration is the highest energy consuming process of the wastewater treatment technology which can account upto 50–60% of all electricity consumption followed by 15–25% of energy by sludge treatment and 15% by secondary sedimentation including recirculation pumps [69].

The energy consumed in UASB+DHS system is through electricity required for pumping [27]. The pumps are used for supplying UASB effluent to the top of DHS system. It is usually estimated on the basis of treatment performance and electricity utilized by pumps. Comparison of energy consumption of UASB+DHS system [27] with UASB+TF [64] is summarized in **Table 3**. From the data, it is evident that the energy consumption of UASB+TF is approximately 5 times higher than that of UASB+DHS system. For UASB+DHS system, 0.05 $kWhm^{-3}$ of energy was consumed by main pumping from UASB unit and 0.07 kWhm-3 for the pump of the DHS system, which sums up the total energy consumption for the system of 0.12 kWh per m^3 of wastewater treated. It is noteworthy that the energy consumption for both these systems was solely by pumping. The UASB+DHS system has likelihood of becoming energy sufficient system. The energy sufficiency of UASB+DHS system can be explained by its minimized energy consumption. In addition, when constructing a UASB+DHS system, the energy sufficiency or neutrality can be achieved if UASB is designed in such a way where the outlet is positioned above the DHS distributor or maintained through gravity.

Considering the overall arguments, environmental indicators suggest that the UASB+DHS system is considerably superior in terms of high treatability, less land requirement and reduced carbon footprint. This information could assist the planners and stakeholders in developing nations for good decision making while selecting WWTPs in future.

4. Economic sustainability

Economic sustainability of WWTPs refers to the economic factors affecting social, environmental and cultural aspects of the treatment systems. Economic efficiency of WWTPs presents the scenario of investments in terms of input and effluent quality as the output [70]. Decision and policy makers in developing countries are challenged with the fact that poor urban residents cannot afford costly conventional sewage treatment systems [71]. Fortunately, a broad range of cost-effective technological options are extensively being studied to cater this category of people. Therefore, economic factors become vital to address these issues. For any WWTPs, economic indicators generally represent the costs associated with the construction and the operation of treatment management during its life time [72]. These are driving factors for decision making while selecting a technology in a practical situation.

For the economic assessment, the two most common indicators called capital expenditure (Capex) and operational expenditure (Opex) were calculated using the equations provided by [73]. The capital expenditure included cost of construction and life cycle costs. Operational expenditure included number of mechanical equipment, skilled workers, power consumption, labor, chemicals, and consumables. The data were taken from the state-of-art literatures [43, 74, 75] for calculating the construction and operating costs for economic evaluation for both the systems. For the comparison in the same scale few considerations were made:

Capex is annualized by taking the means of the initial investment costs by the life period of the project accounting for the time value of money while Opex is total discounted lifetime operational expenses.

- life span of each treatment plant was considered as 20 years.
- All costs were expressed in US dollars ($) /Population equivalence (p.e.).
- The cost of implementing and operating the WWTPs are adjusted to 2015 as the base year and was discounted to 2015 values using equations given in **Table 1**.

The economic costs comparison **Table 4** showed that capex and opex costs of UASB+DHS system of 9740 p.e. is almost equal to the UASB+TF of 50,000 pe.

Process system	Country	Treatment volume (PE)	CAPEX US$/PE	OPEX US$/PE/year	References
UASB+DHS	India	9740	86.2	0.36	[43]
UASB+TF	Egypt	50,000	92.14	2.07	[74]
UASB+TF	Egypt	2337	519.31	8.25	[75]

Table 4.
Economic assessment of UASB+DHS system and UASB+TF system.

While for the UASB+TF of 2337 p.e. UASB+DHS system expressed significantly less capex and opex values. For most of the UASB+TF, among the various costs, cost of personnel is the maximum for opex [50]. This could be also the reason for the decreased economic costs for UASB+DHS system. The another rationale for reduced economic costs of UASB+DHS system could be less manpower required due to the simple O&M processes including cleaning of the mechanical parts, chemical free operation and easy handling of sludge [76]. It is interesting to note that DHS inclination is towards negative value for economic assessment which means it is economically fit for the developing countries as capex and opex costs are negative indicators of sustainability and qualifies the criteria to be considered as self-sustainable WWTP system.

5. Social sustainability

Social assessment based on indicators portrays a big picture of multidimensional issues and facilitate decision making. Many key aspects such as community management aspects, satisfaction and opinions of users, service quality, materials and personnel managements, etc. have to be profoundly analyzed before and after the establishment of WWTPs [45]. In this light, social indicators are rapidly becoming the preferred tools for policymakers and public communicators for disseminating the information on the advantages and disadvantages of the WWTPs [77]. However, societal indicators are generally difficult to quantify and often their meaning and relevance is based on the local stakeholders [78]. The data on social indicators for UASB+TF in this review is lacking since data availability on TF. This does not impair our comparison as it's combinations with other systems is a disadvantage identified at the global level in developing countries [79]. Hence, assessment of the chosen indicators is based on TF studied [43]. Caution should be exercised that all the data for social assessment do not represent any generic weighting. The chosen indicators for this assessment are (i) simplicity of the system (ii) esthetics, and (iii) public acceptance of the technology.

5.1 System manageability

System manageability includes ease of construction, complexity of O&M issues and professional skills required for the troubleshooting of the issues during the O&M of the WWTPs. Studies on the social sustainability of WWTPs in India demonstrated that the UASB+DHS system has fewer mechanical parts than TF. The simple configuration of UASB+DHS system makes it easy for construction and several intensive studies have followed up the ramifications of sponge designs for the easy packaging and enhanced efficiency [76]. For developing countries, simplicity of the system might be a key factor in the selection of the WWTP. Due to simple construction, there are few mechanical parts for O&M issues and UASB +DHS system has already been proven to be no laborious maintenance system [80]. Supplementary to this, the operators do not need to have a high technical knowledge.

5.2 Esthetics

UASB+DHS system has a slightly better stance on esthetics than the other TFs. Hydrogen sulphide (H_2S) and ammonia (NH_3) are the predominant objectionable odors in wastewater [81]. TF has been reported to have limited and passive aeration which might create an anaerobic condition which degrades organic matter and nutrients to release malodorous by-products. Also, TF is very sensitive to temperature change especially high temperatures which generate odors [5]. However, till this date, there are no evidences of odor problems in UASB+DHS system though it was operated at the high temperature in India (~40°C). Nonetheless, both the systems have one common problem i.e. flies and snails which might affect the visual aethetics. Whilst none of the insects found have been found as a nuisance to the surrounding people and there is no evidences of any diseases caused by these insects. However, this issue could be resolved by the installation of nets or covers and cleaning with water sprays [82].

5.3 Public acceptance of the technology

One of the major problems faced during the establishment of WWTPs is its location. Very often resistance and protests from the local people significantly impact on the implementation of any social infrastructure plan [83, 84]. Therefore, "Public acceptance" is a key component when it comes to establishing a new WWTP [85]. In most of the researches, public acceptance of WWTPs facilities are based on concepts such as LULU (locally unwanted land use) and NIMBY (not in my backyard [86, 87]. These studies have highlighted the preferability of TF over activated sludge process. Besides, the result of investigation of Indian WWTPs has also exhibited that moderate value for the public acceptance of TF over UASB+DHS system and suggested WWTPs' location far from the settlement zones. Besides, the sludge and treated water from UASB+DHS system was used by the nearby farmers for agriculture and did not show any social concerns. It is clear that further studies need to be undertaken analyzing local conditions in a stepwise manner towards the acceptance of these technologies.

The above discussions on sustainability indicators assessment support the notion that DHS system could have a wide range of commercial applications for different kinds of wastewater. In most of the developing countries, centralized WWTP are limited to urban areas due to the several financial and social constraints. Likewise, many institutions such as large-scale apartments, complexes, hospitals, hotels, etc. need to maintain their own onsite WWTPs. Until now, the most preferred WWTP option was ASP based treatment plants. However, these systems are expensive due to its high operational and maintenance cost. On this verge, there is a potential use of DHS system as a substitute for ASP which would lead to huge commercial benefit. The technology validation of DHS system over ASP for developing countries has already been reported in several research investigations [12–18, 43]. Another area of commercial application of this system could be in aquaculture industry. Most of the aquariums and fish farms require frequent exchange of water leading to huge financial burden and increased workload. Current progressive researches on the application of DHS system for aquarium water treatment and live seafood transportation by minimizing the exchange of water and decreasing workload have broaden its scalability for aquaculture industry [21, 88]. In line with the environmental, economic and social sustainability of DHS system, it could have prospective applications in industrial wastewater treatment specifically for developing nations. Moreover, DHS reactor has commercial applications in the industries such as food processing industry, beverage industries, rubber processing industries and many agriculture products processing industries. Therefore, it could be concluded that with more researches and real scale implementation of this system, there could be a huge commercial demand for DHS system in near future.

6. Conclusion

From the retrospection of the state-of-art of DHS system, it has always been considered a sustainable system for developing countries. Even though, there are extensive researches being carried out for the performance improvement and application of DHS for other types of wastewater, few efforts have been made for testing and validating the sustainability of DHS system. This chapter introduced and analyzed the sustainability indicators of DHS system based on environment, economic and social indicators. By assessing the range of environmental performance indicators, including treatment performance, land requirement, carbon footprint along with economic costs and social factors, this review provides information on DHS system pursuing positively towards sustainability than UASB+TF. However, the availability of data is still an issue in this context for both the systems. It is recommended to conceptualize the sustainability assessment framework that will also encourage and support data collection for better and more quantitative analysis to ensure the applicability and usefulness of DHS technology. Considering the outcomes from sustainability assessment, regardless of data insufficiency, the DHS system fulfilled the criteria of self-sustaining WWTP to a greater extent. Nevertheless, more comprehensive studies are suggested for understanding the other aspects of self-sustainability which are not discussed in this chapter.

Author details

Namita Maharjan[1]*, Choolaka Hewawasam[2], Masashi Hatamoto[3], Takashi Yamaguchi[3], Hideki Harada[4] and Nobuo Araki[1]

1 Nagaoka National Institute of Technology, Nagaoka, Niigata, Japan

2 University of Sri Jayewardenepura, Colombo, Sri Lanka

3 Nagaoka University of Technology, Nagaoka, Niigata, Japan

4 Tohoku University, Sendai, Miyagi, Japan

*Address all correspondence to: namimaha@nagaoka-ct.ac.jp

References

[1] Sustainable Development Goals Report 2017 (United Nations, 2017).

[2] Online source: World data lab www.worlddata.io/data/forecast

[3] Goffi G, Masiero L, Pencarelli T. Rethinking sustainability in the tour-operating industry: Worldwide survey of current attitudes and behaviors. Journal of cleaner production. 2018 May 10;183:172–82.

[4] U.S.Environmental Protection Agency (EPA) 2002, Onsite Wastewater Treatment Systems Manual (Report). Washington, D.C.

[5] Tchobanoglous G, Burton FL, Stensel HD. Metcalf & Eddy wastewater engineering: treatment and reuse. International Edition. McGrawHill. 2003;4:361–411.

[6] Bressani-Ribeiro T, Almeida PGS, Volcke EIP, Chernicharo CAL. Trickling filters following anaerobic sewage treatment: state of the art and perspectives. Environ Sci Water Res Technol [Internet]. 2018;4(11):1721–38

[7] Von Sperling M, de Lemos Chernicharo CA. Biological wastewater treatment in warm climate regions. IWA publishing; 2005 Sep 30.

[8] Lettinga G, Roersma R, Grin P. Anaerobic treatment of raw domestic sewage at ambient temperatures using a granular bed UASB reactor. Biotechnology and bioengineering. 1983 Jul;25(7):1701–23.

[9] Chernicharo CAL, Nascimento MCP. Feasibility of a pilot-scale UASB/trickling filter system for domestic sewage treatment. Water Sci Technol. 2001;44(4):221–8.

[10] Tandukar M, Machdar I, Uemura S, Ohashi A, Harada H. Potential of a Combination of UASB and DHS Reactor as a Novel Sewage Treatment System for Developing Countries : Long-Term Evaluation. 2006;132(2):166–72.

[11] Von Sperling M. Comparison of simple, small, full-scale sewage treatment systems in Brazil: UASB-maturation ponds-coarse filter; UASB-horizontal subsurface-flow wetland; vertical-flow wetland (first stage of French system). Water Sci Technol. 2015;71(3):329–36.

[12] Tandukar M, Uemura S, Machdar I, Ohashi A, Harada H. A low-cost municipal sewage treatment system witha combination of UASB and the "fourth-generation" downflow hanging sponge reactors. Water science and technology. 2005 Jul;52(1–2):323–9.

[13] Uemura S, Harada H. Application of UASB technology for sewage treatment with a novel post-treatment process. Environmental anaerobic technology: applications and new developments 2010 (pp. 91–112).

[14] Nurmiyanto A, Ohashi A. Downflow Hanging Sponge (DHS) Reactor for Wastewater Treatment-A Short Review. MATEC web of conferences. 2019 (Vol. 280, p. 05004). EDP Sciences.

[15] Design Guidelines for UASB-DHS System and Operation and Maintenance Guidelines for UASB-DHS System in collaboration with Tohoku University. 2016 Submitted to MoEF-NRCD.

[16] Machdar I, Harada H, Ohashi A, Sekiguchi Y, Okui H, Ueki K. A novel and cost-effective sewage treatment system consisting of UASB pre-treatment and aerobic post-treatment units for developing countries. Water Science and technology. 1997 Jan 1;36 (12):189–97.

[17] Tandukar M, Ohashi,Harada H. Performance comparison of a pilot-scale UASB and DHS system and activated sludge process for the treatment of municipal wastewater. Water Res. 2007 Jul;41(12):2697–705.

[18] Onodera T, Okubo T, Uemura S, Yamaguchi T, Ohashi A, Harada H. Long-term performance evaluation of down-flow hanging sponge reactor regarding nitrification in a full-scale experiment in India. Bioresource technology. 2016 Mar 1;204:177–84.

[19] Hewawasam C, Matsuura N, Maharjan N, Hatamoto M, Yamaguchi T. Oxygen transfer dynamics and nitrification in a novel rotational sponge reactor. Biochem Eng J. 2017;128:162–7

[20] Tawfik A, Ohashi A, Harada H. Sewage treatment in a combined up-flow anaerobic sludge blanket (UASB)-down-flow hanging sponge (DHS) system. Biochem Eng J. 2006;29(3): 210–9.

[21] Adlin N, Matsuura N, Ohta Y, Hirakata Y, Maki S, Hatamoto M, Yamaguchi T. A nitrogen removal system to limit water exchange for recirculating freshwater aquarium using DHS–USB reactor. Environmental technology. 2018 Jun 18;39(12):1577–85.

[22] Tanikawa D, Yamashita S, Kataoka T, Sonaka H, Hirakata Y, Hatamoto M, Yamaguchi T. Non-aerated single-stage nitrogen removal using a down-flow hanging sponge reactor as post-treatment for nitrogen-rich wastewater treatment. Chemosphere. 2019 Oct 1; 233:645–51.

[23] Watari T, Mai TC, Tanikawa D, Hirakata Y, Hatamoto M, Syutsubo K, Fukuda M, Nguyen NB, Yamaguchi T. Performance evaluation of the pilot scale upflow anaerobic sludge blanket–Downflow hanging sponge system for natural rubber processing wastewater treatment in South Vietnam. Bioresource technology. 2017 Aug 1;237: 204–12.

[24] Nguyen TH, Watari T, Hatamoto M, Sutani D, Setiadi T, Yamaguchi T. Evaluation of a combined anaerobic baffled reactor–downflow hanging sponge biosystem for treatment of synthetic dyeing wastewater. Environ Technol Innov 2020;19:100913

[25] Ismail S, Tawfik A. Treatment of hazardous landfill leachate using Fenton process followed by a combined (UASB/DHS) system. Water Science and Technology. 2016 Apr 7;73(7):1700–8.

[26] Yoochatchaval W, Onodera T, Sumino H, Yamaguchi T, Mizuochi M, Okadera T, et al. Development of a down-flow hanging sponge reactor for the treatment of low strength sewage. Water Sci Technol. 2014;70(4):656–63.

[27] Okubo T, Onodera T, Uemura S, Yamaguchi T, Ohashi A, Harada H. On-site evaluation of the performance of a full-scale down-flow hanging sponge reactor as a post-treatment process of an up-flow anaerobic sludge blanket reactor for treating sewage in India. Bioresour Technol 2015;194:156–64.

[28] Okubo T, Kubota K, Yamaguchi T, Uemura S, Harada H. Development of a new non-aeration-based sewage treatment technology: performance evaluation of a full-scale down-flow hanging sponge reactor employing third-generation sponge carriers. Water research. 2016 Oct 1;102:138–46.

[29] Cao LT, Kodera H, Abe K, Imachi H, Aoi Y, Kindaichi T, Ozaki N, Ohashi A. Biological oxidation of Mn (II) coupled with nitrification for removal and recovery of minor metals by downflow hanging sponge reactor. water research. 2015 Jan 1;68:545–53.

[30] Kodera H, Hatamoto M, Abe K, Kindaichi T, Ozaki N, Ohashi A. Phosphate recovery as concentrated

solution from treated wastewater by a PAO-enriched biofilm reactor. Water Research. 2013 Apr 15;47(6):2025–32.

[31] Yamaguchi T, Nakamura S, Hatamoto M, Tamura E, Tanikawa D, Kawakami S, Nakamura A, Kato K, Nagano A, Yamaguchi T. A novel approach for toluene gas treatment using a downflow hanging sponge reactor. Applied microbiology and biotechnology. 2018 Jul 1;102(13): 5625–34.

[32] Matsuura N, Hatamoto M, Sumino H, Syutsubo K, Yamaguchi T, Ohashi A. Closed DHS system to prevent dissolved methane emissions as greenhouse gas in anaerobic wastewater treatment by its recovery and biological oxidation. Water Science and Technology. 2010 May;61(9):2407–15.

[33] Li R, Morrison L, Collins G, Li A, Zhan X. Simultaneous nitrate and phosphate removal from wastewater lacking organic matter through microbial oxidation of pyrrhotite coupled to nitrate reduction. Water Res [Internet]. 2016;96:32–41.

[34] Borowy I. Defining sustainable development for our common future: A history of the World Commission on Environment and Development (Brundtland Commission). Routledge; 2013 Dec 4.

[35] Muga HE, Mihelcic JR. Sustainability of wastewater treatment technologies. J Environ Manage. 2008; 88(3):437–47.

[36] Molinos-Senante M, Gómez T, Garrido-Baserba M, Caballero R, Sala-Garrido R. Assessing the sustainability of small wastewater treatment systems: a composite indicator approach. Sci Total Environ. 2014 Nov 1;497–498: 607–17.

[37] Sato N, Okubo T, Onodera T, Ohashi A, Harada H. Prospects for a self-sustainable sewage treatment system: A case study on full-scale UASB system in India's Yamuna River Basin. J Environ Manage. 2006;80(3):198–207.

[38] Chang NB, Pires A. Sustainable solid waste management: a systems engineering approach. John Wiley & Sons; 2015 Mar 16.

[39] Kalbar PP, Birkved M, Hauschild M, Kabins S, Nygaard SE. Environmental impact of urban consumption patterns: Drivers and focus points. Resour Conserv Recycl. 2018;137:260–9

[40] Murphy HM, McBean EA, Farahbakhsh K. Appropriate technology–A comprehensive approach for water and sanitation in the developing world. Technology in Society. 2009 May 1;31(2):158–67.

[41] Feachem RG, Bradley DJ, Garelick H, Mara DD. Appropriate technology for water supply and sanitation. Vol. 3. Health aspects of excreta and sullage management: a state-of-the-art review. Appropriate technology for water supply and sanitation. Vol. 3. Health aspects of excreta and sullage management: a state-of-the-art review.. 1981;3.

[42] Lundin M, Molander S, Morrison GM. A set of indicators for the assessment of temporal variations in the sustainability of sanitary systems. Water Sci Technol. 1999;39(5):235–42.

[43] Maharjan N, Nomoto N, Tagawa T, Okubo T, Uemura S, Khalil N, et al. Assessment of UASB–DHS technology for sewage treatment: a comparative study from a sustainability perspective. Environ Technol 2019;40(21):2825–32.

[44] Molinos-Senante M, Gómez T, Caballero R, Hernández-Sancho F, Sala-Garrido R. Assessment of wastewater treatment alternatives for small communities: An analytic network

process approach. Sci Total Environ. 2015;532:676–87.

[45] Cossio C, Norrman J, McConville J, Mercado A, Rauch S. Indicators for sustainability assessment of small-scale wastewater treatment plants in low and lower-middle income countries. Environ Sustain Indic 2020;6(March):100028

[46] Cornejo PK, Zhang Q, Mihelcic JR. How does scale of implementation impact the environmental sustainability of wastewater treatment integrated with resource recovery?. Environmental Science & Technology. 2016 Jul 5;50 (13):6680–9.

[47] Capodaglio AG. Integrated, decentralized wastewater management for resource recovery in rural and peri-urban areas. Resources. 2017 Jun;6(2): 22.

[48] Gustavsson DJ, Tumlin S. Carbon footprints of Scandinavian wastewater treatment plants. Water Science and Technology. 2013 Aug;68(4):887–93.

[49] IPCC Guidelines for National Greenhouse Gas Inventories, Prepared by the National Greenhouse Gas Inventories Programme, IPCC, 2006 [Eggleston H. S., L. Buendia, K.Miwa, T. Ngara and K. Tanabe K. (eds.)], (IGES), Japan; 5 Chapter 6.Wastewater Treatment and Discharge, 6.1–6.28.

[50] Von Sperling M. Urban wastewater treatment in Brazil. Minas Gerais Brazil [Internet]. 2016;(August):27. Available from: www.iadb.org

[51] Araki N, Ohashi A, Machdar I, Harada H. Behaviors of nitrifiers in a novel biofilm reactor employing hanging sponge-cubes as attachment site. Water Science and Technology. 1999 Apr 1;39(7):23.

[52] Miyaoka Y, Hatamoto M, Yamaguchi T, Syutsubo K. Eukaryotic Community Shift in Response to Organic Loading Rate of an Aerobic Trickling Filter (Down-Flow Hanging Sponge Reactor) Treating Domestic Sewage. Microb Ecol. 2017;73(4): 801–14.

[53] Bitton G. Wastewater microbiology. John Wiley & Sons; 2005 May 27.

[54] He Y, Zhu Y, Chen J, Huang M, Wang G, Zou W, Wang P, Zhou G. Assessment of land occupation of municipal wastewater treatment plants in China. Environmental Science: Water Research & Technology. 2018;4(12): 1988–96.

[55] Delre A, ten Hoeve M, Scheutz C. Site-specific carbon footprints of Scandinavian wastewater treatment plants, using the life cycle assessment approach. Journal of Cleaner Production. 2019 Feb 20;211:1001–14.

[56] Xu J, Li Y, Wang H, Wu J, Wang X, Li F. Exploring the feasibility of energy self-sufficient wastewater treatment plants: a case study in eastern China. Energy Procedia. 2017 Dec 1;142: 3055–61.

[57] Holmes K, Hughes M, Mair J, Carlsen J. Events and sustainability. Routledge; 2015 Mar 24.

[58] Shine KP, Fuglestvedt JS, Hailemariam K, Stuber N. Alternatives to the global warming potential for comparing climate impacts of emissions of greenhouse gases. Climatic Change. 2005 Feb 1;68(3):281–302.

[59] US EPA. ENERGY STAR for wastewater plants and drinking water systems, 2011.Online:http://www.energystar.gov/index .cfm?c=water. wastewater_drinking_water

[60] Stocker TF, Qin D, Plattner GK, Tignor M, Allen SK, Boschung J, Nauels A, Xia Y, Bex V, Midgley PM. Climate change 2013: The physical science basis. Contribution of working group I to the

fifth assessment report of the intergovernmental panel on climate change. 2013 Sep;1535.

[61] Onodera T, Matsunaga K, Kubota K, Taniguchi R, Harada H, Syutsubo K, Okubo T, Uemura S, Araki N, Yamada M, Yamauchi M. Characterization of the retained sludge in a down-flow hanging sponge (DHS) reactor with emphasis on its low excess sludge production. Bioresource Technology. 2013 May 1; 136:169–75.

[62] Noyola A, Paredes MG, Morgan-Sagastume JM, Güereca LP. Reduction of Greenhouse Gas Emissions From Municipal Wastewater Treatment in Mexico Based on Technology Selection. Clean - Soil, Air, Water. 2016;44(9): 1091–8.

[63] Almeida PGS, Marcus AK, Rittmann BE, Chernicharo CAL. Performance of plastic- and sponge-based trickling filters treating effluents from an UASB reactor. Water Sci Technol. 2013;67(5): 1034–42.

[64] Henrich C-D, Marggraff M. Energy-efficient Wastewater Reuse – The Renaissance of Trickling Filter Technology. Proc 9th Int Conf Water Reuse 2013;27–31.

[65] Parker DS, Romano LS, Horneck HS. Making a trickling filter/solids contact process work for cold weather nitrification and phosphorus removal. Water environment research. 1998 Mar; 70(2):181–8.

[66] Matsuura N, Hatamoto M, Sumino H, Syutsubo K, Yamaguchi T, Ohashi A. Recovery and biological oxidation of dissolved methane in effluent from UASB treatment of municipal sewage using a two-stage closed downflow hanging sponge system. J Environ Manage 2015;151:200–9.

[67] Awe OW, Liu R, Zhao Y. Analysis of Energy Consumption and Saving in Wastewater Treatment Plant : Case Study from Ireland. J Water Sustain2016;6(2):63–76.

[68] McCarty PL, Bae J, Kim J. Domestic wastewater treatment as a net energy producer-can this be achieved? Environ Sci Technol. 2011;45(17):7100–6.

[69] Chen H, Liu S, Yang F, Xue Y, Wang T. The development of simultaneous partial nitrification, ANAMMOX and denitrification (SNAD) process in a single reactor for nitrogen removal. Bioresour Technol. 2009 Feb.

[70] Guerrini A, Romano G, Indipendenza A. Energy efficiency drivers in wastewater treatment plants: a double bootstrap DEA analysis. Sustainability. 2017 Jul;9(7):1126.

[71] Tsinda A, Abbott P, Pedley S, Charles K, Adogo J, Okurut K, Chenoweth J. Challenges to achieving sustainable sanitation in informal settlements of Kigali, Rwanda. International journal of environmental research and public health. 2013 Dec;10 (12):6939–54.

[72] Balkema AJ, Preisig HA, Otterpohl R, Lambert FJD. Indicators for the sustainability assessment of wastewater treatment systems. Urban Water. 2002; 4(2):153–61.

[73] Sun Y, Garrido-Baserba M, Molinos-Senante M, Donikian NA, Poch M, Rosso D. A composite indicator approach to assess the sustainability of wastewater management alternatives. Science of The Total Environment. 2020 Apr 3:138286.

[74] Schellinkhout A. UASB technology for sewage treatment: experience with a full scale plant and its applicability in Egypt. Water Science and Technology. 1993 May;27(9):173–80.

[75] Van Lier JB, Vashi A, Van Der Lubbe J, Heffernan B. Anaerobic sewage

treatment using UASB reactors: engineering and operational aspects. InEnvironmental anaerobic technology: applications and new developments 2010 (pp. 59–89).

[76] Tandukar M, Machdar I, Uemura S, Ohashi A, Harada H. Potential of a combination of UASB and DHS reactor as a novel sewage treatment system for developing countries: Long-term evaluation. J Environ Eng. 2006;132(2): 166–72.

[77] Singh RK, Murty HR, Gupta SK, Dikshit AK. An overview of sustainability assessment methodologies. Ecol Indic. 2009;9(2): 189–212.

[78] Padilla-Rivera A, Morgan-Sagastume JM, Noyola A, Güereca LP. Addressing social aspects associated with wastewater treatment facilities. Environ Impact Assess Rev.2016;57: 101–13.

[79] Post T, Medlock J. Wastewater Technology Fact Sheet Tricking Filters. US Environmental Protection Agency, Office of Water Washington DC EPA. 2002:832-F00.

[80] "Design Guidelines for UASB-DHS System" and "Operation and Maintenance Guidelines for UASB-DHS System" (2016) in collaboration with Tohoku University. Submitted to MoEF-NRCD.

[81] Gostelow P, Parsons SA, Stuetz RM. Odour measurements for sewage treatment works. Water Res. 2001;35 (3):579–97.

[82] Hatamoto M, Okubo T, Kubota K, Yamaguchi T. Characterization of downflow hanging sponge reactors with regard to structure, process function, and microbial community compositions. Appl Microbiol Biotechnol. 2018;102 (24):10345–52.

[83] Coppens T, Van Dooren W, Thijssen P. Public opposition and the neighborhood effect: How social interaction explains protest against a large infrastructure project. Land Use Policy. 2018 Dec 1;79:633–40.

[84] Naderpajouh N, Mahdavi A, Hastak M, Aldrich DP. Modeling social opposition to infrastructure development. Journal of Construction Engineering and Management. 2014 Aug 1;140(8):04014029.

[85] Huang Y, Ning Y, Zhang T, Fei Y. Public acceptance of waste incineration power plants in China: Comparative case studies. Habitat International. 2015 Jun 1;47:11–9.

[86] Ohsawa Y, Tamura K. Efficient location for a semi-obnoxious facility. Annals of Operations Research. 2003 Oct 1;123(1–4):173–88.

[87] Li W, Zhong H, Jing N, Fan L. Research on the impact factors of public acceptance towards NIMBY facilities in China-A case study on hazardous chemicals factory. Habitat International. 2019 Jan 1;83:11–9.

[88] Oshiki M, Aizuka T, Netsu H, Oomori S, Nagano A, Yamaguchi T, Araki N. Total ammonia nitrogen (TAN) removal performance of a recirculating down-hanging sponge (DHS) reactor operated at 10 to 20° C with activated carbon. Aquaculture. 2020 Apr 15;520:734963.

Chapter 22

Experimental Investigation of Biomass Attachment to Wastewater Reactors

Renato Benintendi

Abstract

Attached mass bioreactors have extensively been adopted in the last decades when specific needs have suggested this choice. Benefits and advantages of this multi-faceted technology in wastewater treatment processing are well known, along with the kinetic and mass transfer aspects regarding their operation, essentially belonging to the mass transfer with chemical reaction theory applied to enzymatic catalysis, referred to as Languimur-Hinshelwood kinetics, notably Monod/ Michaelis Menten equations. On the other hand, a consolidated literature has dealt with many aspects of the development of strain colonies forming a biofilm. However, a few works have been devoted to the systematic analysis of its physiology, within the framework of the wastewater management of complex substrates and high-loads effluents. This article presents the experimental findings of a research activity covering the junction area between microbiology and bioreactor engineering, against a multifaceted set of operating parameters directly affecting health and stability of the attached biomass. In this respect, important results have been obtained, providing guidance on the attached mass reactor start-up, steady- state operation, impact of xenobiotic substrates, role of nutrients, filaments and foam formation, as well as qualitative aspects of the post-treatment effluent.

Keywords: attached mass, biofilm, wastewater, nutrients, adhesion test

1. Introduction

The adoption of attached mass reactors for the biological treatment of wastewater started in the late 1800s early 1900s [1]. Trickling filters have been used much earlier than the fundamentals of biochemical engineering applied to sewage purification had been established. During the early 1950's, development of plastic media resulted in the introduction of innovative packing and during the early 1960s rotating discs, included in several chemical process, have begun to be considered for oxygen transfer [1]. In the last 40 years submerged attached growth aerobic processes became established options for wastewater treatment, taking the advantage of a reduced space required with respect to activated sludge [2]. Nowadays, the combination of attached growth/activated sludge systems, also referred to as hybrid process or Integrated Fixed-Film Activated Sludge (IFAS) is an optimized alternative, as well as the Moving Bed Biofilm Reactor (MBBR), which typically ensure an increased treatment capacity, reduced sludge production and handling, low spaces and independency of a secondary clarifier. All attached mass systems have to take into account that mass transfer is the rate-limiting-steps of the overall biodegradation process. In addition, biofilm quality, structural properties and stability greatly affect the overall process performance, as well as effluent quality and conditions for foam formation. Consequently, the characterization and the understanding of physiology of the biofilm is an essential key for a correct management of the biodegradative process.

2. Background

Historically, Van Leeuwenhoek is credited as the first scientist who observed microbial biofilm formed on a surface [3]. Important contributions to the understanding of the mechanisms and circumstances according to which adhesion takes place were taken by Characklis [4] and Costerton et al. [5]. In the last decades, kinetic and mass transfer aspects of attached biomass have been well studied and parameterised in the scientific and technologic literature [2, 6]. Diaco and Eramo [7, 8] carried out one of the most complete studies combining diffusional and kinetic aspects with the structure of the attached mass. Oxygen transfer from the bulk liquid to the bacterial colony surface and subsequent kinetics follows the general laws of absorption with chemical reaction on a surface layer, as described in the early works of Sherwood [9] and Astarita [10]. On the other hand, kinetics taking place in a biofilm has extensively been studied according to Michaelis- Menten-Monod scenario and successfully framed within Langmuir-Hinshelwood and Hougen - Watson reaction schemes, according to the formalism of heterogeneous catalysis [11]. Williamson and McCarty [12] provided one of the first approach specifically related to the biofilm, showing the correlation between Fick diffusional phenomena and Monod kinetic theory. Authors such as O′ Toole et al. [13] and many others have studied the biofilm structure with the aim to identify how diverse was the mass microbiological behavior with respect to the suspended structure, as to the metabolic, pathogenic and clinical scenarios. More recently, several authors, such as Muslu [14] and Feng et al. [15] have worked on modeling further biofilm as a chemical reaction site, bringing specific contributions to the definition of the biodegradation rate and of the related conditional factors, within the theoretical frame built up by early authors such as Atkinson and Davies [16] and La Motta [17]. Naz et al. [18] evaluated the biofilm succession on stone media and compared the biochemical changes of sludge in attached and suspended biological reactors operated under aerobic and anaerobic conditions. Ercan et al. [19] have studied the biofilm development conditions on different surfaces, predominantly from the point of view of the relationship between the structural and geometrical features of the supports and the process variables, such as oxygen transfer and chemical reaction rate for production purposes. As a matter of fact, a large and consolidated literature has been produced about the characterization of biofilm and attachment mechanism, predominantly either analyzing the process engineering aspects of the bioreactor schemes or studying the purely microbiological and clinical topics per se. Relatively few efforts have been devoted to the analysis of the physiology of the attached biomass, notably dealing with the effects of nutrients, of substrate origin, of its transient behavior in terms of formation rate and stabilization framed within the biodegradation of complex pollutants contained in the wastewaters.

3. Unknown variables and uncertainties

Jenkinson and Lappin-Scott [20] define biofilm as *the microorganism consortium which develops at the interface solid–liquid or liquid–gas*. Gottenbos et al. [21] have pointed out that *the biofilm is a specific micro-ecosystem inside which different microbial strains effectively cooperate to get protection from ambient stress and to promote nutrients absorption*.

Formation and development of biofilm is assumed to take place according to the following phases, [22–24]:

1. *Adhesion of some planktonic microorganisms to a surface*

2. *Consolidation of a preliminary film layer promoted by Extracellular Polymeric Substances (EPSs) via hydrogen-type bond*

3. *Formation of a monolayer by bacterial strains grown on it*

4. *Development of a three-dimensional biofilm, consisting of a composite structure made of additional planktonic microorganisms and inert materials-EPSs (75 – 95%), cells (5 – 25%)*

5. *Dispersion and expansion of the biofilm*

All these phases are mediated by and are related to specific metabolic factors, which have a specific role in the progression of the process. The definition and characterization of the related parameters are of paramount importance to establish the conditions for a biofilm to be formed and stably persist in an aqueous medium. Parameters and factors affecting biofilm formation, development and effectiveness are very numerous, ranging from metabolic chemistry to genetics, hydrodynamics and transport properties. In the experimental research programme underpinning the present article, a specific focus has been made on those which were expected to have a direct, significant and macroscopic effect on the effectiveness of the wastewater treatment process.

3.1 Nutrients and thermochemical environment

Nutrients, temperature and pH significantly affect the biofilm formation and its behavior. Specifically, effect of temperature and pH is substantially known, whereas, even if a basic understanding of nutrients role has been achieved, often monitored in terms of C/N and C/P ratios, the assessment of biofilm behavior with time and with regards to changing nutrients respectively is important [25].

3.2 Surface and hydrodynamics

Surface and hydrodynamics play an important role. In order for the attachment process to start, contact of cell with solid surface is required, followed by a rapid bond. Strengthen and rapidity of these bonds are rather known, depending on a series of chemical–physical interactions and biological process which lead to a reversible adhesion [26]. This reversibility is to be considered as purely theoretical, as, if a minimum set of bonds has been achieved, the cumulative effect is sufficient to make the adhesion permanent [27]. Once again, nutrients and gram reactivity as well were expected to play an important role both in terms of adjusting the attachment capability to the specific surface and in terms of the transport properties in the aqueous medium, such as viscosity, which, in turn, belongs to the hydrodynamic part of the system.

3.3 Nitrogen contents and filamentous forms

Formation of filamentous forms is a particular drawback of attached growth systems adopted in wastewater treatment. In this respect an important parameter is a high COD value and, connectedly, nitrogen balance [28]. This aspect deserves a specific attention for the possible effects on the correct development and functional characteristic of the biofilm.

3.4 Toxic substances

This is a crucial parameter in wastewater management and possibly is more critical for attached mass bioreactors [29]. In this respect, an indicative approach could be the determination of the difference of the biofilm structure and functional capability against substrates showing an increasing leaving of toxicity.

4. Objectives and novelty of the research

This article presents the findings of an experimental research, conducted at the *Department of Agronomy, Section of Microbiology* of the University of Naples Federico II, Naples (Italy), consisting of the analysis of attached mass formation, based on both mixed industrial-municipal and of the effect of high tannery wastewaters COD, along with the endogenous strains. The research aimed at:

- Studying the main physiological features of the biofilm in relationship to its capacity to settle down and to reduce the COD load of different effluents with different substrates

- Comparing the behavior and the features of the biological structures against different wastewaters

- Appreciating the effects of different nutrients at various dilutions

- Finding out the qualitative aspects of the treated effluents, also considering foam formation, presence of filamentous forms and degree of wastewater clarification, against different parameters and different effluents

- Analyzing the influence of nitrogen substances

- Carrying out a transient analysis from the biofilm formation through its full development to identify any specific growth factors, including CFU/ml

Due to the high level of parameterization related to the high number of variables, experimentally tested, the research is considered a step forward to the understanding of the biomass attached growth mechanism.

5. Materials and research preparation

5.1 Sampling

The study has been carried out on microbial colonies of sludges collected in sterile bottles from Acerra (Italy) mixed industrial-civil wastewater treatment plant and from CoDiSo tannery wastewater treatment plant of Solofra (Italy), subject and not subject to a physical–chemical process. More precisely, the overall treatment process consists of a sequence of a biological and of a physical–chemical process. Samples have been taken downward of the primary settlement and transported with a thermally insulated bag. After laboratory water removal, the sludges have been distributed into 1.5 ml sterile polystyrene Eppendorf pipettes and 50 ml Falcon tubes. Both have been centrifuged at 10,000 rpm for microbial cell separation and fed with PCB nutrient broth at 20% glycerol. PCB composition has been included in **Table 1**. Settled cells have been suspended in a vortex and stored at −20°C. Wastewater has been collected as well with 4 sterile bottles, notably 2 liters upstream and 2 liters downstream of the physical–chemical treatment unit, with the aim to investigate the toxic and substrate diversity associated with the different sampling points. One fraction of this wastewater has undergone centrifugation, suspension of microbial cells in PCB nutrient broth at 20% glycerol and stored at −20°C. The remaining fraction has been frozen for further tests.

5.2 Inoculum preparation

The microbial matrix has been processed after growth induction. Each test has been carried out after defrosting up to ambient temperature and inoculation of 150 ml of 1% PCB broth. The latter has been incubated at 28°C for about 12 hours overnight, in a rotary shaker. This culture medium was the inoculum for the experimental campaign.

5.3 Material list

The following main materials have been adopted for the execution of the experimental study:

- Microbial material and cultures, provided by the Acerra Wastewater Treatment Plant (Italy)
- Nutrients and additional substrates, collected from the inventory of the Laboratory of the Institute of Microbiology of the Faculty of Agronomy of the University Federico II, Portici (Italy)
- Chemicals and reactants, microbiological grade, supplied by Sigma-Aldrich

5.4 Microbial growth and biofilm development with a substrate of given composition

Investigations have preliminary been carried out with the aim to identify the best conditions for the film development with respect to various glucidic forms and different nitrogen concentrations.

Yeast extract	5 g
D-glucose	2 g
Distilled water	1000 ml
pH = 7.0 ± 0.2 a 25°C / Strain storage with sterile 20% glycerol at – 20°C.	

Table 1.
PCB (plate count broth) – liter^{-1}.

5.4.1 Influence of glucidic compounds

Inoculum M9 at different glucidic concentrations has been tested (**Table 2**). Notably: 1 liter at 1% glucose and 0.05% glucose respectively, and 1 liter at 1% saccharose and 0.05% saccharose respectively. These substrates have been selected in light of their ease to be metabolized so that fast results were expected to be attained with respect to the work objectives. The four substrates, after sterilization at 5% inoculum, have been put in a 1.5 liter flask equipped with two air-in / air-out capillaries for sterile air blowing. Each flask has been equipped with a 2 cm2 Petri dish, adopted as attachment surfaces. The culture has been conducted under hood for 24 hours at ambient temperature of around 25°C. Suspended bacteria growth rates have been measured during this time, by means of spectrophotometric readings at 600 nm, and utilizing, as reference, the relevant blank substrates. During this phase, visual checks have been made as to the clarification degree and foam formation for all cases, and dissolved oxygen has been measured. After 24 hours, fragments of Petri dishes have been collected, gently washed with deionized water and attached microorganism have been stabilized on the solid support by using a Bowin fixative (**Table 3**). Afterwards, the fragments have been dried under laminar hood for 24 hours and fixative removal has been accomplished by means a 50% ethanol solution, and Crystal violet Hucker's solution has been used for attached cells colouration (**Table 4**). Microscopic observations have allowed for a qualitative-quantitative evaluation of the attached biomass.

5.4.2 Influence of nitrogen-containing compounds

The same procedure adopted for glucidic substances has been followed for M9 nutritive standard substrate, modified with nitrogen - containing compounds. 3 liters of nitrogen-containing mineral substrate (ammonium chloride) have been prepared, notably 1 liter at 0.1%, 1 liter at 0.2% and 1 liter at 0.4%, and 1 liter of organic nitrogen-containing substance (pancreatic digest of casein) at 0.5%. In

$NaHPO_4$	6 g
KH_2PO_4	3 g
NH_4Cl	1 g
NaCl	0.5 g
D-glucose (20 g/100 ml solution)	10 ml
$MgSO_4 \cdot 7\, H_2O$ (246.5 g/ l solution)	1 ml
Thiamine · HCl (10 mg / 20 ml solution)	1 ml
$CaCl_2$ (14.7 mg/100 ml solution)	1 ml
Deionized water	1000 ml

Table 2.
M9 medium – liter^{-1}.

Saturated solution of picric acid	15 ml
Formalin	5 ml
Glacial acetic acid	1 ml

Table 3.
Bowin fixative.

Solution **A**:	
Crystal – violet	2 g
Ethanol 95°	20 ml
Solution **B**:	
Oxalate ammonium	0.8 g
Deionized water	80 ml

Table 4.
Hucker crystal - violet.

addition to being a fundamental element of the biochemical balance of the microbial cultures, the parametrization of nitrogen content has been considered as a fundamental information to assess the behavior of the biofilm. Also for this inoculum, the microorganism growth rate and the biofilm development has been assessed, in addition to the clarification degree, foam formation and filamentous forms.

5.5 Observation of microbial growth and biofilm development in industrial wastewater

Four microbial cultures have been prepared utilizing tannery wastewater as substrate. Specifically:

- Untreated wastewater (IN)
- Wastewater with chemical–physical treatment (OUT)
- Diluted 1:1 untreated wastewater (IN 1:1)
- Diluted 1:10 untreated wastewater (IN 1:10)

The substrates have been collected in a flask and sterilized at 121°C for 15 minutes, then inoculated with at 5% with a microbial sludge collected from the mixed sludges. Microbial growth has been promoted as described previously, as well as for the procedure for air blowing, fragments collection, evaluation of attachment, clarification and foam formation. The development of microorganism has been checked after 24 hours according to the scalar dilution's method and PCA bacterial count on Petri dish (**Table 5**).

5.6 Isolation and purification

Microorganisms of 3 samples have been isolated and then purified. Specifically:

- Attached biomass of dish fragment of previous test
- Sludges collected from mixed wastewater
- Tannery wastewater treated in the physical–chemical unit

PCA has been utilized as sterile agar culture medium. Isolation has been accomplished through suspension dilutions (up to 10^{-9}) of samples and plating by inclusion of 1 ml for each dilution. After incubation for 24 hours at 27°C, largest colonies have been purified by smear on PCA plating. Purified strains have been stored at

Yeast extract	2.5 g
Pancreatic digest of casein	5 g
Glucose	1 g
Agar	15 g
Deionized water	1000 ml

Table 5.
PCA agar– liter^{-1}.

−20°C in PCB at 20% glycerol and at 4°C in slant test tubes, to be collected for the subsequent identification tests. On the same dishes, for sludges and wastewater, prior to smear purification, a bacterium count of colonies of dilutions showing the greatest growth has been made and the results have been expressed in CFU/ml.

5.7 Identification tests

Three main characteristics have been considered with the aim at systematically framing isolated strains: gram-reaction, catalase and cellular morphology.

5.7.1 Gram-reaction test

Gram reaction has been accomplished according to Gregersen [30], adopting a 3% KOH solution, wherein one colony is dissolved and then the loopful is removed and checked for viscosity. A significant viscosity change is indicative of Gram negativity, whereas an unchanged viscosity reveals the Gram positivity. The rationale is based on Gregersen's observation that Gram+ withstand potassium hydroxide whereas in Gram- cellular wall is destroyed and DNA is released resulting in a viscosity change.

5.7.2 Catalase test

Presence of catalase has been sought through the simple hydrogen peroxide test [31]. Notably, observing the reaction oxygen presence can be detected by formation of bubbles, visible with the naked eye. Catalase test protocol originally proposed by Gagnon [32] has substantially been followed for catalase determination.

5.7.3 Microscopic Observation

Cells shape has been controlled through microscopic observation.

5.8 Test of adhesion

This test has been carried out by the adoption of strain B_1, as it constitutes most part of biomass isolated microorganisms. This test allows one to quantify attached mass through the measurement of the optical density of a mature biofilm on a plastic surface [33]. In this respect, Christensen's method has been applied based on the modifications described by Baldassarri et al. [34]. In this study, 7 dilutions of overnight cultures at 28°C have been made in order to obtain, in addition to an evaluation of the strain attachment characteristics, also a relationship between cellular concentration and bioattachment, at the following dilutions:

- 2.5%
- 5%

- 10%
- 20%
- 40%
- 80%
- Culture broth

5 ml have been withdrawn from each dilution, 1 ml of which has been used to prepare suspension-dilutions with the aim to determine the bacterial load corresponding to each sample. Portions left have undergone spectrophotometric readings (600 nm). All samples, except lowest dilutions, have been centrifuged for 10 minutes at 10,000 rpm and, further to supernatant elimination, cells have been resuspended in sterile broth. 2 ml have been collected from each dilution and distributed into 10 pits (200 μl each) of a 96-wells Enzyme-Linked Immunosorbent Assay (ELISA) plate with flat bottom. The plates, provided with 10 wells filled PCB sterile broth as blank sample, have been incubated at 28°C for 24 hours. The broth has been then gently removed from the wells by means of Pasteur pipettes and wells have been washed three times with sterile deionized water to remove unattached cells. This phase has been conducted very carefully as it often causes removal of attached cells also, resulting in false findings. Unlike proposed by Christensen et al. [33], sterile pipettes have been preferred to direct immersion of plates into clean water, in order to minimize any disturbances and handling. Attached cells have then been fixed by exposure to 60°C for 1 hour and colored by Hucker crystal violet. Plates have been rinsed until rinsing water has been completely colorless, then overturned and dried for 30 minutes at 37°C. Biofilm density has been assessed by a spectrophotometer for ELISA plates, after calibrating the blank sample of the instrument with the wells containing non-inoculated sterile broth. Readings have been taken at 550 nm. According to the adopted protocol, mean of taken readings for each well has been interpreted as:

- Positive: ≥ 0.24
- Weak: ≥ 0.12 and < 0.24
- Negative: < 0.12

5.9 Observation of the growth-curve shown by a pure culture in presence of a complex substrate

A study has been carried out with the aim to investigate the growth of a B_1 strain with a substrate consisting of pure butanol only. In this respect, two tests have been performed:

- Synthetic substrate consisting of 1000 ppm of butanol inoculated with 5% B_1 strain
- ATP substrate inoculated with 5% B_1 strain

The Inoculum has been prepared from a strain stored in a freezer on slant PCA. Sterile broth culture contained by a 20 ml tube has been inoculated with B_1 strain, and incubated for 15 hours overnight at 28°C. The culture obtained was the 10%

Pancreatic digest of casein	12.5 g
Glucose	10 g
Yeast extract	7.5 g
NaCl	5 g
K_2HPO_4	5 g
Sodium citrate	5 g
$MgSO_4 \cdot 7\,H_2O$	0.8 g
$MnCl_2 \cdot 4\,H_2O$	0.14 g
$FeSO_4 \cdot 7\,H_2O$	0.04 g
Tween 80	0.2 g
Na_2CO_3	1.25g
Thiamine · HCl	1 mg
Deionized water	1000 ml
pH = 7.0 ± 0.2 a 25 °C.	

Table 6.
APT broth – liter^{-1}.

inoculum to 200 ml of APT broth (**Table 6**), incubated in a rotary-shaker for 15 hours at 28°C, resulting in the inoculum for the two tests. The two substrates have been kept under chemical hood for 48 hours at ambient temperature. At the same time, a fraction of the following substrates has been stored in sterile flasks:

- Synthetic substrate consisting of 1000 ppm of butanol inoculated with 5% sterile APT
- APT sterile cultural substrate

These substrates have been used as blank samples in the periodical spectrophotometric readings (600 nm) taken for either test.

5.10 C.O.D removal by a pure culture and by a heterogeneous culture (sludge) of tannery wastewater not subject to physical–chemical treatment

Two sterilized samples of untreated tannery wastewater collected in two flasks have been adopted as substrate for B_1 strain. The two samples have been inoculated at 5% with the strain and with sludge respectively. Following exactly the procedure illustrated previously, after 24 hours, 50 ml of substrate have been extracted, transferred into sterile Falcon polystyrene tubes and centrifuged at 10,000 rpm for 10 minutes, to remove microbial cells. Supernatant, collected in sterile medium, has undergone C.O.D test with potassium bichromate 0.0417 M [35]. The protocol utilizes a solution 0.25N of FAS (ferrum(II) ammonium sulphate) as titrant, and C. O.D has been calculated according to the formula:

$$C.O.D. = DF \bullet (V_b - V_s) \bullet M \bullet 800 \qquad (1)$$

where:

- V_b is the average volume (ml) of iron(II) ammonium sulphate solution used in the titration of blank solution

- V_s is the volume (ml) of iron(II) ammonium sulphate solution used in the titration of the samples
- DF is the dilution factor
- M is the molarity of standardized iron(II) ammonium sulphate solution

5.11 C.O.D removal by a pure culture in a pure substrate

The same strain has been tested for the removal of the C.O.D. of two synthetic substrates consisting of 1000 ppm of butanol, inoculated with 5% sterile APT and 5% B_1 respectively. Growth stimulation, post-treatment and determination have been performed as described for previous tests.

6. Results and discussion

6.1 Sampling and storage of microbial material

Microbial materials utilized for the present study have been collected from two wastewater treatment plants, notably:

- Acerra (Italy) industrial and civil wastewater treatment plant
- Solofra (Italy) tannery pole wastewater treatment plant

Acerra wastewater undergoes a traditional activated sludge treatment process, following preliminary removal of gross materials and primary settlement. Activated sludge produced is split between 98% recirculated fraction and 2% fed to thickening and stabilization processes. Colonies populating the sludge flocs have been included in **Table 7**. Samples associated sludge parameters are B.O.D.5 = 20 mg/l and active mass concentration = 2600 mg/l. Solofra plant implements a biological/physical–chemical process over a C.O.D of 7000-8000 mg/l. After a preliminary phase of precipitation with limestone, polyelectrolyte and ferric chloride/aluminum sulphate, wastewater is supplied to a biological plant, with a final activated carbon adsorption stage for the removal or coloring and residual organic matter. Activated sludges of Solofra plant are composed predominantly of proteobacteria, Bacteroidetes, firmicutes with a presence of actinobacteria, planctomycetes and chloroflexi [36].

Aspidisca costata	Lionotus lamella
Colpidium colpodum	*Paramecium caudatum* et petridum
Didinium	Opercularia
Carchesium polypium	*Arcella vulgaris*
Epistylis	*Rotaria rotatoria*
Philodina	Nematodi
Stylonychia mytilus	Crustaceans molds
Vorticella microstoma	
(Courtesy of Acerra wastewater treatment plant)	

Table 7.
Microorganisms of activated sludges.

Inventory of bacterial matters consisted of 1.5 ml 300 Eppendorf, of which 100 of sludge, 100 of untreated water and 100 of physically-chemically treated water. Also, No. 105 50 ml Falcon tubes, of which 35 of sludge, 100 of untreated water and 100 of physically-chemically treated water. Biodegradative inherent capacity of Acerra sludges in a suspended lab scale bioreactor about a specific substrate such as ethyl acetate had preliminarily been tested by the author [37].

6.2 Microbial growth and development of biofilm in sugars-broth

Development of the biofilm is promoted by organic molecules through the formation of a conditioning layer on solid surfaces. M9 has been selected with the aim to prevent false results. Carbon source for this inoculum is glucose only, so that observed attachment could be put in relationship with bacteria capabilities only.

6.2.1 Influence of sugars

Bacteria growth with sugar type and concentration has been obtained through spectrophotometric controls shifted by 1 hour. The findings have been depicted in **Figure 1** [38]. The curves have been obtained by data reparameterization according to Gompertz equation as modified by Zwietering et al. [39]. As expected, growth of microorganisms is characterized by a latency phase, an exponential growth, a steady-state growth and death. Resulting growth with time can be described by the following parameters:

- The maximum growth rate μ_m, coincident with the tangent at the flex point
- The duration of the steady-state phase, λ, that is the intercept of the tangent, on the horizontal axis
- The asymptote A, defined as the maximum attained value:

$$y = A \cdot exp\left\{-exp\left[\frac{(\mu_m \cdot e)}{A} \cdot (\lambda - t) + 1\right]\right\} \tag{2}$$

After 24 hours substrate clarification, foam formation and presence of filamentous forms caused by bacteria growth have been evaluated as shown in **Figure 2**. It

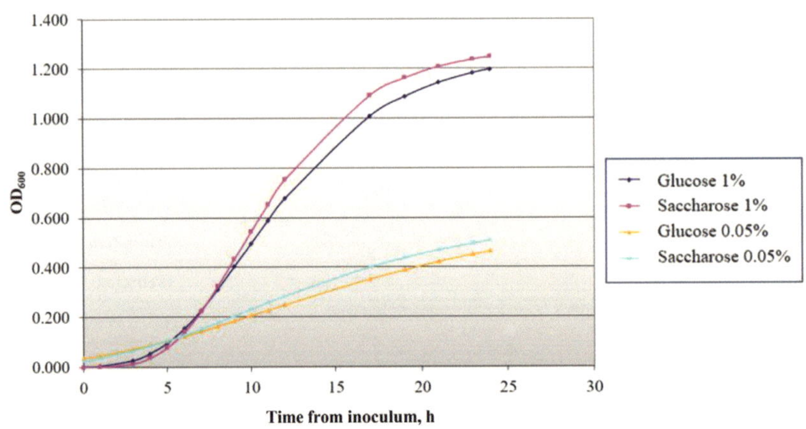

Figure 1.
Microbial growth in sugar-broth.

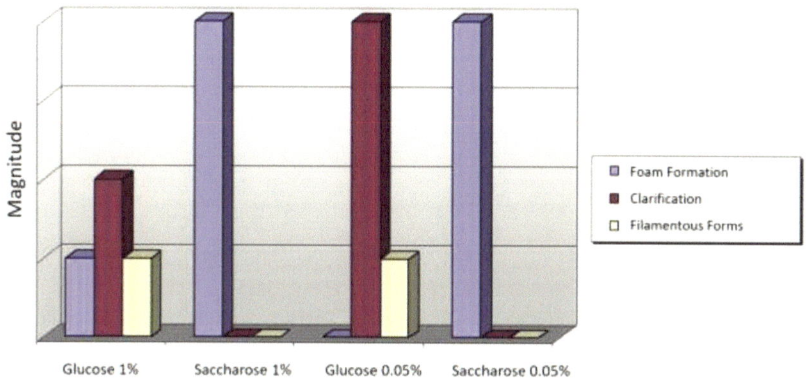

Figure 2.
Effects of substrate of microbial growth in sugar-broth.

can be noted that glucose provides better results than saccharose, which produces both at high and low concentrations, a significant foam formation along with an intense turbidity. Glucose 1% promotes a faster colonies growth than glucose 0.05%: the former can be adopted during the start-up, the latter at the steady-state, where effluent quality matters. The microscopic observation of the attached mass, accomplished on 4 fragments for each test, has revealed the presence of filamentous forms in glucose test, as shown in **Figure 2**, where arbitrary unitless relative values have been assigned. However, this does not affect the quality of the effluent as stated before, as to ideal sugars. The microscopic observation has revealed that the fours tests have shown the same results in terms of biofilm formation, as illustrated in **Figure 3**, based again on relative unitless values. The biofilm was well developed and appeared as a dense, uniform and well-structured cells mass. In some instances, the subsequent cellular layers have been observed.

6.2.2 Influence of nitrogen-substances on microbial growth and development of biofilm

Influence of nitrogen on biofilm development has been assessed with a specific test. **Figure 4**. shows that organic nitrogen promotes cell growth much more than mineral nitrogen, whose different concentration does not significantly affect the

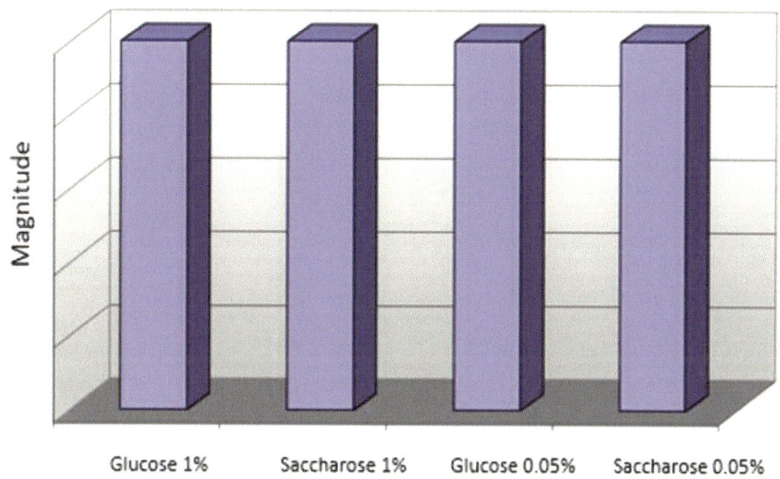

Figure 3.
Biofilm development – Attachment tests in sugar-broth.

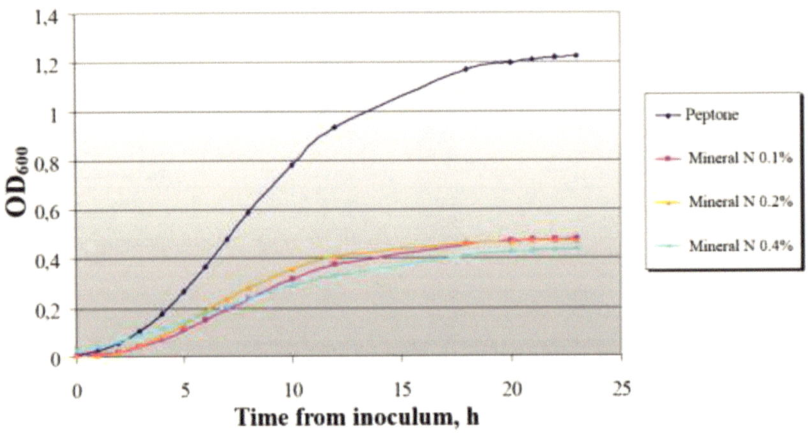

Figure 4.
Microbial growth with nitrogen-compounds.

growth. However, the bigger growth is accompanied by a significant foam formation. This has not been observed when ammonium chloride has been adopted, along with a good clarification, as shown in **Figure 5**. Microscopic observation of the plate fragments collected from the various tests has provided unexpected results (**Figure 6**). Notably, no difference has been detected between biofilms developed with organic and inorganic nitrogen. This disagrees with Mittelman [40] and is in agreement with Abbott et al. [41], according to which the availability of an organic layer next to the attachment surface would be substantially uninfluential, as this would be controlled mainly by electrostatic interactions. Filamentous forms have resulted to be absent in all four tests.

6.3 Microbial growth and development of biofilm in industrial wastewater

Activated sludges collected from Acerra treatment plant have been adopted as microbial matrix, whereas treated and untreated wastewaters coming from Solofra treatment plant (**Table 8**) have been used as substrate. Count of suspended bacteria instead of spectrophotometry was due to the significant TDS and to the related difficulty to calibrate the instrument. Having the count been carried out at irregular intervals during the test, it was not possible to draw a growth curve, but, based on the final bacterial load, it was possible to estimate the growth trend and the

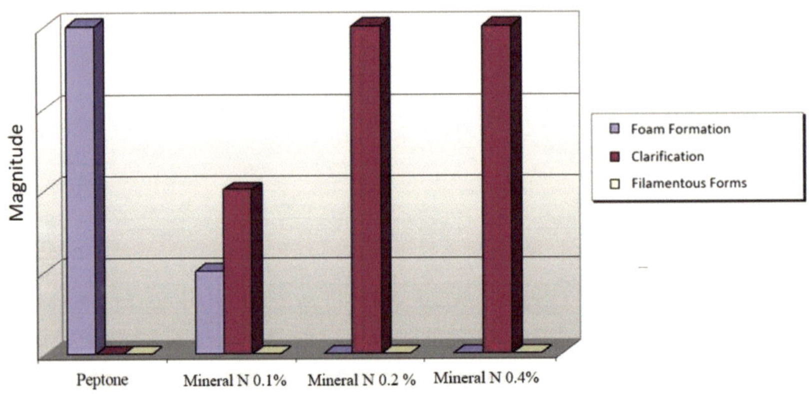

Figure 5.
Effects of substrate of microbial growth with nitrogen-compounds.

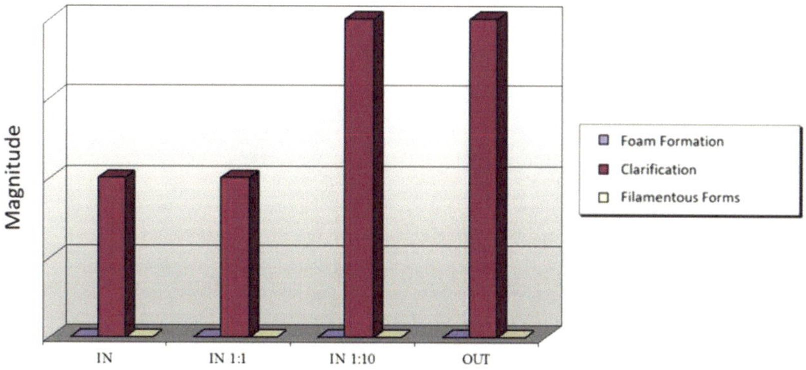

Figure 6.
Tannery wastewater treatment – Quality of the effluent.

COD (mg/l)	8860
TDS (mg/l)	2210
N_4 (mg/l)	100.1
pH	3.6
(Courtesy of Solofra wastewater treatment plant)	

Table 8.
Untreated wastewater characteristics.

survivability of the microbial matrix. Findings of bacterial count are shown in **Table 9**. A longer survivability was observed in wastewaters which underwent chemical settlement and flocculation, as expected from the very high toxic load of tannery effluents. Quality of the effluent in terms of foam formation, clarification and presence of filaments has been shown in **Figure 6** and **Table 10**. Once again, presence or removal of toxic substances are a key driver for the lower or higher degree of clarification. Contrary to the findings of the other tests, which showed same biofilm formation and features, significant even microscopic differences have been identified in this test, as to the attached biomass in the various substrates. A qualitative relative indication of the biomass formation has been shown in **Figure 7**. The layer observed at the microscope even at high toxic concentrations can be

	IN	IN 1:1	In 1:10	OUT
CFU/ml	2E-04	2E-05	4E-05	4E-07

Table 9.
Tannery wastewater. Bacterial count after 24 hours at different dilutions.

	IN	IN 1:1	In 1:10	OUT
Foam formation	Absent	Absent	Absent	Absent
Clarification	Medium	Medium	High	High
Filamentous forms	Absent	Absent	Absent	Absent

Table 10.
Quality of treatment of tannery wastewater.

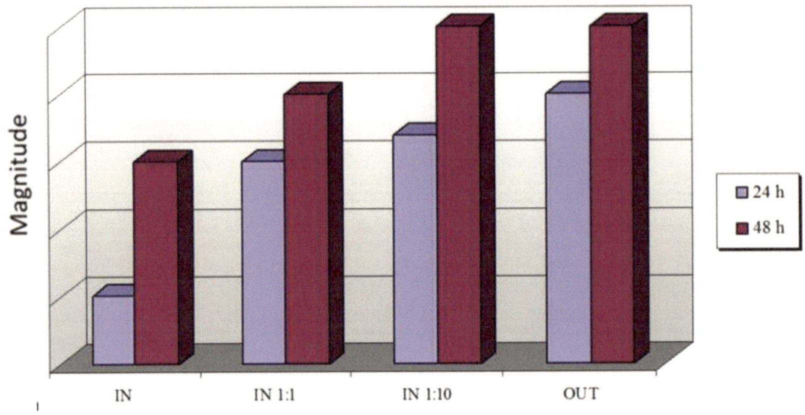

Figure 7.
Biofilm formation in tannery wastewater.

justified by the survival at these concentrations of morphologically different bacterial colonies.

6.4 Isolation and purification

Nineteen strains have been selected from the available samples, notably:

- 5 collected from the attached biomass, indicated with initial B
- 5 collected from the mixed-wastewater treatment plant, indicated with initial S
- 9 from the physically-chemically treatment tannery wastewater, indicated with initial T

It is worth noting that more numerous strains have been taken and untreated tannery effluents have been excluded from the test. Morphology, Gram reaction and catalase test findings have been included in **Table 11**. Morphologically different strains have shown same behavior, and rod shape and $Gram^{+}$ have been predominant. Bacterial counts executed on mixed wastewater sludges, as well as on treated and untreated tannery effluents, have resulted in 2·108, 2·108 and 4·17 CFU/ml respectively.

6.5 Attachment test

This test has consisted of the determination of the relationship between OD600 values and concentration of cells materials per ml of assumed dilution. **Figures 8** and **9** show the relevant spectrophotometric findings. **Table 12** shows the results of the OD_{550} spectrophotometric findings of the observations on ELISA plates adopted for the attachment test. **Table 12** also shows means and standard deviations related to the 10 times repeated tests for each dilution. The test has revealed an excellent attachment capability of the studied strain.

6.6 Statistical analysis

Experimental results have undergone statistical analysis, with the aim to assess the significance of the differences observed between the analyzed samples, notably,

Strain	Morphology	Gram reaction	Catalase
S_1	Cocci	—	+
S_2	Cocci, diplococci	—	+
S_3	Rods	+	+
S_4	Rods	+	—
S_5	Cocci	—	+
B_1	Rods arranged in chains	+	+
B_2	Cocci arranged in chains	+	+
B_3	Cocci arranged in chains	+	—
B_4	Streptobacteria	+	+
B_5	Cocci	—	–
T_1	Rods arranged in chains	+	+
T_2	Rods arranged in chains	+	+
T_3	Rods arranged in chains	+	+
T4	Rods arranged in chains	+	+
T5	Rods arranged in chains	+	+
T6	Rods arranged in chains	+	+
T7	Rods arranged in chains	+	+
T8	Rods arranged in chains	+	+
T9	Cocci	+	—

Table 11.
Differential characteristics of isolated strains.

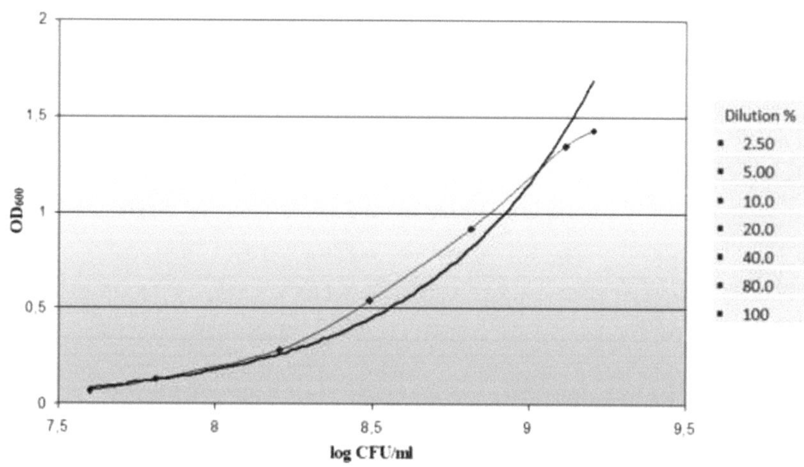

Figure 8.
OD_{600} vs. CFU/ml at different dilution ratios.

through Fischer and Yates variance ratio (F). Based on this ratio, significance of variability due to identified variation sources has been assessed, in relationship with the variability due to error. Here a F has been found to be equal to 380.764, so much greater than the tabulated value (3.14). The analysis has been refined further through Duncan's multiple range test [42]. This was necessary as the means were

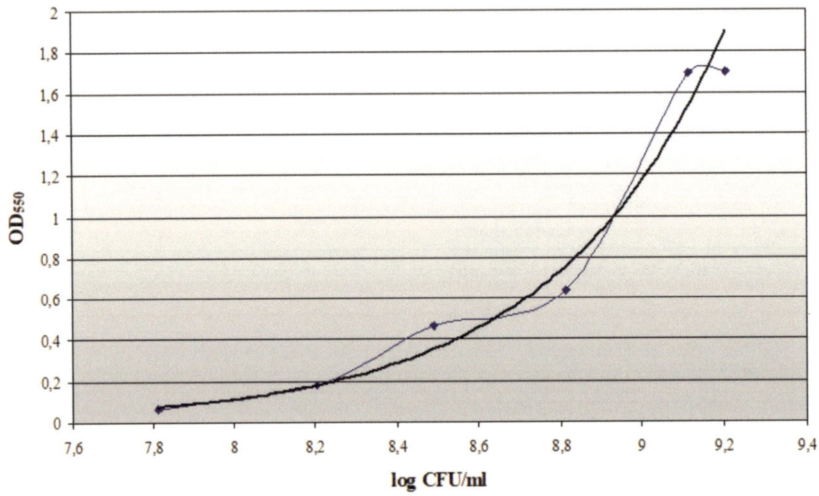

Figure 9.
Attachment behavior vs. CFU/ml.

	A	B	C	D	E	F	Blank
OD_{600}	0.068	0.395	0.505	0.227	1.892	1.763	0.075
OD_{600}	0.141	0.368	0.432	0.859	1.823	1.632	0.085
OD_{600}	0.090	0.495	0.565	0.611	1.632	1.721	0.079
OD_{600}	0.076	0.143	0.357	0.891	1.702	1.963	0.082
OD_{600}	0.134	0.153	0.589	0.695	1.712	1.812	0.060
OD_{600}	0.140	0.231	0.508	0.771	1.638	1.601	0.067
OD_{600}	0.169	0.175	0.571	0.801	1.856	1.897	0.068
OD_{600}	0.129	0.216	0.851	0.685	1.685	1.851	0.074
OD_{600}	0.185	0.220	0.262	0.797	1.886	1.612	0.059
OD_{600}	0.227	0.152	0.362	0.763	1.824	1.843	0.066
OD_{600}	0.135	0.251	0.536	0.710	1.765	1.769	0.0715
Mean	**0.049**	**0.114**	**0.144**	**0.188**	**0.101**	**0.125**	**0.0089**
St.Dev	**0.068**	**0.395**	**0.505**	**0.227**	**1.892**	**1.763**	**0.0750**

A = $6.5{\cdot}10^7$ CFU/ml; B = $1.6{\cdot}10^8$ CFU/ml; C = $3.1{\cdot}10^8$ CFU/ml; D = $6.5{\cdot}10^8$ CFU/ml; E = $1.3{\cdot}10^9$ CFU/ml; F = $1.6{\cdot}10^9$ CFU/ml.

Table 12.
Spectrophotometric check (OD550) on 10-times repeated test – Elisa plate.

different and, in fact, significant differences between the means of the readings of the various cells populations have been found. Specifically, internal variability of the series related to the blank and to the cell concentration of order of magnitude of 107 CFU/ml have resulted to be too high to conclude that means are different. This allows one to state that attachment phenomenon is a direct function of CFU/ml only, within a certain range of cell concentrations: in fact, the growth is weak and independent of the microbial population until around 10^7 CFU/ml, whereas within 10^8 e 10^9 CFU/ml it becomes exponential, after which becomes stable. The results of the statistical analysis have been reported in **Tables 13** and **14**.

	Squares sum	DF	Squares mean	F	Sign.
Between groups	32.030	6	5.338	380.764	0.000
Within groups	0.833	63	$1.402 \cdot 10^2$		
Total	32.913	69			

Table 13.
Results of ANOVA univariate.

	Reiterations	Subgroups for alfa = 0.05 Means for homogeneous subgroups shown				
		1	2	3	4	5
Blank	10	$7.15 \cdot 10^2$				
A	10	0.1359				
B	10		0.2512			
C	10			0.5366		
D	10				0.7100	
E	10					1.7650
F	10					1.7650
Sig.		0.228	1.000	1.000	1.000	0.993

Table 14.
Comparison between means based on variance (Duncan test).

6.7 Growth-curve of a pure culture in presence of a complex substrate

Strain B_1 has been investigated as to its capacity to grow in presence of carbonaceous rich substrate such as ATP and of 1000 ppm of pure butanol in deionized water. For the purpose of the research, complex substrate is meant any substances whose degradation time is longer than those corresponding to municipal wastewater. B_1 shows a rapid growth with ATP reaching the steady state in less than 8 hours, whereas with butanol it is reached in about 3 hours (**Figure 10**). Lack of nutrients

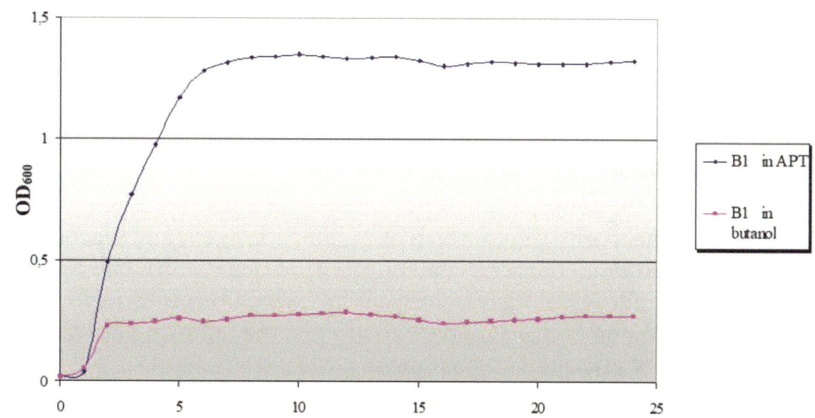

Figure 10.
Spectrophotometric checks for B_1 growth (readings at 600 nm).

plays a specific role in the observed behavior. On the other hand, the exponential growth shows that B_1 tolerates butanol quite well and uses ii for its own development. This confirms Eckenfelder [43] and Benintendi [37] as to the time taken to completely assimilate the organic matter. It can be concluded that the biofilm can be effective in removing COD even in presence of xenobiotic substances.

7. Findings highlights

The Strain B_1 was identified by Marino and Benintendi [44] and Benintendi [37] as *Bacillus alvei*. Based on the evidence described, the following conclusions can be drawn on the attachment mechanism:

- Strain B_1 has shown excellent attachment features according to the parameters indicated by Baldassari et al. [34]. Notably, in correspondence of cell concentration assumed by Baldassarri's protocol (overnight 1:2 culture dilution) a OD_{550} of 0.53 is attained, significantly greater than the minimum value required to defined the strain as positive (OD_{550} = 0.24)

- Attachment depends on a specific cell range, i.e. 10^8 - 10^9 CFU/ml

- Implementing the attachment based on these characteristics could be effective by adopting the strain as inoculum in an attached mass equipment and, complying with the strain needs, the attachment starting phase would be easily promoted.

- Further investigation is required to understand how xenobiotic substances affect strain growth and biodegradative capabilities

8. Conclusions

The experimental research has shown important data related to the knowledge of the physiology of attached mass bioreactors for treatment of complex wastewater.

- Attachment capabilities of bacterial strains have been observed within a broad range of cultural conditions.

- The variation of nutritional conditions results in a selection of microorganisms which, in turn, affects the biofilm structure and thus the quality of the treated effluent, as to clarification and foam formation.

- Contents of mineral nitrogen, instead of organic nitrogen, and glucose instead of saccharose, promote the achievement of a good quality effluent. Specifically, as to saccharose adoption, high concentrations would be required for the start-up, whereas lower amounts should be adopted in the following phases, in order for a good quality effluent to be obtained.

- The presence of toxic substances, even if does not influence significantly qualitative characteristics of the effluent, results in a selective promotion of the microorganisms, which delays the biofilm formation.

- Strains identified as B_1 has undergone specific tests, showing very satisfactory results. This suggests its adoption for the inoculum during the start-up of an attached-mass wastewater treatment plant.

Further studies are recommended to complete the analysis to fully understand the effect of xenobiotic substances on the strain growth, which have been shown to be expectedly well biodegraded, the biofilm formation and stability and the qualitative aspects of the effluent.

The present study offers a good and comprehensive basis for further research phases.

Acknowledgements

The research underpinning the present study has been supported by the work of Prof. Paolo Marino and of Dr. Giusi Iammarino of the former Institution Microbiology of the University of Naples Federico II, faculty of Agronomy. The author is grateful to the Acerra Wastewater Treatment Plant Laboratory, which provided the biological mass, along with its accurate characterization.

Author details

Renato Benintendi
Service de Chimie Physique CP 165/62, Ecole Polytechnique de Bruxelles, Université Libre de Bruxelles, 50 Av. F. Roosevelt Brussels 1050, Belgium

*Address all correspondence to: renato.benintendi@ulb.ac.be

References

[1] Alleman JE. The History of Fixed-Film Wastewater Treatment Systems, Proceedings of the International Conference of Fixed Film Biological Processes. Ohio: Kings Island; 1982

[2] Metcalf and Eddy. Wastewater Engineering: Treatment and Resource Recovery. 5th ed. New York: McGraw-Hill; 2014

[3] Donlan RM. Biofilms: Microbial life on surfaces, Emerg infect dis. Sep. 2002; **8**(9):881-890

[4] Characklis WG. Attached microbial growths-II. Frictional resistance due to microbial slimes. Water Research. 1973; 7:1249-1258

[5] Costerton JW, Geesey GG, Cheng K-J. How bacteria stick. Scientific American. 1978;**238**:86-95

[6] Eckenfelder WW. Industrial Water Pollution Control. 3rd ed. McGraw-Hill Companies Inc.; 1999

[7] Diaco L, Eramo B. Adesione batterica e sviluppo di biofilm su superfici solide (I parte). Ingegneria Sanitaria–Ambientale. pp. 1993:34-76

[8] Diaco, L., Eramo, B. (1994), Adesione batterica e sviluppo di biofilm su superfici solide (II parte), Ingegneria Sanitaria Ambientale., pp. 77-91.

[9] Sherwood TK. Absorption and Extraction. First Edition: McGraw Hill Book Company; 1937

[10] Astarita G. Mass Transfer with Chemical Reaction. Amsterdam, London: Elsevier; 1967

[11] Satterfield, C.N., 1980. Heterogeneous Catalysis in Practice, McGraw-Hill Company

[12] Williamson K, McCarty PL. Verification studies of the biofilm model for bacterial substrate utilization. Journal (Water Pollution Control Federation). 1976;**48**(2) February:281-296

[13] O'Toole G, Kaplan HB, Kolter R. Biofilm formation as microbial development. Annual Review of Microbiology. 2000;**54**:49-79

[14] Muslu Y. Kinetics characteristics of biofilm reactors, water, air, and soil pollution. October. 2002;**140**(1–4):1-20

[15] Feng L, Mu J, Sun J, Kong Y, Wang J, Lv Z, et al. Kinetic characteristics and bacterial structures in biofilm reactors with pre-cultured biofilm for source water pretreatment. International Biodeterioration & Biodegradation. 2017;**121**:26-34

[16] Atkinson B, Davies IJ. The overall rate of substrate uptake (reaction) by microbial films. Part 1. A biological rate equation. Transactions of the Institution of Chemical Engineers. 1974; **52**:248-259

[17] La Motta EJ. Kinetics of growth and substrate uptake in a biological film system applied and environmental microbiology. Feb. 1976;**31**(2):286-293

[18] Naz, I., Seher, S., Perveen, I., Devendra P. Saroj, D.P., Ahmed, S (2015) Physiological activities associated with biofilm growth in attached and suspended growth bioreactors under aerobic and anaerobic conditions. Journal of Environmental Technology, Volume 36 - Issue 13

[19] Ercan, D., Pongtharangkul, T., Demirci, A., Pometto III, A.L., (2015). Applications of Biofilm Reactors for Production of Value-Added Products by Microbial Fermentation, Book Editor(s): Anthony L. Pometto III Ali Demirci

[20] Jenkinson HF, Lappin-Scott HM. Biofilms adhere to stay. Trends Microbiol, Jan. 2001;**9**(1):9-10

[21] Gottenbos B., Van der Mei H. C., Busscher H. J. (2001). Models for studying initial adhesion and surface growth in biofilm formation surfaces. Methods in Enzymology, 2, 38, pp.523-534.

[22] Borenstein SB. Microbiologically Influenced Corrosion Handbook. New York: Industrial Press Inc.; 1994

[23] Geesey GG, Lewandowski Z, Flemming H-C, editors. Biofouling and Biocorrosion in Industrial Water Systems. Ann Arbor: Lewis Publishers; 1994

[24] Kaplan JB, Ragunath C, Ramasubbu N, Fine DH. Detachment of Actinobacillus actinomycetemcomitans biofilm cells by an endogenous β-hexosaminidase activity. Journal of Bacteriology. 2003;**185**(16):4693-4698

[25] Stoodley, P., Dodds, I., Boyle, John, D., Lappin-Scott, Influence of hydrodynamics and nutrients on biofilm structure, Journal of Applied Microbiology, January

[26] Ladd TI, Costerton JW. Methods. Microbiology. 1990;**22**:285

[27] McEldowney S, Fletcher M. Applied and Environmental Microbiology. 1986; **52**:460

[28] Zhao H, Zhang K, Rong H, Zhang C, Yang Z. Effect of carbon nitrogen ratio on simultaneous nitrification and denitrification via nitrite Technology in Sequencing Batch Biofilm Reactor and the process control. Advanced Materials Research. 2013;**777**:232-237

[29] Papadimitriou C, Palaska G, Lazaridou M, Samarasc P, Sakellaropoulos GP. The effects of toxic substances on the activated sludge microfauna. Desalination. 2007;**211**:177-191

[30] Gregersen T. Rapid method for distinction of gram negative from gram positive Bacteria. Applied Micro-Biology and Biotechnology. 1978;**5**:123-127

[31] George, P., (1947). Reaction between catalase and hydrogen peroxide, Nature, Volume 160, pages41–43 (1947)

[32] Gagnon, M., Hunting, W. M., Esselen, W.B., 1959. A new method for catalase determination. Analytical Chemistry31:144.

[33] Christensen, G.D., Simpson, W.A., Younger, J.J., Baddour, L.M., Barrett, F. F., Melton, D.M., Beachey E.H., (`1985). Adherence of Coagulase-Negative Staphylococci to Plastic Tissue Culture Plates: a Quantitative Model for the Adherence of Staphylococci to Medical Devices, Journal of Clinical Microbiology, Dec; 22(6): 996–1006.

[34] Baldassarri L, Simpson WA, Donelli G, Christensen GD. Variable fixation of staphylococcal slime by different histochemical fixatives, European journal of clinical microbiology and infectious diseases. November. 1993;**12** (11):866-868

[35] Canuti, A., (1990). L' Ultima Acqua, Chiriotti Editore

[36] Liang, H., Ye, D., Li, P., Su, T., Wua, J., Luo, L., Evolution of bacterial consortia in an integrated tannery wastewater treatment process, RSC Adv. 6: 87380-87388, Issue 90

[37] Benintendi R. Modelling and experimental investigation of activated sludge VOCs adsorption and degradation. Process Safety and Environmental Protection. 2016;**101**: 108-116

[38] Gompertz B. On the nature of the function expressive of the law of human mortality, and on a new mode of

determining the value of life contingencies. Philosophical Transactions. Royal Society of London. 1825;**1825**(115):513-583

[39] Zwietering, M.H., Jongenburger, I.,. Rombouts, F.M.., Van 't Riet, K., (1990), Modeling of the Bacterial Growth Curve, Applied and Environmental Microbiology, June, p. 1875-1881

[40] Mittelman MW. Biological fouling of purified-water systems: Part 1, bacterial growth and replication. Microcontamination. 1985;**3**(10):51-55 70

[41] Abbott A, Rutters PR, Berkeley RCW. The Influence of Ionic Strength, pH and a Protein Layer on the Interaction between Streptococcus Mutans and Glass Surfaces. Journal of General Microbiology. 1983;**129** (439445)

[42] Duncan DB. Multiple range and multiple F tests. Biometrics. 1955;**11**:1-42

[43] Eckenfelder WW. Activated Sludge Processes Requiring Excess Sludge Disposal; Symposium on Waste Water Treatment for Small Municipalities. Montreal: École Polytechnique; 1965

[44] Marino P, Benintendi R. Research project: Characterization, definition and modelling of inter/intra- phase transport and degradation mechanisms of complex substrates in attached biomass reactors, final report. Funding by Italian Regional Law N. 2002;**41**: 31.12.1994